Handbook on
Dairy Chemistry

(For B.Sc. Dairy Chemistry as per ICAR Syllabus)

THE AUTHOR

Dr. M.K. Srivastava has done his M.Sc (Organic Chemistry) and D.Phil (Micro Analysis of Organic Compounds) from Allahabad University, Allahabad. He has got one year training in Quality Analysis of Milk and Milk Products from National Dairy Development Board (NDDB), Anand and different trainings in analysis of Milk products and Minerals from National Dairy Research Institute (NDRI), Karnal.

He has more than 35 years working experience in Dairy industry and well versed with the Quality and Food Safety Management for Food and Dairy Products and Lab accreditation under NABL. He has good experience in dealing with National (BIS, Agmark, FSSA-2006, NABL etc) and International (Codex, IDF *etc.*) Quality as well as Accreditation agencies.

He has worked as Senior Chemist in Pradeshik Co-operative Dairy Federation Ltd and worked as HOD (Q.A. and R&D), Incharge, Central Lab., Lucknow and undertaken various projects of Council of Science and Technology, U.P. UPDASP etc. He is also academic counsellor in IGNOU for DDT programme and visiting faculty for MSc (Food Processing), Lucknow University. He has published 20 research papers and four Books. He is an approved Food Safety Lead Auditor.

Handbook on
Dairy Chemistry

(For B.Sc. Dairy Chemistry as per ICAR Syllabus)

– Author –

Dr. M.K. Srivastava

M.Sc., D.Phil (A.U.)

Ex Senior Dairy Chemist, PCDF (Parag), Lucknow,

Food Safety Lead Auditor/Consultant (Milk/Food Processing),
Faculty, DDE, IGNOU, Lucknow

Ex Faculty, MSc (Food Processing and Food Technology,
Lucknow University, Lucknow

&

Qualified Master Trainer (Dairy Processing), PMFME

2024

Daya Publishing House®
A Division of

Astral International Pvt. Ltd.
New Delhi – 110 002

ISBN 978-93-5461-436-1 (HB)

Published by : **Daya Publishing House®**
A Division of
Astral International Pvt. Ltd.
– ISO 9001:2015 Certified Company –
4736/23, Ansari Road, Darya Ganj
New Delhi-110 002
Ph. 011-43549197, 8130496929
E-mail: info@astralint.com
Website: www.astralint.com

Printed at : **Replika Press Pvt Ltd.**

Acknowledgements

The author expresses his gratitude and thanks to all Dairy chemists, Dairy/ Food Scientists, Dairy Institutions, Websites and Manufacturers of Processing/ Lab Instruments under references for providing technical information.

The author is also thankful to quality assurance personnel and managers (QA) of PCDF, especially Mrs S. Chopra and Mr H.S. Verma. for supporting and inspiring him for the same.

The author is genuinely indebted to his Mother,Father, Brothers,sisters, wife, kids and other family members for providing continuous moral support without which, it is impossible for author to complete the task.

Last but not Least, author acknowledges his thanks to all his friends and all staff of Central Quality Assurance Lab, PCDF Lucknow and Quality Assurance Lab, Lucknow Milk Union, Lucknow.

Dr. M.K. Srivastava

Preface

The Quality of milk and milk products play a very important role as far as consumer acceptability is concerned and if dairy industry/Food industry is not giving emphasis on quality then it is very difficult to survive in present competitive national and international market. Good quality raw materials, infrastructures for processing, packing, quality analysis, plant hygiene *etc.* are the prime requirement for maintaining the quality of milk, milk products and other foods as per the standard quality norms.

The above cited national and international pressures on the quality of milk/milk products, and other food products compel the well equipped dairy industries/Food industries to equip the laboratories also with latest analytical equipment, skilled man power and technical literatures related to the analysis of quality of milk/milk products and other food products because the aforesaid requirements can only make the dairy/food industries able to evaluate the quality of milk, milk products, food products properly and accurately. Realising the importance of compilation of technical literatures related to analysis of milk and milk products, author has made successful attempts to produce a compiled and completed advanced technical information related to the analysis of quality of milk and milk products in the form of **"Hand Book on Analysis of Milk"(Chemical and Microbiological Analysis of Liquid Milk)** and **"Hand Book on Analysis of Milk Products"(Chemical and Microbiological Analysis of Milk Products**).To provide a compiled and complete advanced technical information related to the instrumental analysis of milk, milk products, other food products *etc.*, author has also written a book entitled **"Instrumentation**

in Chemical and Microbiological Analysis of Food and Dairy Products (Part-1 and 2)"and to provide valuable technical information related to these technologies, author has tried to compile all required technical information as a book **entitled " Hand Book on Alternative, Innovative and Novel Processing Technologies (Milk/Milk Products and other Foods) ".**

To run a dairy industry, qualified and trained man power is highly required. To facilitate dairy graduate students, author now tries to write a book entitled **"Handbook on Dairy chemistry"** as per ICAR syllabus. Author expects that the said publication is able to meet the requirements of dairy graduate students.

Dr. M.K. Srivastava

Contents

— Part 3 —

CHEMISTRY OF DAIRY PRODUCTS

Part 1

Physical Chemistry of Milk

Chapter 1

Introduction

Milk Constituents and Gross Composition

The milk is provided by the nature as source to nourish and provide immunological protection for the mammalian young. Milk has been a food source for humans since prehistoric times. Milk is a white liquid produced by the mammary glands of mammals. Milk is a complex and an important food containing several constituents and is excellent source of nutrients necessary for growth and development of young and old.

- ☆ Milk constituents are important as human food. It is the primary source of nutrition for young mammals before they are able to digest other types of food. Milk is food for mankind and young ones of all mammals as a liquid food. Milk contains water, fat, protein, lactose, vitamins and minerals. Early-lactation milk contains colostrums, which carries the mother's antibodies to the baby and can reduce the risk of many diseases in the baby. The precise components of raw milk vary by species and by a number of other factors, but Milk contains water, fat, protein, lactose, vitamins and minerals

Milk Constituents [1,2,3]

Milk is natural secretion of the mammary gland with very complex composition and have broad components consisting of a mixture of water, fat, proteins, lactose, minerals and vitamins as the nutrient. On keeping at room temperature milk separates into fat as cream, curd or casein and whey. Curd and whey separates on curdling of milk. The **constituents** present in milk are broadly categorized in the following two groups:

a. **Major constituents**- Water, Fat, Proteins,Lactose and Minerals.

b. **Minor constituents**- Phospholipids, Cholesterol, Pigments, Vitamins and Enzymes.

Milk fat, Casein, Lactose are called as True constituents. These constituents are present in following three phases:

i. **True solution,**

ii. **Colloidal dispersion and**

iii. **Fat-in-water type emulsion.**

The above constituents can also be broadly divided in Water, Fat and Solids -not- Fat. The Fat and Solids-not- Fat (constituents other than water) are called Total Milk solids. Milk - fat (skim milk) is called as **Plasma where as Plasma-casein micelles (whey) is called as Serum.** The constituents of milk are listed diagrammatically in following two figures and details of percentages of different constituents are given in following tables. Milk can also be described in terms of the following **three physical states**:

1. **A true solution** (lactose, soluble proteins, minerals, vitamins *etc.* are available in solution),
2. **A colloid suspension** (casein micelles, globular proteins and lipoprotein particles)
3. **An oil-in-water emulsion** (fat globules dispersed in the continuous serum phase).

There are always a state of equilibrium between colloidal state and true solution in normal milk.

Composition of Milk [1,2,3]

Milk is an emulsion or colloid of butterfat globules within a water-based fluid that contains dissolved carbohydrates and protein aggregates with minerals. Milk provides essential nutrients and is an important source of dietary energy, high-quality proteins and fats. Milk can make a significant contribution to the required nutrient intakes for calcium, magnesium, selenium, riboflavin, vitamin B_{12} and pantothenic acid.

- Milk and milk products are nutrient-dense foods and their consumption can add diversity to plant-based diets. It is the primary source of nutrition for young mammals, including breastfed human infants before they are able to digest solid food.
- The composition of milk varies with the species to ideally suit the need of new born infants. However, milk from cow, buffalo and goat ideally meet the requirement of human beings as liquid milk or its products. The nutritional value of milk is high. Milk is also a very complex food with over 100,000 different molecular species found.

☆ Milk mainly contains water, Fat and Solids not fat (SNF).

1. **Milk Fat portion-** This portion contains Milk Fat, Cholesterol, Phospholipids and Fat soluble vitamins.
2. **The SNF portion**- This portion largely consists of proteins, lactose and minerals. These solids are also referred to as 'serum solids'.

Milk contains water, Milk fat, Proteins (Casein and Whey proteins),Lactose, Minerals, Enzymes and Pigments as shown in Figure and Tables.

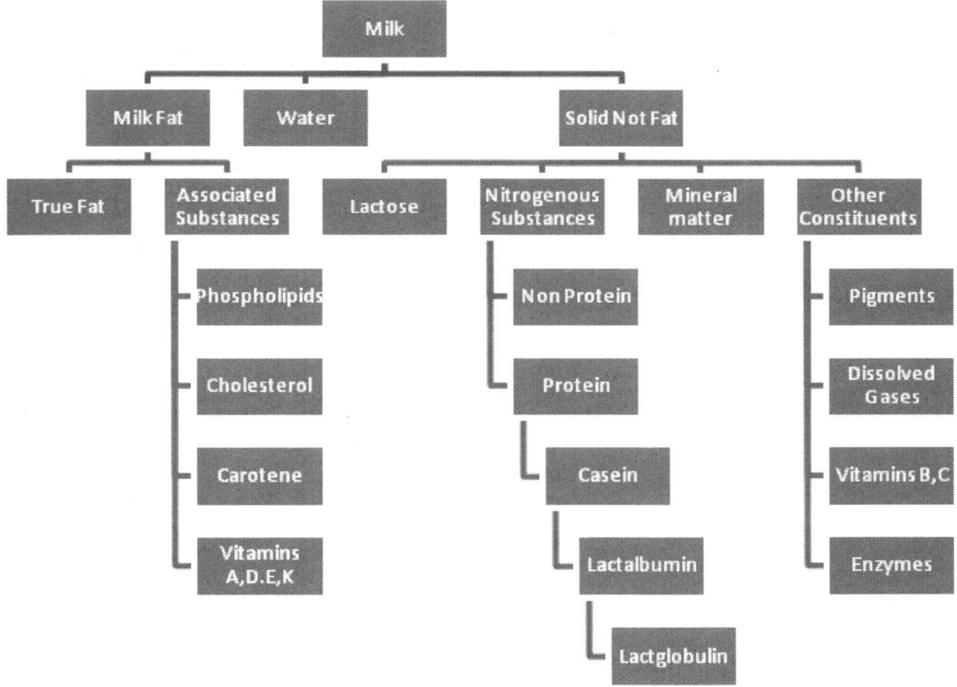

Figure 1.1: Flow Chart Showing Milk Composition [1,2,3].

Table 1.1: Detailed Composition of Milk [2, 3, 4, 4a]

Sl. No.	*Constituents/Groups of Constituents*	*Percentage*	*Approx. Concentration (Weight per litre of milk)*
1	Water	86-88	860-880
2	Lipids in emulsion phase		
	i. Milk Fat (a mixture of mixed triglycerides), Phospholipids	03-05	30-50gms
	ii. Sterols	0.03	0.30gms
	iii. Carotenoids	0.01	0.10gms
	iv. Vitamins A and E, D and K	—	—

Sl. No.	Constituents/Groups of Constituents	Percentage	Approx. Concentration (Weight per litre of milk)
3	Proteins in colloidal dispersion		
	i. Casein (alfa, beta and gamma)	2.50	25 gms
	ii. Beta -Lactalbumins	0.30	3 gms
	iii. Alfa Lactalbumins	0.07	0.7gms
	iv. Enzymes (catalase, peroxidase, phosphatase *etc.*)	—	—
4	Dissolved Materials		
	a. Carbohydrates		
	i. Lactose	4.5-5.0	45-50 gms
	ii. Glucose	0.005	0.05gms
	b. Inorganic and organic ions and salts		
	i. Calcium	0.125	1.25 gms
	ii. Phosphate (as Po_4)	0.210	2.10 gms
	iii. Citrates (as citric acid)	0.20	2.0 gms
	Partly in colloidal dispersion		
	iv. Chlorides	0.10	1.0 gms
	v. Sodium, Potassium, Magnesium	—	—
	c. Water soluble vitamins		
	(Thiamine, Riboflavin, Niacin, Pyridoxin, Pentothenic acid, Biotin, Folic acid, Vita. B_{12}, Ascorbic acid).	—	—
	d. Nitrogenous Materials (Not protein or vitamins) Ammonia, Amino acids, Creatine/creatinine, uric acid *etc.* Urea.	—	—
	e. Gases- CO_2, O_2, N_2	—	—
	f. Trace Elements- Cu, Fe, *etc.*	—	—

The details of aforesaid constituents of milk are discussed in respective chapters but short descriptions of milk constituents are given in Table 2 [1a,3]

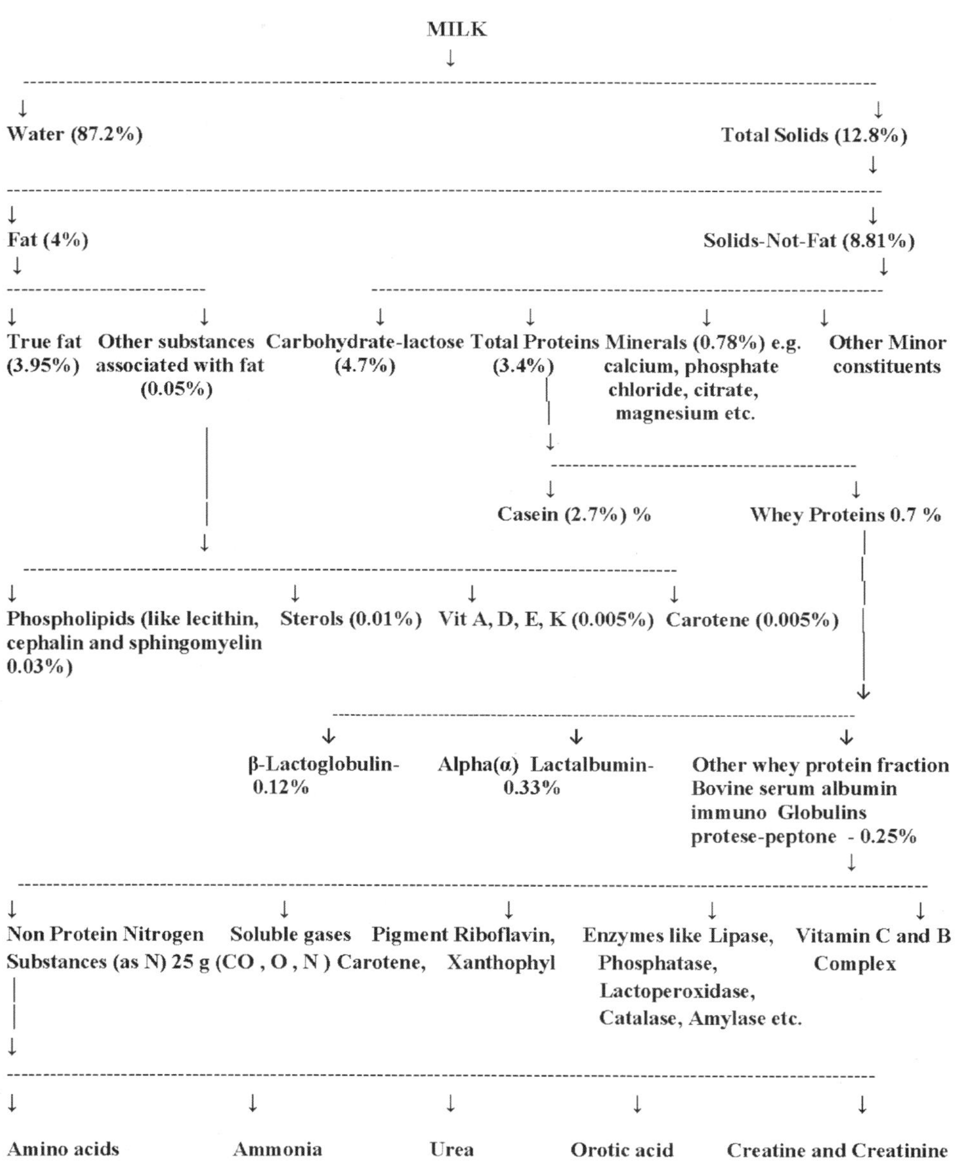

Figure 1.2: Distribution of Constituents of Cow Milk [3].

Table 1.2: Short Descriptions of Milk Constituents

Sl.No.	Milk Constituents	Short Description
1	**Water**	It is the medium in which all the milk constituents are dispersed and plays an important role in the physical properties of milk. It helps in maintaining the much desirable equilibrium and keeps the constituents of the milk in their native state. It acts as a solvent and assists in keeping the constituents in soluble, colloidal or emulsion form. It influences the colligative properties of milk.
2	**Fat Portion**	
	Milk Fat	It is the costliest constituent of milk and present as an emulsion. It is rich in saturated fatty acids and mono unsaturated fatty acids and also acts as a carrier for the fat soluble vitamins. The milk fat has short chain saturated fatty acids in substantial quantities and imparts the unique flavour to the milk and milk products. The long chain unsaturated fatty acids reduce the melting point of milk fat. In order to maintain the emulsion stability several surface active substance are also associated with the milk fat and present in higher proportion in fat globule membrane.
3	**SNF Portion**	
	Lactose	Lactose is a disaccharide of glucose and galactose, reducing sugar and a major soluble component present in milk. Milk contains approximately 4.6 per cent carbohydrate (predominately lactose with trace amounts of monosaccharides and oligosaccharides). It is first carbon source for the microbes gaining entry into milk. It is a good source of energy for the young ones. Milk also contain several other carbohydrates in trace levels.
	Proteins	Milk contains unique protein namely the casein and whey proteins. It is present in the form of colloidal dispersion in milk and is responsible for several physical properties of milk. Milk protein provides all the essential amino acids and is easily digestible. Milk is a good source for several enzymes which have both desirable and also undesirable effect on preservation and processing of milk.
	Minerals	Milk contains minerals which are mainly inorganic salts, partly ionized and partly present as complex salts. Some inorganic matter is bound covalently such as calcium phosphate groups in casein. Minerals are responsible for the ionic balance of milk and helps in maintaining the equilibrium between the soluble and colloidal state.
	Vitamins	Milk contains both fat soluble and water soluble vitamins (essential vitamins). Milk is not a good source of Vitamin A,K and Vitamin C. Vitamin A is existing in the cow milk in its precursor form *i.e.* carotene a provitamin A imparts golden yellow colour to milk.
	Miscellaneous Compounds	Milk contains many other components in low concentrations. These are gases, alcohols, carbonyl compounds, carboxylic acids, conjugated compounds, non protein nitrogenous compounds, phosphate esters, nucleotides, nucleic acids, sulfur containing compounds, hormones *etc.*

The details related to the composition of milk of different milch animals are given in Table 1.3.

Table 1.3: Composition of Milk of Different Milch Animals [1,2,3,5]

Species	Water per cent	Total Solids per cent	Fat per cent	Proteins per cent			Lactose per cent	Mineral/ Ash per cent	SNF per cent Min.	Energy kcal
				Casein	Whey	Total				
Goat	87.00	13.0	4.0-4.5	2.6	0.6	3.2-3.7	4.4-4.6	0.8	8.5-9.0	69
Sheep	81.50	18.5	7.6-8.6	3.9	0.7	4.6	4.8	08	9.9-10.9	108
Horse/Mare	88.3-90.1	9.9-11.7	1.6	1.3	1.2	2.5	5.7-6.2	0.4	8.3-10.1	-
Ass/Donkey	88.5-91.5	8.5-11.5	0.6-1.5	1.0	1.0	1.4-2.0	6.1-7.4	0.4	-	-
Camel	86.00	14.0	3.0-4.0	2.7	0.9	3.6-4.0	5.5	0.7	10.0-11.0	-
Buffalo	83.50	16.5	7.2-7.8	3.2	0.6	3.8	4.9	0.7	8.7 - 9.3	97
Cow (Indian)	87.20	12.8	4.0-4.8	2.8	0.6	3.4	4.7-4.95	0.7	8.0-8.8	65
Cow (Foreign)	88.00	12.0	4.6	2.8	0.6	3.4	4.9	0.7	7.4	61
Human	87.50	12.5	4.38	1.03-1.25*			6.89 **	0.2	8.12	70
Dog	75.50	24.50	9.5	11.1			3.3	0.6	15.0	-
Cat	82.10	17.90	3.3	9.0			4.9	0.6	14.6	-
Soy	93.30	6.70	1.92	2.75 *			1.81 **	0.27	4.78	33
Elephant	67.80	32.20	19.5	3.2			8.8	0.6	12.7	-

* Total proteins; ** Carbohydrates; SNF: Solid Not Fat.

Early-lactation milk is called colostrum, which contains antibodies that strengthen the immune system and thus reduces the risk of many diseases. During approximately the first three days of lactation, the buffalo/cow secretes colostrum. This is very vital for the new born and its composition reflects the calf's need. Colostrum contains the important proteins; the immuno-globulins [3,6]. The difference between the composition of cow milk and colostrum is tabulated in Table 1.4.

Table 1.4: Differences between Cow Milk and Colostrum [3,6]

Sl.No.	*Constituents*	*Cow Milk*	*Colostrum of Cow*
1	Water per cent	87.27	75.09
2	Fat per cent	3.68	3.97
3	Casein	3.39	4.18
4	Albumin and Globulin	0.44	12.97
5	Lactose	4.95	2.28
6	Minerals	1.53	7.0
7	Gravity (SP)	1.032	1.046
8	Others Carotene	9 Times more than normal	
	Iron	17 Times more than normal	

Differences in the Composition of Milk from various Species [1a]

The composition of milk changes with the change of milch animals. The following table showing the composition of milk from various species is given for understanding the variation in the composition of the milk secreted by them.

- ☆ The buffalo milk is having the maximum fat while the fat content in sheep milk is also much close to that. Similarly the fat per cent in goat milk is much similar to cow milk.
- ☆ Variation among the protein per cent is also very less among these species.
- ☆ The lactose content in Human milk is characterized by having higher per cent of lactose and fat while the protein and ash content is much less when compared with other species.
- ☆ The energy supplied through the milk is highest in buffalo and sheep while the difference is very much less between the milk from the remaining species.

Table 1.5: Composition of Milk from different Milch Animals [1a,8]

Sl. No.	*Milch Animals*	*Composition (g/100g)*						
	Species/Scientific Name	*Energy Kcal/100g*	*Water*	*Fat*	*Lactose*	*Casein*	*Whey Protein*	*Ash*
1	Human/*Homo sapiens*	72	87.1	4.5	7.1	0.4	0.5	0.2
2	Cow/*Bos taurus*	66	87.3	3.9	4.6	2.6	0.6	0.7
3	Zebu/*Bos indicus*	74	86.5	4.7	4.7	2.6	0.6	0.7
4	Goat/*Capr hircus*	70	86.7	4.5	4.3	2.6	0.6	0.8
5	Sheep/*Ovis aries*	102	82	7.2	4.8	3.9	0.7	0.9
6	Horse/*Equus caballus*	52	88.8	1.9	6.2	1.3	1.2	0.5
7	Rabbit	202	67.2	15.3	2.1	9.3	4.6	1.8
8	Camel/*Camelus dromedaries*	70	86.5	4.0	5.0	2.7	0.9	0.8
9	Buffalo/ *Bubalus bubalis*	**102**	**82.8**	**7.4**	**4.8**	**3.2**	**0.6**	**0.8**

1. Cow Milk v/s Buffalo Milk [3]

Cow milk differs from buffalo milk in composition, Fat characteristics and physico chemical properties as given below:

i. Composition [3]

The buffalo milk contains more fat, solids not fat and total solids and hence yield of products prepared from buffalo milk, be it cream, butter, cheese or condensed milk will be always higher. The inherent properties of buffalo milk like high total solids content, superior whiteness and viscosity render it eminently suitable for the manufacture of traditional (indigenous) milk products. Average and gross composition of cow and buffalo milk are tabulated in Table 1.6.

Table 1.6: Gross Composition of Cow Milk [7]

Sl.No.	*Component*	*Average Content Per cent w/w*	*Component Range Per cent w/w*	*Average Per cent of Dry Matter*
1	Water	87.3	85.5 -88.7	-
2	Solids not fat	8.8	7.9-10.0	69
3	Lactose	4.6	3.8-5.3	36
4	Fat	3.9	2.4-5.5	31
5	Protein	3.25	2.3-4.4	26
6	Casein	2.6	1.7-3.5	20
7	Mineral substances	0.65	0.53-0.80	5.1
8	Organic acids	0.18	—	1.4
9	Miscellaneous	0.14	—	1.1

Table 1.7: Average Composition of Milk of different Cows [3a]

Sl.No.	Species	Water	Fat	Protein	Lactose	Ash
1	Friesian cow[1]	87.92	3.40	3.13	4.86	0.69
2	Sindhi cow[2]	86.07	4.90	3.42	4.91	0.70
3	Gir cow[2]	86.44	4.73	3.32	4.85	0.66
4	Tharparkar cow[2]	86.58	4.55	3.36	4.83	0.68
5	Sahiwal cow2	86.42	4.55	3.33	5.04	0.66
6	Crossbred cow[2]	86.54	4.50	3.37	4.92	0.67

Note: 1: *Bos taurus* and 2: *Bos indicuis*. Values in per cent.

Table 1.8: Composition of Buffalo Milk [3b]

Constituent	Per cent	Constituent	Per cent
Water	83.18	Calcium	0.18
Fat	6.71	Magnesium	0.02
Protein	4.52	Sodium	0.04
Lactose	4.45	Potassium	0.11
Total solids	16.82	Phosphorous	0.10
Solids Not Fat	10.11	Citrate	0.18
Ash	0.80	Chloride	0.07

ii. Milk Fat Difference [3]

Milk fat of cow and buffalo milk differs in following properties-

- **a.** Buffalo milk fat has a higher Reichert value, Saponification value, Kirschner value and melting point but lower Polenski value, Iodine and Butyro refracto- meter index than cow milk fat.
- **b.** Buffalo milk fat is higher than cow milk fat in Butyric, palmitic, stearic and oleic acids but lower in caproic, caprylic, capric, lauric and mysteric acids.

iii. Physico-chemical Properties [3,4]

Buffalo milk normally has a higher pH, acidity buffer value, density, viscosity and fat globule size than cow milk. The detailed differences in physico-chemical properties of Cow and Buffalo milk are given in Table 9.

Benefits of Buffalo Milk [3,3b]

As per the information available, the following benefits of buffalo milk are listed:

Table 1.9: Physico-Chemical Properties of Milk [3,4,4a]

Sl. No.	Properties – Physico-Chemical	Cow Milk		Buffalo Milk	
		Mean	Range	Mean	Range
1	Density (absolute) at 20°C	1.0287	1.028-1.030	1.0305	1.030-1.032
2	Specific gravity) at 20°C	1.0330	—	1.0323	—
3	Viscosity (centipoises) at 20°C	1.86	1.5-2.0	2.245	2.086-2.538
4	Curd tension (g)	23.3	28-54	40.2	32-85
5	Surface tension (dyne/cm.) at 20°C	55.9	40-60	55.4	45.5-48.2
6	Freezing point	0.540	0.525-0.565	0.544	0.5454-0.5443
7	pH at 20°C	6.64	6.5-6.7	6.73	6.63-6.87
8	Titratable acidity (as per cent Lactic Acid)	0.15	0.14-0.18	0.16	0.10-0.28
9	Oxidation reduction potential (+volts)	0.270 at 20°C	0.18-0.38	0.310 at 30°C	0.129*-0.469
10	Electrical conductivity (milli mho/cm.) at 30 °C	4.67	4. 14-5.03	3.60	3.28-3.81
11	Thermal Conductivity Kcal/ h/m/degree celsius	0.4572 at 36.6 °C	—	0.5689 at 42.5 °C	0.5487-0.5937
12	Refractive Index	—	03449-1.3448	1.0	—
13	Heat Capacity at 40°C (cal/g/degree Celsius)	0.930	—	0.8191	0.7402-0.9261
14	Fat globule size (mm)	3.85	—	5.01	—
15	Phosphatase (units)	83	—	28	—
16	Ultra violet fluoresence	Pale bluish	—	Greenish -yellow	—

- ☆ Lower cholesterol content -0.65 mg/g in buffalo milk as compared to the 3.14 mg/g in cow milk.
- ☆ More protein -2.74 protein efficiency ratio (PER)of buffalo milk as compared to 2.49 PER of cow milk (11 per cent higher protein in buffalo milk than cow milk). Greater proportion and larger size particles of misceller casein in buffalo milk than cow milk.
- ☆ More important minerals –Ca,Fe and P are 92 per cent,37.7 per cent and 118 per cent respectively higher in buffalo milk than cow milk. There are more cations (Ca and Mg) in buffalo milk but fever anions (phosphate and citrate). The calcium phosphorous ratio is higher in buffalo milk (2.26) than cow milk (1.96). Due to high salt balance ratio buffalo milk has a low heat stability.

- ☆ More Vitamin –A in Buffalo Milk as buffalo metabolizes all the carotene into vitamin –A.
- ☆ No difference in nutritive value and digestibility of milk and milk products.
- ☆ Better whey protein –proteins of buffalo milk particularly whey proteins are more heat resistant to heat denaturation as compared to cow milk proteins.
- ☆ Better health foods–Higher levels of various bio- protective factors, such as immunoglobulins, lactoferin, lysozyme, lactoperoxidase as well as bifidogenic.
- ☆ More viable commercially –commercially more viable than cow milk for the manufacture of fat based and SNF based milk products because of low water content, high fat content and white color of buffalo milk.

2. Goat Milk [1b]

The goat milk has following characteristics:

- ☆ Fat globules: Mean 2.76 μM, Range 0.73 – 8.58 μM versus 3.5 and 0.92 – 15.8 μM for cow milk.
- ☆ More lipolysis especially greater release of short chain fatty acids relative to cow milk.
- ☆ Low heat stability at native pH relative to cow milk.
- ☆ Casein reduced in Alpha-S1 null milk.
- ☆ More fat relative to protein.
- ☆ Plasmin activity.

3. Sheep Milk [1b]

The sheep milk has following characteristics:

- ☆ Much higher solids.
- ☆ More fat relative to protein.
- ☆ Traditionally valued for hard cheese.

Source

1. *Edgar R. Ling, A textbook of dairy chemistry, Chapman and Hall publication, 1948.*

1a. *K.S. Sharma, Chemistry of milk, Agrimoon.*

1b. *Art Hill, Milk Composition and Properties. University of Guelph, www.foodscience.ca.*

2. *Sukumar De, Practical dairy technology.*

3. *M.K. Srivastava, Handbook on Analysis of Milk, CBS Publications, New Delhi, 2010. ISBN 13: 9788123926681.*

3a. *Narendra K Nayak, Milk composition and its constituents, http://ndvsu.org.*

3b. *Cow milk vs Buffalo milk, www.dairy for all.com.*

4. *Robert Jenness and Stuart Patton, Principles of Dairy Chemistry, John Wiley and Sons, Inc., NY, 1959.*

4a. *B.H. Webb, A.H. Johnson, and J.A, Alford, 1965. Fundamentals of dairy chemistry. AVI Publ. Co., New York.*

5. *Davies, Chemistry of milk.*

6. *Dr. Nanak Singh, Dairy chemistry.*

7. *P. Walstra and Robert Jenness, Dairy Chemistry and Physics, 1984, John Wiley and Sons, New York, ISBN 9780471097792 - 467*

8. *Robert Jenness and Sloan, Dairy Science abstracts, 1970, 32, 599-612.*

Chapter 2

Colloidal and Emulsion State

Colloidal State [1]

Colloids (also known as colloidal solutions or colloidal systems) are mixtures in which microscopically dispersed insoluble or soluble particles of one substance are suspended in another substance [2]. For a mixture to be classified as a colloid, its suspended particles must not settle (in the manner that the particles of suspensions settle at the bottom of the container if left undisturbed).

- ☆ A colloidal system may be solid, liquid or gaseous condition. Sometimes the dispersed substance alone is called the colloid [2], the term **colloidal suspension** refers unambiguously to the overall mixture (although a narrower sense of the word *suspension* is distinguished from colloids by larger particle size).

  The IUPAC definition of the colloidal state can be written as follows: **"The colloidal state is the state of subdivision in which molecules or polymolecular particles having at least one dimension in the range of 1 nanometre and 1 micrometre, are dispersed in some medium" [3].**

  Considering the degree of dispersion, colloids will be between the coarse dispersions and molecular dispersions or true solutions. Unlike a solution, whose solute and solvent constitute only one phase, a colloid usually consist of at least two phases- **dispersed phase or internal phase** (the suspended particles) and **a continuous phase** (the medium of suspension) that arise by phase separation. Colloids usually feature substances that are evenly dispersed in another. In such mixtures, the substance that is dispersed is referred to as the **dispersed phase** whereas the substance through which it is dispersed is called the **continuous phase**.

☆ To qualify as a colloid, the mixture must be one that does not settle or would take a very long time to settle appreciably. The colloids are **micro heterogeneous** in nature and in these colloids dispersed particles comprise a separate phase that is divided by an interface from the dispersion medium. In a colloid system, the particle size is larger than dissolved molecules but smaller than settleable solids since the particles are having fine dimensions and greater surface area

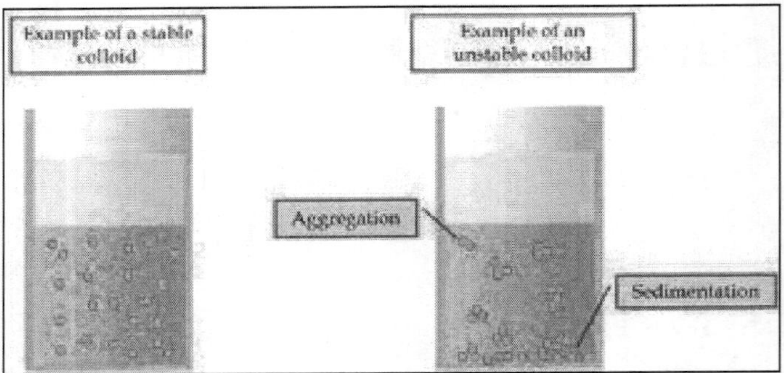

Figure 2.1: Colloids [1].

The dispersed-phase particles have a diameter between approximately 1 and 1000 nanometers [4]. Such particles are normally easily visible in an optical microscope, although at the smaller size range ($r < 250$ nm), an ultramicroscope or an electron microscope may be required. Homogeneous mixtures with a dispersed phase in this size range may be called **colloidal aerosols, colloidal emulsions, colloidal foams, colloidal dispersions, or hydrosols.** The dispersed-phase particles or droplets are affected largely by the surface chemistry present in the colloid.

Some colloids are translucent because of the Tyndall effect, which is a phenomenon in which beams of light incident on colloids are scattered due to the interactions between the light and the colloidal particles.Other colloids may be opaque or have a slight color. The cytoplasm of living cells is an example of a colloid, containing many types of biomolecular condensate.

Colloidal suspensions are the subject of interface and colloid science. This field of study was introduced in 1845 by Italian chemist Francesco Selmi [5] and further investigated since 1861 by Scottish scientist Thomas Graham [6].

Source

1. *Wikipedia.*
2. *"Colloid". Britannica Online Encyclopedia. Retrieved 31 August 2009.*

3. *Richard G. Jones; Edward S. Wilks; W.Val Metanomski; Jaroslav Kahovec; Michael Hess; Robert Stepto; Tatsuki Kitayama, eds., 2009, Compendium of Polymer Terminology and Nomenclature (IUPAC Recommendations 2008) (2nd ed.). RSC Publ. p. 464. ISBN 978-0-85404-491-7.*

4. *Levine, Ira N. (2001). Physical Chemistry (5th ed.). Boston: McGraw-Hill. p. 955. ISBN 978-0-07-231808-1.*

5. *Francesco Selmi, Studi sulla dimulsione di cloruro d'argento, Nuovi Annali delle Scienze Naturali di Bologna, fasc. di agosto 1845.*

6. *Graham coined the term "colloid" in 1861, See: Graham, Thomas, 1861, "Liquid diffusion applied to analysis", Philosophical Transactions of the Royal Society of London, 151, 183-224.*

Classification [1,2]

The colloids are sometimes identified and characterized by their physico-chemical and transport properties because the size of dispersed phase may be difficult to measure and colloid has appearance of solution.

For example, if a colloid consists of a solid phase dispersed in a liquid, the solid particles will not diffuse through a membrane, whereas with a true solution the dissolved ions or molecules will diffuse through a membrane.

Because of the size exclusion, the colloidal particles are unable to pass through the pores of an ultrafiltration membrane with a size smaller than their own dimension. The smaller the size of the pore of the ultrafiltration membrane, the lower the concentration of the dispersed colloidal particles remaining in the ultrafiltered liquid.

The classification of colloids can be made on the basis of different properties. The details related to the classification of colloid, are given below:

1. On the Basis of the Types of Dispersed and Dispersion Medium (Physical states), colloids can be classified as follows [1,2]

Medium/Phase		*Dispersed Phase*		
		Solid	*Liquid*	*Gas*
Dispersion Medium	**Gas**	**Solid aerosol** Examples: smoke, ice cloud, atmospheric particulate matter (Dust).	**Liquid aerosol** Examples: fog, clouds, condensation, mist, hair sprays.	No such colloids are known. Helium and xenon are known to be immiscible under certain
	Liquid	**Sol** or Suspension Examples: milk (protein fraction), pigmented Ink, blood, Paint, sediment.	**Emulsion** Examples: milk (fat fraction), mayonnaise, hand cream; latex	**Foam** Example: whipped cream, shaving cream
	Solid	**Solid sol** **Example:** cranberry glass, Coloured gem stone.	**Gel** Examples: agar, gelatin, jelly	**Solid foam** Examples: aerogel, styro foam.

2. Based on the Nature of Interaction between the Dispersed Phase and the Dispersion Media (The affinity of dispersed molecules with dispersing media), colloids can be classified as [1,2]:

i. Hydrophilic Colloids

The colloid particles are attracted toward water. They are also called reversible sols. Hydrophilic - Molecules that are water soluble and do not readily cross the plasma membrane.

ii. Hydrophobic Colloids

These are opposite in nature to hydrophilic colloids. The colloid particles are repelled by water. They are also called irreversible sols. Hydrophobic - Molecules are lipid soluble and can easily cross the plasma membrane.

The terms hydrophilic and hydrophobic are used in case of aqueous colloid solution.

- ☆ If the affinity on dispersion medium is more it is **lyophillic or reversible sol**. The lyophilic (solvent attracting) colloid is one whose particles have a strong attraction for the molecules of dispersion medium and binding large number of them into so called solvent shells. Reversible colloids are thermo dynamically stable. They are either macromolecules (such as proteins) or association of colloids (micelles). Lyophilic colloidal systems with aqueous continuous phase are known as hydrophilic colloidal system.
- ☆ If the affinity on dispersion medium is less it is **lyophobic or irreversible sol**. In lyophobic (solvent repelling) colloid, the particles do not interact so strongly with the molecules of the surrounding medium [2]. Irreversible colloids constitute a true phase in aqueous medium such as oil in water emulsions. The lyophobic systems with aqueous dispersion medium are known as hydrophobic.

Lipophilicity [1]

Its refers to the ability of a chemical compound to dissolve in fats, oils, lipids, and non-polar solvents such as hexane or toluene. Thus lipophilic substances tend to dissolve in other **lipophilic** substances, while **hydrophilic** substances tend to dissolve in water and other **hydrophilic** substances.

Difference between Lipophilic and Lipophobic Colloids

The differences between Lipophilic and Lipophobic colloids, are tabulated in Table 2.1 [2].

Table 2.1: Difference between Lipophilic and Lipophobic Colloids

Sl.No.	Parameters	Lipophilic	Lipophobic
1	Reversible/Irreversible	Reversible.	Irreversible.
2	Tyndall Effect	Exhibit weak tyndall effect.	Exhibit Strong tyndall effect.
3	Preparation	Can be prepared by simple methods of mixing.	Needs special methods for preparation.
4	Surface Tension	Surface tension is less than dispersion medium.	Surface tension of colloid sol is same for dispersion medium.
5	Viscosity	Viscosity is more for sol than for dispersion medium.	Viscosity of sol is similar to dispersion medium.
6	Stability	More solvated because of which it is more stable.	Not solvated hence less stable but can be made stable by adding small quantity of lyophillic colloid.

Source: 1: Wikipedia and 2: K.S. Sharma, Physical Chemistry of milk, Agrimoon.com.

Definition of True Solution [1]

The homogenous mixture of two or more substances, where the solute is dissolved in the solvent is called the true solution. Here the size of the particles is less than 1 nm.

- ☆ The example of the true solution is when sugar or salt dissolved in water. The particles cannot be filtered or separated through filter paper or parchment paper. Even the particles are invisible through naked eyes.
- ☆ As the mixture is in the liquid phase and transparent, it allows the light to pass through the solution without getting scattered.
- ☆ When the solution is said as homogenous, it means the particles are evenly distributed in the solution and do not get settled at the bottom of the container. As the amount of particles present in per unit volume of the solution is equal everywhere, the particle density is higher.
- ☆ The Brownian effect is not observed in true solutions, and even the Tyndall effect is absent.

Definition of Colloidal Solution [1]

The heterogeneous mixture of two or more substances, where the size of the particles lies between 1- 1000 nm, is known as a colloidal solution.

- ☆ The colloidal solution is the intermediate between true solution and suspension, though it is also in the liquid phase. When starch dissolved in water or Gelatin mixed in the water are the examples of the colloidal solutions, here the tiny particles will float instead of getting dissolved.

- ✫ Likewise, the true solution, the particles of the colloidal solution are invisible through naked eyes but can be observed electron microscope.
- ✫ The separation of the colloidal particles can be done through parchment paper, but not through the filter paper.
- ✫ The particles can be obtained by the process of centrifugation, where they (particles) will get settled at the bottom.
- ✫ As the mixture is the heterogeneous type, the particles are not evenly distributed in the solutions.
- ✫ As the colloidal solutions are translucent, they allow the light to pass through the liquid, but due to the presence of particles, the light gets scattered. The colloidal particles were observed to be in constant motion in a zig zag path in all possible directions. This motion is known a Brownian motion. It arises because of the impact of solvent molecules on the colloidal particles. The forces are unequal from different directions. **Brownian motion and Tyndall effect is observed in Colloidal solution.**
- ✫ Emulsion, Foam, Sol, Hydrocolloid, Reversible or Irreversible Colloids are the various types of colloids.

Definition of Suspension [1]

Suspensions are the mixture, where the size of the particles is more than the 1000 nm. When the soil is dissolved in water, which is stirred strongly, after some time the particles of the solution gets settle at the bottom of the container due to the gravity; This is the example of the suspension.

- ✫ The particles in the true solution are visible through naked eyes.
- ✫ Brownian motion and Tyndall effect is observed in suspension.

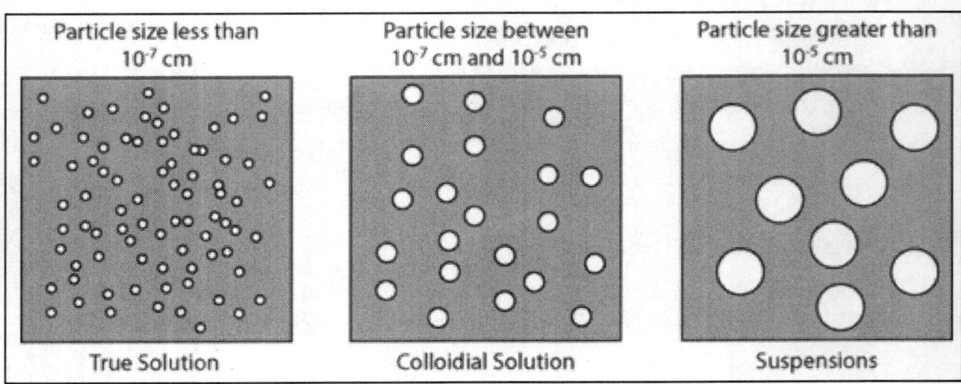

Figure 2.2:: True Solution, Colloidal Solution and Suspensions [2].

Differences Between True Solution, Colloidal Solution, and Suspension [1]

The following are the major differences between True Solution, Colloidal Solution, and Suspension:

1. True solutions are the type of mixtures, where the solute and solvents are properly mixed in the liquid phase, while Colloidal solutions are the type of mixture in the liquid phase, where the solute (tiny particles or colloids) is uniformly distributed in the solvent (liquid phase). The suspension is the mixture, where the solute does not get dissolved, rather get suspended in the liquid and float freely.
2. Sugar solution in water is the example of the true solution. Starch dissolved in water is the example of the colloidal solution and Soil dissolved in water is the suspension.
3. True solutions are homogenous and are transparent in appearance, while colloidal solutions are heterogeneous and appear to be translucent, whereas suspension is also heterogenous but appear to be opaque.
4. As the size of the particles is less than 1nm (10^{-7}cm), the particles easily get pass through parchment paper and filter paper, but the particles size in colloidal solution is between 1-100 nm(10^{-5}-10^{-7}cm), the particles of the colloidal solutions do not diffuse or pass through parchment paper but it is easy through filter paper, in the suspension the particle size is more than the 1000 nm, the particles of the suspension do not pass through parchment or filter paper.
5. In the true solutions particles are invisible through the naked eye and even not under powerful microscope, while the particles in the colloidal solution are visible through the electron microscope, but not through naked eye and the particles in the suspension are visible through naked eyes as well as under electron microscope.
6. Tyndall effect and Brownian effect is not observed in true solutions, while these features are observed in colloidal solutions and suspension.

Source

1. *C. Rachna, Differences Between True Solution, Colloidal Solution, and Suspension, BioDifferences,2019.*
2. *https: //chemistry.stackexchange. com.*

Properties of Colloidal System [1,2]

The colloidal system will have two phases (a dispersed phase and a dispersion medium). The fundamental characteristics of the colloid system,are determined by the state of aggregation of dispersed phase and the intensity of its interaction with dispersion medium.

1. Lyophillic colloids have a strong attraction for the molecules of dispersion medium and thus binding large number of them into so called solvent shells.
2. Lyophillic colloids are in fact true solutions which mean that molecular species disperse homogenous in the system and not colloids,
3. Lyophobic colloids are heterogeneous and are highly disperse colloidal systems,
4. In lyophobic colloids the particles do not interact so strongly with the molecules of the surrounding medium,
5. The great difference in the size of the solvent and solute particles and to the structure of these particles. However to the large size of the molecules solutions of such substances are in many respects similar to colloid system.

Source

1. *Physical chemistry of milk, milk as colloidal System, http://ecourseonline.iasri.res.in.*

2. *K.S. Sharma, Physical Chemistry of milk, Agrimoon.com.*

Gels [1]

The word gel was coined by 19th-century Scottish chemist Thomas Graham by clipping from gelatine [5].

- ☆ Gels play a very significant role in the food system and also in the industrial application [1a]. A gel is a semi-solid that can have properties ranging from soft and weak to hard and tough [2,3].

Definition

The following definitions of gel are given-

- **i.** Gels are defined as a substantially dilute cross-linked system, which exhibits no flow when in the steady-state [4].
- **ii. IUPAC definition of Gel**- Nonfluid colloidal network or polymer network that is expended throughout its whole volume by a fluid [6,7].
  - ☆ By weight, gels are mostly liquid, yet they behave like solids due to **a three-dimensional cross-linked network** within the liquid. It is the crosslinking within the fluid that gives a gel its three dimensional structure (hardness) and contributes to the stickiness (adhesive ness). This structure of gel has increased their usefulness in various fields of application.
  - ☆ Gels are a dispersion of molecules of a liquid within a solid medium. A gel is formed when particles aggregate to such an extent forming a continuous network throughout the liquid. **A colloidal system**

in which liquid is a dispersed phase and solid is dispersion medium is known as Gel. Example-Cheese, jellies.

- It is immobile semi solid and has honey comb like structure. Gels do not show Tyndall effect, Brownian movement and electrophoresis.
- A gel has a finite, usually rather small, yield stress. A gel can contain followings:
  - i. A covalent polymer network, *e.g.*, a network formed by cross-linking polymer chains or by nonlinear polymerization.
  - ii. A polymer network formed through the physical aggregation of polymer chains, caused by hydrogen bonds, crystallization, helix formation, complexation, *etc.*, that results in regions of local order acting as the network junction points. The resulting swollen network may be termed a "thermo-reversible gel" if the regions of local order are thermally reversible.
  - iii. A polymer network formed through glassy junction points, *e.g.*, one based on block copolymers. If the junction points are thermally reversible glassy domains, the resulting swollen network may also be termed a thermo-reversible gel.
  - iv. Lamellar structures including mesophases. Sing *et al.* [8] define lamellar crystal and mesophase), *e.g.*, soap gels, phospholipids, and clays.
  - v. Particulate disordered structures, *e.g.*, a flocculent precipitate usually consisting of particles with large geometrical anisotropy, such as in V_2O_5 gels and globular or fibrillar protein gels.

Types of Gels [1,1a]

There are different types of gels depend upon the swelling agent present and elasticity etc.

a. Gels with Swelling Agent [1]

Hydrogel and Xerogel are the examples of this types of gel. The details of hydrogel and xerogel are given below:

i. Hydrogel

Gel in which the swelling agent is water. The network component of a hydrogel is usually a polymer network. A hydrogel in which the network component is a colloidal network may be referred to as an aqua gel [7]. Hydrophilic gels that are usually referred to as hydrogels are networks of polymer chains that are sometimes found as colloidal gels in which water is the dispersion medium [9].

ii. Xerogel

Open network formed by the removal of all swelling agents from a gel. Examples of xerogels include silica gel and dried out, compact macromolecular structures such as gelatin or rubber.

b. Gels with Elastic/Non Elastic Properties [1a]

There are two types of gels:

i. Elastic Gels

Elastic gel after gradual removal of water, becomes elastic solids and by adding water they can again be transformed into a gel, *e.g.* gum used for pasting. The linkages between the particles of a gel are due to electrical attraction and they are not rigid or strong.

ii. Non Elastic Gels

This type of gels are rigid and upon dehydration they set in to become glossy powder and thus lose their elasticity and the glossy powder cannot be converted back to gel by addition of water.

- ☆ These are prepared by suitable chemical action such as addition of concentrated hydrochloric acid to sodium silicate solution of correct concentration makes silica gel. Initially silicic acid is formed which gradually polymerizes to silica gel. This has a strong rigid covalent structure.

Food/Edible Gels [1,1a]

Edible gels are used widely in the food industry and mainly refer to gelling polysaccharides (*i.e.*Hydrocolloids) [1,10].Foods gels may be formed by proteins and polysaccharides, which may participate in gel formation in the form of solutions, dispersions, micelles,or even in disrupted tissue structures [1].

A food gel consists of a continuous phase of inter connected particles and/ or macro molecules intermingled with a continuous liquid phase such as water. Gels possess various degrees of rigidity, elasticity and brittleness depending on the type and concentration of the gelling agent, salt content and pH of the aqueous phase and temperature.

- **i.** Gelling agents present at levels of 10 per cent or less may be polysaccharides, proteins or colloidal complex particles such as casein micelles.
- **ii.** Firm gels can be prepared by few types of gums, pectins, and gelatin at levels of 1 per cent or lower.
- **iii.** Gels prepared with colloidal particles are generally not very rigid even when the solids content is considerably higher than 1 per cent.

iv. Some of the gels can be melted (liquefied) and reset with the addition or removal of thermal energy and these have been designated as thermo reversible.

v. Gels with covalent bonds between the molecules or complex particles however are generally thermo-irreversible.

Transformations of a Sol into a Gel [1a]

The conditions for conversion of a sol into gel are given below:

1. Temperature change
2. Chemical alteration of gelling agent
3. Reduction in number of charged groups by adjustment of pH or addition of salt and addition of a water competitive compound such as sugar.

During the sol gel transformation a three dimensional network is formed involving interaction of groups of polymer chains or particles to form cross linkages at the site of junction zones. The aqueous phase is entrapped in the interstitial areas of the structure.

a. In some gelling agents, the junction zones consist of microcrystallites involving specific chain units arranged in a crystal like fashion. Bonds in the junction zones are electrostatic, hydrophobic, covalent and hydrogen bonds.

b. Thermo reversible gels have preponderance of intermolecular hydrogen bonds whereas in a protein gel a few disulfide linkages per polymer chain may be sufficient to render them thermo irreversible.

Visco-Elastic Behavior of Gel [1a,1b]

Viscoelastic behaviour is a combination of elastic and viscous behaviour where the applied stress results in an instantaneous elastic strain followed by a viscous, time-dependent strain.

- A gel has an elastic behavior similar to a solid and is deformed when stress is applied to it but recovers to its original shape after removal of the stress as deformation has left the bonds between the particles intact.
- A gel also has viscous properties because part of the deformation is not recovered after the stress is removed and the gel flows since the bonds are broken and new bonds are formed as in a liquid when stress is applied.
- If the stress is applied for a very short duration the elastic deformation will be predominant *i.e.* recoverable. If the stress lasts long the viscous deformation is most conspicuous *i.e.* flow.

- ☆ A gel thus shows the **visco -elastic behavior** which is characterized by the following two rheological parameters:

  Gels exhibit an elastic or storage module (G′) and the viscous or loss module (G″).

  i. If the G′ is greater than the G″ the elastic properties prevail, and if the G′ is less than the G″ the viscous properties prevail.
  ii. The overall resistance to deformation is expressed in the combined modulus G*. The moduli give the ratio of the stress to the relative deformation and thus have the dimension of stress *e.g.* N.M-2 or Pa.

- ☆ They often strongly depend on the timescale of deformation. The above reasoning holds when the deformation is about proportional to the stress (so called linear behavior).

  i. Gels unlike solids often can be deformed considerably (*e.g.*by 10 per cent) and still show linear behavior.
  ii. At larger deformation this relationship breaks down and as the bonds in the net work are broken that do not reform within the time scale of the experiment.
  iii. A still larger deformation soon causes yielding or breaking of the gel and the net work is locally destroyed. The stress needed to do this may be called 'yield stress'.

- ☆ It is not a well defined quantity as it depends much on conditions like geometry of the measuring instrument and time scale.
- ☆ The rheological properties will be altered when the deformation of the gel results in permanently breaking of the bonds.

Source

1. *Wikipedia.*

1a. *K.S. Sharma, Physical Chemistry of milk, Agrimoon.com.*

1b. *JA Yanez, E Laarz and L Bergstrom, Viscoelastic Properties of Particle Gels, J Colloid Interface Sci, 1999 Jan 1; 209(1): 162-172.*

2. *Khademhosseini A, Demirci U (2016), Gels Handbook: Fundamentals, Properties and Applications. World Scientific Pub Co Inc. ISBN 978981465 6108.*

3. *Seiffert S, ed., 2015, Supramolecular Polymer Networks and Gels. Springer. ASIN B00VR5CMW6*

4. *Ferry JD, 1980, Viscoelastic Properties of Polymers. New York: Wiley. ISBN 0471048941.*

5. *Harper D."Online Etymology Dictionary: gel". Online Etymology Dictionary. Retrieved, 2013.*

6. *Jones RG, Kahovec J, Stepto R, Wilks ES, Hess M, Kitayama T, Metanomski WV (2008). IUPAC. Compendium of Polymer Terminology and Nomenclature, IUPAC Recommendations, 2008, (the "Purple Book") (PDF). RSC Publishing, Cambridge, UK.*
7. *Slomkowski S, Alemán JV, Gilbert RG, Hess M, Horie K, Jones RG, et al. (2011). "Terminology of polymers and polymerization processes in dispersed systems (IUPAC Recommendations 2011)" (PDF). Pure and Applied Chemistry. 83 (12): 2229-2259.*
8. *Sing KS, Everett DH, Haul RA, Moscou L, Pierotti RA, Rouquérol J, Siemieniewska T, 1985, "Reporting physisorption data for gas/solid systems with special reference to the determination of surface area and porosity". Pure Appl. Chem. 57: 603*
9. *Enas M. Ahmed, Fatma S. Aggor, Ahmed M. Awad, Ahmed T. El-Aref, An innovative method for preparation of nanometal hydroxide superabsorbent hydrogel, Carbohydr Polym, 91 (2013), pp. 693-698.*
10. *Phillips G. O. Williams P. A. 2000 Handbook of hydrocolloids. In Starch, (ed P.Murphy), Woodhead Publishing limited, Cambridge.*

Formation of Hydrogel [1,2]

Gelation (gel transition) is the formation of a gel from a system with polymers. Branched polymers can form links between the chains, which lead to progressively larger polymers. As the linking continues, larger branched polymers are obtained and at a certain extent of the reaction links between the polymer result in the formation of a single macroscopic molecule. At that point in the reaction, which is defined as **gel point,** the system loses fluidity and viscosity becomes very large. The onset of gelation, or gel point, is accompanied by a sudden increase in viscosity [3]. This "infinite" sized polymer is called the gel or network, which does not dissolve in the solvent, but can swell in it [4].

- ✰ Hydrogels may be synthesized in a number of "classical" chemical ways. These include one-step procedures like polymerization and parallel cross-linking of multifunctional monomers, as well as multiple step procedures involving synthesis of polymer molecules having reactive groups and their subsequent cross-linking, possibly also by reacting polymers with suitable cross-linking agents [2].
- ✰ The polymer engineer can design and synthesize polymer networks with molecular-scale control over structure such as cross-linking density and with tailored properties, such as biodegradation, mechanical strength, and chemical and biological response to stimuli [5].

Source

1. *Wikipedia.*

2. *Enas M.Ahmed, Hydrogel-Preparation, characterization, and applications, A review, Journal of Advanced Research, Volume 6, Issue 2, March 2015, Pages 105-12.*
3. *Odian, George, 2004, Principles of Polymerization. ISBN 0471274003.*
4. *Chanda, Manas, 2006, Introduction to Polymer Science and Chemistry, ISBN 9781420007329.*
5. *Sina Burkert, Thomas Schmidt, Uwe Gohs, Helmut Dorschner Karl-Friedrich Arndt cross-linking of poly(N-vinyl pyrrolidone) films by electron beam irradiation, Radiat Phys Chem, 76, 8-9, 2007, pp. 1324-1328.*

Properties of Gels [1,2]

A gel is a semi-solid that can have properties ranging from soft and weak to hard and tough. Gel has following properties:

1. **Swelling or Imbibition [1,2]-** In swelling process,partially dehydrated gels when dipped in water absorb water resulting in increase in the volume of gel. The tendency of gel to take up the large quantity of water or liquid and go on increasing in volume is called swelling or smiling of gel or imbibition of gel. Only elastic gels show this property.
2. **Hydration [1]-**The non elastic gels cannot be hydrated while elastic gels after complete dehydration can be regenerated by adding water.
3. **Synersis [1,2]-**Most of the inorganic gels on standing shrink and exude water and discharge water in the form of tear. This is known as synersis. This process is reverse of imbibition.
   - ☆ During storage of a gel,the system may shift to a more stable state with changes in the junction zones and solvent solute relationships. This process leads to synersis which involves spontaneous release of water and contraction of gel volume.
4. **Thixotrophy [1,2]-** In thixotrophy, there is reversible sol gel transformation. Some gels liquefy on shaking and reset when allowed to stand and this is known as thixotrophy.
5. **Fragile nature [2]-**In newly opened container of yoghurt and sour cream (gel) appears to be smooth and firm but once some of the material has been spooned out, little puddles of liquid appears in hollowed portion. This is because of the fact that as the spoon is plunged in to the gel, it pulls the nearby layers of the gel along with it, creating a shearing action that breaks it apart, releasing the liquid entrapped inside

Source

1. *K.S. Sharma, Physical Chemistry of milk, Agrimoon.com*
2. *Gels, Physical chemistrty, https: //thefactfactor.com.*

Milk as a Colloidal System [1,2]

Milk, when secreted by the mammary glands is a mixture of various substances which exist in different states and to maintain the stability of substances in native state,the maintenance of similar conditions continuously is essential.

- The constituents like casein micelles and minerals exist in the form of a colloidal suspension and exists in that condition as long as favorable conditions prevail.
- Milk is not a homogeneous fluid and at the same time milk is also not a pure colloidal system. It could be even observed that in some milk products, the fat globules exist as hydrophobic colloids along with air bubbles, and crystals of fat and lactose. The hydrophobic colloids are thermodynamically unstable.
- ***Milk* is an emulsified colloid of liquid butterfat globules dispersed within a water-based solution.**
- Milk has various components in the form of emulsions, colloids, molecular and ionic solutions. **Lactose in milk is present in form of true solution and milk fat as an emulsion whereas milk proteins are dispersed as a colloid.**
- The term colloid is not applicable to a given substance but rather to a specific state of a matter endowed with certain characteristic properties. Milk is a colloidal system in which liquid is dispersed in liquid. Whipped cream is a foam is which gas is dispersed in liquid.

Stability of the Colloidal System in Milk [1,2]

The capability of the colloidal system to remain as it, is called as stability of a colloidal system. The aggregation and sedimentation phenomena that determine phase separation, cause hindrance in stability of colloidal system.

Ostwald's Ripening

A phenomenon in solid solutions or liquid sols which describes the change of an inhomogeneous structure over time during which small crystals or sol particles dissolve, and redeposit onto larger crystals or sol particles is known as **Ostwald's ripening**. This happens due to the difference in solubility larger particles which tend to grow at the expense of smaller ones and the latter disappears.

- Generally the radii should be below 0.1µm to exhibit this effect for solids or liquids or below 100 µm for gases. Further, it is necessary that the substance in the particles be at least slightly soluble in the surrounding medium.

☆ In milk the hydrophobic colloids are subjected to this Ostwald's ripening particularly when they are very small. Brownian motion prevails when the particles are so smaller than < 1 μm. The smaller particles show a tendency to flocculate which could effectively be prevented by the steric repulsion caused by their surface material.

i. If the hydrophobic colloid particles are fluid (fat globules at normal temp and air bubbles) which may coalesce with each other if the thin film of continuous phase between closely approaching globules or bubbles is ruptured.

ii. Two colloidal particles of equal composition always attract each other because of van der Waals forces. If the attraction energy is not too strong the Brownian motion will be sufficient to keep the particle apart and maintain the colloidal stability.

iii. There will also be repulsion due to hydration on close contact of individual molecular chains when the protruding chains are hydrophilic. This is important for the stability of fat globules.

Source

1. *Physical chemistry of milk, milk as colloidal System.*

2. *K.S. Sharma, Physical Chemistry of milk, Agrimoon.com.*

Emulsion State [1]

The word "emulsion" comes from the Latin emulgere "to milk out," from *ex* "out" + mulgere "to milk", as milk is an emulsion of fat and water, along with other components, including colloidal casein micelles (a type of secreted biomolecular condensate) [1a].

☆ **Emulsion**, in physical chemistry, mixture of two or more liquids in which one is present as droplets, of microscopic or ultramicroscopic size (dispersed phase), distributed throughout the other (dispersion medium).

☆ An **emulsion** is a colloid or mixture of two or more liquids that are normally immiscible owing to liquid-liquid phase separation. For an emulsion to be stable, the two liquids must practically be insoluble in each other or only slightly soluble.

☆ Emulsions are part of a more general class of two-phase systems of matter called colloids. Although the terms colloid and emulsion are some times used interchangeably, emulsion should be used when both phases, dispersed and continuous, are liquids.

☆ Examples of emulsions include homogenized milk/milk, liquid biomolecular condensates, butter and margarine, cream. and some cutting fluids for metal working [1].

Emulsions, being liquids, do not exhibit a static internal structure. The droplets dispersed in the continuous phase (dispersion medium) are usually assumed to be statistically distributed to produce roughly spherical droplets. When molecules are ordered during liquid-liquid phase separation, they form liquid crystals rather than emulsions. Lipids, used by all living organisms, are one example of molecules able to form either emulsions (*e.g.*, spherical micelles; Lipoproteins) or liquid crystals (lipid bilayer membranes) [1].

Types of Emulsion [1,4]

A emulsion is mixture of oil and water (Two liquids). An emulsion has two phases, a continues phase (large quantity) and a discontinues phase particle size 0.1um 100um (present in small quantity).*e.g.* Milk. Two liquids can form different types of emulsions. Now based on the quantity of these liquids there are following two types of emulsion:

i. Oil in Water (O/W)

This is system in which consists of water is in large quantity hence is the continues phase (dispersion medium) and small quantity of oil (non-polar organic liquid,) which is the discontinues phase (dispersed phase). This is denoted by O/W. *e.g.* milk. In oil-in-water emulsion the lipid droplets are dispersed in water. Cream is also an example of this type of emulsion.

- ☆ Milk is an emulsion of this type with fat particles (globules) dispersed in an aqueous (watery) environment. The fat globules do not coalesce and form a separate layer (oil off or churn) because they are protected by a membrane layer which keeps the fat particles separate from the water phase.
- ☆ Homogenized milk is an emulsion of milk fat in water, with milk proteins as the emulsifier.

ii. Water in Oil (W/O)

This system consists of oil in large quantity (continues or dispersion phase) and water in small quantity (discontinues phase or dispersed phase) *e.g.*

- ☆ Butter, Margarine. In water in oil emulsion, the water droplets are dispersed in continuous oil phase.
- ☆ Mayonnaise is oil-in-water emulsions stabilized with egg yolk lecithin, or with other types of food additives, such as sodium stearoyl lactylate.

iii. K/A Multiple Emulsion

A third type of emulsion also exists **K/A multiple emulsion** which is (w/o/w). Multiple emulsions are also possible, including a "water-in-oil-in-water" emulsion and an "oil-in-water-in-oil" emulsion [2].

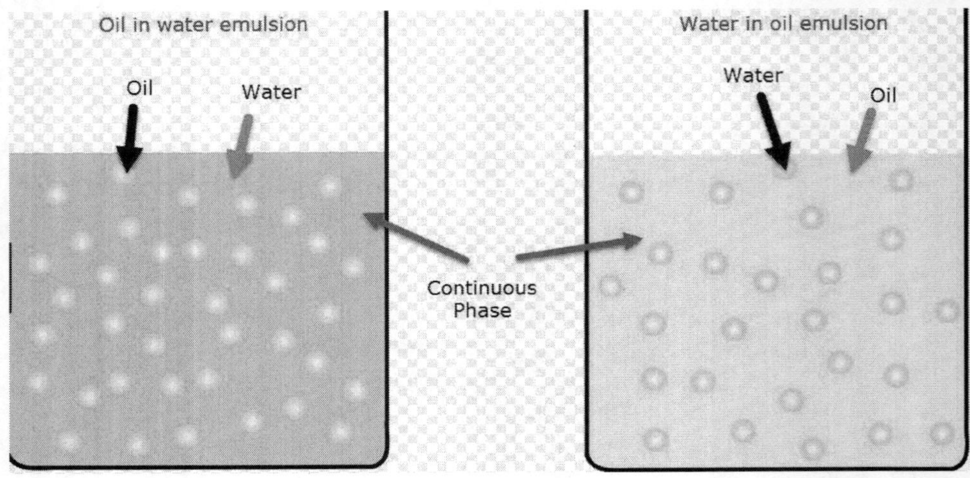

Figure 2.3: Types of Emulsions [1b].

Formation and Separation of Emulsion [3,4]

The two liquids that involve in the formation of emulsion must be insoluble in each other. The formation of the interface always requires work to be done and as such greater is the interfacial tension, greater is the amount of work required.

- Addition of emulsifying agents will reduce the interfacial tension by being adsorbed on the inter phase. Due to the stabilizing effect of these emulsifiers, substances with much larger size than that of ordinary colloids can also be stabilized.
- Emulsions are formed from the component liquids either spontaneously or, more often, by mechanical means, such as agitation, provided that the liquids that are mixed have no (or a very limited) mutual solubility.
- Unstable emulsions eventually separate into two liquid layers.
- Stable emulsions can be destroyed by inactivating or destroying the emulsifying agent *e.g.*, by adding appropriate third substances or also by freezing or heating.

The following **different methods are used to separate emulsions** into its constituent liquids -

i. Heating
ii. Centrifuging
iii. Freezing, *etc.*

Food Emulsions [4]

Food emulsions may possess solid particles (coagulated protein clumps) and gas bubbles in addition to the two liquid phases. Example for food emulsions are Butter (W/O), margarine (W/O), mayonnaise (O/W), salad dressing (O/W), milk (O/W), cream (O/W), non dairy creamer (O/W) and ice cream (O/W) [4]. The acceptability of food emulsions is dependent on attributes such as appearance (colour and opacity), texture (viscosity, plasticity, and oiliness) and flavour. Certain features of food emulsions prove to be advantageous for modifying some these attributes.

1. It is possible to incorporate oil, soluble flavouring and coloring matter or vitamins in an oil- in- water emulsion.
2. By dispersing droplets of particles with a diameter of 0.05 to 1.0µm into the continuous phase, it is possible to create opacity in the fluid product.
3. By increasing the concentration of the dispersed liquid to about 60 per cent or more it is possible to introduce the oil into an emulsion without imparting any oily sensation *e.g.* **mayonnaise**.

Source

1. *Wikipedia.*

1a. *Harper, Douglas, "Online Etymology Dictionary", www, etymonline.com. Etymonline, Retrieved 2 November 2019.*

1b. *Making an emulsion, https: //cwsimons.com.*

2. *Khan, A. Y., Talegaonkar, S, Iqbal, Z., Ahmed, F. J., Khar, R. K., 2006, "Multiple emulsions: An overview". Current Drug Delivery, 3 (4): 429-43.*

2a. *Emulsion, BYJU'S, The Learning App.*

3. *Emulsion chemistry, Britannica.*

4. *K.S. Sharma, Physical Chemistry of milk, Agrimoon.com.*

Appearance of Emulsion [1]

Emulsions contain both a dispersed and a continuous phase, with the boundary between the phases called the "interface" [2].

- ☆ Emulsions tend to have a cloudy appearance because the many phase interfaces scatter light as it passes through the emulsion.
- ☆ Emulsions appear white when all light is scattered equally. If the emulsion is dilute enough, higher-frequency (low-wavelength) light will be scattered more, and the emulsion will appear bluer – this is called the **"Tyndall effect"** [3].
- ☆ If the emulsion is concentrated enough, the color will be distorted toward comparatively longer wavelengths, and will appear more yellow. This phenomenon is easily observable when comparing skimmed milk, which contains little fat, to cream, which contains a much higher concentration of milk fat. One example would be a mixture of water and oil.

Source

1. *Wikipedia.*
2. *Loi, Chia Chun; Eyres, Graham T.; Birch, E. John, 2018, "Protein-Stabilised Emulsions", Reference Module in Food Science, Elsevier, ISBN 9780081005965.*
3. *Joseph Price Remington, 1990, Alfonso R. Gennaro (ed.). Remington's Pharmaceutical Sciences. Mack Publishing Company (Original from Northwestern University) (Digitized 2010). p. 281.*

Physical Properties of Emulsions [1,1a]

The important physical properties of emulsion which play significant roles, are given below:

i. Rheological Properties [1a]

These properties of emulsions play an important role in the practical utility of the emulsions. The rheological property helps in removing the product from the containers. The pouring out property of the emulsions depends on the size distribution of the suspended particles. Butter (an emulsion of water- in-oil type) has to be modified to get a product, butter spread by modifying its rheological property.

ii. Droplet Size Distribution [1a]

a. In coarse low concentration of emulsions (have with differences in their densities of the dispersed and dispersion phases), the phases either tend to rise to the surface or settle at the bottom of the vessel.

b. In highly concentrated emulsions the dispersion medium remains as a thin film between the particles of the emulsion. In such systems, it is the strength of this film that prevents the coalescence of the particles. The geometrical form of the particles in highly concentrated emulsions may not be spherical in shape.

iii. Optical Properties [1,1a]

The emulsion has optical property (Light scattering property). This property of the emulsion systems is the greatest for the shorter wave lengths *i.e.* for the blue and violet rays. Light scattering property is effectively studied by using the spectrophotometers [1a].

a. Two special classes of emulsions [1] *e.g.* **microemulsions and nano-emulsions**, with droplet sizes below 100 nm – appear translucent [2]. This property is due to the fact that light waves are scattered by the droplets only if their sizes exceed about one-quarter of the wavelength of the incident light. Since the visible spectrum of light is composed of wavelengths between 390 and 750 nanometers (nm), if the droplet sizes in the emulsion are below about 100 nm, the light can penetrate through the emulsion without being scattered [3].

b. Due to their similarity in appearance, translucent nanoemulsions and microemulsions are frequently confused. Unlike translucent nanoemulsions, which require specialized equipment to be produced, microemulsions are spontaneously formed by "solubilizing" oil molecules with a mixture of surfactants, co-surfactants, and co-solvents [2].

c. The required surfactant concentration in a microemulsion is, however, several times higher than that in a translucent nanoemulsion, and significantly exceeds the concentration of the dispersed phase. In addition, the stability of a microemulsion is often easily compromised by dilution, by heating, or by changing pH levels.

iv. Stability of Emulsion [1]

Common emulsions are inherently unstable and, thus, do not tend to form spontaneously. Energy input through shaking, stirring, homogenizing, or exposure to power ultrasound [4] is needed to form an emulsion. Over time, emulsions tend to revert to the stable state of the phases comprising the emulsion.

- ☆ An example of this is seen in the separation of the oil and vinegar components of vinaigrette, an unstable emulsion that will quickly separate unless shaken almost continuously.
- ☆ Emulsions are stabilized by agents that form films at the surface of the droplets (*e.g.*, soap molecules) or that impart to them a mechanical stability (*e.g.*, colloidal carbon or bentonite)]4a].

There are important exceptions to this rule. Microemulsions are thermodynamically stable, while translucent nanoemulsions are kinetically stable [2]. Whether an emulsion of oil and water turns into a "water-in-oil" emulsion or an "oil-in-water" emulsion depends on the volume fraction of both phases and the type of emulsifier (surfactant) present [1].

The following factors influence the properties of emulsions [1a]:

1. The continuous phase
2. Temperature
3. Average droplet size
4. Distribution of the droplets
5. Volume fraction of the oil dispersed in water
6. The oil that is dispersed
7. Various additives which help in the stability of the emulsion.

Source

1. *Wikipedia.*

1a. *K.S. Sharma, Physical Chemistry of milk, Agrimoon.com.*

2. *Mason TG, Wilking JN, Meleson K, Chang CB, Graves SM, 2006, "Nanoemulsions: Formation, structure, and physical properties" (PDF). Journal of Physics: Condensed Matter, 18 (41): R635-R666.*

3. *Leong TS, Wooster TJ, Kentish SE, Ashokkumar M, 2009, "Minimising oil droplet size using ultrasonic emulsification" (PDF), Ultrasonics Sono-chemistry. 16 (6): 721-7.*

4. *Kentish, S, Wooster, T.J., Ashokkumar, M.; Balachandran, S.; Mawson, R.; Simons, L. (2008). "The use of ultrasonics for nanoemulsion preparation", Innovative Food Science and Emerging Technologies. 9 (2): 170-175.*

4a. *Emulsion chemistry, Britanica.*

Chapter 3

Physical Properties of Milk

Physico-chemical Properties

Milk possess characteristic physico-chemical properties which are specific in nature. These are due to the constituents present in milk. The Physico–chemical Properties of milk depends upon the composite effects of its different constituents.

- ☆ Certain physical and physico-chemical properties of milk like density/specific gravity,viscosity, acidity, pH, buffering, surface tension, freezing point, oxidation-reduction, boiling point, electric conductivity,specific heat, *etc.*, are important characteristics of milk with a very narrow range of variation in these properties.
- ☆ Knowledge of physico-chemical properties of milk is essential for identification and effective quality control of milk *e.g.*
  - **i.** Adulteration of milk can be detected by specific gravity, freezing point and refractive index.
  - **ii.** Adulteration of ghee can be detected by refractive index.
- ☆ Such properties also help in selection and modification of processing parameters for manufacturing purposes *e.g.*in the processing of milk for ice cream, condensed milk, dried milk, butter, whey protein concentrate, *etc.*

The details description of the Important physico-chemical properties of milk are given below:

1. Density and Specific Gravity [1]

The definition and other details of Density and Specific Gravity are given below:

a. Density

Density has a direct relationship with weight and volume of a substance. The knowledge about density is helpful in understanding the behavior of substance when placed in a different environment [1a]. The density of a substance is its mass (weight) per unit volume. Density is expressed in gm/ml at a given temperature.

Density = weight/volume

Density is expressed in $kg.m^{-3}$ (SI units) or $g.ml^{-1}$. The symbol ρ used to express density. Since density closely depends on temperature it is usually denoted with temperature ρ 20 meaning the density at 20°C.

- ☆ Density is a characteristic property of milk. Though the composition of milk is variable but the density of milk remain within a very short range. Same is true for various constituents of milk *e.g.* milk fat [1]. The density of milk and milk products is used to convert volume into mass and vice versa. The relationship of this property with the temperature will also be helpful in developing various processing techniques of milk products [1a].
- ☆ Density is an absolute value *i.e.* absolute density. The density is dependant on the:
  - i. Temperature history of the material.
  - ii. Temperature at the time of measurement
  - iii. Inclusion of air (a complication with more viscous products).
  - iv. Composition of the material (especially the fat content).

The density of milk and milk products is used for the:

- **i.** To determine solid contents.
- **ii.** To calculate other physical properties (*e.g.* kinematic viscosity.
- **iii.** To estimate the solids content).
- **iv.** To convert volume into mass and vice versa.

Table 3.1 gives the density of various Milk and fluid dairy products as a function of fat and solids-not-fat (SNF) composition.

Table 3.1: Density of Various Fluid Dairy Products [1,2]

Product	*Composition*		*Density (kg/L) at*			
	Fat (Per cent)	*SNF (Per cent)*	*4.4°C*	*10°C*	*20° C*	*38.9°C*
Producer milk	4.00	8.95	1.035	1.033	1.030	1.023
Homogenized milk	3.6	8.60	1.033	1.032	1.029	1.022
Skim milk, pkg	0.02	8.90	1.036	1.035	1.033	1.026
Fortified skim	0.02	10.15	1.041	1.040	1.038	1.031
Half and half	12.25	7.75	1.027	1.025	1.020	1.010
Half and half, fort.	11.30	8.90	1.031	1.030	1.024	1.014
Light cream	20.00	7.20	1.021	1.018	1.012	1.000
Heavy cream	36.60	5.55	1.008	1.005	0.994	0.978

Source

1. *M.K. Srivastava, Handbook on Analysis of Milk, CBS Publications, New Delhi, 2010. ISBN 13: 9788123926681.*

1a. *K.S. Sharma, Physical Chemistry of milk, Agrimoon.com*

2. *Manual of University of Guelph.*

b. Specific Gravity [1,2]

It is the ratio of density of the substance to density of a standard substance (water) [1] equivalently, it is the ratio of the mass of a substance to the mass of a reference substance for the same given volume.

- ☆ Apparent specific gravity is the ratio of the weight of a volume of the substance to the weight of an equal volume of the reference substance. The reference substance is nearly always water at its densest (4°C) for liquids, for gases it is air at room temperature (21°C). Nonetheless, the temperature and pressure must be specified for both the sample and the reference [2].

  Specific Gravity= Density of substance/Density of water

  The mass or weight of a certain volume of the product divided by the mass of an equal volume of water is known as specific gravity. Thus specific gravity (s.g) of a product can be obtained by using the formula [1a].

$$s.g = \rho product/\rho water$$

- ☆ Generally density of water at 4°C is used as the standard for specific gravity for liquid and solids. Since the absolute value of water at 4°C is unity, the numerical values for absolute density and sp.gr. are identical [1].

- ☆ The density of substance varies with temperature and pressure so it is necessary to specify the temperatures and pressures at which the densities or weights are determined. Usually the density measurement is made nominally at 1 atmosphere (1013.25 millibars/ hectopascal) pressure. If the substances are incompressible liquids, the effect of atmospheric pressure on this property is negligible during measurement of density. As such the variations caused by changes in pressure are neglected. For true (in vacuo) specific gravity calculations, air pressure must be considered. Temperatures are specified by the following notation [1a]-

  Ts/Tr

  where, Ts - the temperature at which the sample's density was determined and Tr- the temperature at which the reference (water) density is specified. For example, s.g (20°C/4°C) would be understood to mean that the density of the sample was determined at 20°C and of the water at 4°C.

- ☆ The specific gravity of milk is usually expressed at 60°F(15.5°C). Milk is heavier than the water. The average specific gravity ranges from 1.028-1.030 for cow milk, 1.030-1.034 for buffalo milk and 1.035-1.037 for skim milk.
- ☆ The specific gravity of milk also depends upon the constituents because each constituent has different specific gravity *e.g.* water -1.000, Fat -0.93, Protein- 1.346, Lactose -1.666 and Salt – 4.12(SNF- 1.616).
- ☆ More SNF in buffalo milk results in a higher specific gravity than cow milk. The specific gravity of milk is lowered by addition of water and cream and increased by addition of skim milk [1].
- ☆ It is customary to represent specific gravity or relative density of any liquid as below:

  $$8.155\,^{20^{\circ}C}_{4^{\circ}C}$$

  where, The superscript indicates temperature at which the density of the material is measured, and the subscript indicates temperature of the reference substance to which it is compared.

- ☆ There is appreciable variation in specific gravity with the change in temperature of both the fluids. As such it is necessary to know both the temperatures to get more reliable information.

Source

1. *M.K. Srivastava, Handbook on Analysis of Milk, CBS Publications, New Delhi, 2010. ISBN 13: 9788123926681.*

1a. *K.S. Sharma, Physical Chemistry of milk, Agrimoon.com.*

2. *Specific Gravity determination, Impact Analytical, https: //www.impactanalytical.com.*

Differences between the Density and Specific gravity

The differences between Density and Specific gravity are shown in Table 3.2 [1]:

Table 3.2

Sl.No.	Specific Gravity (i.e. relative density)	Density
1	It is a dimensionless quantity and is mostly useful as a comparison but not in absolute units. It is a measure of a relative term.	It can measured in specific units. It is a measure of an absolute term. The dimensional formula of density is [M L^{-3}].
2	It does not provide the actual difference in the mass of two substances since it is only on relative basis. It tends to be rather vague.	Density will help in giving the exact mass of the product per unit volume.
3	It is more suitable for only liquids.	Density is for both solids and liquids.
4	Measurement of the specific gravity is simple and more convenient, less laborious and does not require any sophisticated equipment and quick results can be obtained.	Density needs specific equipment like pycnometer, sensitive balance and other materials.
5	It can be measured by people with minimum skill	Density measurement requires specially trained individual.
6	In a given,a large number of samples could be tested for specific gravity.	In a given time, only limited samples for density could be determined.
7-	Determination of specific gravity is more economical in comparison to the density determination.	Determination of density is less economical in comparison to the specific gravity determination.
8	It has no unit.	Unit of density is Kg/m^3
9	The calculation of specific gravity has applications in industries for the measurement of the solutions concentration.	It has several domestic as well as commercial applications.

Source

1. *K.S. Sharma, Physical Chemistry of milk, Agrimoon.com.*

2. *Difference Between Density and Specific Gravity, https: //www.vedantu.com.*

Effect of Various Processing Variables on the Density and Specific Gravity of Milk [1,2]

Milk is often subjected to various processes before being packed and sent to consumers. During processing of milk there will be change in the density or the specific gravity of milk. Density or specific gravity of milk is influenced by following factors:

A. Components of Milk [1,3]

Milk density is influenced by the combined effect of densities of various components present in milk. It is dependent on the amount of dissolved or suspended matter, changes in the constituents and the physical state of components in milk. Thus, milk density is effected by various factors such as temperature history of samples, biological differences of micelles, and processing of milk. Milk fat content is known to be the main cause for variation in the density of milk [3,4]. The physical state of the milk fat was observed to have greater influence on density.

- ☆ The types of milk and breed are the two factors that influence the density/specific gravity of milk since there is variation in the composition which would otherwise influence this property. Milk fat is not of consistent density.
- ☆ Milk fat is lighter constituent, the milk with higher fat content will have lower the specific gravity and vice versa. Buffalo milk contains more fat than cow milk and hence its specific gravity is higher than the cow milk. This is because buffalo milk contains more solids-not-fat as well, which ultimately results in a higher specific gravity [1,2].

The specific gravity of milk is decreased by the followings [1]:

i. Addition of water.
ii. Addition of cream (fat) and
iii. Increased temperature.

The specific gravity of milk is increased by the followings:

i. Addition of separated milk,
ii. Removal of fat
iii. Reduction of temperature.

B. Recknagel Phenomenon [1,3]

Increased density of cold stored milk has been known for a long time. Recknagel (1883), observed that the density or specific gravity of milk as soon as milking is lower than the same milk held for long periods of time, especially of milk under cold storage. Thus, this increase in density during storage was named Recknagel's phenomenon.

- ☆ The increase in the hydration of the protein(casein) at low temperature as the major cause for such a phenomenon rather than the escape of the air bubbles. Subsequent work carried out by other scientists attributed this phenomenon to the ratio between the liquid and solid.
- ☆ It was concluded that the fat caused the variations because fat-free milk showed no such variation and the variation in whole milk was linearly related to the fat content.

☆ Richmond stated that most of the rise in density was caused by the solidification of the fat. The study suggested that the specific gravity used in calculating the composition of milk should be determined when the fat is in a definitely known and reproducible physical state, which was the liquid state [3].

C. Temperature [1,2,3]

Density and specific gravity decreases with increase of temperature. This is the reason for obtaining corrected lactometer reading while calculating the specific gravity of milk at designated temperature of the lactometer.

Sp. gravity of milk is = 1 + CLR/1000

where, CLR is Corrected lactometer reading at a particular temperature expression.

☆ For attaining the maximum density and to avoid the Recknagel phenomenon it is suggested [British Standards Institution - specification No. 734 (1955)] that milk should be heated to a temperature of 40°C and specific gravity is determined when cooled to 20°C to ensure reproducible state of milk fat. The temperature history is necessary to secure high precision and accuracy with density determination.

D. Processing and other Factors [1,3]

Milk density is increased when it is stored under refrigerated conditions and this is due to the slow crystallization of the fat and change in the hydration of the globule membrane.

i. Fat content and temperature have been related to the density of creams. The specific gravity of cream decreases regularly with increase in the fat content.

ii. **Pasteurisation, Homogenization and Sterilization**- Pasteurisation has very small effect [3]. Homogenization slightly increases the density of whole milk but not of skim milk. Sterilization decreases the density of both milks. These changes are very small and negligible due large variations observed from sample to sample [1].

iii. **Removal of water from milk** – Condensation, Drying *etc.* are generally used to remove water from milk. Skim milk has specific gravity of 1.0320 to 1.0365 at 15°C/15°C.

a. Condensation

The removal of water in the manufacture of concentrated milk products raises the specific gravity. Baume hydrometer is widely used for this purpose. The Baume's reading is related to specific gravity as follows:

- ☆ Sp. Gravity at 60°F/60°F = 145/145-Be Where Be = Baume scale reading at 60°F.
- ☆ The concentrated milk products will have high viscosity as such the Baume's reading is often taken at 120°F instead of 60°F.

b. Drying

The milk with higher SNF content will have higher density. Similarly the processes in which water is removed (drying *etc.*) will have similar effect on the density of milk.

iv. Liquid Dairy Products [1]

Densities of the liquid dairy products such as whey, evaporated milk, sweetened condensed milk and freshly frozen ice cream vary in similar way as to milk.

Source

1. *Physical chemistry of milk, http://ecourseonline.iasri.res.in.*
2. *M.K. Srivastava, Handbook on Analysis of Milk, CBS Publications, New Delhi, 2010. ISBN 13: 9788123926681.*
3. *Ayako Ueda, Thesis on Relationship among milk density, composition and temperature, 1999, Faculty of Graduate Studies, The University of Guelph.*

Determination of Density of Fluid milk [1,2,2a]

The density or specific gravity of liquid/milk is usually measured by either determining the weight of a known volume or volume of known weight with a pycnometer/hydrostatic balance and with lactometer respectively. Another instrument used to determine the density of a liquid or a gas is the digital density meter - based on the oscillating U-tube principle.

i. Pycnometer (Pyknometer)/Specific Gravity Bottle (Typical ASTM Test method D1840) [1,2,2a]

A pycnometer (from Greek: πυκνός (puknos) meaning "dense") is usually made of glass, with a close-fitting ground glass stopper with a capillary tube through it, for facilitating the escape of air bubbles from the apparatus. This device enables accurate measurement of a liquid's density in reference to an appropriate working fluid, such as water or mercury, using an analytical balance [1a].

a. Specific Gravity of Liquids

The specific gravity of the liquid can easily be calculated if the weight of empty flask, flask with full of water, and flask with full of an experimental liquid are known [1a]. Density is determined by weighing a predetermined volume of the Milk/Material taken in a special container, after careful adjustment of the temperature of material (20°C).

- ☆ Pyknometer is a simplified form of specific gravity bottle used commonly for measuring sp gr of liquids like fluid milk. Density is determined by weighing milk/liquid in the pycnometer with a specific volume generally at 20°C.
- ☆ **Specific gravity bottle method [1b]**- It uses a specific-gravity bottle, *i.e.*, a flask made to hold a known volume of liquid at a specified temperature (usually 20°C). The bottle is weighed, filled with the liquid whose specific gravity is to be found, and weighed again. The difference in weights is divided by the weight of an equal volume of water to give the specific gravity of the liquid.

b. Specific Gravity of Powder [1a]

The particle density of a powder, can also be determined with a pycnometer. The powder is added to the pycnometer and is then weighed, to get weight of the powder sample. The pycnometer is then filled with a liquid of known density, in which the powder is completely insoluble. The weight of the displaced liquid can then be determined, and hence the specific gravity of the powder.

The pycnometric method gives the most accurate results but consume considerable time. It is thus not used generally in routine analysis, but used out for reference purposes [1].

ii. Hydrometer [1a,1b] and Lactometer [1]

a. Hydrometer [1a,1b,2a]

It is an instrument used to measure the specific gravity (or relative density) of liquids.

Principle

The hydrometer operates on the Archimedes principle that a solid suspended in a fluid will be buoyed up by a force equal to the weight of the fluid displaced. Thus, the lower the density of the substance, deeper the hydrometer will sink.

Construction of Hydrometer

A hydrometer is usually made of glass and consists of a cylindrical stem and a bulb filled with mercury or lead shot to make it float upright. Hydrometers usually contain a paper scale inside the stem, so that the specific gravity can be read directly. The scales may be Plato, Oechsle, or Brix, depending on the purpose.

Procedure

The liquid to be tested is poured into a tall jar or measuring cylinder and the hydrometer is gently inserted/lowered into the liquid until it floats freely. The point at which the surface of the liquid touches the stem of the hydrometer is noted.

Hydrometers may be calibrated for different uses, such as a lactometer for measuring the density (creaminess) of milk, a saccharometer for measuring the density of sugar in a liquid, or an alcohol meter for measuring higher levels of alcohol in spirits.

b. Lactometer (or Galactometer) [1]

Rapid determination of density of liquid is also made using the principle of floatation and standardized plunger called hydrometer, which is termed as lactometer while determining the specific gravity of milk.

Lactometer is used as an instrument for rapid determination of density of milk. The specific gravity of milk does not give a conclusive indication of its composition since milk contains a variety of substances that are either heavier or lighter than standard substance *i.e.* water. The average lactometer reading for normal cow whole milk is between 26-30. For buffalo milk the range is between 28-32. The lactometer reading can be changed to sp. gr. by dividing the LR by1000 and prefexing 1.0. Thus a reading of 32 indicates a specific gravity of 1.032.

Construction of Lactometer

It consists of a long, slender glass stem of uniform diameter connected to a larger glass air chamber. This chamber causes the instrument to float. Beneath this chamber at the lowest end of the body is a glass bulb which is filled with lead shots or mercury which causes the lactometer to sink to a proper level and also to float in an upright position in milk. Of the various types of lactometers in use, the Quevenne type is mostly followed in this country. The lactometer may or may not contain a built in thermometer. The lactometer is usually graduated at equal intervals from 15 at the top to 40 at the bottom, each division is being known as lactometer degree.

Principle

It is based on the principle of floatation which displaces specific volume of milk on floatation. Lactometer provides quick result with fair accuracy. In routine analysis density of fluid milk is expressed in terms of specific gravity, compared with density of water as unity at 15.5°C (60° F). Specific gravity of milk is usually determined for the purpose of calculating the solids contents in milk.

Procedure

i. Warm the milk sample to 40° C to 45°C and maintain at this temperature for 5 minutes.

ii. Mix the contents by rotating and inverting the bottle, taking care to avoid the formation of air bubbles and froath.

iii. Cool the sample to 15.5°C.

iv. Invert the sample bottle two or three times, pour enough milk into the lactometer jar taking care to avoid the formation of air bubbles, so that some milk overflows when the lactometer is inserted.

v. Insert the lactometer gently to wet the stem not more than a short length, about 3 mm beyond the position of equilibrium. The lactometer should float freely and not touch the sides of the cylinder.

vi. Allow the lactometer to remain steady in the milk. Take the reading within about 30 seconds. Note the reading of the lactometer corresponding to the top of the Meniscus of the stem without the error of parallax. The density/specific gravity of milk is determined by observing the Corrected Lactometer reading (CLR) of milk and putting the observed CLR in the following formula:

Specific gravity/Density = 1+ CLR/1000

Note: The CLR of milk is determined at the same temperature at which lactometer is calibrated. If the temperature of milk is different from the calibrated temperature then apply the correction to CLR as per the following:

Correction Factor = T- 80 or 60/5

where, T - Temperature of milk.

Figure 3.1: Hydrometer

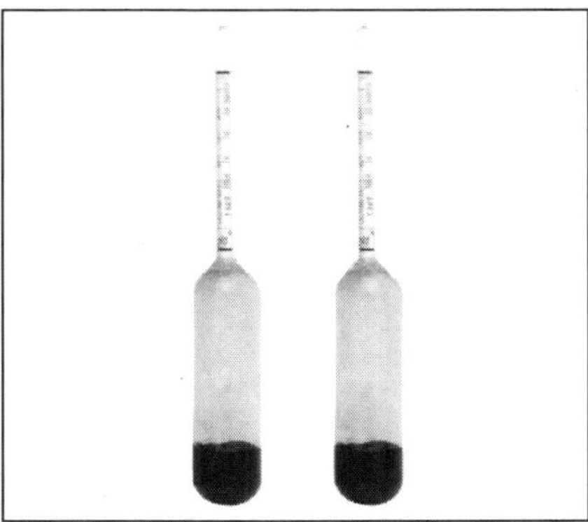

Figure 3.2: Lactometer

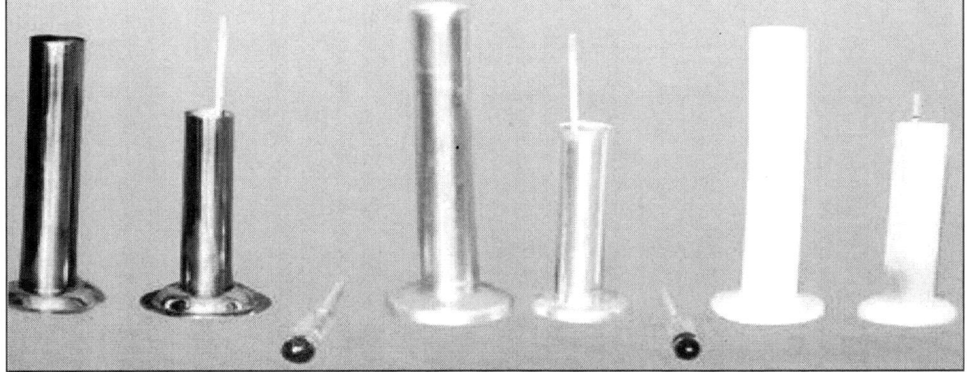

Figure 3.3: Lactometer and Jars.

If temperature of milk is more than the calibrated temp, then the observed correction factor is added to lactometer reading to get corrected LR and vice versa. Generally specific gravity of milk(having higher temp.) comes from village level Dairy societies or vendor, is determined by Lactometer calibrated at 84°C where as specific gravity of City supply/Tanker milk(having lower temp.) is determined by Lactometer calibrated at 60°C.

iii. Westphal Balance [1,1a]

A Westphal balance (also known as a Mohr balance) is a scientific instrument for measuring the density of liquids and it was devised by a German pharmacist, Karl Friedrich Mohr, and improved by Georg Westphal (1836-1902). It measures sp. gravity of liquids quickly and with better accuracy than lactometer.

Instrument

This is usually supplied in its box, along with the characteristic U-shaped balancing masses and the plummet in the upper right-hand corner. The sliding top of the box has the decorative molding attached to it. The arm is first balanced with the plummet totally immersed in water at 15.5°C. The plummet has a builtin thermometer and has a known volume (equal to 5 gms of water at 15.5°C) and mass (15 g).

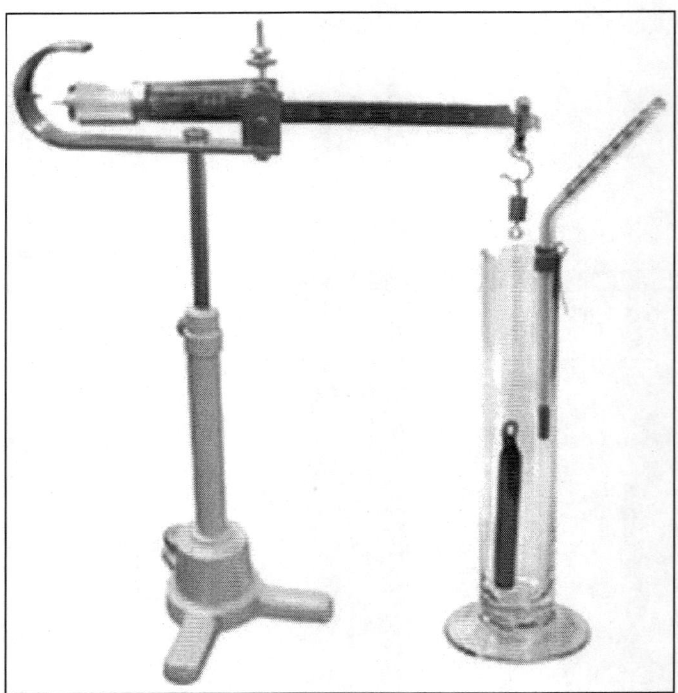

Figure 3.4: Westphal Balance.

Procedure

A 5 g mass is placed on the hook holding the plummet, and screw on the foot is adjusted until the index pointer on the end of the beam lines up with the point on the frame.

- ☆ The plummet is then completely immersed in the unknown liquid, and the system is rebalanced, using a series of riders on the nine equally paced notches on the beam, thus giving the value of the added mass for each decimal place.
- ☆ This gives the buoyant force of the liquid relative to water, and hence the specific gravity, which may be obtained to four decimal places.

iv. Digital Density Meters (Typical ASTM Test method D4052) [2]-

a. This type of instrument (Vibrating Element Transducers) requires a vibrating element to be placed in contact with the fluid of interest. The resonant frequency of the element is measured and is related to the density of the fluid by a characterization that is dependent upon the design of the element.

- ☆ In modern laboratories precise measurements of specific gravity are made using oscillating U-tube meters. These are capable of measurement to 5 to 6 places beyond the decimal point and are used in the brewing, distilling, pharmaceutical, petroleum and other industries.
- ☆ The instruments measure the actual mass of fluid contained in a fixed volume at temperatures between 0 and 80 °C but as they are microprocessor based can calculate apparent or true specific gravity and contain tables relating these to the strengths of common acids, sugar solutions, hydrocarbons *etc.*

b. Anton Par Model DMA 45 [3] digital density meter is also used, which has an accuracy of ±0.0001 g/cm^3 when it is used with a circulating thermostat having an accuracy of ±0.05°C.

- ☆ The rneasuring principle of the instrument is based on the change of the natural frequency of a hydrogen filled hollow oscillator, when filled with different liquids. The mass, and thus the density of the liquid, changes this natural frequency due to a gross mass change of oscillator caused by the introduction of the liquid.
- ☆ The electronic measurement of the time period is used to automatically calculate the density by a built-in arithmetic processor.

Source

1. *M.K. Srivastava, Handbook on Analysis of Milk, CBS Publications, New Delhi, 2010. ISBN 13: 9788123926681.*

1a. *K.S. Sharma, Physical Chemistry of milk, Agrimoon.com.*

1b. *Methods of Determining Specific Gravity, https: //www.infoplease.com.*

2. *Specific Gravity determination, Impact Analytical, http://www.impactanalytical.com.*

2a. *Wikipedia- Relative density.*

3. *Ayako Ueda, Thesis on Relationship among milk density, composition, and temperature, 1999, Faculty of Graduate Studies, The University of Guelph.*

2. Surface Tension [1,2]

The knowledge of surface tension is helpful in understanding the mechanisms of interfacial interactions such as adsorption, wetting or adhesion and more specifically in milk and food packaging industry in relation with wettability of systems.

- ✰ Surface tension is a property of the surface of a liquid that allows it to resist an external force,due to the cohesive nature of the water molecules [1b,2].
- ✰ In other words surface tension is the property of a liquid that makes it stretch like elastic [1b,2].
- ✰ Surface tension is defined as the force acting at right angles to any line of unit length on the surface of a liquid or as the force along a line of unit length where the force is parallel to the surface but perpendicular to the line [1,1a,2].
- ✰ Surface tension is **the energy, or work, required to increase the surface area of a liquid due to intermolecular forces**. Since these intermolecular forces vary depending on the nature of the liquid (*e.g.* water vs. gasoline) or solutes in the liquid (*e.g.* surfactants like detergent), each solution exhibits differing surface tension properties [1a].
- ✰ An equivalent definition is work done per unit area and is measured in SI system as joules per square meter and in the cgs system as ergs per cm^2. Although these two expressions are equivalent, but it is customary to refer the term **surface energy** when the energy per unit of area is to be mentioned. The term surface energy is more of a general term in the sense that it also applies to solids and not just liquids [2].
- ✰ The **cohesive forces** between liquid molecules are responsible for the phenomenon known as surface tension [1b,2].

Cohesion is the attraction between molecules that are like each other. Cohesive forces are intermolecular forces which causes liquid to resist separation. For example, in water, there are strong cohesive forces, hydrogen bonds, that causes the rain to come down as a water drop rather than a mist [3].

Adhesion is the interaction between unlike molecules. Adhesion force is caused by forces acting between two substrates, such as mechanical and electrostatic forces. When two immiscible liquids are brought into contact, the adhesion forces play a role. Adhesion forces are also important when liquid is brought in contact with solid [3].

Mechanism [1b,2]-A molecule in the interior of a liquid is attracted in all directions because it is surrounded by molecules equally in all directions resulting in a net force of zero. However, the molecules at the surface of the liquid are attracted sideways and downwards but not upwards creating some internal pressure and thus forcing liquid surfaces to contract to the minimal area [1b].

In terms of energy, a molecule in contact with a neighbor is in a lower state of energy than when it is not in contact with a neighbor. The molecules in the interior of liquid have as many neighbors. The molecules at boundary are having some missing neighbors and therefore have a higher energy. For a liquid to minimize its energy state the number of higher energy boundary molecules must be minimized and this results in a minimized surface area [2].

- ☆ The cohesion or tension of molecules at the surface, resulting from imbalance of forces acting on them, convert the surface to act as though covered with a film or skin. This, phenomenon is known as surface tension. Surface tension is responsible for the shape of liquid droplets [1,2].
- ☆ The area of contact between two phases is called the interface and is known as surface when one of the two phases is gaseous phase. The properties of interfaces and surfaces are determined by the number, kind, and orientation of the molecules located in them [2].

Unit of Surface Tension

Surface tension has the dimension of force per unit length or of energy per unit area. Surface tension, represented by the symbol γ. Surface tension is expressed as newton per meter in SI unit. Dynes per centimeter (dynes/cm) in cgs unit is also used [1,2].

Surface Tension of Milk [1,2]

The Values of surface tension of cow and buffalo milk at 20°C are 52 dyne/cm and 50.5dyne/cm respectively. Surface tension of milk falls in the range of 40 to 60 dynes/cm measured at 20°C and the mid value of this range will give the average value.

- ☆ A value of 50 dynes is commonly taken as surface tension of milk at 20°C. Water has a surface tension of 72.75 dynes/cm at 20°C. The lower value of milk compared to water is due to substances which lowers the surface tension of milk.

- ☆ The milk proteins, fat, phospholipids and free fatty acids are the principal surface active components that contribute to the surface properties [2]. Fat significantly lowers the surface tension of milk as cream has a surface tension value of 39-40 dynes/cm [1].
- ☆ Lipolysis also lowers down the surface tension of milk. Surface tension varies with the temperature and Total solids in milk.

Surface Tension Depends on Temperature and Pressure [3]

- ☆ One of the main factors affecting surface tension of pure liquids is the temperature. Experimentally it has been found that the surface tension decreases almost linearly with temperature. When temperature increases, the molecular thermal activity increases causing a decrease in cohesive interaction. This causes decrease in surface tension.
- ☆ Another factor affecting the surface tension is pressure. In most applications the pressure does not play a role, but needs to be considered when processes that happen under high pressures are studied. In general, dissolution of the gas in liquid increases with the increasing pressure, decreasing the surface tension.

Examples of Surface Tension [1b,2]

Some examples of surface tension are given below:

i. **Floating a needle-**Surface tension can be demonstrated by carefully horizontally placing needle at the surface of the water where it will float due to the forces of surface tension [1b,2].

ii. **Why bubbles are round-** The surface tension of water provides the necessary wall tension for the formation of bubbles with water. The tendency to minimize that wall tension pulls the bubbles into spherical shapes [1b].

iii. **Soaps and detergents-** These help the cleaning of clothes by lowering the surface tension of the water so that it more readily soaks into pores and soiled areas [1b].

iv. **Washing with cold water-**The major reason for using hot water for washing is that its surface tension is lower and it is a better wetting agent. But if the detergent lowers the surface tension, the heating may be unnecessary [1b].

Interfacial Tension [2,3]

Surface tension is the property of the liquid in contact with gas phase (usually air). Interfacial tension, on the other hand, is the property between any two substances. It could be liquid-liquid, liquid-solid or solid-air. Interfacial tension is somewhat similar to surface tension in that cohesive forces are also involved. However the main forces involved in interfacial tension are adhesive forces (tension) between the liquid phase of one substance and either a solid, liquid or gas phase of another substance. The interaction occurs at the surfaces of the substances involved, that is at their interfaces [3].

The tension caused across the two phases in a liquid is known as interfacial tension. The more effective depressants of interfacial tension tend to be concentrated at the interface to the exclusion of other substance present there. In many cases the amount of material concentrated at the interface is greater than would be predicted.

- ☆ In milk the important interfaces are those between the liquid product and air and between the milk plasma and the fat globules.
- ☆ Studies on the surface tension (liquid/air) have been made to ascertain the relative effectiveness of the milk components as depressants, to understand the release of the surface active components as a result of processing and the release of the free fatty acids during lipolysis and to explain the characteristic foaming phenomenon of the milk.
- ☆ Interfacial tensions between milk fat and solutions of milk components have been measured during the stabilization of fat globules in raw and processed milks [2].

Determination of Surface Tension [1,2,3]

The capillary-rise method was the first method used to measure surface tension. Today, force and optical tensiometry are the most widely accepted methods to measure static surface and interfacial tension due to automation and disposable sample vessels enabling faster measuring cycles [3]. The methods used for measuring the surface and interfacial tension may be grouped as dynamic and static. Majority of the information on surface and interfacial phenomena in milk has been obtained with various static methods. The major principles involved in the determination are given below [2]:

i. Rise in the height of liquid in a capillary tube.
ii. Weight of drops formed by specific volume of liquid flowing from a capillary tip. Some times this method is considered to be semi dynamic method,
iii. Force required to pull a ring or plate out of the surface.
iv. Maximum pressure required to force a bubble of gas through a nozzle immersed in the liquid and.
v. Shape of a drop hanging from a capillary.

Surface Tension Measuring Methods and Principle [1,2,3]

Several methods of measuring surface tension in milk, are available. The details related to the methods and principle on which these methods work, are given on next page:

i. Ring Detachment or Tensiometer Method [1,3,4]

The surface tension of a liquid can be measured by the duNouy ring method which involves slowly lifting a ring, often made of platinum, from the surface of a liquid. The force required to pull the ring from the liquid's surface is measured and related to the liquid's surface tension.

These instruments are based on measuring the forces exerted on a probe that is positioned at the liquid-gas or liquid-liquid interface. The probe is connected to a very sensitive balance and the liquid interface of interest is brought into contact with the probe. The forces measured by the balance as the probe interacts with the surface of the liquid can be used to calculate the surface tension [3].

- ✰ This method is rapid, simple, more reliable and give an accurate result ± 0.25 per cent. The result could be more accurate when the force is measured with an analytical balance [2].
- ✰ Probes are typically made of platinum which helps to ensure a zero degree contact angle between the probe and the liquid to be studied. Two configurations of probes are commonly used; the duNouy ring(Platinum ring) and the Wilhelmy plate [3]. Apparatus employing tension balance with various degrees of sensitivity (du Nouy balances) are also available [2].

ii. Drop Weight Method [1,2]

The principle is to measure the weight of drops of a fluid of interest falling from a capillary glass tube, and thereby calculate the surface tension of the fluid.

- ✰ The weight of the falling drops is determined by counting them and from it determination of surface tension can be done. Here the number of drops formed when a given amount of liquid is allowed to fall from a pipette is measured.

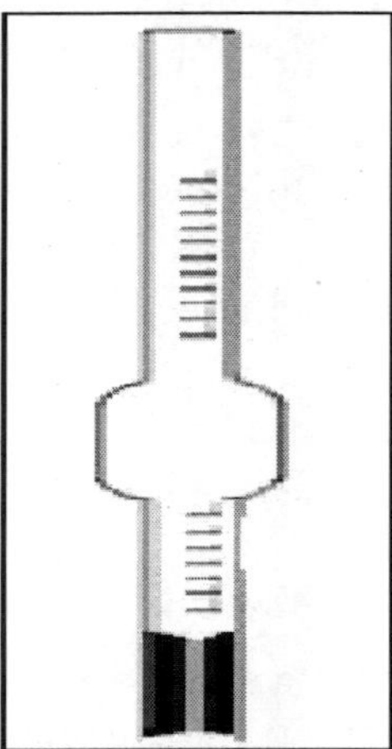

Figure 3.5: Stalagmometer, Straight Form.

☆ The instrument used to measure drop weight is known as **Stalagmometer**. It consists of a glass tube with a uniform diameter of 1-3 mm with a small bulb which is sharply grounded so that the liquid drops from the tube in an almost spherical form. The tip of the tube can be enclosed in a circulating water bath to control the temperature. Droplets falling from the tube are collected and weighed.

iii.

The pressure required to force a bubble through a nozzle immersed in the liquid. A measurement technique for determining surface tension at short surface ages. Maximum pressure of each bubble is measured.

Source

1. *M.K. Srivastava, Handbook on Analysis of Milk, CBS Publications, New Delhi, 2010. ISBN 13: 9788123926681.*

1a. *Surface Tension, Chemistry, Libretexts, https: //chem. libretexts.org.*

1b. *Surface Tension and Water, USGS, https: //www.usgs.gov.*

2. *K.S. Sharma, Physical Chemistry of milk, Agrimoon.com.*

3. *Susanna Laurén, Difference between Surface tension and Interfacial tension, Biolin Scientific, 2018.*

4. *Du Nouy, P.L., "A new apparatus for measuring surface tension", Journal of genetic physiology 1, 1919, 521.*

Factors Influencing Surface Tension Determination of Milk [1,2]

The factors affecting the measurement of surface tension are given below:

i. Temperature history of the milk
ii. Age of the milk
iii. Time required for measurement and.
iv. Correction factors required for individual instruments.

The surface tension of milk is influenced by several factors and it is difficult to specify with any certainty an average value.

The surface active substances tend to concentrate in the surface in proportion to their concentration and ability to lower the surface tension and. Soluble constituents like salts and sugars do not tend to concentrate at the surface because they do not lower the surface tension. Various dilution studies of milk, skim milk, whey, and solutions of milk proteins reveal that-

i. **Casein and whey proteins** of lactalbumin fraction (β-lactoglobulin, α-lactoglobulin and bovine serum albumin) are powerful depressants while immunoglobulin fraction are not having greater influence on this property of milk.

ii. **The shorter chain fatty acids-** These contribute more to the rancid flavour and the larger ones are more effective surface tension depressants. The fatty acids released by lipolysis are very effective depressants of surface tension. Butyric acid has little action on reducing the surface tension, acids with longer chain lengths particularly capryllic, capric and possibly mono and di glycerides are responsible for lowering of surface tension due to lipolysis.

iii. **Homogenization [2]**–The expanded fat surface in homogenized milk withdrew surface-active materials (protein) from the plasma and raised the surface tension. If lipase was not inactivated by heat, it was activated by homogenization to produce fatty acids which lowered the surface tension.

In milk containing less than 5 to 7 per cent fat the fatter effect predominated; whereas, with a higher per cent fat, aging was required to overcome the initially predominating former effect. The latter (enzyme) effect was destroyed by preheating the milk to 150°F. (64.6°C.) but not to 140 ° F. (60.0 ° C.)

iv. **Heat treatment [1]** -It has little effect on the surface tension except that sterilization treatments cause an increase by few dynes cm-1 coinciding with grain formation. This effect undoubtedly results from denaturation and coagulation of the proteins so that they are no longer effective surface active agents.

v. **Cooling of milk [2]-** Cooling milk for 45 min. or longer to less than 10°C. caused all average decrease in surface tension of 2.12 dynes/cm. An average additional drop of 3.64 dynes/cm occurred on standing 12 hr, Rewarming to 20 or 37°C. caused only a small fraction, of the reverse change, whereas rewarming to 50°C. restored the original surface tension value.

Table 3.3: Surface Tension of Various Milk Products [1]

Sl.No.	*Milk Products*	*Surface Tension (dyne/cm)*
1	Rennet whey	51-52
2	Skim milk	52-52.5
3	Whole milk	46-47.5
4	Cream 25 per cent	42-45
5	Sweet cream butter milk	39-40

It could be revealed from the above data that [1]:

- ☆ The proteins of whey are not effective surface tension depressants.
- ☆ The fat globules in milk and cream (as well as traces of free fat on the surface) serve to reduce the surface tension significantly below that of skim milk.

- ✰ The fat globule membrane material released from the fat globules during churning of cream are exceedingly surface active and are responsible for the very low surface tension of butter milk.

Significance of Studies of Surface Tension in Milk [2]

i. Detection of Adulteration

The simplest objective in studying the surface tension of milk has been to detect adulteration with water, starch, foreign fat. Other features of change in surface tension clearly rule out surface tension as a simple test for adulteration.

ii. Bacterial Growth

Bacterial growth as affected by the surface tension of milk should be considered in relation to older studies on effects of surface tension in other media. The nature of the depressant, the kind of bacteria, and sometimes the salt concentration of the medium were found to be important factors affecting the relation of growth to surface tension. Surface tension depressed with Tween 40 did not retard growth of *S. lactis* in milk. Toxic specific fatty acids, rather than lowering surface tension, were suggested as the cause of the above effect. Oleic acid lowered surface tension but not growth rate, whereas caprylic acid lowered growth rate but not surface tension.

iii. Others

It was early shown that the surface tension of milk increased with decreased concentration of fat or casein caused, respectively, by skimming or acid precipitation. Replacement of the constituent again lowered the surface tension. Effects of udder inflammations, pregnancy, stage of lactation, and feed also were noted.

Source

1. *K.S. Sharma, Physical Chemistry of milk, Abgrimoon.com.*
2. *C. H. Whitnak, The surface tension of milk. A review, Department of chemistry, Kansas state university, Manhattan, https: //www.journalofdairyscience.org.*

3. Viscosity [1,2]

Viscosity of a fluid (liquid or gas) is a measure of its resistance towards flow. This resistance is due to internal friction within a liquid as they slide each other [1]. Viscosity, resistance of a fluid to a change in shape, or movement of neighbouring portions relative to one another. Viscosity denotes opposition to flow.

- ✰ Viscosity is termed as "thickness" or "internal friction". Thus, water is "thin", having a lower viscosity, while milk is "thick", having a higher viscosity. The nature, concentration and type of milk components have a direct bearing on the viscosity [2].

- ✰ In others words viscosity could simply be defined as the resistance of liquid to flow or pour, meaning the less viscous the fluid is, the greater its ease of movement or fluidity [2].
- ✰ Viscosity of liquids can be defined as the resistance of a liquid to the motion of its layers relative to one another [1,2].
- ✰ Viscosity is also defined as ratio of shearing stress (S) to the shear rate which pertains to simple shear flow (*i.e.* laminar flow with parallel streamlines). In such fluids the shear rate is then equal to the velocity gradient G which is perpendicular to the direction of flow [2].
- ✰ The usual symbol for dynamic viscosity is the Greek letter μ. The symbol η is also used by chemists.

Units of Viscosity [2,2a]

The details related to the units of viscosity in cgs and SI systems are given below:

Unit in cgs

The c.g.s physical unit for dynamic viscosity is the poise(named after Poiseuille). It is more commonly expressed as centi poise (cP).

- ✰ A poise is the force of one dyne acting on area of one square centimeter between two parallel planes one centimeter apart, to produce a difference in flow rate between the planes of one centimeter per second.
- ✰ It is commonly expressed, particularly in ASTM standards, as **centipoise** (cP), because it is more convenient (for instance the viscosity of water at 20 °C is about 1 cP), and one centipoise is equal to the SI millipascal second (mPa·s).
- ✰ In milk centipoise is commonly used to express viscosity, which is one hundredth of a poise.

Unit in SI System

The SI physical unit of dynamic viscosity is the pascal-second (Pa·s) which is equivalent to N·s/m^2 (newton-second per square meter), or kg/m·s (Kg per meter per second).

Viscosity of Milk [1,2]

The viscosity of a heterogeneous substance such as milk at a given temperature depends upon its composition and the physical state of its colloidally dispersed substances, including milk fat.

- ✰ The viscosity of fluids is influenced by various factors and knowledge of which is of immense value in understanding the behavior of these fluids during processing [2]. It ranges between 1.5 to 2.0 centipoise at

20°C.Due to fat emulsion and colloidal particles milk is viscous than water.

- Any alternation in the physical nature of fat or protein hydrolysis, cooling or heating of milk affects proteins and fat and thus the viscosity. Clustering of fat globules affects viscosity *e.g.* cream where viscosity increases due to clustering of fat globules.
- Likewise homogenization of milk results in the state of sub-division of dispersed constituents *e.g.* fat. Thus homogenization of milk increases the viscosity.
- Viscosity of milk and milk products is important in determining the followings:
  - i. The rate of creaming.
  - ii. Rates of mass and heat transfer.
  - iii. The flow conditions in dairy processes [1].

Viscosity of Concentrated Milk and Evaporated Milk

- Viscosity increases also due to heating and concentration *e.g.* condensed milk due to increased total solids and changes in milk constituents [1].
- The viscosity of evaporated milk passes through well-defined changes arising from the manufacturing processes and the storage conditions to which it is subjected. There is first a thickening during the heat of sterilization, then a thinning during early storage and finally, after long periods of storage, the viscosity may again increase until a gel has formed.
- The magnitude of the decrease in the viscosity of evaporated milk during early storage was dependent upon the storage temperature.
  - i. At temperatures below 16°C (60.8 ° F.) the loss of body was small.
  - ii. At temperatures above 30° C. (86 ° F.) evaporated milk sometimes lost 40 per cent of its original viscosity during the first 10 days of storage [4].
  - iii. Increase in time and/or temperature of the process results in higher viscosity of the evaporated milk or in more coagulation.
- The viscosity of evaporated milk decreases rapidly on vigorous agitation and slowly during quiescent aging. "Age thinning" of evaporated milk is retarded significantly by cool storage temperatures. These facts indicate that the suspending medium in evaporated milk does not exhibit true viscous flow characteristics.

Viscosity of other Products

The viscosity of milk and other products is tabulated on next page [2]:

Table 3.4: Viscosity of Milk and Milk Products

Sl.No.	Type of Milk	Viscosity (centi poise)
1	Whey	1.2
2	Skim Milk	1.5
3	Whole milk	2.0

Factors affecting Viscosity of Milk [1,2]

Viscosity of milk and milk products is important in determining the flowing rate of cream, rates of mass and heat transfer, the flow conditions in dairy processes. Milk and skim milk, excepting cooled raw milk, exhibit Newtonian behavior in which the viscosity is independent of the rate of shear [2]. Important factors that influence the viscosity of milk are as follows [1,2]:

pH

An increase or decrease in pH of milk also causes an increase in casein micelle voluminosity.

Temperature [1,2,2a]

The viscosity of milk and milk products depends up on the temperature and on the amount and state of dispersion of the solid components.

i. The viscosity of liquids decreases rapidly with an increase in temperature, and the viscosity of gases increases with an increase in temperature.

ii. Cooler temperatures increase viscosity due to the increased voluminosity of casein micelles,

iii. Temperatures above 65°C increase viscosity due to the denaturation of whey proteins [1]. The effect of temperature on the viscosity of normal fluid milk and in comparison to the various other fluids is given in following table [2].

Table 3.5: Effect of Temperature

Temperature °C	Viscosity (cp)			
	Water	5 per cent Lactose	Skim Milk	Whole Milk
5	1.519	1.76	3.96	3.254
10	1.308	1.50	2.47	2.809
15	1.140	-	2.10	2.463
20	1.005	1.15	1.79	2.127
25	0.894	1.03	1.54	1.857
30	0.894	0.91	1.33	1.640
35	0.723	-	1.17	-
40	0.656	0.74	1.04	-

By going through the aforesaid table, followings important points are observed:

- The viscosity of 5 per cent lactose solution at a given temperature is not much greater than that of water but skim milk has an appreciable viscosity approaching to that of whole milk.
- The difference in the viscosity between 5 per cent lactose and skim milk indicates the role of milk proteins on the viscosity of milk.
- Lactose, whey proteins and milk salts have relatively small contribution towards the viscosity of milk whereas the casein alone is contributing substantially towards this property.
- The tendency of milk to increase the viscosity upon heating as it will approaches to a point of coagulation of proteins. This tendency of milk is the basis for producing high viscosity super heated condensed milk.
- Conditions and treatments that affect the stability of casein are very significant in the viscosity of milk.
- Acidity, salt balance, heat treatment and the action of various enzymes and bacteria are some of the factors which affect the stability of caseins.

Concentration of Fat and Agitation [1,2]

The effects of concentration of fat and agitation on viscosity of milk, are discussed below:

1. Viscosity increases with increasing concentration of fat and solids not -fat, but consistent general relationship could not be established [2].
   - Viscosity varied exponentially with temperature and linearly with fat content of milk [3]
   - Viscosity for the fat and temperature range measured varied from 1 to 60 centipoise [3].
2. Agitation cause partial coalescence of the fat globules (partial churning) which increases viscosity. Fat globules that have under gone cold agglutination, may be dispersed due to agitation, causing a decrease in viscosity.
   - Milk and skim milk (except cooled raw milk) exhibit Newtonian behavior, in which the viscosity is independent of the rate of shear. Cooled raw milk and cream exhibit **non-Newtonian** behavior in which the viscosity is dependant on the shear rate.

State and Concentration of the Protein [2]

The viscosity of colloidal systems depends upon the volume occupied by the colloidal particles. Changes in the caseinate micelles produced by either raising or lowering the pH results in increased viscosity.

- ☆ The viscosity is approximately doubled by the addition of 10 ml of 1.4 to 3.8 N ammonia to 90 ml milk.
- ☆ Addition of alkali (pH up to 11.7), urea (up to 4.8M) and calcium complexing agents to concentrated (22.7 per cent solids) skim milk causes a marked transient increase of several folds in viscosity followed by a sharp decline. This is due to the swelling of the micelles followed by their disintegration.

Source

1. *M.K. Srivastava, Handbook on Analysis of Milk, CBS Publications, New Delhi, 2010. ISBN 13: 9788123926681.*

2. *K.S. Sharma, Physical Chemistry of milk, Agrimoon.com.*

2a. *Wikipedia- Viscosity.*

3. *A. S. Bakshi and D. E. Smith, Effect of Fat Content and Temperature on Viscosity in Relation to Pumping Requirements of Fluid Milk Products, 1984, J Dairy Sci 67: 1157-1160.*

4. *E. F. Deysher, B. H. Webb, and G. E. Holm, The Viscosity of Evaporated Milks of Different solids concentration,Division of Dairy Research Laboratories, Bureau of Dairy Industry, Agricultural Research Administration, U. S. Department of Agriculture.*

Types and Forms of Viscosity [1,2]

Newton's law of viscosity defines the relationship between the shear stress and shear rate of a fluid subjected to a mechanical stress [2]. Newton's law of viscosity is not a fundamental law of nature but an approximation that holds in some materials and fails in others. Thus there exist following forms of viscosity:

1. Newtonian Fluids

Newtonian fluids obey Newton's law of viscosity. The viscosity is independent of the shear rate. Fluids for which the viscosity coefficient depends only on temperature and pressure and is independent of the rate of shear are called " Newtonian".

- ☆ Newtonian fluid is the fluid in which the viscosity remains constant for all rates of shear if constant conditions of temperature and pressure are maintained. In common terms, this means the fluid continues to flow, regardless of the forces acting on it.

Examples of Newtonian Fluids

Some examples are discussed below:

1. Water is Newtonian, because it continues to exemplify fluid properties no matter how fast it is stirred or mixed.

2. Other examples may be aqueous solutions, emulsions. Alcohol, glycerol, and thin motor oil are also examples of this type of fluids.
3. Gases, pure liquids, and solutions of materials of low molecular weight exhibit behavior of this type.

Skim milk and whole milk do not differ appreciably from Newtonian behavior.

For a Newtonian fluid, the viscosity depends only on temperature and pressure not on the forces acting upon it if the fluid is a pure substance or else it will also be influenced by the chemical composition if it is not a pure substance.

2. Non Newtonian Fluids [1,2]

Non-Newtonian fluids do not follow Newton's law and, thus, their viscosity (ratio of shear stress to shear rate) is not constant and is dependent on the shear rate [2].

- ☆ A non-Newtonian fluid is a fluid whose flow properties differ in any way from those of Newtonian fluids.
- ☆ Generally speaking, a non-Newtonian fluid is defined as one in which the relationship between shear stress and shear rate (S/R) is not constant. Most commonly the viscosity of non-Newtonian fluids is dependent on shear rate or shear rate history.
- ☆ The viscosity of non-Newtonian fluids changes as the shear rate is varied. Non-Newtonian fluids exhibit a more complicated relationship between shear stress and velocity gradient than simple linearity.

Examples [2]

Newtonian liquids such as water do not change viscosity when force is applied. Non-Newtonian fluids such as **ketchup** change viscosity when force is applied. Soap solution, Toothpaste, butter and cheese are the another examples of this type of fluids.

Non Newtonian Behaviours and Fluids [1,3,4]

There are several types of Non newtonian flow behaviour, characterized by the way a fluid's viscosity changes in response to variations in shear rate [1,3].

Non-Newtonian fluids change their viscosity or flow behaviour under stress. If you apply a force to such fluids (say you hit, shake or jump on them), the sudden application of stress can cause them to get thicker and act like a solid, or in some cases it results in the opposite behaviour [4].

i. **Shear Thickening [1,4]:** Viscosity of the substance increases with the shear rate is increased. Some liquids behave differently when stress is applied (application of force). Shear thickening liquids increase in viscosity as stress increases.

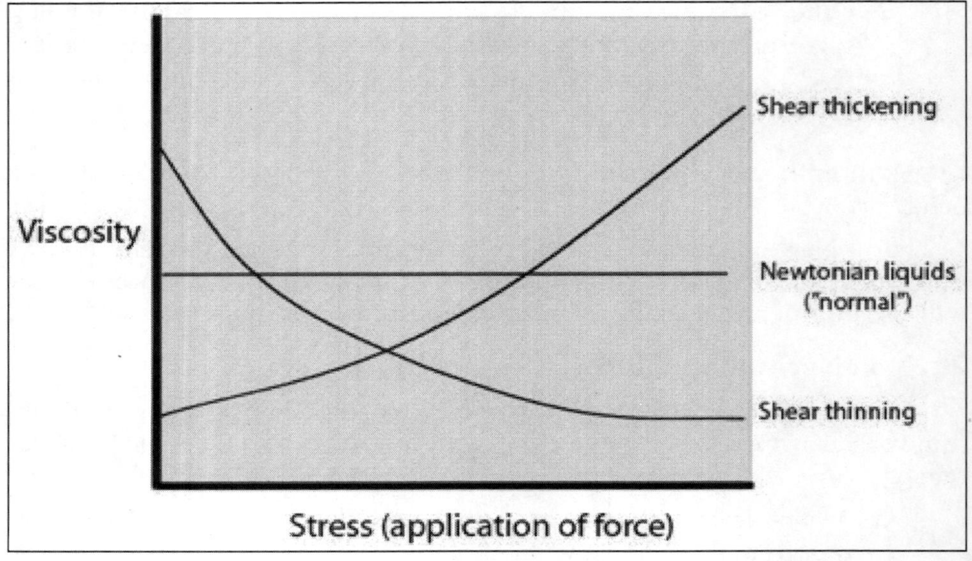

Figure 3.6: Relationship between Stress and Viscosity [4].

ii. **Shear thinning [1,4]:** Viscosity of the substance decreases with the shear rate is increased. Shear thinning liquids decrease in viscosity as stress increases.

iii. **Thixotropic:** A substance that is gel-like when still, but liquid when agitated. It is the property of certain gels or fluids that are viscous under normal conditions, but become thin, less viscous and flow over time when shaken, agitated, or otherwise stressed. Shear thinning liquids are very commonly, but misleadingly, described as thixotropic.

iv. **Rheopectic:** It is the rare property of some non-Newtonian fluids to show a time-dependent change in viscosity; the longer the fluid undergoes shearing force (shaken, agitated, or otherwise stressed) the higher its viscosity.

v. **A Bingham plastic** [1,1a1b,]**:** It is a viscoplastic material that behaves as a solid at low stresses but flows as a viscous fluid at high stresses. Mayonnaise, chocolate and Toothpaste are Bingham plastic. Blood is considered as Bingham plastic fluid.

vi. **A magnetorheological fluid (MR Fluid or MRF) [1,5,6]-** A magnetorheological (MR) fluid is a fluid that has good magnetic properties. MR fluid responds to the magnetic field and changes its properties when a magnetic field is presented [6].

   ✰ It is a type of "smart fluid" in a carrier fluid which, when subjected to a magnetic field, greatly increases its apparent viscosity, to the point of becoming a viscoelastic solid [1,5].

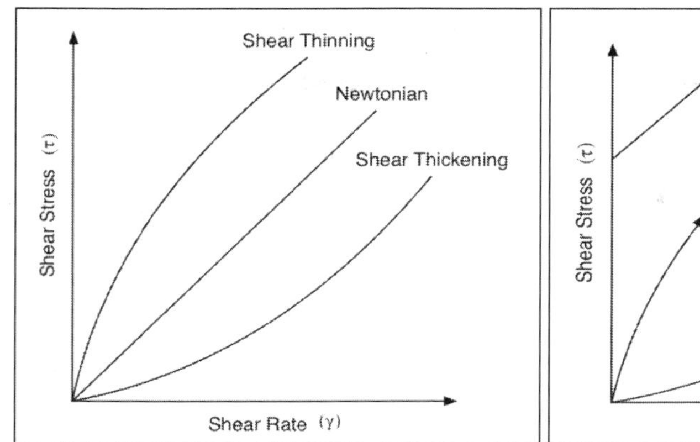

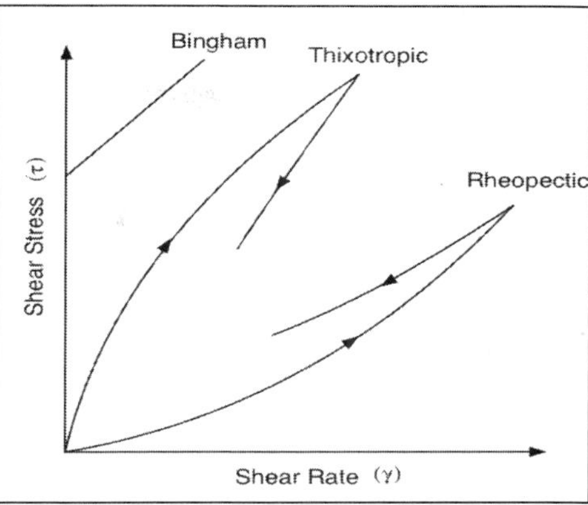

Figure 3.7: Relationship between Shear Rate and Shear Stress [3].

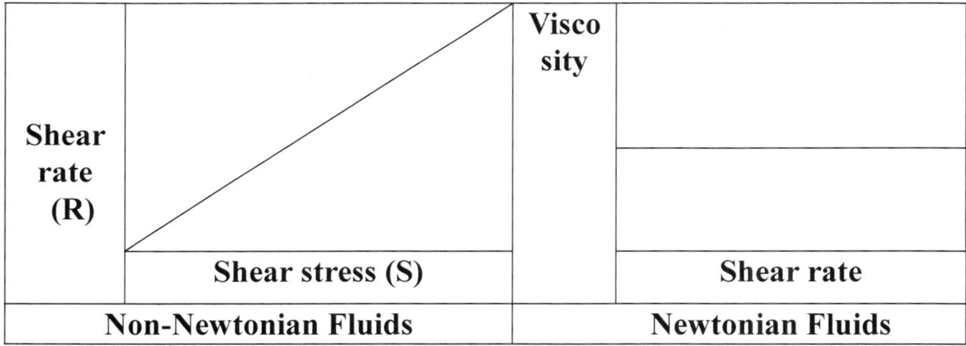

Figure 3.8: Relationship between Shear Stress (S) and Shear Rate (R) [1].

- ☆ **Hysteresis** -Many materials also exhibit **hysteresis**, where by the coefficient of viscosity at a particular shear rate depends upon whether the shear rate is being decreased or increased. The cream, concentrated milks, butter and cheese exhibit varying degree of non-Newtonian behaviour.
- ☆ Not all non-Newtonian Fluids behave in the same way when stress is applied – some become more solid, others more fluid. Some non-Newtonian fluids react as a result of the amount of stress applied, while others react as a result of the length of time that stress is applied [4].
- ☆ The table below summarises four types of non-Newtonian fluids, behaviour and examples [4]-

Table 3.6: Non-Newtonian Fluids, Behaviour and Examples

Sl.No.	Types of Behaviour	Description	Examples
1	Shear thinning	Viscosity decreases with increased stress	Tomato sauce
2	Dilatant or shear thickening	Viscosity increases with increased stress	Oobleck
3	Thixotropic	Viscosity decreases with stress over time	Honey – keep stirring, and solid honey becomes liquid
4	Rheopectic	Viscosity increases with stress over time	Cream – the longer you whip it the thicker it gets

3. Apparent Viscosity [1,7]

It is the viscosity of a fluid measured at a given shear rate at a fixed temperature. In order for a viscosity measurement to be meaningful, the shear rate must be stated or defined [1].

- ☆ For a Newtonian fluid, the apparent viscosity is constant, and equal to the Newtonian viscosity of the fluid, but for non-Newtonian fluids, the apparent viscosity depends on the shear rate [7].
- ☆ This type of viscosity is ordinarily met in concentrated fluid milk products particularly in ice cream mix [1].
- ☆ It refers to a thickened condition of the product which can be dispelled by agitation. It results from the formation of gel structure in the medium [1].

4. Plasticity [1]

The property exhibited by a complex, non-newtonian fluid in which the shear force is not proportional to the shear rate. This property is ordinarily differentiated from viscosity on the basis of the force necessary to cause flow.

- ☆ Plasticity, ability of certain solids to flow or to change shape permanently when subjected to stresses of intermediate magnitude between those producing temporary deformation, or elastic behaviour, and those causing failure of the material, or rupture.
- ☆ Liquids of substantial fluidity start to flow and continue to do so by driving force of their own weight. However, certain relatively non fluid substances will only start to flow after application of initial external pressure. These substances are said to exhibit plastic flow [1].

Source

1. *K.S. Sharma, Physical Chemistry of milk, Agrimoon.com.*

1a. *Wikipedia-Bingham plastic.*

1b. *E.C. Bingham, "An Investigation of the Laws of Plastic Flow". Bulletin of the Bureau of Standards. 13 (2): 309–353, 1916.*

2. *Herman F. George and Farrukh Qureshi, Newton's Law of Viscosity, Newtonian and Non-Newtonian Fluids, Encyclopedia of Tribology, 2013, https: //link.springer. co.*

3. *Viscosity, Neutrium, https: //neutrium.net*

4. *Non- Newtonian fluids, https: //www.sciencelearn.org.nz.*

5. *N.M. Wereley, Magneto-rheology: Advances and Applications, 2014, Ed., Royal Society of Chemistry, RSC Smart Materials, Cambridge, UK.*

6. *Maher Y. Salloom, Modeling Behavior of Magnetorheological Fluids, Reference Module in Materials Science and Materials Engineering, 2020.*

7. *Wikipedia- Apparent viscosity.*

Stoke's Law [1,2,3]

The law(or the frictional force – also called drag force), first set forth by the British scientist Sir George G. Stokes in 1851, is derived by consideration of the forces acting on a particular particle as it sinks through a liquid column under the influence of gravity [2].

- ✰ It is an expression describing the resisting force on a particle moving through a viscous fluid and showing that a maximum velocity is reached in such cases, *e.g.* for an object falling under gravity through a fluid [2].
- ✰ When small spherical bodies move through a viscous medium, the bodies drag the layers of the medium that are in contact with them. This dragging results in relative motion between different layers, which are away from the body. Therefore, a viscous drag comes into play, opposing the motion of the body. It is found that this backward force or viscous drag increases with increase in velocity of the body. According to Stoke, the viscous drag 'f ', depends on the coefficient of viscosity 'η' of the medium, the velocity (v) of the body and radius (r) of the spherical body [1,3].

$$f \propto \eta v r$$

$$f = k \eta v r$$

where k is found to be 6π.

Stokes's law finds application in several areas, particularly with regard to the settling of sediment in fresh water and in measurements of the viscosity of fluids [2].

Source

1. *K.S. Sharma, Physical Chemistry of milk, Agrimoon.com.*
2. *Stokes's law, physics, written by The Editors of Encyclopaedia Britannica.*
3. *Viscosity, Neutrium, https: //neutrium.net.*

Principle and Determination of Viscosity [1,2,3,4]

There are two kinds of viscosity commonly reported, kinematic and dynamic. Dynamic viscosity is the relationship between the shear stress and the shear rate in a fluid. The Kinematic viscosity is the relationship between viscous and inertial forces in a fluid. Viscosity can be measured both in absolute or relative terms.

- ☆ **Absolute viscosity** is the viscosity in poise or centipoise. Relative viscosity is the rate of flow of liquid. It is either volume flow during a fixed period of time or time for a fixed volume under specified conditions. The absolute viscosity of water is 1.005 centipoise at 20°C. Thus centipoise is the viscosity exhibited by water at 20°C [1].
- ☆ The unit of viscosity is poise. It is defined as the force required to maintain a relative velocity of 1cm/sec between two parallel planes placed 1 cm apart. The unit commonly used for milk is centi poise (10^{-2}poise).
- ☆ A useful quantity in fluid flow calculations is the **kinematics viscosity** or viscosity/density. In dealing with the solutions and colloidal dispersion the following quantities are often used.

  Relative viscosity: $\eta rel = \eta soln/\eta solv$

  Specific viscosity: $\eta sp = \eta rel - 1$

  Reduced viscosity: $\eta red = \eta spC$ where c is the concentration of the solute.

  Intrinsic viscosity: $[\eta] = \lim (\eta sp/C)$ as c goes to zero.
- ☆ **Viscometers**-Stokes law is the basis of the falling-sphere viscometer, in which the fluid is stationary in a vertical glass tube. A sphere of known size and density is allowed to descend through the liquid. If correctly selected, it reaches terminal velocity, which can be measured by the time it takes to pass two marks on the tube. Electronic sensing can be used for opaque fluids. Knowing the terminal velocity, the size and density of the sphere, and the density of the liquid, Stokes' law can be used to calculate the viscosity of the fluid [3].
- ☆ Viscosity can be determined by the following three (Three Viscometers) methods [1,2]-

  i. **Ostwald pipette (capillary tubes)** which is based on the principle of time of flow under a fixed pressure.

ii. **Falling Ball Viscometer**-by measuring the fall of a ball through a column of liquid *e.g.* Hoeppler viscometer.

Falling sphere and capillary tube viscometers are not suitable for measurements of non-Newtonian fluids because of the correction for non uniform shear rates. These methods are not only tedious but the results obtained are not accurate.

iii. **MacMichael Viscometer**- MacMichael Viscometer that consists of an outer cylinder (coaxial) rotated by a variable-speed motor; the measured torque exerted on an inner cylinder is a function of the viscosity of the test fluid.

☆ It is suitable for non-Newtonian suspensions. Measuring the force required to move two layers of liquid past each other. **Brookfield Viscometer** is also used for determination of viscosity.

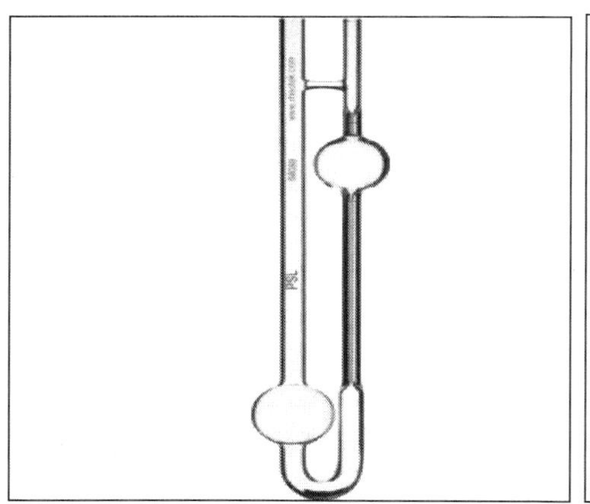

Ostwald U-Tube Viscometer

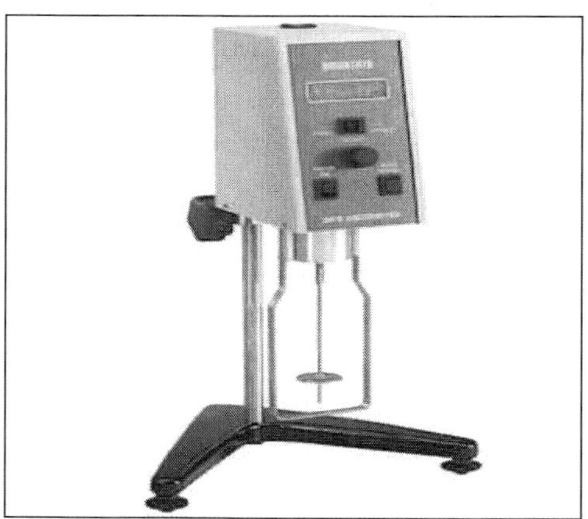

Brookfield Low Range Viscometer

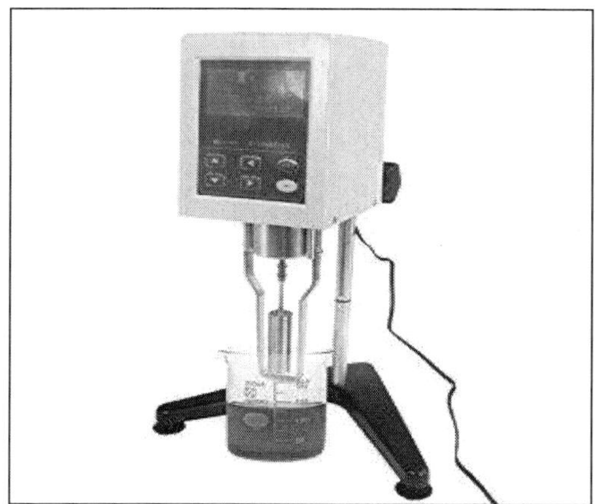

Vibrational Viscometer

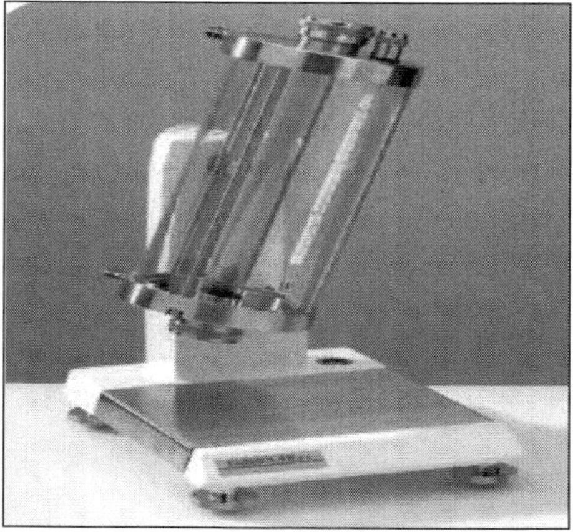

Falling Ball Viscometer

Figure 3.9: Different Types of Viscometers.

- ☆ **A Mobil meter**, which has some features of both the coaxial cylinder and falling sphere viscometers, has been used to measure viscosities of evaporated milk.
- ☆ **A sealed micro viscometer** of the falling sphere type in which the specimen can be sterilized has been proposed for studies of changes occurring in sterilization of concentrated milk products.

Source

1. *M.K. Srivastava, Handbook on Analysis of Milk, CBS Publications, New Delhi, 2010. ISBN 13: 9788123926681.*
2. *K.S. Sharma, Physical Chemistry of milk, Agrimoon.com.*
3. *Wikipedia.*
4. *Viscosity, Neutrium, https: //neutrium.net.*

4. Refractive Index (RI) [1-5]

The optical properties of milk are also very important not only in understanding the composition/Adulteration of milk but also in determining the appearance of milk. Optical properties provide the basis for following rapid, indirect methods of analysis [2,3]:

- **i.** Proximate analysis by infrared absorbency or light scattering.
- **ii.** Also helps in determining the appearance of milk and milk products.
  - ☆ The Refractive Index of milk is very important optical property. Refractive index also called index of refraction is the measure of the bending of a ray of light when passing from one medium into another [4].
  - ☆ The velocity of light varies with density of the medium through which it passes. Thus when light passes from a less denser medium such as air into a more dense medium water it is bent or refracted. The magnitude of this bending when expressed as a ratio of sines of the angles of incidence and refraction of the light, is the refractive index. It is designated by the letter *n* [2,3,4].

$$n = \frac{\sin i}{\sin r}$$

where- i -angle of incidence, r - angle of refraction and *n*- refractive index constant.

  - ☆ Refractive index is also equal to the velocity c of light of a given wavelength in empty space divided by its velocity v in a substance, or n = c/v [4].
  - ☆ Since the refractive index varies with the sample temperature and the wave length of the light these must be critically controlled and specified.

- **Refractive index** (RI) is normally determined at 20° C with the D line of the sodium spectrum. Thus n^{20}_{D} refers to the index at 20°C with D line of the sodium spectrum (589.0 and 589.6nm) [3,4].
  i. The refractive index of water is $n^{20}_{D} = 1.33299$.
  ii. The value of n^{20}_{D} for cow's milk generally falls in the range of 1.3440 to 1.3485 [1,4].
  iii. Buffalo milk is similar to that of cow's milk while human, goat, and ewe milk appear to have higher refractive index values.
- RI can be used to estimate
  i. Concentration of dissolved solids (total solids) in milk and condensed milk.
  ii. Sugar content
  iii. Purity of ghee.
  iv. Adulteration of milk with water.
- The Refractive Index of milk is the refractive index of the solvent and indices of the solutes. The refractive index of material such as milk is higher than water due to dissolved material in milk. The dissolved material includes salts, lactose, *etc.* A butyro refractometer (B.R. Index) reading between 40-43 is the index of purity of ghee. Ghee exhibiting a refractive index of 1.4545 corresponds to a B.R. index of 43 at 40°C [2,3].
- Since refractive index increments contributed by each solute in a solution are additive, much consideration has been given to the possible use of refractive index as a means of determining total solids or added water in milk [4].
- The relation between solids content and refractive index is linear and the contributions of the several components are additive. The individual components of milk differ in specific refractive increment $\Delta n(\rho c)$ where 'ρ' is the density of the sample and 'c' is the weight/weight concentration specific refractive increment 'n') of the component. Thus the relation between per cent solids and refractive index will vary between lot of milk [4].
- Researchers have revealed the following specific refractive increments (mng-1):
  i. Casein complex 0.207
  ii. Soluble proteins 0.187
  iii. lactose 0.140

  The total contribution to the refractive index for milk containing 2.34 per cent casein complex, 0.83 per cent serum proteins and 4.83 per cent lactose becomes 0.00500 +0.00159 + 0.00695 or 0.01354.

The residue 0.95 per cent contributes 0.00166 to the total difference between the refractive indices of water and the milk [4].

- ✰ The refractive index of milk fat is 1.4537 to 1.4578 at 40°C.
- ✰ Fat does not contribute to the refractive index of whole milk because refraction occurs at the interface of air and the continuous phase [3,4].
- ✰ Sterilization and storage does not alter the refractive index.

Determination of Refractive Index [2]

The following methods are being used to determine the Refractive Index:

i. **Immersion refractometer:** It is used for measuring refractive index of milk. Milk serum is prepared with the help of copper sulphate. The proteins are removed by filtration. A drop or two is placed on the prism of refractometer and the refraction is measured.

ii. **Abbe's refractometer:** The refractive index of milk itself is somewhat difficult to determine because of the opacity but by using a refractometer such as Abbe's instrument which employs a thin layer of sample, it is possible to make satisfactory measurement particularly of skim milk products and sweetened condensed milk [4].

Abbe's refractometer is a modified version of refractometer. It is provided with a narrower range for measuring RI for fats and oils. There is an arrangement for circulation of water so that fat remains in a melted state and temperature can also be controlled.

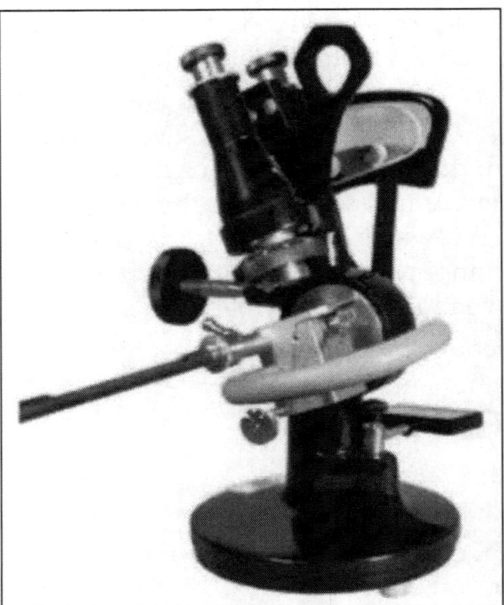

Figure 3.10: Abbe Refractometer

The scale is usually graduated in numbers (*e.g.* 43 which corresponds to n = 1.4545). Abbe's refractometer is used to measure purity of ghee. A drop or two of ghee droplets are placed on the surface of Abbe's refractometer prism, which is heated at 40°C through water circulation. The refractometer reading called as butyro refractometer reading (B.R) is recorded on the scale. A B.R reading between 40-43 is taken as the criteria for purity of ghee. This is commonly known as the B.R. index of ghee. An increase in B.R. shows the adulteration of ghee vegetable oil/animal body fat [2].

Brix value of glass distilled water at 40 ° C. It should be zero. Dry the prism and put few drops of melted ghee and take the R.I. reading at 40 ° C and convert it into B.R. reading. Some digital refractometers also give both R.I. and B.R. readings.

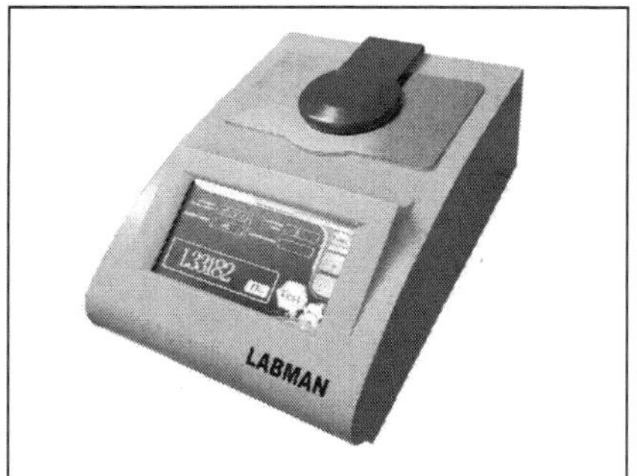

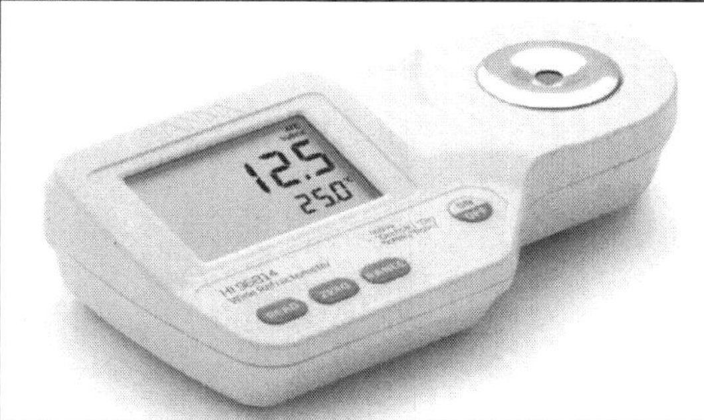

Figure 3.11: Automatic Digital Refarctometers.

Source

1. *Physical properties of milk, Food Science, University of Guelph.*
2. *Refractive index, dairy technology, dairy-technology.blogspot.com*
1. *M.K. Srivastava, Handbook on Analysis of Milk, CBS Publications, New Delhi, 2010. ISBN 13: 9788123926681.*
4. *K.S. Sharma, Physical Chemistry of milk, Agrimoon.com.*
5. *M.K. Srivastava, Handbook on Analysis of Milk and Milk Products, Write and Print Publications, New Delhi, 2016. ISBN: 9789384649258.*

Thermal Conductivity [1,2]

It determines how fast milk is cooled or heated. The Values of Thermal conductivity of cow and buffalo milk at 30°C are 0.486 Kcal/hr.m^2 and 0.569 Kcal/hr.m^2 respectively. It increases with increase in temperature and decreases as concentration level increases. For a given temperature and concentration, higher the fat content decreases the thermal conductivity.

Electrical Conductivity [1,2]

Solutions of electrolytes can conduct electric current due to the presence of electrolytes under the influence of an electric field. As milk contains electrolytes and hence it has Electrical conductivity.

Electrical conductivity may be **defined as a measure of the ability of a solution to carry electric current.** This conductance is obeyed by ohm's law and is measured by specific resistance. Specific resistance is defined as resistance in ohms of a column of solution 1 cm long and 1sq cm in cross-section. The conductivity or specific conductance is the reciprocal of specific resistance. It is expressed as reciprocal ohms (*i.e.*, ohms 1 or mhos). The Values of Electrical conductivity of cow and buffalo milk at 30°C are 4.67 milli mhos/cm and 3.67 milli-mhos/cm respectively. It increases with increase in temperature. In milk, the fat and colloidally dispersed substances obstructs the ions in their migration hence lowers the electrical conductivity.

Determination of Electrical Conductivity

The electrical conductivity is measured in terms of **specific conductance which is reciprocal of specific resistance.** The specific resistance of milk is low with an average value of 0.005 ohm-1. During **mastitis** the value increases due to the presence of various ions of milk. The various ions responsible for the conductivity of milk are sodium, potassium and chloride ions. In mastitis and colostrum milk these ions are abnormally high. The specific conductance is estimated from the resistance obtained from the filled solution in the cell. The electric resistance of milk is measured by placing milk in a cell, between two platinum electrodes fixed at a distance. The electrical resistance is measured with the help of Wheatstone bridge. The specific conductance is calculated from the measured resistance of the cell filled with solution *e.g.* milk.

Thus Specific conductance = K/R

where, K-cell constant determined with cell of solutions of known conductance and R - measured resistance in ohms.

Optical Properties [1,2]

Light scattering by fat globules and casein micelles causes milk to appear turbid and opaque. Light scattering occurs when the wave length of light is near the same magnitude as the particle. Thus, smaller particles scatter light of shorter wavelengths. Skim milk appears slightly blue because casein micelles scatter the shorter wavelengths of visible light (blue) more than the red. The carotenoid precursor of vitamin A, β -carotene, contained in milk fat, is responsible for the 'creamy' colour of milk. Riboflavin imparts a greenish colour to whey.

Colour [1,2]

The Colour of milk is another important optical property. The colour has three aspects *e.g.* tint, intensity and uniformity. The colour of an opaque material is the colour it reflects, the colours of visible spectrum are absorbed. A White object reflects all colours of light that fall on it where as a black object absorbs all colours. Thus an object is yellow because more yellow light is reflected to the eye than any other colour. The colour of the milk depends upon the different constituents of milk and it is a blend of the following individual effects:

- **i.** Light scattering colloidal casein particles and dispersed fat globules.
- **ii.** Carotene, which imparts yellowish tint.

Cow milk is yellowish creamy white where as buffalo milk is creamy white. Certain breeds of cow imparts a deeper yellow tint. The greater the intake of green feeds, the deeper yellow the colour of cow milk. The intensity of yellow colour increases due to increase in fat per cent age and size of fat globules. The skim milk has a bluish where as whey has a greenish yellow colour.

Specific Heat of Milk [1,2]

It is the amount of heat required to raise the temperature of a given mass of the substance one degree centigrade, compared with the amount of heat required to raise an equal mass of some standard substance one degree centigrade. The specific heat of Water is 1 and it is taken as a standard.

- ☆ The specific heat of milk is 0.9454(Compared to water) which is lower than that of water. Thus it requires less heat to raise the temperature compared to same quantity of water. It also takes less ice to cool a certain volume of milk one degree than it does to cool the same quantity of water through one degree.
- ☆ The specific heat of skim milk is lower than milk because of the absence of fat from skim milk. Skim milk has a value of 0.933 to 0.954 cal g-1C-1.
- ☆ Fat has a higher specific heat of about 0.52 cal g-1C-1. The specific heat of milk and cream depend strongly upon the fat content.
- ☆ Specific heat of milk varies with the temperature.
    - i. Specific heat of warm milk is the same as that of normal milk. This is due to the fact that fat is in the liquid state.
    - ii. This value is however, lower of milk as milk is cooled below 19°C. At this temperature some of the heat supplied to the milk system at a temperature near the melting point of the fat is used by the fat for its melting.
- ☆ Specific heat of milk and milk products is given in Table 3.7.

Table 3.7: Specific Heat of Milk and Milk Products

Milk and Milk Products	*Specific Heat*	
	at 60 °F	*at 40 °F*
Milk	0.94	0.93
Whey	0.98	-
Butter	0.53	-
Cream (30 per cent fat)	0.98	0.85
Cream (60 per cent fat)	1.05	0.72
Cheese	**0.64**	-

☆ Specific heat is measured easily with the help of a calorimeter with an electric heater. With the help of calorimeter energy used to raise the temperature can be easily measured.

Source

1. *M.K. Srivastava, Handbook on Analysis of Milk, CBS Publications, New Delhi, 2010. ISBN 13: 9788123926681.*

2. *M.K. Srivastava, Handbook on Analysis of Milk and Milk Products, Write and Print Publications, New Delhi, 2016. ISBN: 9789384649258.*

Chapter 4

Colligative Properties of Dilute Solutions

Dilute Solution [1,3]

A dilute solution is one which contains very small amount of the solute in comparison to the amount of the solvent.

- A **dilute** solution is one in which there is a relatively small amount of solute dissolved in the solution.
- The dilute solutions show more or less ideal behavior as the heat and volume changes.
- Dilute solutions containing non-volatile solute exhibit some special properties which depend only upon the number of solute particles present in the solution irrespective of their nature and these properties are termed as colligative properties.
- The colligative properties have been one of the means for determining the molecular weight of unknown compounds.

Definition of Colligative Properties [1,1a,2]

The meanings of colligative in latin and Greek are as follows:

Colligative-Latin- tied together [2] or
Greek colligatus = Collected together [1a]

- A colligative property may be defined as one which depends on the number of particles in solution and not in any way on the size or chemical nature of the particles [1a].

- ✰ These properties are properties of solutions that depend on the number of molecules in a given amount of solvent and not on the properties/identity (*e.g.* size or mass) of the molecules or those properties of solutions that depend on the number of dissolved particles in solution, but not on the identities of the solutes [1].
- ✰ Colligative properties are the physical changes that result from adding solute to a solvent.
- ✰ Colligative Properties depend on how many solute particles are present as well as the solvent amount, but they do not depend on the type of solute particles.

For Example [1,1a]

1. The freezing point of salt water (solution) is lower than that of pure water, due to the presence of salt dissolved in it. It does not matter whether the salt dissolved in water is sodium chloride or potassium nitrate.
2. If the molar amounts of solutes are the same and the number of ions is the same, the freezing points will be the same. Aluminium chloride ($AlCl_3$) and potassium phosphate (K_3PO_4) would exhibit essentially the same colligative properties, since each compound dissolves to produce four ions per formula unit.
   - ✰ Consequent to the above definition, each colligative property is exactly related to any other. Thus if one property is measured, the other can be calculated. The colligative properties of dilute solutions are particularly important as these provide valuable methods for finding the molecular weights of the dissolved substances [1,1a]. If the following three conditions are satisfied then we get excellent results:
     - ✰ The solution should be very dilute
     - ✰ The solute should be non-volatile and
     - ✰ The solute does not dissociate or associate in solution.
   - ✰ Colligative properties play an important role in the processing of milk and preparation of milk products.

Common Colligative Properties [1,1a,2]

Molecular solutes have only one particle per formula, but ionic materials come apart into their ions and have almost as many particles in the solution as there ions are available.

Sometimes re-association of the ions decreases the number of particles. The effect of the ionic pair depends upon the properties of the species dissolved and the concentration of solute. The more concentrated the solute, the greater percentage of ion pairing takes place.

The four commonly studied colligative properties are as follows:

1. **Freezing point depression**-The decrease in the freezing point (FP) of the solution, compared to pure solvent at the same pressure. FP decreases if solute in solution increases.
2. **Boiling point elevation**-The increase in the boiling point of a solution containing nonvolatile solutes, compared to pure solvent at the same pressure. The solution shows an increase in boiling point as the amount of solute in it is increased. The boiling point of a liquid is just the point at which the vapor pressure of the liquid equals the surrounding pressure. **If the vapor pressure decreases, it will take a greater temperature to boil the liquid.** The solution shows a decrease in melting point as the amount of solute is increased.
3. **Lowering of vapor pressure**-The decrease in the vapor pressure of a solution containing nonvolatile solutes, compared to the vapor pressure of the pure solvent at the same temperature. Solution of a solid (non-volatile) solute in a liquid solvent shows a decrease in vapor pressure above the solution as the amount of solute is increased.
4. **Osmotic pressure**-The increase in the osmotic pressure of the solution that places the solvent in transfer equilibrium with pure solvent at the same temperature and pressure as the original solution. The solution shows an increase in osmotic pressure between it and a reference solution as the amount of solute is increased.

Semi permeable membranes are an important part of any living thing. Cell membranes are semi permeable. The membranes on the outside of eggs are semi permeable. A semi permeable membrane is one which allows only water from the solution but not the solutes. When two solutions are separated by such a membrane osmotic pressure will develop across the membrane, consequently due to the process of osmosis the water from the solution having lesser solute will pass through the semi permeable membrane and dilutes the solution with more solutes. Trees pull up water from their roots by a process of osmosis.

All four properties are defined by an equilibrium between the liquid solution and a solid, liquid, or gas phase of the pure solvent [2,3].

Sources

1. *K.S. Sharma, Physical Chemistry of milk, Agrimoon.com.*

1a. *Theory of Dilute Solutions,chapter-15, http: //icvcollege.edu.in.*

2. *http:/Chemo.libretexts.org*

3. *M.K. Srivastava, Handbook on Analysis of Milk, CBS Publications, New Delhi, 2010. ISBN 13: 9788123926681.*

1. Vapor Pressure and Lowering of Vapor Pressure [1,2,3]

That vaporization process is reversible. Gaseous molecules coming into contact with the surface of a liquid can be trapped by intermolecular forces in the liquid. Eventually the rate of escape will equal the rate of capture to establish a constant, equilibrium vapor pressure above the pure liquid.

- ✰ The vapour pressure of a solution of a non-volatile solute is equal to the vapour pressure of the pure solvent at that temperature multiplied by its mole fraction.
- ✰ Vapour pressure can be defined as pressure formed by the vapor of the liquid (or solid) over the surface of the liquid. This pressure is formed in a thermodynamic equilibrium state in a closed container at a certain temperature. vapour pressure increases with the temperature [1].
- ✰ When a nonvolatile solute is added to a liquid to form a solution, the vapor pressure above that solution decreases. When a solute is present in a solvent, the vapor pressure is lowered because fewer solvent molecules are present at the top of the solution.
- ✰ Liquid molecules at the surface of a liquid can escape to the gas phase when they have a sufficient amount of energy to break free of the liquid's intermolecular forces. If we add a nonvolatile solute to that liquid, the amount of surface area available for the escaping solvent molecules is reduced because some of that area is occupied by solute particles. Therefore, the solvent molecules will have a lower probability to escape the solution than the pure solvent. That fact is reflected in the lower vapor pressure for a solution relative to the pure solvent.
- ✰ That statement is only true if the solvent is nonvolatile.
- ✰ If the solute has its own vapor pressure, then the vapor pressure of the solution may be greater than the vapor pressure of the solvent [2].

The identification of the nature of the solvent or the solute (except for its lack of volatility) is not needed to derive that the vapor pressure should be lower for a solution relative to the pure solvent. **That is what makes vapor pressure lowering a colligative property it only depends on the number of dissolved solute particles(solute) [2].**

Factors on which Vapor Pressure of Liquid Depends [3]

The vapour pressure of liquid mainly depends on temperature and nature of the liquid.

1. **Temperature:** The vapour pressure of liquid increases with increase of temperature. The increase in temperature causes more molecules to have larger kinetic energies and larger number of molecules will escape from the surface of liquid to the vapour phase resulting in higher vapour pressure.

2. **Nature of liquid:** Each liquid has a characteristic vapour pressure because each liquid has different magnitude of intermolecular forces. A liquid having weaker intermolecular force, has more vapour pressure.

Lowering of Vapor Pressure-Raoult's Law [2,4]

The law that mathematically describes the vapor pressure lowering phenomenon. Raoult's law is given below:

$$P = c_{solvent} P^o$$

where,

P - Vapor pressure of a solution,$c_{solvent}$ - Vapour pressure of the solvent above a solution and P^o - Vapor pressure of the pure solvent.

On the surface of the pure solvent (shown on the left) there are more solvent molecules at the surface than in the right-hand solution flask. Therefore, it is more likely that solvent molecules escape into the gas phase on the left than on the right. Therefore, the solution should have a lower vapor pressure than the pure solvent.

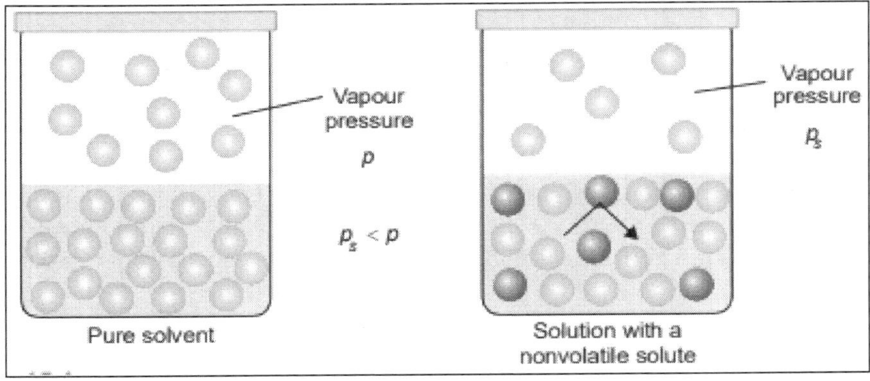

Figure 4.1: Lowering of Vapour Pressure by a Nonvolatile Solute [4].

- ✰ The vapour pressure of the pure solvent is caused by the number of molecules evaporating from its surface. When a nonvolatile solute is dissolved in solution, the presence of solute molecules in the surface blocks a fraction of the surface where no evaporation can take place. This causes the lowering of the vapour pressure.

The vapour Pressure of a given solution is lower than that of the pure solvent. The Raoult's law helps in explaining the aforesaid fact and the explanation is given below:

Raoult's law says that the vapor pressure of a solution(P) equals the mole fraction of the solvent($C_{solvent}$) multiplied by the vapor pressure of the pure solvent(P^o).

Ideal and Non Ideal Solutions [2,4]

On the basis of the Raoult's Law, solutions can be categorised in to Ideal and Non ideal solutions.

Ideal Solutions

Solutions that obey Raoult's law are called ideal solutions because they behave exactly as we would predict.

Non-Ideal Solutions or Real Solution

Solutions that show a deviation from Raoult's law are called non-ideal solutions because they deviate from the expected behavior.

Very few solutions actually approach ideality, but Raoult's law for the ideal solution is a good enough approximation for the non- ideal solutions that we will continue to use Raoult's law.

Deviation from Raoult's Law [2,4]

This "law" is approximately obeyed by most solutions, some show deviations from the expected behavior. Deviations from Raoult's law can either be positive or negative [2].

A Positive Deviation

A molecule will escape from the solution surface more readily and the vapour pressure of the solution will be higher than predicted by Raoult's law. If the solute and solvent are not as tightly bound to each other as they are to themselves, then the solution will show a positive deviation from Raoult's law because the solvent molecules will find it easier to escape from solution into the gas phase.

A Negative Deviation

A molecule will escape less readily and the vapour pressure will be less than that predicted by Raoult's law. If the solute is strongly held by the solvent, then the solution will show a negative deviation from Raoult's law because the solvent will find it more difficult to escape from solution.

Measurement of Lowering of Vapour Pressure [4]

The following methods are being used to determine the lowering of vapour pressure:

i. Barometric Method

Raoult measured the individual vapour pressure of a liquid and then the solution by this method and introduced the liquid or the solution into Toricellian vacuum of a barometer tube and measured the depression of the mercury level.

- ☆ This method is neither practicable nor accurate as the lowering of vapour pressure is too small.

ii. Manometric Method

The vapour pressure of a liquid or solution can be conveniently measured with the help of a manometer. The bulb B is charged with the liquid or solution. The air in the connecting tube in then removed with a vacuum pump. When the stopcock is closed, the pressure inside is due only to the vapour evaporating from the solution or liquid.

- ☆ This method is generally used for aqueous solutions. The manometric liquid can be mercury or n-butyl phthalate which has low density and low volatility. Evacuated Scale space Manometer Solution Vapour pressure to vacuum pump

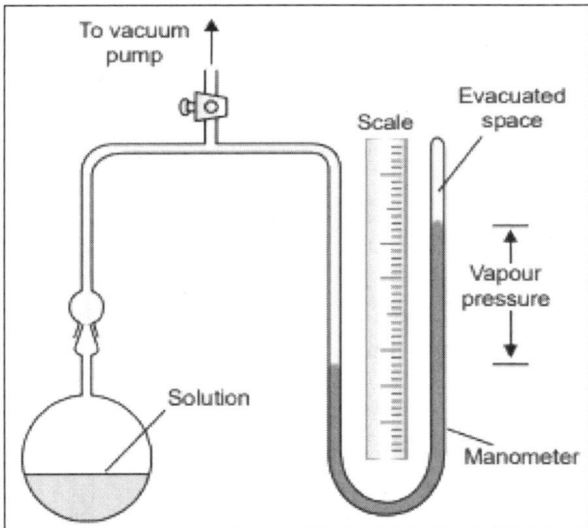

Figure 4.2: Measurement of Vapour Pressure of Aqueous Solutions with a Manometer [4].

iii. Ostwald and Walker's Dynamic Method (Gas Saturation Method)

In this method the relative lowering of vapour pressure can be determined straightway. The measurement of the individual vapour pressures of a solution and solvent is thus eliminated.

Procedure

The apparatus used by Ostwald and Walker is shown in Figure 4.3.

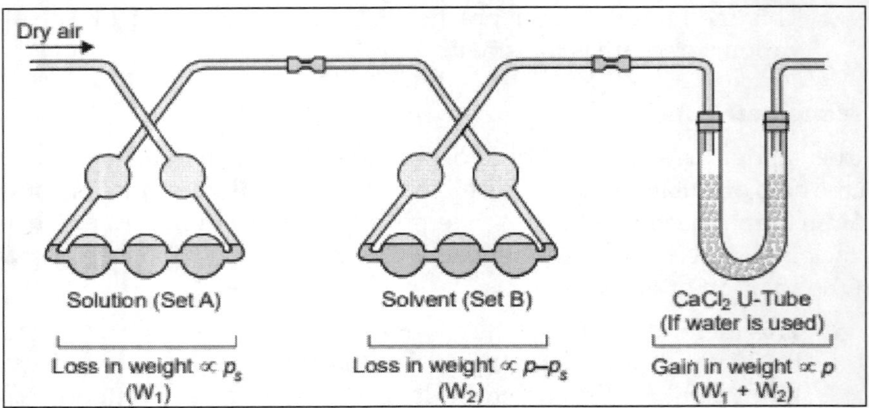

Figure 4.3: Ostwald-Walker Method of Measuring the Relative Lowering of Vapour Pressure [4].

It consists of following two sets of bulbs:

a. Set A containing the solution

b. Set B containing the solvent.

Each set is weighed separately. A slow stream of dry air is then drawn by suction pump through the two sets of bulbs. At the end of the operation, these sets are reweighed. From the loss of weight in each of the two sets, the lowering of vapour pressure is calculated. The temperature of the air, the solution and the solvent must be kept constant throughout.

Source

1. *M.K. Srivastava, Handbook on Analysis of Milk, CBS Publications, New Delhi, 2010. ISBN 13: 9788123926681.*

2. *K.S. Sharma, Physical Chemistry of milk, Agrimoon.com.*

3. *Vapor pressure of liquid, https: //classnotes.org.in.*

4. *Theory of Dilute Solutions, Chapter-15, http: //icvcollege.edu.in.*

2. Freezing Point and Depression in Freezing Point

The freezing point is the temperature at which the liquid solvent and solid solvent are at equilibrium, so that their vapour pressures are equal [1].

☆ The freezing point of any aqueous systems(milk *etc.*) depends on the concentration of water soluble components. The mathematical relationship between depression of freezing point and concentration of the solute was determined by Raoult and is expressed in the below mentioned equation [3]:

$$T_f = K_f M$$

where,

T_f is the difference between the freezing points of the solvent and the solution, K_f is the molal depression constant (1.86°C for water) and M is the molar concentration of the solute.

- ☆ This relationship is valid only for dilute solution of undissociated solutes. The freezing point is a property controlled by the number of particles rather than the kind and nature of the particles in the solvent.

Freezing-Point Depression

It is the decrease of the **freezing point** of a solvent on the addition of a non-volatile solute. Examples include salt in water, alcohol in water, or the mixing of two solids such as impurities into a finely powdered drug [1].

- ☆ The increase of solute percentage in solvent decreases the freezing point (**a colligative property**) of solution where as Freezing point of water is **a constitutive property** (depends upon the nature of the molecules of the liquid) [2].
- ☆ The phenomenon of freezing-point depression is analogous to boiling point elevation. However, the magnitude of the freezing point depression is larger than the boiling point elevation for the same solvent and the same concentration of a solute. Because of these two phenomena, the liquid range of a solvent is increased in the presence of a solute [1].
- ☆ When ice and water are in equilibrium with one another that is called freezing. At this temperature they have the same vapour pressure. The addition of solute to water lower the vapour pressure. The reason is that it delays the escape of the vapours from the surface. Thus the freezing of a given solution is always lower than that of water. The amount of depression of the freezing point is proportional to the fraction of the total number of solute particles, which are solute molecules. Also freezing point depression is a function of the osmotic pressure of the solution [1,2,3].
- ☆ In the dairy field, the objective of freezing point measurement is virtually restricted to determination of the water content of the product in order to detect the illegal addition of water. The solvent of milk is water while solute consists of molecules like sugars, minerals, proteins, milk fat in suspension, *etc.* As milk contains several dissolved substances hence freezing point of milk is lower than water [1,2,3].
    i. The Fat, protein and colloidal salts do not contribute to Freezing point. Fat, protein, colloidal calcium phosphate, casein colloids and fat globules have negligible effect on the freezing point of milk because of their high molecular weight.

ii. Lactose,dissolved gases and salts are the major contributors to Freezing point. 75 to 80 per cent of the depression of the freezing point is due to these two constituents.

iii. The freezing point of milk is usually in the range of -0.512 to -0.550° C with an average of about **-0.522° C**. In order to obviate negative sign the term freezing point depression is used. The average freezing point depression of Indian cow milk and buffalo milk are 0.547°C (31.02°F) and 0.549°C (31.04°F) respectively with an average of bulk milk is 0.530°C (31.05° F) [2]. Sheep milk and Goat milk have depression of the freezing point -0.588°C and -0.575°C respectively.

Since milk is having same osmotic pressure as that of the blood on physiological grounds and has fairly constant value. Freezing point depression of milk is inversely proportional to the amount of water added [3].

iv. Colostrums is having a freezing point -0.605°C which is slightly less than normal milk.

☆ In the dairy industry, freezing point is mainly used to determine **added water** but it can also been used to determine lactose content in milk, estimate whey powder contents in skim milk powder, and to determine water activity of cheese. The freezing point depression value lower than the aforesaid values indicates the addition of water in milk.

Factors affecting Freezing Point of Milk [2,3]

The following factors are responsible for change in freezing point of milk-

☆ Mastitis milk shows normal freezing point [2]. The composition of milk is altered by the udder diseases like mastitis, but in order to maintain the osmotic pressure in equilibrium with blood the lowering of the lactose content is compensated by increase in the sodium chloride. As such the milk obtained from the animal affected with mastitis will not alter the freezing point [3].

☆ The boiling and sterilization increases the freezing point depression where as pasteurization does not have any effect.

☆ The freezing point depression test is not able to detect the addition of skim milk or removal of fat from the sample and it also not gives clear picture about the addition of soured water to milk.

☆ Large temporary increases in freezing point occur after consumption of large amounts of water because milk is iso-osmotich blood.

☆ With respect to interpretation of freezing points for added water determination, the most significant variables are the nutritional status of the herd and the access to water.

- ☆ The primary sources of non-intentional added water in milk are residual rinse water and condensation in the milking system.
- ☆ Skim milk also has the same freezing point as milk adulteration of skim milk cannot be detected by this method.
- ☆ Souring of milk lowers the freezing point as the number of molecules increases due to breakdown of lactose, which affect the osmotic pressure of milk. This results in an increase in freezing point depression.
- ☆ The fortification of milk with lactose or non fat solids or addition of sugar will increase the depression of the freezing point.
- ☆ Environmental factors also cause the variation in freezing point. Some of the variations have been attributed to seasonal effects, feed, water intake, stage of lactation, breed of the cow, heat stress and time of the day (*i.e.* morning or evening milk). If the net number of osmotically active particles are changed due to the handling of milk from the time it is drawn from the udder and the freezing point depression will also be changed [3].

Determination of Freezing Point Depression of Milk [2,3]

Freezing point is a fairly sensitive and constant property of milk. It is principally used to detect adulteration of milk with water. Since the freezing point depression of milk differs from water with only 0.5°C, highly sensitive and accurate thermometers are required for its measurement, which can read up to 0.001°C.

- ☆ Regardless of the method used, determination of the freezing point of solutions is empirical.
- ☆ The general principle employed for the determination of the freezing point is to super cool a sample slightly to induce crystallization and then to observe the maximum temperature attained.
    - i. The temperature of cooling bath must be controlled otherwise the rate of heat loss will be greater than the rate of heat transfer to the bath by the heat of fusion of the solution and the observed freezing point will be too low.
    - ii. Control of temperature of super cooling and seeding techniques is extremely important. If it varies the amount of solvent that crystallizes out and consequently the observed freezing point also varies.
- ☆ The depression of freezing point can be measured more correctly and with less difficulty. Two simple methods commonly used are outlined below. The following methods are used to determine freezing point of milk:

1. *Cryoscope Methods [2,3]*

i. *Hortvet Cryoscope Method [2,3]*

It was one of the earliest method to determine freezing point of milk by an instrument called Hortvet cryoscope. It was developed and used as early as 1923.

- ☆ Although the Hortvet method has been accepted as the official method, a number of modifications have been developed such as replacing the ether cooling system with mechanical refrigeration and adding mechanical stirring and tapping devices.
- ☆ The Hortvet method involves measurement of the differences between the freezing point of a standard solution and the freezing point of milk.
- ☆ This instrument has a number of drawbacks. Major difficulty being the cumbersome of the operation of cryoscope and limitations in getting reproducibility of results. This instrument uses ordinary thermometer called freezing point thermometer.

ii. *Thermistor Cryoscope Method [2,3]*

Due to the development of special type of thermometers called thermistor probes, thermistor cryoscope has been developed which has replaced Hortvet cryoscope in determining freezing point of milk. Thermistor probes are special type of thermometers based on the measurement of changes in electrical resistance with variation in temperatures.

- ☆ Thermistor cryoscope are vary popular in determining freezing point depression of milk. Fiske Cryoscope and Advanced Milk Cryoscope are the important cryoscope instruments which are used for determining freezing point of milk. Official methods of Analysis of the Association of Official Analytical Chemists (AOAC) has recommended the method for determining freezing point depression of milk using thermistor probes. In order to determine percentage of water added it can be calculated by the formula:

$$\text{Minimum percentage of added water by mass} = \text{F.P.} = \frac{0.530 - \Delta T}{0.530} \times (100 - \text{SNF})$$

2. *Beckmann's Method [4]*

It is invented in 1903 and details of this method are given below:

Apparatus- It consists of:

i. A freezing tube with a side-arm to contain the solvent or solution, while the solute can be introduced through the side-arm.

ii. An outer larger tube into which is fixed the freezing tube, the space in between providing an air jacket which ensures a slower and more uniform rate of cooling.

iii. A large jar containing a freezing mixture *e.g.*, ice and salt, and having a stirrer.

Procedure

15 to 20 g of the solvent is taken in the freezing point of the solvent by directly coding the freezing point tube and the apparatus set up as shown in following figure so that the bulb of the thermometer is completely immersed in the solvent.

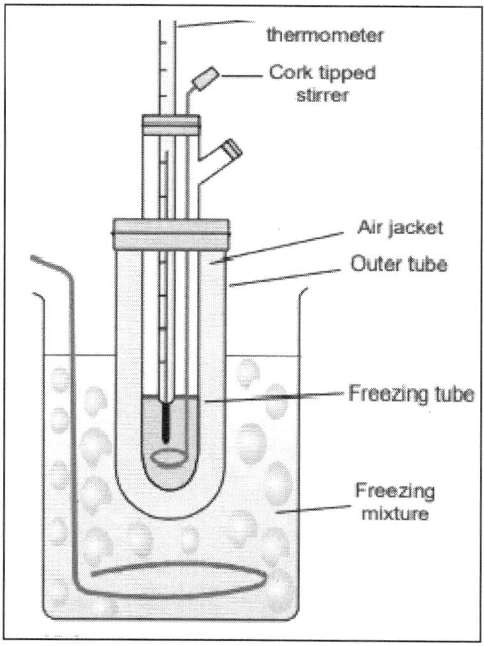

Figure 4.4: Apparatus for Beckmann's Method.

- ☆ First determine the approximate freezing point of the solvent by directly cooling the freezing point tube in the cooling bath. When this has been done, melt the solvent and place the freezing-point tube again in the freezing bath and allow the temperature to fall. When it has come down to within about a degree of the approximate freezing point determined above, dry the tube and place it cautiously in the air jacket. Let the temperature fall slowly and when it has come down again to about 0.5° below the freezing point, stir vigorously. This will cause the solid to separate and the temperature will rise owing to the latent heat set free. Note the highest temperature reached and repeat the process to get concordant value of freezing point.
- ☆ The freezing point of the solvent having been accurately determined, the solvent is remelted by removing the tube from the bath, and a

weighed amount (0.1–0.2 g) of the solute is introduced through the side tube. Now the freezing point of the solution is determined in the same way as that of the solvent

Source

1. *Wikipedia*
1. *M.K. Srivastava, Handbook on Analysis of Milk, CBS Publications, New Delhi, 2010. ISBN 13: 9788123926681.*
3. *K.S. Sharma, Physical Chemistry of milk, Agrimoon.com.*
4. *Theory of Dilute Solutions, Chapter-15, http: //icv college. edu.in.*

3. Boiling Point and Elevation of Boiling Point [1,2,3]

Boiling point of a solution is the temperature at which the vapour pressure of the liquid equals to the external (Atmospheric) pressure [1,2].

- ☆ The increase of solute percentage in solvent increases the boiling point of solution. A solution (solvent + solutes) always boils at a higher temperature compared to a pure solvent. The increase in boiling points depends on the concentration of the dissolved substances (or the solute concentration) [2].
- ☆ Since the vapour pressure of a solution is always less than the vapour pressure of the pure solvent it follows that the boiling point of a solution will always be higher than that of the pure solvent. Pure water boils at 100°C. The normal Ionic ↔ molecular ↔ colloidal equilibrium is altered due to heating [2].

Elevation of Boiling Point

Boiling-point elevation describes the phenomenon that the **boiling point** of a liquid (a solvent) will be higher when another compound is added, meaning that a solution has a higher **boiling point** than a pure solvent.

- ☆ The boiling point elevation is a colligative property, which means that it is dependent on the presence of dissolved particles and their number, but not their identity. The boiling point elevation happens both when the solute is an electrolyte, such as various salts, and a nonelectrolyte.
- ☆ In thermodynamic terms, the origin of the boiling point elevation is entropic and can be explained in terms of the vapor pressure or chemical potential of the solvent [3].

  The result is that in dilute ideal solutions, the extent of **boiling-point elevation** is directly proportional to the molal concentration (amount of substance per mass) of the solution according to the following equation [3]:

$$\Delta T_b = K_b \times m$$

The **boiling point elevation**, is defined as $T_{b\ (solution)} - T_{b\ (pure\ solvent)}$.
where, ΔT_b - Boiling point elevation, Kb- the boiling point elevation constant, m - the molality (mol/kg solvent) of the solute.

Boiling Point Elevation in Milk [1,2]

The water content in fluid milk is highest and hence the boiling point of milk will be close to that of water. But being a colligative property it is influenced by the dissolved substances like lactose, minerals *etc.* and hence it is slightly higher than that of water [1].

- As milk contains several dissolved substances such as protein, fat, minerals, *etc.* hence boiling point of milk is greater than water. Milk boils more quickly than water. The reason is not that milk boils at a lower temperature but it requires less heat to raise its temperature than water does. Milk contains a number of easily heated solids. Therefore, if the same amount of heat is applied to equal quantities of milk and water the temperature of the milk will be raised more than that of water.
- Milk constituents are responsible for the elevation of the boiling point of milk. The boiling point of both buffalo and cow milk ranges from 100.2°C to 101°C with an average of 100.5 °C.Value of 100.15°C is taken as the boiling point of milk but actual boiling points is 100.45°C.
- The reason for this discrepancy is an alteration of the normal ionic and molecular colloidal equilibrium as a result of heating. As a conductor of heat milk is poorer than water. Addition of water lowers the concentration of the dissolved substances responsible for elevating the boiling point of milk.
- Though boiling point is lowered by addition of water it can be used as a method of water adulteration in milk. As such adulterated milk with water boils at a lower temperature than the normal milk. However, there are practical difficulties in determining the boiling point of milk [1,2].

Determination of Elevation of Boiling Point [1,4]

It can be determined by the following two methods:

i. Landsberger-Walker Method [4]

This method was introduced by Landsberger and modified by Walker.

Apparatus

The apparatus used in this method is shown in following figure and consists of:

i. An inner tube with a hole in its side and graduated in ml.

ii. A boiling flask which sends solvent vapour in to the graduated tube through a 'rosehead' (a bulb with several holes)

iii. An outer tube which receives hot solvent vapour issuing from the side-hole of the inner tube and.

iv. A thermometer reading to 0.01 K, dipping in solvent or solution in the inner tube.

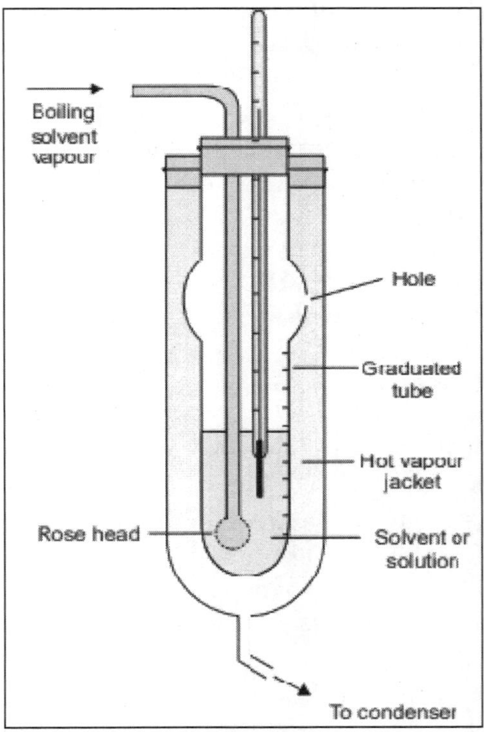

Figure 4.5: Landsberger-Walker Apparatus.

Procedure

Pure solvent is placed in the graduated tube and vapour of the same solvent boiling in a separate flask is passed into it. The vapour causes the solvent in the tube to boil by its latent heat of condensation. When the solvent starts boiling and temperature becomes constant, its boiling point is recorded.

- Now the supply of vapour is temporarily cut off and a weighed pellet of the solute is dropped into the solvent in the inner tube. The solvent vapour is again passed through until the boiling point of the solution is reached and this is recorded.
- The solvent vapour is then cut off, thermometer and rosehead raised out of the solution, and the volume of the solution read.

ii. Cottrell's Method [1,4]

A method better than Landsberger-Walker method was devised by Cottrell (1910). The details related to the apparatus and method are given below:

Apparatus

The apparatus consists of followings:

i. **a boiling tube (a)-**It is graduated and contains weighed amount of the liquid under examination.

ii. **An inverted funnel tube (b)-** It is placed in the boiling tube collects the bubbles rising from a few fragments of a porous pot placed inside the liquid.

iii. **Beckmann thermometer (f)-** It is held a little above the liquid surface.

iv. **Condenser (C)** - It has circulation of water through (d) and (e).

Method

Determine boiling point of pure solvent. When the liquid starts boiling, it pumps a stream of a liquid and vapour over the bulb of the Beckmann thermometer held a little above the liquid surface. In this way, the bulb is covered with a thin layer of boiling liquid which is in equilibrium with the vapour. This ensures that the temperature reading is exactly that of the boiling liquid and that superheating is minimum.

☆ After determining the boiling point of the pure solvent, a weighed amount of the solute is added and procedure is repeated for another reading. The vapours of the boiling liquid is cooled in a condenser which has circulation of water through (d) and (e). The cooled liquid drops into the liquid in (a).

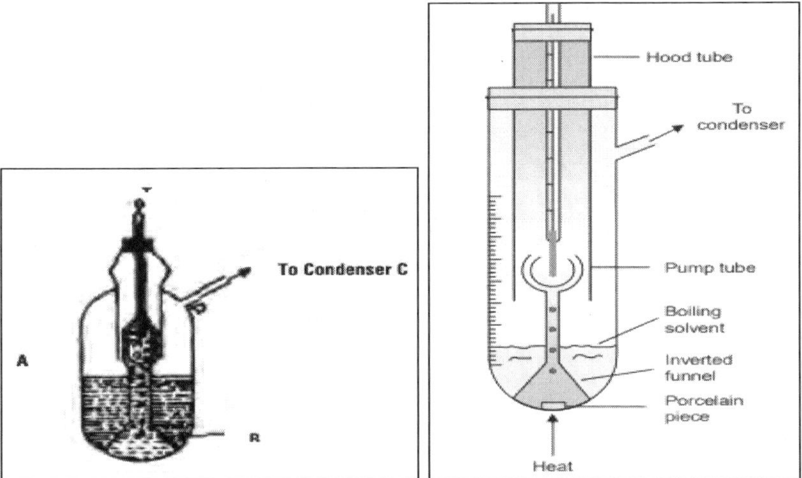

Figure 4.6: Apparatus for Determination of Elevation of Boiling Point [1,2,4].

Sources

1. *M.K. Srivastava, Handbook on Analysis of Milk, CBS Publications, New Delhi, 2010. ISBN 13: 9788123926681.*
2. *K.S. Sharma, Physical Chemistry of milk, Agrimoon.com.*
3. *Wikipedia.*
4. *Theory of Dilute Solutions, Chapter-15, http: //icv college. edu.in.*

c. Osmosis and Osmotic Pressure [1-4]

Osmosis is a vital process in biological systems, as biological membranes are semipermeable. In general, these membranes are impermeable to large and polar molecules, such as ions, proteins and polysaccharides, while being permeable to non-polar or hydrophobic molecules like lipids as well as to small molecules like oxygen, carbon dioxide, nitrogen, and nitric oxide. Permeability depends on solubility, charge, or chemistry, as well as solute size [2,3].

- ✰ **Osmosis** – The term "osmosis", from Greek *`óμüò* ("a push"), has a different definitions.
- ✰ It is the spontaneous net movement of solvent molecules through a selectively permeable membrane from a less concentrated solution into a region of higher solute concentration, aiming to equalize the solute concentrations on the two sides [1,2].
- ✰ It is a process of net movement of water caused by a concentration difference of water [1a].
- ✰ Osmosis is a special type of diffusion, namely the diffusion of water across a semipermeable membrane [3,4].
  1. Osmosis is a passive process and happens without any expenditure of energy. Any solvent can undergo the process of osmosis including gases and supercritical liquids.
  2. Some good examples of osmosis are- The raisin gets puffed when keep raisin in water, Movement of salt-water in animal cell across our cell membrane. Plants take water and mineral from roots with the help of Osmosis.
- ✰ **Osmotic pressure**- It is defined as the external pressure required to be applied so that there is no net movement of solvent across the membrane. It can also be defined as pressure required to stop solvent (water) from diffusing through a membrane by osmosis [2,3]. The osmotic pressure is the pressure required to counter, not sustain, osmosis.

  Osmotic pressure is a colligative property, meaning that the osmotic pressure depends on the molar concentration of the solute but not on its identity [2].

- ☆ When a solution and the pure solvent used in making that solution are placed on either side of a semi permeable membrane, it is found that more solvent molecules flow out of the pure solvent side of the membrane than solvent flows into the pure solvent from the solution side of the membrane. That flow of solvent from the pure solvent side makes the volume of the solution rise.
- ☆ When the height difference between the two sides becomes large enough, the net flow through the membrane ceases due to the extra pressure exerted by the excess height of the solution chamber. Converting that height of solvent into units of pressure gives a measure of the osmotic pressure exerted on the solution by the pure solvent. 'P' stands for 'pressure,'r' is the density of the solution, and 'h' is the height of the solution [1].
- ☆ Osmosis is influenced by the pressure exerted by the liquid on the membrane. It is helpful in isolating some of the constituents of milk.
- ☆ More solvent molecules are at the membrane interface on the solvent side of the membrane than on the solution side. Therefore, it is more likely that a solvent molecule will pass from the solvent side to the solution side than vice versa. That difference in flow rate causes the solution volume to rise. As the solution rises, by the pressure depth equation, it exerts a larger pressure on the membrane's surface. As that pressure rises, it forces more solvent molecules to flow from the solution side to the solvent side. When the flow from both sides of the membrane is equal, the solution height stops rising and remains at a height reflecting the osmotic pressure of the solution [1,2].

  The equation relating the osmotic pressure of a solution to its concentration has a form quite similar to the ideal gas law [1,2]- **PV=inRT.**

  This form of the equation has been derived by realizing that n/V gives the concentration of the solute in units of molarity, **P = iMRT**.

  Or Π = **iMRT**

  The osmotic pressure can be approximated by using the aforesaid formula(Morse equation).

  where, Π denotes for osmotic pressure, i - the dimensionless van 't Hoff index, M- the molar concentration of the solute, R- the gas constant and T- the temperature.

Raoult's Law [1,2]

The French chemist Francois Henricus Raoult discovered a law in 1887 to determine the vapor pressure of a solution when a solute has been added to it. Raoult's law is a chemical **law** that states that the vapor pressure of an ideal solution is dependent on the vapor pressure of each chemical component and the mole fraction of the component present in the solution. **Raoult's Law** is expressed by the following **formula**:

$$P_{solution} = \chi_{solvent} P^{o}_{solvent}.$$

It is a law stating that the freezing and boiling points of an ideal solution are respectively depressed and elevated relative to that of the pure solvent by an amount proportional to the mole fraction of solute.

Limitations of Raoult's Law [2]

There are a following limitations to Raoult's law-

- **i.** Raoult's law is apt for describing ideal solutions. **Raoult's Law only work for ideal solutions.** However, ideal solutions are hard to find and they are rare. Different chemical components have to be chemically identical equally.
- **ii.** Since many of the liquids that are in the mixture do not have the same uniformity in terms of attractive forces, these type of solutions tends to deviate away from the law.

Significance of Osmosis in Dairy Industry

Osmosis plays important role in milk but Reverse osmosis is widely used in dairy industry for processing especially in concentration of milk **or volume reduction of** milk and whey, milk solids recovery and water reclamation.

- ☆ Some examples related to the importance of osmosis in milk industry are given below:
  - i. Sweetened condensed milk is a good ingredient for sweet recipes because of all the added sugar about 25 per cent by weight. That added sugar kills bacteria that would otherwise digest the milk and spoil it. The sugar kills not by poisoning the bacteria, but by a more direct physical process. It draws water out of the bacteria so the bacterial cells shrivel and die.
  - ii. A bacterial cell in a can of sweetened condensed milk finds itself immersed in an extremely concentrated solution of sugar. Water inside the cell will, therefore, pass out through the cell membrane into the sugar solution. The bacterial cell dehydrates and dies in a sea of sugary water.

Sources

1. *K.S. Sharma, Physical Chemistry of milk, Agrimoon.com.*

1a. *Arthur C. Guyton and John E. Hall, Textbook of Medical Physiology, ISBN: 978-1-4160-4574-8.*

2. *Wikipedia.*

3. *M.K. Srivastava, Handbook on Analysis of Milk, CBS Publications, New Delhi, 2010. ISBN 13: 9788123926681.*

4. *William Stillwell, Membrane Transport in An Introduction to Biological Membrane Second Edition, 2016*

5. *Steve Fentress, Milk Osmosis, 2019. Indianapublicmedia.org.*

Chapter 5

Aqueous Solution of Electrolytes

It is observed that the electrical current will pass through the medium when certain substances are dissolved in it but in certain situations it will not allow the current to pass through. The study on the nature of these substances is needed to understand such phenomena substances.

Electrolytes and Non-Electrolytes Substances [1,2,3]

The definitions and examples of electrolytes and nonelectrolytes are discussed below:

Electrolytes [1,2,3]

These are salts or molecules that ionize completely in solution. Substances which cause increase in electrical conductivity in solutions and which dissociate in the process of conducting the electric current are known as electrolytes.

- ☆ Electrolyte is an electrical conductor in which current is carried by ions rather than by free electrons (as in a metal).
- ☆ An **electrolyte** is a substance that produces an electrically conducting solution when dissolved in a polar solvent, such as water. The dissolved electrolyte separates into cations and anions, which disperse uniformly through the solvent [1a].

For example, when table salt, NaCl, is placed in water, the salt (a solid) dissolves into its component ions, as per the following dissociation reaction-

$$NaCl(s) \rightarrow Na^{+}(aq) + Cl^{-}(aq)$$

The compounds that readily form ions in solution are known as ***strong*** **electrolytes** and a compound dissociates to a small extent, the solution will be a weak conductor of electricity; a compound that only dissociates weakly, therefore, is known as a **weak electrolyte**.

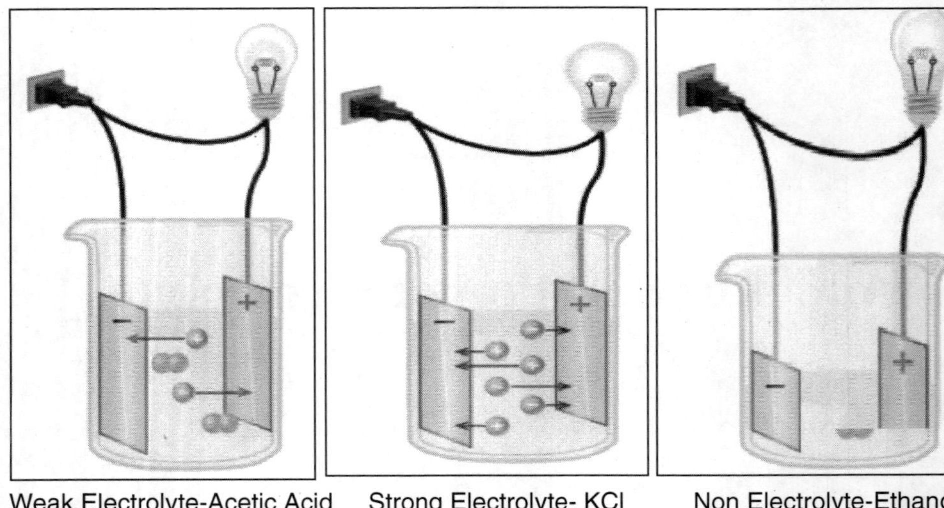

Weak Electrolyte-Acetic Acid Strong Electrolyte- KCl Non Electrolyte-Ethanol

Figure 5.1: Electrolyte and Non-electrolyte Solutions [4].

Nonelectrolytes [1,2,3]

They do not dissociate into ions in solution. A non-electrolyte, however, is a compound composed of molecules that does not conduct electricity when molten or in aqueous solutions. If the substances which give the expected result to osmotic pressure, freezing point and boiling point but do not significantly increase the electrical conductivity such substances are known as non electrolytes.

- ☆ Nonelectrolytes are primarily held together by covalent rather than ionic bonds. Glucose, or $C_6H_{12}O_6$ is a common example of a nonelectrolyte. Glucose (sugar) readily dissolves in water, but because it does not dissociate into ions in solution and hence does not conduct electricity.Urea, Benzene, Ethanol, Chloroform, ether *etc.* are the other examples of non electrolytes.

Sources

1. *Electrolytes and nonelectrolytes solution, https: //courses. Lumen learning.com,1a-Wikipedia-Electrolytes.*
2. *K.S. Sharma, Physical Chemistry of milk, Agrimoon.com.*
3. *Electrolytes and Non electrolytes, Ionic mobility, Lesson 14, http: //ecoursesonline. iasri.res.in.*
4. *Electrolytes-chemistry, https: //open textbc.ca.*

Electrolytic Dissociation [1,2,3]

In electrolytic, or ionic, dissociation, the addition of a solvent or of energy in the form of heat causes molecules or crystals of the substance to break up into ions. Water is a neutral substance, which does not conduct electricity. By dissolving some substance in water, it conducts electricity. These substances are called electrolytes and the process is known as "Electrolytic dissociation.".

- ☆ Non electrolytes behave in a definite manner with respect to osmotic pressure, freezing point and boiling point and this behavior can be calculated and predicted with a fair degree of accuracy.
- ☆ The electrolytes when subjected to similar calculations, the predicted results were always observed to be less than their effect as observed by actual trial.
- ☆ In case of very dilute solutions, their effect comes close to being even 2, 3, or 4 times greater than as predicted.

To explain such behavior, electrical conductivity of solutions and the facts of electrolysis, Arrhenius proposed the following theory of electrolytic dissociation or ionization:

Arrhenius Theory [1,2,3]

This theory is introduced in 1887 by the Swedish scientist Svante Arrhenius.

- ☆ According to this theory, molecules of an electrolyte on going into solution dissociate into ions. In electrolytic dissociation, molecules or crystals breaks into ions by addition of energy or solvent.
- ☆ The ions are dissimilar as to the elements or elements represented in them and as to the electrical charge which they carry. In each case there are ions that carry positive charge and ions that carry negative charges in such numbers that the two types just compensate each other so that the solution is electrically neutral or balanced. This is illustrated by giving a few chemical dissociation equations.

$$NaCl(s) \rightarrow Na^{+}(aq) + Cl^{-}(aq)$$

$$K_2SO_4 \rightarrow 2K^{+} + SO_4^{2-}$$

The above reactions are in equilibrium and the dissociation is more complete, the more dilute the solutions are made. At complete dissociation each molecule gives rise to 2, 3 or 4 ions as the case may be and this accounts for the fact that in very dilute solutions the effect on the osmotic pressure, freezing point and boiling point is nearly 2, 3 or 4 times as great as anticipated on the basis of calculations that holds for non electrolytes.

- ☆ According to the modern views, we consider the electric current to be a stream of electrons. The electron is the ultimate unit of electricity its charge is negative its mass is approximately 1/1000 the mass of the hydrogen atom.
  - i. A positive ion has 1, 2 or 3 positive charges because it lacks 1, 2 or 3 electrons.
  - ii. A negative ion has 1, 2 or 3 negative charges because it contains an excess of 1, 2 or 3 electrons.
  - iii. A limited number of ions with a still larger number positive or negative charges are also there.
  - iv. To understand the difference between atoms and ions more fully, the example of sodium atom and sodium ion is given below:

    Na (sodium atoms)as compared to Na^+ (sodium ion).
  - v. The atom consists of a positive nucleus surrounded by electrons in such number that the positive nucleus is just satisfied.
- ☆ The atoms of some elements have a strong tendency to lose electrons such elements yield positively charged ions whereas other elements have a strong tendency to take on additional electrons such elements yield negatively charged ions. When an electric current is passed through a solution it is conducted by the migration of ions.
- ☆ The positive ions migrate to the negative electrode and then take on the requisite number of electrons to neutralize the charge.
  - i. The discharged ion is then an atom (if the ion contained only one element) and is deposited on the electrode (electro-plating) or
  - ii. it may escape as a gas (two atoms combining to form molecule) or
  - iii. it may react with water of the solution, depending upon the chemical nature of the element involved.
- ☆ Similarly negative ions migrate to the positive electrode and there give up the excess electrons.

Sources

1. *K.S. Sharma, Physical Chemistry of milk, Agrimoon.com.*
2. *Britanica.*
3. *Electrolytes and Non electrolytes, Ionic mobility, Lesson 14, http: //ecoursesonline.iasri.res.in.*

The Theory of Electrolytic Dissociation [1,2]

The theory of electrolyte dissociation explains the followings [1]-

i. The abnormal behaviors of electrolytes with respect to osmotic pressure, freezing point and boiling point.

ii. The electrical conductivity of solutions, and

iii. The facts of electrolysis.

- ☆ The degree of dissociation can be measured in various ways. The extent, to which electrolytes deviate from the accepted rules for osmotic pressure, freezing point, and boiling point, can be a basis for computing the degree of dissociation. Since electrical conductivity is dependent upon ions which can also be used to measure the dissociation.
- ☆ The equilibrium conditions follow the law of mass action only in a qualitative sense in the case of electrolytes that dissociate very readily, but in strict mathematical sense in the case of electrolytes they dissociate only to a slight extent. Weak acids and bases fall under the latter class.

Limitation of Classical Theory of Electrolytic Dissociation [2]

The classical theory of electrolytic dissociation, developed by S. Arrhenius and W. Ostwald (1880's), is based on the assumption of the incomplete dissociation of the solute, characterized by the degree of dissociation α, that is, the fraction of the electrolyte molecules that dissociate.

- ☆ The dynamic equilibrium between the undissociated molecules and the ions is described by the law of mass action.

  For example, the electrolytic dissociation of a binary electrolyte CA is described by an equation of the type $CA \rightleftarrows C^+ + A^-$. The dissociation constant K_d is determined by the activities of the cations a_{c+}, anions a_{A},and undissociated molecules a_{CA} as follows:

$$K_d = \frac{a_{c+} + a_{A-}}{a_{cA}}$$

  The value of K_d depends on the nature of the solute and solvent, as well as on the temperature, and may be determined by several experimental methods.

- ☆ The classical theory of electrolytic dissociation is applicable only to dilute solutions of weak electrolytes. Strong electrolytes in dilute solutions are virtually completely dissociated, and consequently the idea of an equilibrium between ions and undissociated molecules is meaningless.

Sources

1. *K.S. Sharma, Physical Chemistry of milk, Agrimoon.com.*

2. *Electrolytic dissociation, https: //encyclopedia2. Thefree dictionary. com.*

Ionic Mobility [1]

In the absence of external electrical field, the ions of a solution are in random motion since all directions are equivalent. When an external electrical field is applied although the randomness of motion will basically remain, one of the directions becomes preferred, and more so the higher the potential gradient *i.e.* the greater drop of potential per cm.

- ☆ Ionic mobility (μ) is the ability of charged particles to move through a medium in response to an electric field that is pulling them at a specified temperature and pressure.
- ☆ Ionic mobility is the ability of ions to move through a medium whereas ionic velocity defines how fast the ions move through the medium.
- ☆ The velocity of an ion is the value of this preferential migration towards one of the electrode expressed in cm/s. The potential gradient for the purpose of comparing the ion velocity is 1volt/cm and these mobilities are known as absolute mobilities. Since the mobilities are taken as proportional to the potential gradient, the absolute mobility is expressed in cm^2/volt.
- ☆ In gaseous electric conduction, the average velocity with which a given ion drifts through a specified gas under the influence of an electric field of unit strength.
- ☆ The mobilities are very low having values of only 0.0005 – 0.003 cm^2/volt. Ionic conductance is the contribution of a given type of ion to the total equivalent conductance in the limit of infinite dilution.

$$\text{Ionic mobility} = \text{Ionic conductance}/96500$$

Ostwald's Dilution Law and Kohlrawsch Law [1,2,4]

The electrolytes dissociate into charged particles known as ions and they play an important role in various chemical reactions. Their behaviour at different dilutions needs to be assessed to understand the degree of ionization and their mobility.

- ☆ The Ostwald's dilution law and Kohlransch's Law explain the behaviour of these ions and help in calculating their ionic mobility.

Ostwald's Dilution Law [1,1a,2,4]

The law was first put forward by Wilhelm Ostwald to account for electrical conductivities of electrolyte solutions.

- ☆ This law is concerned with the relationship between the dissociation constant and the degree of dissociation of a weak electrolyte (acids or bases) [1,2,4].

☆ For any weak electrolyte, Ostwald's dilution law states that the degree of dissociation is inversely proportional to square root of the molar concentration and is directly proportional to the square root of the volume containing one mole of electrolyte [1,2,4].

$$Kp \text{ or } kd = \frac{C(K^+).C(A^-)}{C(KA)} = \frac{\alpha^2}{1-\alpha} \cdot C_O$$

where, Kp - constant of protolysis/dissociation, α - degree of dissociation (or degree of protolysis), $C(A^-)$ - concentrations of anions, $C(K^+)$ - concentration of cations, C0 - overall concentration and C(KA) - concentration of associated electrolyte.

The value of α can be calculated by measuring conductance of the solution as-

$$\alpha = \frac{\lambda_v}{\lambda_{infty}}$$

where, λ_v is the equivalent conductance at a particular dilution and λ_∞ is equivalent conductance at infinite dilution.

☆ If a weak acid dissociates in water $HA \rightleftharpoons H^+ + A^-$ the dissociation constant K_a is given by the following equation- [1a]

$$K_a = \alpha^2 n/(1-\alpha)V$$

where, α -is the degree of dissociation, n -the initial amount of substance (before dissociation), and V -the volume.

If α is small compared with 1, then $\alpha^2 = KV/n$.

☆ According to Arrhenius theory of electrolyte dissociation, the molecules of an electrolyte in solution are constantly splitting up into ions and the ions are constantly reuniting to form unionized molecules. Therefore, a dynamic equilibrium exists between ions and unionized molecules of the electrolyte in solution. It was pointed out by Ostwald that like chemical equilibrium, law of mass action can be applied to such systems also.

Consider a binary electrolyte AB which dissociates into A^+ and B^- ions and the equilibrium state is represented by the equation [1,2,4]-

$$AB \rightarrow A^+ + B^-$$

	AB	A^+	B^-
Concentration, when initially t=0	C	0	0
At equilibrium	C	αc	αc

So dissociation constant may be given as-

$$K = \frac{[A^+][B^-]}{[AB]} = \frac{[\alpha c][\alpha c]}{[1-\alpha]c} \qquad \textit{Eq.-1}$$

For very weak electrolytes, $\alpha <<< 1$, implying that $(1 - \alpha) \approx 1$.

$$K = c\alpha^2$$

$$\alpha = \sqrt{\frac{K}{c}} \qquad \textit{Eq.-2}$$

Thus, degree of dissociation of a weak electrolyte is proportional to the square root of dilution [1,1a,2,4].

Ostwald's dilution law states that only at infinite dilution the weak electrolyte undergoes complete ionization.

Limitations of Ostwald's Dilution Law [1,2,4]

The law holds good only for weak electrolytes and fails completely in case of strong electrolytes.

- ☆ The value of 'α' is determined by conductivity measurements by applying the formula $\Lambda/\Lambda\infty$. The value of α' determined at various dilutions of an electrolyte when substituted in Eq. 1 gives a constant value of K only in the case of weak electrolytes like CH_3COOH, NH_4OH, *etc.* [1,2].
- ☆ Lewis and Randall recognized that the law fails badly since the supposed equilibrium constant is actually far from constant [2]. The cause of failure of Ostwald's dilution law in case of strong electrolytes is due to the following factors"
  - **i.** The law is based on the fact that only a portion of the electrolyte is dissociated into ions at ordinary dilution and completely at infinite dilution. Strong electrolytes are almost completely ionized at all dilutions and $\Lambda/\Lambda\infty$ does not give accurate value of 'α'.
  - **ii.** When concentration of the ions is very high, the presence of charges on the ions appreciably affects the equilibrium. Hence, law of mass action its simple form cannot be strictly applied in case of strong electrolytes.

Kohlransch's Law [1,2,4]

Friedrich Kohlrausch, a german physicist, gave this law in 1875-1879.

- ☆ This law is popularly known as **Kohlrausch law of independent migration of ions.** Kohlransch's Law of independent migration of ions states that the limiting molar conductivity of an electrolyte, is equal to the sum of the amount of its each constituent ion's limiting molar conductivity [1b].

- ✰ This law states that the equivalent conductivity of an electrolyte at infinite dilution is equal to the sum of the conductance of the anions and cations [1,4].
- ✰ At infinite dilution, when dissociation is complete, each ion makes a definite contribution towards equivalent conductance of the electrolyte irrespective of the nature of the ion with which it is associated and the value of equivalent conductance at infinite dilution for any electrolyte is the sum of contribution of its constituent ions, *i.e.*, anions and cations.

$$\Lambda_\infty = \lambda a + \lambda c$$

where, λa- is the equivalent conductance of the anion and λc-that of the cations.

The ionic conductances are proportional to their ionic mobilities. Thus, at infinite dilution,

$$\lambda c = kuc \text{ and } \lambda a = kua$$

where, uc and ua are ionic mobilities of cation and anion respectively at infinite dilution. The value of k is equal to 96500 c, *i.e.*, one Faraday.

- ✰ Thus, assuming that increase in equivalent conductance with dilution is due to increase in the degree of dissociation of the electrolyte, it is evident that the electrolyte achieves the degree of dissociation as unity when it is completely ionized at infinite dilution.
- ✰ Therefore, at any other dilution, the equivalent conductance is proportional to the degree of dissociation. Thus,

Degree of dissociation $\alpha = \Lambda/(\Lambda_\infty)$ = (Equivalent conductance at a given concentration)/(Equivalent conductance at infinite dilution).

Applications of Kohlrausch Law [1b,3]

Some important applications of Kohlrausch law of independent migration of ions are given below:

i. **Calculation of molar conductivity:** This law helps in the determination of limiting molar conductivities for any electrolyte. When concentration of electrolyte is almost zero at that point **molar conductivity is called as limiting molar conductivity.**

   Weak electrolytes have lower molar conductivities and lower degree of dissociation at higher concentrations. The molar conductivity of weak electrolyte increases steeply at lower concentrations. Therefore, limiting molar conductivity, $\ddot{E}_m^{\,o}$ cannot be obtained by extrapolation of molar conductivity to zero concentration.

Hence, we use the Kohlrausch law of independent migration of ions for the determination of limiting molar conductivity, $\ddot{E}_m^{o}$ for weak electrolytes.

ii. Value of dissociation constant- This law also helps in determining the value of dissociation constant from the value of molar conductivity and limiting molar conductivity for a weak electrolyte at a given concentration by following equation-

$$\alpha = \Lambda \ddot{E}m^{o}$$

where, α -dissociation constant, Λ - molar conductivity and $\ddot{E}_m^{o}$ - limiting molar conductivity;

iii. Degree of dissociation-The degree of dissociation of a weak electrolyte is proportional to its molar conductance. It is also used for calculation of degree of dissociation. If at any concentration C the value of molar conductance is λme then

$$d = \lambda m^{e} / \lambda_m^{\infty}$$

d = degree of dissociation.

iv. Determination of the solubility of a sparingly soluble salt - Kohlrausch's law is used for calculation of solubility of moderately or sparingly soluble salt. The solubility of a sparingly or moderately soluble salt in a solvent is quite low. The solubility of silver chloride, barium sulphate, lead sulphate *etc.*,in a solvent is quite low. Even a saturated solution of such a salt is so dilute that it can be assumed to be at infinite dilution.

v. Determination of ionic product of water.

vi. It is also used in the calculation of the cell potential in various electrochemical cells [1b].

Sources

1. *K.S. Sharma, Physical Chemistry of milk, Agrimoon.com.*

1a. *Ostwald dilution law, Overview, Oxford Reference, https: //www.oxfordreference.com.*

1b. *https: //protonstalk.com/electrochemistry/kohlrausch-law.*

2. *Wikipedia- Law of dilution.*

3. *Application of Kohlrausch law, https: //www.emedicalprep.com.*

4. *Kohlrawsch Law and Ostwald's Dilution Law and Electric Conductance of Milk, Lesson-15, http: //ecoursesonline.iasri.res.in.*

Equilibria, Ionic Equilibria and Dissociation of Ionic Product of Water [1-4]

Equilibrium is the state of a process in which the properties like temperature, pressure, and concentration *etc.* of the system do not show any change with passage of time. In all processes which attain equilibrium, two opposing processes are involved. Equilibrium is attained when the rates of the two opposing processes become equal. If the opposing processes involve only physical changes, the equilibrium is called **Physical Equilibrium.** If the opposing processes are chemical reactions, the equilibrium is called **Chemical Equilibrium [1]**.

- ☆ Chemical reactions mostly take place in solutions. All chemical substances are made up of either polar units (called ions) or non-polar units. The activity of these entities is more evident and pronounced in solution. The behaviour of these substances depends upon their nature and conditions of the medium in which they are added [2].

  Ionic Equilibrium [2,4] - The equilibrium established between the unionized molecules and the ions in the solution of weak electrolytes is called ionic equilibrium. For example, take acetic acid breaking up into acetate ions and hydrogen ions [3]-

$$CH_3\,COOH \rightarrow CH_3\,COO^- + H^+$$

- ☆ This type of equilibrium is observed in substances that undergo ionization easily, or in polar substances in which ionization can be induced. Ionic and polar substances are more easily soluble in polar solvents because of the ease of ionization taking place in the solvent medium. With the dissolution of ionic and polar substances in the solvent, these solutions become rich in mobile charge carriers (ions) and thus can conduct electricity [2].
- ☆ Substances in Ionic Equilibrium can be classified into two categories on the basis of their ability to conduct electricity given as under [2,4]-
  - i. **Electrolytes**: Substances, that dissociate into their constituent ions in their aqueous solution and thus these substances are capable of conducting electricity, are called as electrolytes
  - ii. **Non electrolytes**: Substances that consist of molecules that bear no electric charge, do not dissociate into their constituent ions and not capable of conducting electricity,are called as non-electrolytes.
- ☆ In a water molecule the electrons are shared unevenly between the oxygen and hydrogen making it a polar molecule. Because of this water molecules have a partial negative charge on Oxygen, and a partial positive charge on Hydrogen. Accordingly the water molecules can form strong electrostatic attractions with other water molecules, polar molecules and ions [2].

Sources

1. *Chemical and Ionic eqilibria, www.askiitians.com.*
2. *K.S. Sharma, Physical Chemistry of milk, Agrimoon.com.*
3. *Ionic equilibrium,https: //study. com/academy/lesson/ionic-equilibrium*
4. *Wikipedia*

Dissociation of Water [1-4]

Water is frequently regarded as a substance that practically does not dissociate into ions. It is observed that it will dissociate to a small extent according to the following equation [1]:

$$H_2O \rightleftharpoons H^+ + OH^-$$

- The hydrogen ion in water always forms hydroxonium H_3O^+. **H_2O can act as both a proton donor and acceptor for itself. A proton can be transferred from one water molecule to another, resulting in the formation of one hydroxide ion (OH^-) and one hydronium ion (H_3O^+).** This is called the **autoionization (self-ionization of water)** or **dissociation**/auto dissociation of water **[2,4].**

$$H_2O + H_2O \rightleftharpoons H_3O^+ + OH^-$$

- In solutions containing acids, the hydrogen ions formed will affect the equilibrium in the above equation. A similar effect could also be seen in solutions containing bases which form the hydroxyl ions (OH-). The equilibrium is determined by the dissociation constant which is expressed in the following equations [1-4]-

$$K_{da} = \frac{a_{H^+} \times a_{OH^-}}{a_{H2O}}$$

- For pure solutions or for pure water the activities (a) can be replaced by concentration (c), then the above equation could be expressed as-

$$K_d = \frac{c_{H^+} \times c_{OH^-}}{c_{H2O}}$$

Equilibrium (in the water dissociation reaction)which greatly shifts the direction of undissociated water. Dissociation of water being very slight is considered that either the activity of undissociated water molecule or its concentration is regarded as constant. We can now combine the dissociation constant and rearrange the above two formulae as follows [1,4]-

$$a_{H^+} \times a_{OH^-} = Kw$$

$$c_{H^+} \times c_{OH^-} = Kw$$

where, the constant $Kw = K_{d.a} \times a\, H_2O$ or

$$Kw = K_d \times C\, H_2O$$

☆ This is called the ionic product of water. In pure water or in an aqueous solution, the product of concentrations of hydrogen and hydroxyl ions is a constant at a given temperature. This is called **ionic product of water** and is conventionally designated by *Kw*.

Ionic product kw, concentration c_{H+} and c_{OH-} of ions in water and pH of water at various temperatures are given in following table [1]-

Table 5.1: Ionic Product, Temp and pH

Temperature (°C)	Ionic Product (kw)	$c_{H+} = c_{OH^-}$	pH
0	1.139×10^{-15}	3.38×10^{-8}	7.472
18	5.702×10^{-15}	7.64×10^{-8}	7.117
25	1.008×10^{-14}	1.004×10^{-7}	6.998
50	5.474×10^{-14}	2.339×10^{-7}	6.631
100	5.9×10^{-13}	7.7×10^{-7}	6.12

At 25°C the ionic product of water kw = 1.008×10^{-14}

a. *In Neutral Solution*

Since pure water as in any other neutral medium $a_{H+} = a_{OH-}$ (or $c_{H+} = c_{OH-}$) for 25° C.

$a_{H+} = a_{OH-} = \sqrt{Kw} = 1.004 \times 10^{-7}$ g ions per litre and hence the degree of dissociation.

$$\alpha = \frac{a^{H+}}{55.5} = \frac{1.004 \times 10^{-7}}{55.5} = 1.8 \times 10^{-9}$$

where, 55.5 = C_{H2O} *i.e.* the number of moles of water per litre at 25°C.

b. *In Non Neutral Media*

In non neutral media,the a_{H+} and a_{OH-} are not equal to each other.

☆ As it could be observed from the above equations these two ions are intimately connected and are inversely proportional to each other. Thus upon adding acid to water we increase the concentration of hydrogen ions and therefore augment the value of a_{H+} but this accelerates the opposite side of the reaction, some of the added H^+ ions bind an equal amount of OH^- to form H_2O and reduction in acid a_{OH} ions occur.

- **Equilibrium is established again when the product of ion activities once more acquires the value it had prior to addition of the acid. Hence, any increase in hydrogen ion concentration causes a corresponding decrease in the hydroxyl ion concentration and vice versa [1].**
- The **dissociation of water** molecules leads to acidic and basic conditions that can affect life.

Effect of Temperature, Pressure and Ionic Strength on dissociation [2]

The dependence of the water ionization on temperature and pressure has been investigated thoroughly. The value of pK_w decreases as temperature increases from the melting point of ice to a minimum at 250 °C, after which it increases up to the critical point of water 374 °C. It decreases with increasing pressure.

Sources

1. *K.S. Sharma, Physical Chemistry of milk, Agrimoon.com.*
2. *Wikipedia.*
3. *Ionic equilibrium,https: //study.com/academy/lesson/ionic-equilibrium.*
4. *Dissociation of water, https: //groups.chem.ubc.ca/courseware.*

pH and pOH Scales [1,2]

When water molecule dissociates and attains ionic equilibrium the concentration of both [H^+] and [OH^-] ions will be equal. Increase of either H^+ or OH^- would result in the imbalance of the equation. As such this phenomenon is used to measure the acidic or alkaline condition of the solution. This is referred as pH scale.

pH

Danish botanist S.P.L. Sorensen, in 1909, proposed a logarithmic scale (called pH scale) for expressing the concentrations of H+ ions. He defined pH as the negative logarithm of the molar concentration of hydrogen ions [3]. The pH scale of acidity defines pH as the negative common logarithm of the concentration of H+:

$$pH = -\log_{10} [H^+]$$

Now a days above is represented as:

$$pH = -\log_{10} [H_3O^+]$$

- A common logarithm is a function that computes what exponent would be on 10 to obtain the input number. For example, the logarithm of 100 is 2 because $10^2 = 100$.

- ☆ Water is frequently regarded as a substance that practically does not dissociate into ions it is in fact always dissociated to a small extent according to the following equation [1]-

$$H_2O \leftrightarrow H^+ + OH^- \quad 1$$

- ☆ In solutions containing acids the hydrogen ions will affect the equilibrium of the above reaction. Similarly in solutions containing bases the hydroxyl ions will affect the equilibrium.
- ☆ The details related to the equilibrium, dissociation constant and ionic products of water are discussed under Dissociation of Water.

Acidity and Alkalinity [1]

The simple reaction between these two quantities (H^+ OH-)may be used to characterize the acidity and alkalinity of media. Any increase in hydrogen ion concentration causes a corresponding decrease in hydroxyl ion concentration and vice versa.

- ☆ The acid properties are due to the hydrogen ion and can be expressed quantitatively by the activity of aH^+.
- ☆ Similarly the alkalinity of a medium can be expressed by the activity of a OH^-. Although both are equivalent in this respect it is customary to use the activity of the hydrogen ions as the parameter utilized by convention.
- ☆ In acid solution aH^+ is greater than in pure water, the difference increasing as the acidity increases. In alkaline media the value of aH + is less than for pure water again the difference increasing as the alkalinity increases. In practice instead of aH^+ the so called hydrogen ion exponent of pH is used which is defined as

$$pH = -\log_{10} aH^+$$

For a dilute solutions $pH = -\log_{10} cH^+$.

Hence pH is a quantity characterizing the activity (or concentration) of hydrogen ion and is numerically equal to the negative common logarithm of the activity of concentration expressed as g. ions per litre [1]. Mathematically, $pH = -\log_{10} CH^+$. pH is an index of the true acidity or alkalinity of the system [4]. pH is usually (but not always) between 0 and 14. Knowing the dependence of pH on $[H_3O^+]$.

- ☆ **Neutral solution [1,2,4]-** At 25°C the pH of a neutral medium is 7. This is known as neutral pH. This value is not randomly selected, but rather it comes from the fact that the $[H^+]$ in pure water is 10^{-7}.
- ☆ **Acidic solution [1,2,4]-** It has a pH value less than 7 because $[H^+]$ is greater than that of pure water.

☆ **Basic solution [1,2,4]**-A basic solution has a pH greater than 7 because there is a lower [H^+] than that of pure water (and consequently a larger OH]).

pOH [1,2,3]

The **pOH scale** is similar to the **pH scale.** pOH is defined as the negative common logarithm of the concentration of OH-

$$pOH = -\log [OH-]$$

pOH scale analogous to the pH acidity scale is used to gauge hydroxide ion concentrations of aqueous solutions. In the pOH scale, pOH of 7 is indicative of a neutral solution. A basic solution has a pOH less than 7, while an acidic solution has a pOH of greater than 7. The pOH is convenient to use when finding the hydroxide ion concentration from a solution with a known pH.

A useful relationship is that pH + pOH = 14. This formula will allow to readily convert values of pH and pOH. Note that because Kw is constant, the product of [H+] and [OH-] is always equal to 10-14.

The notation "p" in front of a symbol means "take the negative log". We have defined pH and pOH as values to describe the acidic or basic strength of solutions. Now pKa (- log Ka) and pKb (- log Kb) are used as measures of acidity and basicity, respectively, for standard acids and bases [1,3].

Strong acid-An acid with a pKa less than zero is called a strong acid because it almost completely dissociates in water, giving an aqueous solution a relatively low pH.

Weak acid-Acids with pKa's greater than zero are called weak acids because they only partially dissociate in water, to make solutions with larger pH's than those strong acids produce at the same concentration.

Strong and weak bases-Similarly, bases with pKb's less than zero are strong bases and bases with pKb's greater than zero are called weak bases [1].

Difference between pH and pOH [5]- pH refers to the "potential of hydrogen". It can be used to determine whether a solution is acidic, basic or neutral. In contrast, pOH is a measure of hydroxide ion (OH^-) concentration. The

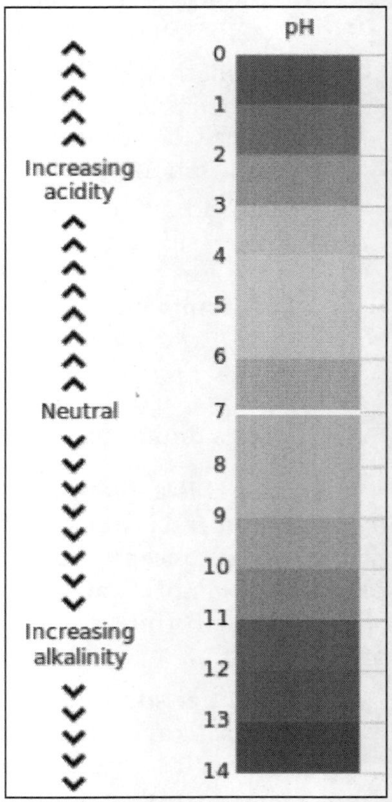

Figure 5.2: pH Scale [5].

key difference between pH and pOH is that pH is a measure of hydrogen ions whereas pOH is a measure of hydroxide ions.

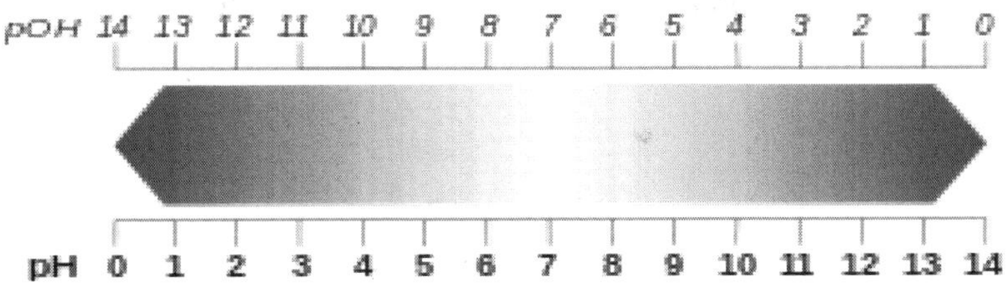

Figure 5.3: Relation between pH and pOH [2].

Sources

1. *K.S. Sharma, Physical Chemistry of milk, Agrimoon.com.*
2. *Wikipedia-pH.*
3. *Ionic equilibrium, Chemical Dynamics, https: //nios.ac.in.*
4. *M.K. Srivastava,Handbook on analysis of milk, ISBN: 978-8123926681,2010*
5. *Difference Between pH and pOH, https: //www.differencebetween.com.*

Chapter 6

Acids and Bases

Lewis Concepts of Acids

Various scientists have attempted to explain the concept of acids and bases.

- ☆ Lewis tried to give his conceptual definition to these acids and bases basing upon their electronic configuration.
- ☆ Similarly Arrhenius and Bronsted and Lewis have also explained the concept of acids and bases taking in to consideration several other parameters of the reactions between an acid and a base.
- ☆ The concepts proposed by Arhenius and Bronsted and Lowry are limited to such substance which contain protons where as the Lewis proposed more general and broader concept for acids and bases.

Lewis Concept [1,1a,2]

The Bronsted- Lowry concept does not depend on the nature of the solvent. However, like Arrhenius concept it does not explain the acidity of the substances which do not have a hydrogen. In 1923, G.N. Lewis proposed a yet another concept of acids and bases that includes such substances also [2].

- ☆ According to this concept an acid is any species (any atom, molecule or ion) which is capable of accepting a pair of electrons and a base is a species(any atom, molecule or ion) having lone pair of electrons for donation [1].
- ☆ A **Lewis acid** is a chemical species that contains an empty orbital which is capable of accepting an electron pair from a Lewis base to form a Lewis adduct. A **Lewis base**, then, is any species that has a filled orbital

containing an electron pair which is not involved in bonding but may form a dative bond with a Lewis acid to form a Lewis adduct [1a].

i. Thus all the Bronsted - Lowry Acids and bases are also Lewis acids and Lewis bases but all the Lewis acids and bases are not Bronsted - Lowry Acids and bases respectively [1].

ii. Water and some other compounds are considered as both Lewis acids and bases since they can accept and donate electron pairs based on the reaction.

iii. Lewis acids also include some other compounds like BF_3, Al Cl_3, Zn Cl_2, Fe Cl_3, SO_3 *etc.* According to Lewis an acid may also be a species without any proton.

$$AlCl_3 + NH_3 \rightarrow Cl_3Al \leftarrow NH_3$$

In the above example $AlCl_3$ is an electron deficient species. It accepts an electron pair from a molecule of NH_3 which has a lone pair of electrons on N atom. Thus, $AlCl_3$ is a Lewis acid and NH_3 is a Lewis base [2].

Types of Lewis Acids [1,1a]

Lewis acids have been grouped based on certain criteria. The criteria for such grouping are discussed here under.

☆ Compounds whose central atom has an incomplete octet. In other words Lewi's acids of this type are the electro deficient molecules such as BF_3, Al Cl_3, Ga Cl_3 *etc.*

BF_3	+	$:NH_3$	$\rightarrow F_3B{:}NH_3$
Acid (Electron donor)		Base (Electron acceptor)	

☆ Compounds in which the central atom has available d-orbital and may acquire more than that of an octet of valence electrons.

SiF4	+	$:2F^-$	$\rightarrow SiF_6^{2-}$
Lewis acid		Lewis base	Complex

☆ Some other examples of this type are PF_3, PF_5, SF_4, Se F_4, Te Cl_4 *etc.*

☆ All simple cations *e.g.*, Na^+, Ag^+, Cu^{2+}, Al^{3+}, Fe^{3+}. They can combine with an electron pair.

$$Ag^+ + 2{:}\ NH_3 \rightarrow (H_3N{:}\ Ag{:}\ NH_3)^+$$

☆ Molecules with a multiple bond between atoms of dissimilar electro negative charge.

$$O = C = O, O = S = O$$

Lewis Bases [1,1a]

Bases are also being grouped similar to acids on the basis of certain criteria. The criteria and the groups(Types) of bases are discussed below:

- ☆ All the simple negative ions such as Cl^-, F^- *etc.*
- ☆ Molecules with one or two unshared pair of electrons - H_2O,NH_3, R OH, R_2O, R_2S Pyridine *etc.*
- ☆ Multiple bonded compounds which form coordination compounds with transition metals *e.g.* CO, NO, Ethylene, Acetylene.

Sources

1. *K.S. Sharma, Physical Chemistry of milk, Agrimoon.com.*

1a. *Wikipedia-Lewis acid and bases.*

2. *Bronsted and Lowry concept, Chemical dynamics, modules-5, G: /Acid per cent 20 bases per cent 20theories.pdf.*

Bronsted-Lowry Concept [1,1a,2]

In 1923, J.N. Bronsted and T.M. Lowry pointed out independently that acid-base reactions can be interpreted as proton-transfer reactions. They have given following definitions for acid and base-

- ☆ An acid as a substance which gives up proton (H^+) or proton donor.
- ☆ A base as a substance which accepts proton or proton acceptor [1,1a,2].

In the Bronsted–Lowry theory acids and bases are defined by the way they react with each other, which allows for greater generality. The definition is expressed in terms of an equilibrium expression [1a]

$$\text{Acid} + \text{Base} \rightleftharpoons \text{Conjugate base} + \text{Conjugate acid.}$$

The following cases are helpful in understanding the aforesaid definition-

i. The species B accepts a proton and thus behaves as a base while the species BH^+ gives up a proton and thus acts as an acid.

$$\underset{\text{Base}}{B} + \underset{\text{Proton}}{H^+} \rightarrow \underset{\text{Acid}}{B^+H}$$

ii. In a reaction between ammonia (NH_3) and hydrogen fluoride (HF); ammonia acts as a base (accepts a proton) while HF behaves as an acid (donates a proton).

$$\underset{\text{Base}}{NH_3} + \underset{\text{Acid}}{HF} \rightarrow NH_4^+ + F^-$$

Such related pairs of an acid and a base is said to conjugate to one another *i.e.* in the above example B is the conjugate base of the acid B^+H and B^+H is the conjugate acid of the base B.

- ☆ On the whole,it can be said that every base has its conjugate acid and every acid has a conjugate base.
- ☆ It is generalized that in a conjugate acid base pair the weaker an acid is, stronger is its conjugate base. Likewise, the weaker a base is, the stronger is its conjugate acid. Hence, strong acid has a weak conjugate base and a weak acid has a strong conjugate base [1,2].
- ☆ Some examples of conjugate acids and bases are given in following Table [1,1b]:

Table 6.1: Conjugate Acids and Bases

Sl.No.	Acid	Base		Conjugated Acid	Conjugated Base
1	H_2SO_4	$+ H_2O$	→	H_3O^+	HSO_4^-
2	NH_4	$+ H_2O$	→	H_3O^+	NH3
3	HCl	OH^-	→	H_2O	Cl^-
4	H3O+	OH-	→	H_2O	H_2O
5	HCl	$H_2PO^-_4$	→	H_3PO_4	Cl^-

- ☆ Thus according to Bronsted -Lowry concept neutralization is a process in which a proton is transferred from an acid to a base. A substance acts as an acid only when another substance capable of accepting a proton (*i.e.* Base) is present.

The following examples help in understanding the aforesaid concept [1,1a]-

i. HCl or CH_3COOH solution in benzene is not acidic because benzene is not in a position to take up protons.

ii. On the other hand HCl or CH_3COOH solutions in water are definitely acidic because HCl or CH_3COOH can ionize in water.

iii. The reaction between ammonia and Hydronium ion is also an example of Bronsted -Lowry acid and base-

$$NH_3\,(g) + H_3O^+\,(aq) \rightarrow H_2O(l) + NH_4^+\,(aq)$$

Here, in aqueous solution, a proton is transferred from the hydronium ion, H_3O^+, to the NH_3 molecule, giving H_2O and NH_4^+. In this case H_3O^+ acts as proton donor or an acid and NH_3 as proton acceptor.

- ☆ Water (the solvent) playing the role of mediator in the transfer of proton from hydrochloric acid to ammonia. **It should be noted that in the Bronsted- Lowry Concept, acids and bases can be either ions or molecular substances [2].**

☆ Water (common solvent) can act both as an acid as well as a base because it can give up a proton as well as can receive a proton. The compound (as water) which can function both as an acid and base is **known as amphoteric or amphiprotic.** Its dual behaviour may be represented by the following equation [1,1a]-

$$\underset{\text{Acid1}}{H_2O} + \underset{\text{Base1}}{H_2O} \rightleftharpoons \underset{\text{Acid2}}{H_3O^+} + \underset{\text{Base2}}{OH^-}.$$

Another example is furnished by substances like aluminium hydroxide, $Al(OH)_3$ [1a].

Sources

1. *K.S. Sharma, Physical Chemistry of milk, Agrimoon.com.*

1a. *Wikipedia - Bronsted and Lowry acid - base theory.*

1b. *Wikipedia-Conjugated acid.*

2. *Bronsted and Lowery concept,Chemical dynamics, modules-5, G: /Acid per cent 20 bases per cent 20theories.pdf.*

Dissociation Constants of Acids and Bases [1]

The dissociation constant quantifies the tendency of a compound or an ion to dissociate.

☆ Dissociation is the process by which the compound or ion is split into two components that are also ions and or compounds.

Dissociation Constant- A general example is the dissociation of the compound AB into A+B which can be written as-

$$AB \rightarrow A + B$$

☆ It could be noticed that when a solution is prepared by adding AB, it is possible to calculate from the dissociation how much of the material added is in the form of AB, A and B which are constant. However, before calculation it is important to know that dissociation of AB into A and B is a reversible process. A and B can assemble back to AB. The dissociation constant can be written as-

$$Ka = \{ [A] \times [B]\} / [AB]$$

☆ If the dissociation constant of this process is *e.g.* 10^{-7}, 1 mol of AB dissolved in to a solution of water will result in a concentration of Both (a) and (b) of approximately 0.000316 moles and a concentration of AB of 0.999684. This gives us an idea that AB is not very soluble. This equation does not give much information unless we take into consideration the concentration. We have to set a second degree equation to understand this aspect of the equation. The process is as follows:

$$10^{-7} = \{ [A\} \times [B]\} / [AB]$$

☆ When AB dissolves x concentration is lost from AB and x concentration of A and B. So the equation is rewritten to-

$10^{-7} = (x \text{ X } x)/1- x$ this could be rewritten as-

$10^{-7} \times (1- x) = (x. x)$

$10^{-7} - 10^{-7}. x = x^2$

$10^{-7} = x2 + 10^{-7}. X$

$0 = x2 + 10^{-7}. x - 10^{-7}$

The result obtained after calculating using this second order reaction is approximately 0.000362 which is x or [a] and [b]. The concentration of AB is 1-x or 1- 0.000362. The above equation is actually a simplification of a general equation. The general equation is written as follows:

$$A_X B_Y \rightleftharpoons A_X + B_Y$$

In this case the dissociation constant is written as:

$$Ka = ([A]x. [B]y) / [AxBy]$$

The summary of all three acid and base theories (Lewis, Bronsted and Lowery and Arrhenius) is tabulated below:

Table 6.2: Summary of Acid and Base Theories [2]

Sl.No.	*Type of Theories*	*Acid*	*Base*
1	Lewis	Electron-pair acceptor	Electron-pair donor
2	Brønsted-Lowry	H^+ donor	H^+ acceptor
3	Arrhenius	H^+ ions in solution	OH^- ions in solution

Sources

1. *K.S. Sharma, Physical Chemistry of milk, Agrimoon.com.*

2. *Lewis Acids and Bases, https: //courses.lumenlearning.com.*

Definitions for Acids and Bases [1]

Arrhenius has proposed that acids are the substances that produce protons H^+ in aqueous solutions while bases are the substances which produce OH^-.

☆ The behaviour of some acids and bases in aqueous solutions was different.

☆ The definition given by Bronsted and Lowry for acids was that the compounds which are capable of donating protons. Similarly bases are the substance which accepts protons. This definition being independent of the solvent explains the behavior of most acids and bases.

Dissociation Constants for an Acid [1,1a,2]

The dissociation constant of an acid (also known as **acidity constant**, or **acid-ionization constant**)is used to study the dissociation of weak or strong acids.

☆ An **acid dissociation constant**, K_a is a quantitative measure of the strength of an acid in solution [1a].

☆ Dissociation constants are often denoted as Ka-values. A simple example on how to use Ka-values is given here-

i. Consider ammonium (NH_4^+) dissociating reversibly into ammonia (NH_3)

$$NH_4^+ \rightleftharpoons NH_3 + H^+$$

ii. In case of strong acids the ionisation is almost complete or close to 100 per cent or the equilibrium lies far to the right. In such cases the sign of equilibrium may be replaced by a single arrow (→) [2].

$$HA\ (aq) + H_2O \rightarrow H_3O^+\ (aq) + A^-\ (aq)$$

$$Keq = \frac{[H_3O^+]\ [A^-]}{[H_2O]\ [HA]}$$

The concentration of a pure liquid or a solid is taken as 1, hence above reaction can also be written as follows:

$$Keq = \frac{[H_3O^+]\ [A^-]}{[HA]} = Ka$$

where, Ka is a new constant called acid dissociation constant or ionisation constant of the acid [2].

The magnitude of the equilibrium constant is a measure of the strength of the acid [1a,2]. An acid dissociation constant is a particular example of an equilibrium constant.

☆ Higher the value of the equilibrium constant the stronger is the acid. For all strong acids the values of the equilibrium constants is quite high and does not help much in suggesting their relative strengths. However, for a weak acid, this constant is quite useful [2].

☆ The values of Ka show the extent to which the acids are dissociated in water. Relative strength of acids is judged on the basis of dissociation constant. Like other equilibrium constants the dissociated constant, Ka also depends on the temperature. Therefore, the ionisation constants or dissociation constants are compared at the same temperature [2].

Reactions of Acids and Bases with Water [1,2]

In the equations showing the reaction of acids and bases, we show the arrows in both the directions indicating that they are in equilibrium process. The proportion of the reagents and products at equilibrium are described by equilibrium constant.

☆ For the reaction of an acid HA with water is shown as:

$$HA(aq) + H_2O(l) \rightleftharpoons A^-(aq) + H_3O^+(aq)$$

$$Keq = \frac{[A-]\,[H_3O^+]}{[HA]\,[H_2O]}$$

We can observe in the above equation that water is the reactant and belongs in the equilibrium constant. Its value is 55.5M in aqueous solution. This is very large when compared with the change in water concentration at equilibrium. So it is assumed that the value of $[H_2O]$ with this assumption we will define the dissociation constant Ka as follows:

$$HA(aq) + H_2O(l) \rightleftharpoons A^-(aq) + H_3O^+(aq)$$

$$Keq\,[H_2O] = \frac{[A-]\,[H_3O^+]}{[HA]}$$

☆ **From this equation we can see that the acids which dissociate to a greater extent will have larger value of Ka are stronger acids while those which have a smaller values of Ka are weaker acids.**

☆ The dissociation constants for acids (Ka) range from 10^{-12} to 10^{13} this shows that the weakest acid is having the lowest equilibrium constant while the strongest acid will have the highest value of equilibrium constant. Further by deriving the equilibrium constant for any acid we will be in a position to know whether a given acid is a strong or a weak acid.

Ionisation/Dissociation of Bases [1,2]

The base dissociation constant, K_b, is a measure of basicity or the base's general strength. It is related to the acid dissociation constant, K_a, by the simple relationship $pK_a + pK_b = 14$, where pK_b and pK_a are the negative logarithms of K_b and K_a, respectively [1b].

☆ The equilibrium constant for bases Kb with the following equation [1,1b]-

$$B(aq) + H_2O(l) \rightleftharpoons BH^+(aq) + OH^-(aq)$$

$$Keq\,[H_2O] = \frac{[BH^+]\,[OH^-]}{[B]} = K_b$$

☆ Higher equilibrium constant of a base indicates that it is a strong base while the smaller value a weak base. These values range from 10^{-11} to 10^3.

☆ The constant Kb is called dissociation constant of the base. Similar to values of Ka, Kb values also give us the idea about the relative strengths of weak bases. Higher the value of Kb the stronger is the base.

Sources

1. *K.S. Sharma, Physical Chemistry of milk, Agrimoon.com.*

1a. *Wikipedia-Acid dissociation constant.*

1b. *Base Dissociation Constant, https: //courses. lumenlearning.com.*

2. *Bronsted and Lowery concept, Chemical dynamics, modules-5, G: /Acid per cent 20 bases per cent 20theories.pdf.*

Acidity of Milk [1,1a,2]

Milk being a biological fluid secreted from mammary cells (gland) has to maintain isotonicity with the blood plasma.

☆ Further the proteins in milk exist in its native colloidal state only when the pH of milk is maintained constantly within the given range. Any change in the pH would destabilize the proteins and result in precipitation and gelation. The determination of acidity of milk is a rapid measure to understand the stability of milk during heat processing

☆ The hydrogen ion concentration of milk is about 10^{-66} per litre which means that pH value for milk is 6.6. Freshly drawn milk is amphoteric to litmus paper that it will turn the red litmus to blue and blue litmus to red. The pH of normal healthy cow milk will range between 6.4 to 6.6 while the pH of buffalo milk is in the range of 6.7-6.8.

☆ If the animal is suffering from udder disease like mastitis the pH would be higher than the normal milk. Due to the variation in the composition which is the result of the mammary gland activity a difference in the pH and buffering capacity of individual lots of fresh milk was observed [1,1a,2].

- ☆ The variation in the composition which is the result of the mammary gland activity causes a difference in the pH and buffering capacity of individual lots of fresh milk
    - i. In general the pH is lower in colostrum (pH 6.0) and higher in cases of mastitis (Up to pH 7.5) than normal milk of mid lactation.
    - ii. Colostrums and mastitis milk differs radically from normal milk in the proportion of proteins, and certain salts. (A radical difference in the colostrum and mastitis milk from normal milk could be observed in which the proportion of proteins and certain salts were observed to be different) [1,1a,2,3].
- ☆ Milks of low phosphorus, casein, and Ca2+ tend to be low in titratable acidity while excessive acidity is related to hyperketonemia, inadequate calcium and excessive concentrates in the ration.
- ☆ The slight lower side of pH is due to natural acidity as a function of natural milk constituents contributing towards acidity. Skimming and dilution with water increases the pH where as sterilization decreases it [2].

The relationship between acidity and pH and relationship among acidity,pH and quality of milk are is given in following tables [2,5]-

Table 6.3: Relationship between Titratable Acidity and pH of Milk [4,5]

Sl.No.	*Titratable Acidity*	*pH*	*Sl.No.*	*Titratable Acidity*	*pH*
01	0.50	6.0	08	0.145	6.7
02	0.43	6.1	09	0.125	6.8
03	0.36	6.2	10	0.115	6.9
04	0.30	6.3	11	0.105	7.0
05	0.25	6.4	12	0.095	7.1
06	0.205	6.5	13	0.090	7.2
07	0.165	6.6	14	-	-

Table 6.4: Animal Breed, Acidity and pH of Milk [1,3]

Sl.No.	*Breed*	*Titratable Acidity (per cent Lactic acid)*		*Mean pH*
		Range	*Mean*	
1	Ayrshire	0.08-0.24	0.160	-
2	Holstein	0.10-0.28	0.161	6.71
3	Guernsey	0.12-0.30	0.172	6.65
4	Jersey	0.10-0.24	0.179	6.66
5	Pooled	0.12-0.21	0.134	6.66

- ☆ Perusal of above Table reveal that there appears to be a reduction in the mean titatratable acidity over a period of study. The drop in the maximum values is consistent with an improvement in the microbiological quality of milk supply [1].
- ☆ There are many components in milk which provide a buffering action. The major buffering groups of milk are caseins and phosphate. During titration of milk with alkali to measure developed acidity induces shifting of pH through hydrogen ion equilibrium due to the buffering action of milk. Buffer action simply indicates a state of resistance to a change in hydrogen ion concentration of a solution *i.e.* milk. The buffer compounds present in milk are the acids, proteins, salts of acids such as phosphate, citrate, carbonates and dissolved CO_2 [2].

Determination of pH [2]

pH of milk can be determined mainly by the following two methods:

i. Indicator Paper Strip Method

With indicator paper strip method pH can be easily determined within few seconds.

- ☆ However, this method gives less accurate results compared to electrometric method.
- ☆ In this method help of pH paper strip is taken. The pH of the strip is so chosen so that it is nearest to the expected pH. On dipping the pH paper strip colour of strip changes. The pH of milk is equal to that of standard pH colour scale to which the dipped paper compares.

ii. Electrometric Method

Electrometric method is based on the principle of potentiometer, which measures electromotive force (emf) of the system, the difference of emf is measured with the help of two electrodes.

- ☆ One of the electrode is known as reference electrode which has potential independent of the solution. The other electrode is pH dependent. It generates emf when dipped in a solution. This emf is generated by H+ ion concentration in the solution, which causes emf generation. It is directly related to pH. Due to emf generation by two electrodes and along with the emf of the reference electrode causes electric current to flow.
- ☆ The magnitude of the electric current flow can be measured accurately and rapidly with modern electromotive devices.
- ☆ For determination of pH of milk, **pH meter** is standardized with a standard pH solution *e.g.* pH 4.0.6.0 or 9.0 and after standardization,

milk is taken in a beaker and the electrodes are dipped in milk. This results in pH measurement, which can be obtained directly from the pH meter.

Source

1. *K.S. Sharma, Physical Chemistry of milk, Agrimoon.com.*

1a. *Acidity of Milk, Lesson 20, http: //ecoursesonline.iasri.res.in.*

2. *M.K. Srivastava, Handbook on Analysis of Milk, CBS Publications, New Delhi, 2010. ISBN 13: 9788123926681.*

3. *Wong et al., 1988, Fundamental of Dairy Chemistry.*

4. *H.V. Atherton and J.A. Newlender, 1982, Chemistry and testing of dairy products. The AV1 Publishing company,Inc. Westport Connecticut, 265.*

5. *N.K.Roy and D.C.Sen, Text book of practical dairy chemistry, Kalyani publishers, Ludhiana.*

Acidity and Types of Acidity [1,1a,2,3]

There are two types of acidity observed in milk. Fresh Milk has Natural or Apparent acidity (caused by the presence of casein, acid –phosphate, citrates *etc.*) and stored milk also has Developed or Real or True acidity (caused due to the bacterial action on lactose) in addition to natural acidity.

i. Natural Acidity or Apparent Acidity

Freshly drawn milk drawn from the udder of a cow/buffaloes is slightly acidic in nature due to its normal constituents. Such acidity is known as Natural acidity or apparent acidity. The constituents that contribute to this acidity are Casein, acid phosphates and citrate and to some extent the albumins and globulins, carbon dioxide, non protein nitrogenous compounds and various acids.

- Colostrum is having high natural acidity due to its protein content.
- The normal acidity of individual milk varies depending on stage of lactation. In early lactation also this value is higher which slowly returns to normal level during the second month of lactation. This level is maintained upto the last month of lactation. At that stage of lactation a slight decline in the acidity is observed. This is the normal phenomenon during the lactation of an animal.
- The variation in the natural acidity of fresh milk is also observed between the species, breed, individual animal, physiological condition of the udder *etc.*
- An increase in the solids non fat content of milk would also increase the natural acidity.

ii. Developed Acidity or True Acidity

During storage of milk acidity develops due to the fermentation(microbial multiplication) of milk lactose to lactic acid and other acidic components, primarily due to microbial effect which would result in increase in the acidity of milk and decrease in the pH value. This acidity is known as developed acidity.

- ✰ The titratable acidity (determined by titration with Sodium hydroxide using phenolphthalein indicator) of stored milk is equal to sum of natural and developed acidities. The titratable acidity of cow milk varies from 0.13 to 0.14 and buffalo milk from 0.14 to 0.15 per cent.
- ✰ Because of the large inherent variation, the measure of titratable acidity has little practical value except to measure changes in acidity (eg. during lactic fermentation) and even for this purpose, pH is a better measurement. The developed acidity along with natural acidity, which is referred as total acidity, gives positive reaction on boiling milk. A platform test named COB test called clot-on-boiling test is used to detect coagulation of milk. Formation of clot on boiling is an index for acidic abnormal milk [2].
- ✰ The normal acidity of individual cow milk ranges from 0.10 to 0.26 per cent lactic acid. Milk having titratable acidity more than 0.18 per cent Lactic acid is not suitable to prepare heat treated products as the milk will coagulate at or above that acidity [1].
- ✰ Heated milk will show an increase in acidity due to the changes in the casein complex and formation of acids by the degradation of lactose.
- ✰ Dilution of milk will decrease the acidity and increases the pH. The acidity of milk cannot be increased through feeding.

Table 6.5: Tentative Relationship between Acidity, pH and Quality Milk [2,3]

Range of Titratable Acidity (as Lactic acid per cent)	*Range of pH*	*Quality pf Milk*
0.14-0.16	6.5-6.7	Normal Fresh Milk
0.16-0.18	6.4-6.5	Limit for receiving milk
0.25-0.27	6.0-6.2	Acid odour detectable
0.26-0.30	5.6-6.0	Acid Flavour detectable
0.50-0.60	4.4-4.8	Coagulation of casein

- ✰ **The total titratable acidity of milk is expressed as per cent lactic acid. The pH of the milk does not truly represent the titratable acidity** because of the non availability of the ionic groups to titration in the normal milk.

☆ It is also a matter of interest that the titratable acidity of a milk sample will vary depending upon the buffering capacity of the milk. The natural constituents protect the milk from developing acidity. Mastitis, even in mild or subclinical form, causes the acidity of the milk to be lower. In rare cases mastitis causes a high acidity in the milk [1].

Determination of Acidity [1,1a,2,4]

The milk is being titrated using the standard sodium hydroxide and the acidity is determined by calculating the acidity using the volume of the standard sodium hydroxide used for the titrating to the phenolphthalein end point.

☆ Acidity of milk can be determined against a standard alkali such as 0.09 N(N/9) NaOH or 0.1N NaOH solution in the presence of phenolphthalein indicator. On complete neutralization of acidity phenolphthalein indicator changes its colour in milk to faint light pink. The following reaction occurs during titration.

$$\underset{\text{Lactic acid}}{CH_3CHOHCOOH} + NaOH \rightarrow \underset{\text{Sodium lactate}}{CH_3CHOHCOONa} + H_2O$$

Method

Mix the milk sample thoroughly by avoiding incorporation of air. Transfer 10 ml milk with the pipette in porcelain dish/conical flask. Add equal quantity of glass distilled water. Add 3-4 drops of phenolphthalein indicator and stir with glass rod. Rapidly titrate the contents with N/10 NaOH solution continue to add alkali drop by the drop and stirring the content with glass rod till first definite change to pink colour which remains constant for 10 to 15 seconds. Complete the titration within 20 seconds. Note the titre value.

Calculation

The titratable acidity is expressed as lactic acid per 100 ml of milk. Acidity is expressed as per cent lactic acid. Since 1 ml of 0.1 N lactic acid contains 0.009 grams of lactic acid, the number of ml. of 0.1 N NaOH required to neutralize the lactic acid in the sample, multiplied by 0.009 will give the amount of lactic acid (grams) in the sample, when the result is divided by weight of milk sample and multiplied by 100 the per cent lactic acid will be obtained. The calculation of titratable acidity is done by using the following formula:

$$\text{Per cent Lactic acid} = \frac{\text{No of ml. of 0.1 N NaOH solutions required for neutralization x 0.009}}{\text{Weight of sample (Weight of sample = Volume of milk x specific gravity)}} \times 100$$

$$= \frac{\text{Titratable acidity (TA)}}{\text{Volume of sample}} = 9 \text{ V N}$$

If 10 ml milk is taken for titration then

Titratable acidity (TA) = 0.9 V N

where, V-Volume in ml of the standard NaOH solution required for titration and N - Actual normality of the NaOH solution

Note: The calculation becomes simpler when N/9 NaOH is used for titration as titre value is direct reading of acidity.

Importance of Acidity of Milk [1,1a]

Titratable acidity of milk could be used in several stages during milk processing. The details are given below:

- ☆ **In grading of raw milk**-It is mostly used as a method to grade raw milk at the plant while it is received at the milk reception dock (RMRD). The conditions of storage of milk would also influence the initial acidity which means that if the milk is held under room conditions for long time and there is higher microbial count in the milk the initial increase in the acidity cannot be detected. It is necessary that higher SPC count should also be associated with high acidity. The determination of acidity of milk is a **rapid measure to understand the stability of milk during heat processing**.
- ☆ **Heat stability of milk**-The acidity of milk could also be used as a measure to study the heat stability of milk during processing. Acidity of more than 0.18 per cent lactic acid would result in the curdling during heat processing.
- ☆ **Monitoring of Udder health-** Titratable acidity could help us in monitoring the udder health of an animal. A change in the titratable acidity indicates the change in the udder health.

Sources

1. *K.S. Sharma, Physical Chemistry of milk, Agrimoon.com.*

1a. *Acidity of Milk, Lesson 20, http: //ecoursesonline.iasri.res.in.*

2. *M.K. Srivastava, Handbook on Analysis of Milk, CBS Publications, New Delhi, 2010. ISBN 13: 9788123926681.*

3. *N.K. Roy and D.C.Sen, Text book of practical dairy chemistry, Kalyani publishers, Ludhiana.*

4. *IS: 1479 (Part 1) – 1960 (Reaffirmed 2003).*

Salts and their Hydrolysis [1,2]

Salt is formed when an acid comes in contact with a base. Salts are ionic compounds containing cations and anions except hydrogen and hydroxide ions. All ions in aqueous solution are hydrated (surrounded by water molecules in aqueous solution).Some ions undergo only hydration while other ions react with the water molecules and result in the formation of acidic and alkaline solutions.

Salt [1,1a,2]

Salts are defined as ionic compounds that result from the neutralization reaction of an acid and a base. Salts are composed of positively charged cations and negatively charged anions in such a proportion that the product formed is electrically neutral (without a net charge) [1,1a]. These component ions can be inorganic, such as chloride (Cl^-), or organic, such as acetate ($CH_3CO^-_2$), and can be monatomic, such as fluoride (F^-), or polyatomic, such as sulfate ($SO^{2-}{}_4$).

Types of Salts [1,1a,2]

While it may seem that salt solutions would always be neutral, they can frequently be either acidic or basic [2]. The following varieties of salts are available-

i. **Acid salts**-Salts that hydrolyze to produce hydronium ions in water are acid salts. Salts that produce acidic solutions are acid salts.

ii. **Basic/Alkali salts**-Salts that hydrolyze to produce hydroxide ions when dissolved in water are basic salts.

iii. **Neutral salts**- These salts are those that are neither acid nor basic salts.

Hydrolysis of Salts

Salt, when dissolved in water it gets dissociated and forms ions. This process is known as hydrolysis. Dissolving salts in water would result in the interaction of its ions *viz.* cations and anions with water (Hydrolysis).

- ☆ This interaction would affect the pH of the solution.
- ☆ The cations (*e.g.*, Na^+, K^+, Ca^{2+}, Ba^{2+}, *etc.*) of strong bases and anions (*e.g.*, Cl^-, SO4, Br^-, $NO3^-$, $ClO4^-$ *etc.*) of strong acids simply get hydrated but do not hydrolyse. Therefore the solutions of salts formed from strong acids and bases are neutral *i.e.*, their pH is 7 for *e.g.*. Na_2S0_4, KC1, $KN0_3$ *etc.*
- ☆ Salts formed by reactions of strong acid and weak base will give acidic aqueous solutions and for *e.g.*, NH_4CI, $CuSO_4$, $Al(N0_3)_3$ *etc.* This is due to the interaction of the ions produced from the salt with the ions formed from water [1].
- ☆ **Salt hydrolysis** is a reaction in which one of the ions from a salt reacts with water, forming either an acidic or basic solution [2 [.Hydrolysis is quite opposite of neutralization process.

Neutralization [1,2]

Reaction between acid and base would result in salt formation. This reaction is known as neutralization. In this reaction water is also formed along with salt This reaction could be represented as follows:

$$\underset{\text{Acid}}{HA} + \underset{\text{Base}}{BOH} \rightarrow \underset{\text{Salt}}{BA} + \underset{\text{Water}}{H_2O}$$

Depending on the solubility of the salt, it may remain in ionized form in the solution or it may precipitate out of solution. Neutralization reactions usually proceed to completion.

Types of Combinations for the Salt Formation [1,2,3]

The interaction between acid and base could be grouped in to following four categories basing on their type:

i. Strong Acid and Strong Base

Interaction between a strong acid and a strong base would result in the formation of salt.

- When strong acid like hydrochloric acid (HCl) reacts with an equally strong base like sodium hydroxide(NaOH) sodium chloride (NaCl) would be formed.

$$HCl + NaOH \rightarrow NaCl + H_2O$$

- The acid and base neutralize each other with the resulting solution will be neutral (pH=7) and there is no interaction between the ions and water [1,2].
- A salt that is derived from the reaction of a strong acid with a strong base forms a neutral solution [2].

ii. Strong Acid and Weak Base

The interaction between a strong acid and a weak base would also result in the formation of salt. The following equation expresses this reaction:

$$HCl + NH_4OH = NH_4^{+} + Cl^{-} + H_2O$$

In the solution, ammonium ion reacts with water and forms ammonia and Hydronium ion as shown in following reaction:

$$NH_4^{+} + H_2O \rightarrow NH_3 + H_3O^{+}$$

- Ammonia ions formed in this reaction would result in the lowering of pH since they react with water. As such the resulting salt which is in the form of a solution is acidic.

☆ Similarly when hydrochloric acid (HCl) reacts with a weak base like ammonia would form ammonium chloride which is a salt, but in this reaction water molecule is not formed because weak bases tend not to be hydroxides [1]. A salt that is derived from the reaction of a strong acid with a weak base forms an acidic solution [2].

☆ The solvent water interacts with the ammonia which is the cation of the salt to reform the weak base:

$$HCl + NH_3 \rightarrow NH_4Cl$$

iii. Strong Base and Weak Acid

Salts that are derived from the neutralization of a weak acid (HF) by a strong base (NaOH) will always produce salt solutions that are basic [2]. When a weak acid reacts with a strong base the resulting solution will be basic. The salt will be hydrolyzed to form the acid, together with the formation of the hydroxide ion from the hydrolyzed water molecules.

a. **Reaction between Bleaching powder and Sodium hydroxide-** Interaction between a weak acid (bleaching powder) with a strong base(sodium hydroxide) would result in the formation of sodium hypochloride which is a salt. The basicity of this salt is due to the hydrolysis of the conjugate base of the (weak) acid used in this reaction [1].

b. **Reaction between acetic acid and Sodium hydroxide**-Similarly sodium acetate formed by the interaction between acetic acid(weak acid) and sodium hydroxide (NaOH- strong base) will be formed which is a basic salt [1].

When salt is dissolved, ionization takes place- The bleaching powder being a weak acid when it reacts with sodium hydroxide-

$$HClO + NaOH \rightleftharpoons NaClO + H_2O$$

iv. Weak Base and Weak Acid

The reaction between the bleaching powder with a weak base like ammonia would form ammonium hypochloride

$$HClO + NH_3 \rightleftharpoons NH_4ClO$$

☆ pH of the solution formed from the reaction of a weak acid with a weak base depends on the relative strengths of the reactants.

For example, if the acid HClO has a Ka of 3.4×10^{-8} and the base NH_3 has a Kb = 1.6×10^{-5}, then the aqueous solution of HClO and NH_3 will be basic because the Ka of HClO is less than the Ka of NH_3. When a salt is formed between a weak acid and a weak base it could be neutral, acidic, or basic depending on the relative strengths of the acid and base.

Characteristics of Buffer [3]

Buffer have following characteristics:

i. Has a definite pH value.
ii. pH value does not change on keeping for long time.
iii. pH value does not change even with addition of small amount of a strong acid or a base and.
iv. pH value does not change on dilution.

Definition of Buffer [1,2,2a]

A buffer solution (more precisely, pH buffer or hydrogen ion buffer) is an aqueous solution consisting of a mixture of a weak acid and its conjugate base, or vice versa [1].

- ✰ Buffers are the solutions with stable hydrogen ion concentration and therefore usually with no change in pH which is almost independent of dilution and changing very little with small additions of a strong acid or alkali [2].
- ✰ In simple way definition of buffer can also be given as a solution which resists any change of pH when a small amount of a strong acid or a strong base is added to it, is called a **buffer solution or a buffer.** Thus, buffers have acidity and reserve alkalinity [2].

Buffer solutions usually consist of a mixture of a weak acid and its salt with a strong base *e.g.*, CH_3COOH and CH_3COONa, or that of a weak base and its salt with a strong acid *e.g.*, NH_4OH and NH_4Cl.

Buffer Action [1,2,3]

The solution of any salt of a weak acid and a weak base *e.g.*, ammonium acetate, also shows buffering property. Certain solutions, such as that of ammonium acetate, have a tendency to resist any change in its hydronium ion concentration or pH, whenever a small amount of a strong acid or a strong base is added to it. This property of a solution is **known as buffer action [2].**

- ✰ In a mixture of acetic acid and sodium acetate, acetic acid is weakly ionized while sodium acetate is almost completely ionized. If a drop of a strong acid like HCl is added, then the H^+ ions from HCl combine with CH_3COO^- to give feebly ionized CH_3COOH. Thus, there is a very slight change in the pH value [1,3].

Types of Buffers [1,2,2a,3]

Buffers are broadly divided into following two types of buffer solutions:

i. **Acid buffer solution**- A buffer solution containing large amounts of a weak acid, and its salt with a strong base, is termed as an acid buffer.

☆ Such buffer solutions have pH on the acidic side *i.e.*, pH is less than 7 at 298 K *e.g.* a mixture of acetic acid (CH_3COOH) and its salt sodium acetate (CH_3COONa) acts as a buffer solution with a pH of about 4.75. The pH of an acid buffer is given by the following equation:

$$pH = pKa + \ln \frac{[Salt]}{[acid]}$$

where, Ka is the acid dissociation constant of the weak acid.

ii. **Basic Buffer solution**- A buffer solution containing relatively large amounts of a weak base and its salt with a strong acid is termed as a basic buffer. Such buffers have pH on the alkaline side *i.e.*, pH is higher than 7 at 298 K. *e.g.* a mixture of ammonium chloride(NH_4Cl) and ammonium hydroxide (NH_4OH) acts as a buffer solution with a pH of about 9.25 and the pH of a basic buffer is given by the following equation:

$$pOH = pKb + \ln \frac{[Salt]}{[acid]}$$

where, Kb is the base dissociation constant of the weak base. **These equations are called Henderson Hasselbalch equations.**

Buffer-Capacity and Buffer-Range [1,2,2a]

The effectiveness of any buffer is described in terms of its buffer capacity.

☆ Buffer capacity is a quantitative measure of the resistance to change of pH of a solution containing a buffering agent with respect to a change of acid or alkali concentration [1].

☆ Buffer capacity is defined as, 'the number of equivalents of a strong acid (or a strong base) required to change the pH of one litre of a buffer solution by one unit, keeping the total amount of the acid and the salt in the buffer constant [2,2a].

i. The buffer capacity of a buffer is maximum when acid to salt or base to salt ratio is equal to 1 *i.e.*, it contains equal number of moles of acid (or base) and the salt.

ii. All buffer solutions remain effective over a small pH range: this pH-range is characteristic of the buffer and is termed as the buffer-range [2,2a].

Mechanisms [2,2a,3]

i. Action of an Acid Buffer

An acid buffer contains relatively large amounts of a weak acid (HA) and it's salt with a strong base (say NaA). The buffer solution thus contains large concentration of HA and A^- (due to the dissociation of the salt), apart from H_3O^+ and OH^-. A mixture of acetic acid and Sodium acetate can be taken as example to understand the buffering action [3].

- ☆ **An addition of small amount of a strong acid** causes the reaction, to proceed in such a direction that an equivalent amount of A- combines with H_3O^+ to give the same amount of undissociated weak acid, HA. Thus, the added acid is picked up by the anions (from the salt) present in large concentrations in the buffer. As long as the added strong acid is in smaller amounts, the changes in the concentrations of salt and that of the weak acid, (HA) are small.
- ☆ Therefore, the acid to salt ratio does not change appreciably by the addition of strong acid to the buffer solution. As a result, no noticeable change is seen in the pH value of the buffer.

$$\underset{\text{From added acid}}{H_3O^+ + A^-} \rightleftharpoons \underset{\text{From buffer}}{H_2O + HA}$$

This resistance to the change in pH up on addition of strong acid is called **reserve basicity [3].**

- ☆ **Similarly addition of a strong base to an acid buffer** on the other hand causes the reaction to proceed in the forward direction, resulting in the formation of an equivalent amount of the salt at the cost of the buffer acid. As long as the added base is in small amounts, the ratio of weak acid to salt remains virtually unchanged. As a result, no observable change in the pH value is seen

$$\underset{\text{From added base}}{OH^- + HA} \rightleftharpoons \underset{\text{From buffer}}{H_2O + A^-}$$

This resistance to the change in pH up on addition of strong base is called **reserve acidity [3].**

ii. Action of Basic Buffer

A basic buffer contains a weak base (BOH), and its salt with strong acid (BX). The buffer solution thus contains large amounts of the weak base BOH, and the cation B^+ (coming from the dissociation of the salt BX), in addition to H_3O^+ and OH^-. Here a mixture of ammonium hydroxide and ammonium chloride can be taken as example to understand the buffering action [3].

☆ The addition of an acid or a base to the basic buffer causes the following reactions-

$$\underset{\text{From added acid}}{H_3O + BOH} \rightleftharpoons \underset{\text{In buffer}}{B^+ + 2H_2O}$$

$$\underset{\text{From added base}}{OH^- + B^+} \rightleftharpoons \underset{\text{In buffer}}{BOH}$$

The reactions proceed in the forward direction.

☆ It is clear that the addition of an acid or a base to any buffer solution does cause a change in the concentrations of the buffer acid (or base) and the salt. But, because of the relatively much larger concentrations of these in the buffer solution, for all practical purposes, the ratio, [Salt]/ [Acid] or [Salt]/ [Base] remains constant. Hence, the pH does not change.

Importance of Buffers [2,2a,4]

Chemistry plays an important role in our surrounding environment, daily lives and biological systems. So buffers being an integral part of inorganic chemistry also prove the importance of applied chemistry in environment and other sectors [4].

☆ Many chemical reactions are affected by the acidity of the solution in which they occur. In order for a particular reaction to occur or to occur at an appropriate rate, the pH of the reaction medium must be controlled. Such control is provided by buffer solutions, which are solutions that maintain a particular pH [2].

Some examples of buffers in physiological system,industry and nature are given below:

i. Importance of Buffers in Physiological Systems [2,2a,4]

Biochemical reactions are especially sensitive to pH. Most biological molecules contain groups of atoms that may be charged or neutral depending on pH, and whether these groups are charged or neutral has a significant effect on the biological activity of the molecule [2].

☆ In all multi cellular organisms, the fluid within the cell and the fluids surrounding the cells have a characteristic and nearly constant pH. This pH is maintained in a number of ways, and one of the most important is through buffer systems [2].

☆ Processes that take place in living organisms are called physiological processes [4].

i. pH plays an important role in almost all biological processes. Small change in pH *i.e.* deceased or high pH can cause metabolic implications in human body like acidosis and alkalosis.

ii. The pH of human blood is 7.4. A blood pH of below 7 or above 7.8 can cause death within minutes. So buffering of blood pH is very important to stabilize it around 7.4. Important buffers that are dominant in human body are:

   a. **Bicarbonate buffers** (Buffering in blood),

   b. **Phosphate buffers** (Buffering of internal cell fluids) and

   c. **Protein buffers** (Buffering in Cells and Tissues).

ii. Importance in Industries [1,4]

Buffers are of prime importance in different kind of industrial processes especially in pharmaceuticals, fermentation/food and textile dyeing industries. It also finds its role in many other industries as described below:

a. **In Pharmaceuticals Industries-**Most of the medicines are prepared in aqueous solution of different chemicals so these aqueous solutions require a constant pH in order to assure the stability and clinical effectiveness of a medicines and this is done through buffers.

b. **In fermentation and food industry-**Each **fermentation process** requires a specific pH for the best results. The pH during a fermentation process changes by itself due to fermentation process so to control this pH change buffers play an important role.

c. **In dyeing industry-**Dyes in textile industries play an important role in giving color to different fabrics. Color strength of dyes is closely related to narrow pH range which is maintained by using different buffer systems.

d. **In other Industries-** Buffers of different types are majorly used in aforesaid three types of industries. However, buffers also play an important role in following industries-

Printing industries, Electroplating industries, Leather industries, Glue and gelatine manufacturing industry.

iii. Importance of Buffers in Nature [1,4]

Buffers are critical to different kinds of natural systems like water and soil systems.

☆ Water bodies like lakes, streams, rivers are important habitat for aquatic life forms like fish and amphibians. Like all other living organisms aquatic life also need stable pH to survive in the water bodies. Buffer helps in maintaining the required pH of water bodies.

☆ Like water bodies pH also plays a critical role in soils also since plants grow best within a narrow pH range. Buffer helps in maintaining the required pH of soil.

Sources

1. *Wikipedia-Buffer solution.*
2. *K.S. Sharma, Physical Chemistry of milk, Agrimoon.com.*

2a. *Buffers, Types of Buffer Solutions, Solutions of Single Substance of Mixture, Lesson 22, http: //ecoursesonline.iasri.res.in.*

3. *Buffer solution, https: //thefactfactor.com/facts/pure_science/ chemistry/physical-chemistry/buffer-solution.*
4. *Importance of buffers, Eco Hurdle, http: //environmental-realm.blogspot.com.*

Henderson and Hasselbalch Equation [1-4]

Lawrence Joseph Henderson (American biological chemist),in 1908, derived an equation to calculate the pH of a buffer solution [1,2] and Karl Albert Hasselbalch (Swedish physiologist), in 1917, re-expressed that formula in logarithmic terms [3], resulting in the Henderson–Hasselbalch equation.

i. Henderson–Hasselbalch equation is a useful expression for buffer calculations and describes the derivation of pH as a measure of acidity using the acid dissociation constant (pKa,) in biological and chemical systems.

ii. The equation is also useful for estimating the pH of a buffer solution and to ascertain the equilibrium pH in acid-base reactions which will be of use to calculate the isoelectric point pI of proteins.

iii. The Henderson–Hasselbalch equation relates the pH of a solution containing a mixture of the two components to the acid dissociation constant, K_a, and the concentrations of the species in solution [1,4].

Derivation of Henderson-Hasselbalch Equation [1,4,4a]

Bronsted- Lowry theory states that an acid (HA) is capable of donating a proton (H^+) and a base (B) is capable of accepting a proton. After the acid (HA) has lost its proton, it is said to exist as conjugate base (A^-). Similarly, a protonated base is said to exist as conjugate acid (BH^+). The dissociation of an acid can be described by an equilibrium expression-

HA	$\rightleftharpoons$	H^+	+	A^-
Acid		Proton		Conjugated base
CH_3COOH	$\rightarrow$	CH_3COO^-	+	H^+
Acetic acid		Acetate (Conjugate base)		Proton

In case of acetic acid (CH_3COOH) and acetate anion ($CH3COO^-$), acetate is the conjugate base of acetic acid. Acetic acid and acetate is a conjugate acid/ base pair. This relationship can be described with equilibrium constant-

$$Ka = \frac{[H^+]\,[A^-]}{HA}$$

Here Ka is used for the acid dissociation constant. Taking the negative log of both sides of the equation gives-

$$-\ln K_A = -\ln \frac{[H^+]\,[A^-]}{HA}$$

This can be rearranged-

$$-\ln K_A = -\ln [H^+] + (-\ln \frac{[A^-]}{HA})$$

By definition, $pK_a = -\ln K_a$ and $pH = -\ln [H^+]$,

so for acid solutions the equation is

$$pK_a = pH - \ln \frac{[A^-]}{HA} \qquad \textit{Eq-1}$$

For solutions of a weak base the equation below can be used-

$$pOH = pKb\ Ilog \frac{B}{BOH}$$

Both equations are perfectly equivalent and interchangeable. So this equation can be rearranged to give the Henderson-Hasselbalch equation

$$pH = pK + \ln [A\text{-}]\,[HA] = pK + \ln \frac{\text{Conjugated base}}{\text{Conjugated acid}} \qquad \textit{Eq.-2}$$

$$pH = pK + \log \frac{\text{Conjugated base}}{\text{Conjugated acid}} \qquad \textit{Eq.-3}$$

Limitations of Henderson-Hasselbalch Equation [4,4a,5]

Henderson-Hasselbalch equation is used mostly to calculate pH of solution which is obtained by mixing known amount of acid and conjugate base (or neutralizing part of acid with strong base). It has following limitations:

1. The Henderson-Hasselbalch equation fails to predict accurate values for the strong acids and strong bases because it assumes that the concentration of the acid and its conjugate base at chemical equilibrium will remain the same as the formal concentration (the binding of protons to the base is neglected).

   Henderson–Hasselbalch equation reveals that when the pH of the solution equals the pK2 of the buffer, [conjugate base] = [acid], and the buffer can therefore respond equally to both added acid and added base.

   i. It also follows from Equation-1, the solution contains approximately 9 per cent unprotonated or protonated species, respectively.

   ii. Similarly, if the pH of the solution is two units above or below the pK2 value, the solution contains almost entirely (99 per cent) unprotonated or protonated species [5].

   - ✰ When a solution is prepared by mixing 0.1 M of acetic acid and 0.05 M NaOH half of the acid is neutralized by the alkali and resulting in the equal concentrations of acid and conjugate base, thus quotient under logarithm is 1, logarithm is 0 and pH=pKa.
   - ✰ However, this approach is justifiable in many cases, it creates false conviction that the equation could be applied in all the situations which is not true Henderson-Hasselbalch equation is valid only when it contains equilibrium concentrations of acid and conjugate base.
   - ✰ In the case of solutions containing not-so-weak acids (or not-so-weak bases) equilibrium concentrations can be far from those predicted by the neutralization stoichiometry. When acetic acid is replaced in the above example with a stronger acids like dichloroacetic acid, with pKa=1.5. Repeating the same reasoning we used earlier we will arrive at pH=1.5 - which is wrong. By determining the pH using either an equation or by pH meter we find the pH will be 1.78. It is because dichloroacetic acid is strong enough to dissociate on its own and equilibrium concentrations of acid and conjugate base are not 0.05 M but 0.0334 M and 0.0666 M respectively.
   - ✰ As a thumb rule it may be remembered that acids with pKa value below 2.5 dissociate too easily and use of Henderson-Hasselbalch equation for pH prediction can give wrong results, especially in case of diluted solutions.

   i. For solutions above 10 mM and acids weaker than pKa>=2.5, Henderson-Hasselbalch equation gives results with acceptable error.

ii. The same holds good for bases with pKb>=2.5.

iii. However, the same equation will work perfectly regardless of the pKa value if it is asked to calculate ratio of acid to conjugate base in the solution with known pH.

2. Since the Henderson-Hasselbalch equation does not consider the self-dissociation undergone by water, it fails to offer accurate pH values for extremely dilute buffer solutions. **Similar problem is present in calculation of pH of diluted buffers as in strong acid. Let's see what happens when acetic buffer 50/50 is diluted:**

Table 6.7: pH of Diluted Solutions of Acetic Buffer 50/50 (Ionic strength ignored) Ca (M) [4,4a]

	Buffer pH (pH calculator Value)
0.1	4.76
0.01	4.76
10^{-3}	4.79
10^{-4}	4.95
10^{-5}	5.47
10^{-6}	6.31
10^{-7}	**6.89**

- ☆ The more diluted the solution is, the more solution pH is dominated not by the presence of acetic acid and its conjugate base, but by the water auto dissociation. pH of 1 mM solution is close enough to the expected (from pKa) value, more diluted solutions deviate more and more. It is worth noting here that 1 mM buffer solution has so low capacity, that it has very limited practical use [4,4a].
- ☆ Henderson-Hasselbalch equation can be also be used for pH calculation of polyprotic acids, as long as the consecutive pKa values differ by at least 2 (better 3). Thus it can be safely used in case of phosphoric buffers (pKa_1=1.963, pKa_2=7.199, pKa_3=12.35),but not in case of citric acid (pKa_1=3.128, pKa_2=4.761, pKa_3=5.40). It is necessary to determine the pH of the solution directly rather than depending on the above equation [4,4a].

Application [1,4,4a]

The Henderson–Hasselbalch equation can be used to calculate the pH of a solution containing the acid and one of its salts(buffer solution). With bases, if the value of an equilibrium constant is known in the form of a base association constant, K_b the dissociation constant of the conjugate acid may be calculated by the following equation-

$$pK_a + pK_b = pK_w$$

where, K_w is the self-dissociation constant of water. pK_w has a value of approximately 14 at 25°C.

Sources

1. *Wikipedia-Henderson-Hasselbalch equation.*
2. *Lawrence J.Henderson, 1908, "Concerning the relation-ship between the strength of acids and their capacity to preserve neutrality", Am.J. Physiol. 21 (2): 173-179.*
3. *Hasselbalch, K. A.,1917,"Die Berechnung der Wasserstoffzahl des Blutes aus der freien und gebundenen Kohlensäure desselben, und die Sauerstoffbindung des Blutes als Funktion der Wasser-stoffzahl", Biochemische Zeitschrift. 78,112-144.*
4. *K.S. Sharma, Physical Chemistry of milk, Agrimoon.*

4a. *Derivations of Henderson and Hasselbalch Equation, Lesson-23, http: // ecoursesonline. iasri.res.in.*

5. *N.V. Bhagavan, Water, Acids, Bases, and Buffers in Medical Biochemistry (Fourth Edition), 2002, https: //www. Science direct.com.*

Buffering Capacity and Buffering Index [1,1a2,3]

In 1937, Kolthoff introduced the term of buffer capacity or buffer index as a name for the differential ratio dB/dpH [1b].

- ✰ The constituents in any biological fluid (milk *etc.*) can not retain their native state when there is change in pH of solution. The constituents in biological fluid can retain their native state so long the pH of the solution is maintained unchanged.
- ✰ For effective utilization and to perform the designated functions by that fluid the constituent exist in different forms *e.g.*, in milk the fat exists as an emulsion while the major protein casein exists in the form of micelles.
- ✰ The maintenance of uniform pH of fluid/solution becomes possible when buffers are present in that solution.

Buffering Capacity [1,1a-3]

The buffer capacity is numerically expressed to be equal with the minimum concentration of strong acid or strong base which causes the variation of buffer's pH with one unit [1b].

- ✰ The more concentrated the buffer solution, the greater its buffer capacity. If the buffer capacity is 10 times larger, then the buffer solution can absorb 10 times more strong acid or base before undergoing a significant change in pH.

- ✰ In milk aggregation or dispersion of protein micelles and fat globules is dependent up on its buffering capacity, pH, acidity and electrical conductivity.
- ✰ The natural range of variation of this property in milk and milk products will determine several technological properties of milk.
- ✰ Stability of milk suspension is mainly controlled by the acid base equilibrium in milk. Substances which liberate protons (acid) and of those that combine with them (Bases) are responsible for maintaining this stability of milk constituents.
- ✰ The ionized and ionizable constituents of milk are in a state of rather delicate physical balance.

The treatments such as heating or changing in milk compositional qualities alter the state of dispersion of proteins and salts which is reflected in protons. Buffers resist changes in pH.

- ✰ The Buffering Capacity increases as the molar concentration (molarity) of the buffer salt/acid solution increases. **The closer the buffered pH is to the pKa, the greater the Buffering Capacity.** Buffering Capacity is expressed as the molarity of Sodium Hydroxide required to increase pH by 1.0 unit/litre of solution.

$$\beta = dn/dpH$$

where, n is number of equivalents of added strong base. Note that addition of dn moles of acid will change pH by exactly the same value but in opposite direction.

Buffer Index [1,1a,2]

Buffer Index can be defined as the differential ratio of the increase in the amount of strong acid or strong base added, to pH variation.

- ✰ The field of application of both Buffer capacity and Buffer Index is different.
  - i. The buffer capacity is used in the quantitative chemical analysis.
  - ii. The buffer index in studying biological systems.

Buffering Capacity of Milk [1,1a-3]

Milk has some normal constituents (having buffering action) which prevents the changes due to variations in the milk such as microbial population in it or atmospheric changes environment in the pH of milk within a given pH range [1].

- ✰ Several constituents of milk act as a buffer *i.e.* they give resistance towards change in pH. These include carbon dioxide, proteins, phosphate, citrate and a number of minor constituents. Due to microbial action additional components are introduced in milk, which include lactate and many other organic anions [2,3]

- The buffering capacity of milk and milk products is an important physico-chemical characteristic that corresponds to the ability of the product to be acidified or alkalinized [1,5].
  i. The parameters of this value depend on several compositional factors including small constituents (inorganic phosphate, citrate, organic acids) and milk proteins (caseins and whey proteins) [1,5].
  ii. Natural and induced variations in the composition of milk affect this physico-chemical parameter. Thus, some processing treatments and physico-chemical changes, such as heat treatment, membrane separation technology, high-pressure treatment and salt addition, results in a buffering capacity specific to the transformed product (heated milk, retentate, fermented milk) [1,5].
  iii. In cheese manufacture, the preparation of cheese milk and the cheese making parameters have major influences on the buffering capacity of the curd at mould removal and during ripening. Cheese quality is therefore affected during ripening, via enzymatic activities and microbial growth [1,5].

  The two most important buffer components of milk are caseins (buffer maximum near pH 4.6) and phosphate (buffer maxima near pH 7.0). The buffer maximum near pH 5.0 is extremely important to cheese manufacture because the optimum pH for most cheese is in the range of 5.0 - 5.2 [5a].
  iv. The maximum buffer index of skim milk, standardized milk (4.5 per cent fat) and whole milk are 0.0290, 0.0283 and 0.0270 for buffalo and 0.0333, 0.0283 and 0.0277 for cow respectively [1].
  v. Both cow and buffalo milks exhibit a maximum buffering at pH 5.3 to 5.4 and show the same buffer intensity curve between pH 4.5 and 10.0 [1].
  vi. Goat milk, unlike cow or human milk, has unique characteristics, such as its high digestibility, distinct alkalinity, high buffering capacity (BC). The protein, primarily casein and phosphate systems within milk, influences its BC [6].
  vii. Preheating conditions (low-, medium-, and high heat-) did not significantly affect the buffering capacity (BC) of skim milk powder (SMP), whereas the level of demineralization significantly affected the BC of whey powders (WP).
  viii. Sodium citrate and pyrophosphate are favourable stabilizers shifting the maximum buffering of milk towards the normal pH of the milk. These additives cause an increase in milk pH and higher rate of dispersion.

Contribution of Constituents to Buffering Capacity of milk [1,1a,2,3]

Several acidic and basic groups are present in milk and because of the presence of these acidic and basic groups the pH changes much less than when titrating with water, these groups have a buffering action.

- ☆ The buffering capacity of solution can be expressed in the buffer index which is the molar quantity of acid or base needed to change the pH of 1 litre of solution by one unit; it is thus the derivative of the titration curve. The buffering index of an ionizable group is given by the following formula:

$$pH- pK = \pm 2.\ ^{dE/dpH\ 2.3cKa}\ {}^{H+}_{(ka\ +aH+)}$$

where, C is the total molar concentration of the group (dissociated plus associated). Again, it concerns intrinsic constants, but the above equation also holds for monoprotic acids if Ka is the stoichiometric dissociation constant.

- ☆ This equation implies that the buffering capacity is maximal (0.58 c) for the pH=pK; it is reduced to 33 per cent of the maximal values for pH- pK= ± 2 [1]. The buffering groups of milk are given in following table [1,1a,2]-

Table 6.8: Buffering Groups In Milk [1,1a]

Sl.No.	Group	Sub group	Concentration (mM)	pKa Value
A	**Protein-Bound residues**			
1		Aspartic acid	19	4.1
2		Glutamic acid	50	4.6
3		Histadine	6	6.5
4		Tyrosine	12	9.7
5		Lysine	20	10.5
6		Ester phosphate	7	2.6
7		Acetyl Neuraminic Acid	0.5	5
8		Terminal group	1.5	3.7, 7.9
B	**Salts**			
9		Phosphate 21	3,5.8,6.6	6.6
10		Citrate	9	3,4.1,4.8
11		Phosphate ester	2.5	1.7,5.9
12		Carbonate	2	6.4,10.1
13		Various carboxylic acids	2	4.8
14		Various amines	1.5	7.6
15		Lactic acid	50-120	3.95

Constituents of Milk Responsible for Buffering [2-4,5a]

Buffering of milk varies between samples and breeds. Some of the constituents responsible for buffering of milk are given below [2,3,4]-

i. **Carbon dioxide**- CO_2 present in milk is responsible for buffering of milk. Milk contains 20 mg CO_2 per 100 ml(10 per cent by volume) of milk. CO_2 behave like an acid in the form of carbonic acid. This on titration offers a titration value of 0.5 meq/100 ml between pH 6.6 and 8.3 out of the total value of 1.3 to 2.0

ii. **Citrate**- Citric acid as citrate ions has very little buffering capacity of 0.1 meq per 100 ml. However, it complexes with calcium as calcium citrate. This delays the precipitation of calcium phosphate and affect the titration indirectly.

iii. **Lactate**-Lactic acid also delays the titration as it also forms a complex with phosphate ion.

$$CH_3CHOH\ COOH + HPO_4^{--} = D\ H_2PO^-_4 + CH_3\ CHOHCOO^-$$

Lactic acid Lactate ions

iv. **Phosphate**- Presence of phosphate in milk has three buffering ranges. In the presence of calcium it forms a complex of calcium phosphate. During titration precipitation of calcium phosphate occurs as the pH is raised *e.g.* pH 6.0 About 0.6 meq of alkali is needed around pH 6.0

$$3Ca^{++} + 2H_2PO_4^- \rightarrow D\ Ca_3\ (PO_4)\ 2\ F + 4H^-$$
$$4H^+ + 4OH^- \rightarrow D\ 4H_2O$$

v. **Proteins**- Milk contain around 2.5 per cent casein and 0.6 per cent whey proteins. Casein contributes to a titration value of 0.8 meq/100 ml and whey proteins 0.1 to 0.2 meq/100ml.

- ✰ Titration of milk in the pH range of 4.8 to 8.3 has been used to assess the behaviour of milk during titration in the presence of phenolphthalein indicator. When milk is titrated against an alkali maximum buffering occurs at a pH range of 4.8 to 6.8 but at pH 8.0 to 8.3 very little alkali is required resulting in lower buffering.
  i. Titration of fresh milk over pH 6.6 to pH 8.3 requires 13 to 20 ml of 0.1 NaoH per 100 ml (1.3 to 2.0 meq. per 100 ml).
  ii. Most fresh milk samples fall in the range of 1.5 to 1.8 meq. per 100 ml [2,3].
- ✰ The ionizable groups of the proteins, the phosphates, and the citrates mainly determine the acidity and the buffering capacity of milk. The considerable variation in the concentration of these substances will cause variation in the titration curves. Some of the ionizable groups in a protein may even be unavailable for titration, as they are buried

in the hydrophobic interior of the molecules. Ionizable groups of the whey proteins and casein molecules are not exposed [1,1a].

- ☆ The presence of Ca^{2+} ions strongly modifies the ionization of di and triprotic acids such as phosphoric acid and citric acids they will behave as if they are titrated by $Ca(OH)_2$ instead of NaOH or HCl. Lowering of the pH of milk will make calcium phosphate soluble because it is not soluble in normal pH of milk whereas colloidal phosphate will increase when the pH is increased [1,1a].
- ☆ The Magnesium citrate and protein affect the composition and quantity of colloidal phosphate. Consequently the buffering index of various groups is not strictly additive [1,1a].

Sources

1. *K.S. Sharma, Physical Chemistry of milk, Agrimoon.*

1a. *Buffering capacity and Buffering Index, Lession-24, http: //ecoursesonline.iasri.res.in.*

1b. *Buffer Index and Buffer Capacity for a Simple Buffer Solution, https: //ui.adsabs.harvard.edu.*

2. *Wong, et al., 1988. Fundamental of Dairy Chemistry.*

3. *N.K.Roy and D.C.Sen, Text book of practical dairy chemistry, Kalyani publishers, Ludhiana.*

4. *M.K. Srivastava, Handbook on Analysis of Milk, CBS Publications, New Delhi, 2010. ISBN 13: 9788123926681.*

5. *Francoise Michel Salaun, Bernard Mietton and Frédéric Gaucheron, Buffering capacity of dairy products, 2005, International Dairy Journal 15(2): 95-109.*

5a. *Concepts of Acidity and pH, Food Science, https: //www.uoguelph.ca.*

6. *Young W. Park, Relative Buffering Capacity of Goat Milk, Cow Milk, Soy-Based Infant Formulas, and Commercial Non prescription Antacid Drugs, https: //www.journalofdairyscience.org.*

Electrolyte Equilibrium [1,2,2b]

Substances that give ions when dissolved in water are called **electrolytes**. They can be divided into acids, bases, and salts, because they all give ions when dissolved in water. These solutions conduct electricity due to the mobility of the positive and negative ions, which are called **cations** and **anions** respectively.

- ☆ **Strong electrolytes** completely ionize when dissolved, and no neutral molecules are formed in solution [1].
- ☆ The number of dissociated and undissociated ions attain equilibrium which necessary for the stability of the solution [2,2b].
- ☆ Non electrolytes although dissolve in water but cannot pass electric current.

Dissociation of Ions in solutions [2,2b]

Some of the compounds like hydrochloric acid (HCl), sodium chloride (NaCl), acetic acid (CH_3COOH) *etc.* dissolve in water by dissociating into ions. To begin this process the ions require some energy which they obtain by the process known as hydration [2,2b].

Strong Electrolytes

For example, NaCl, HNO_3, $HClO_3$, $CaCl_2$ *etc.* are strong electrolytes. An ionization can be represented by the following equation [1]-

$$NaCl_{(s)} \rightarrow Na^{+}_{(aq)} + Cl^{-}_{(aq)}$$

Since NaCl is an **ionic solid** (s), which consists of cations Na^+ and anions Cl^-, no molecules of NaCl are present in NaCl solid or NaCl solution. The ionization is said to be complete. The solute is one hundred per cent (100 per cent) ionized [1].

Weak Electrolytes

Small fractions of **weak electrolytes'** molecules ionize when dissolve in water. Some neutral molecules are present in their solutions. For example, H_2CO_3(Carbonic acid),CH_3COOH(Acetic acid), NH_4OH(Ammonia), and most organic acids and bases are weak electrolytes. The following ionization is not complete-

$$H_2CO_{3\,(aq)} \rightleftharpoons H^{+}_{(aq)} + HCO_3^{-}{}_{(aq)}$$

Hydration [2,2b,3]

Hydration may refer to hydrate, a substance that contains water. It is the act or process of combining or treating with water. This is a process in which opposite charges attract one another.

- ☆ In the present example the partial positive charge on the hydrogen molecule of H_2O is attracted to the anion and the partial negative charge on oxygen is attracted to cations.
- ☆ In this way during hydration each ion is surrounded by a group of opposite charges there by lowering the overall energy of the solution. It may be mentioned that these substances in their pure state are not separated ions, are electrically neutral and do not conduct electricity.

Conductivity [2,2b,4]

Conductivity is the measure of the ease at which an electric charge or heat can pass through a material. A conductor is a material which gives very little resistance to the flow of an electric current or thermal energy (Electrical and Thermal conductivity).

- The molecular compounds such as pure acetic acid do not conduct electricity. Under the influence of water and during the process of hydration these compounds can be broken down into ionic species due to which the solution conducts electricity.
- Ionic compounds are also poor conductors of electricity when not in solution.

Reversible Reactions and Equilibrium [2,2b,5]

Reversible reactions that happen in a closed system eventually reach equilibrium. At equilibrium, the concentrations of reactants and products do not change. But the forward and reverse reactions have not stopped - they are still going on, and at the same rate as each other. This is an example of equilibrium.

- The situation or the stage in which the number of ion pairs and molecular particles stay the same is called Equilibrium. Equilibrium should not be thought of as the end of a reaction but rather a stage or point at which no net change occurs.

  Hydration is an example of a reversible reaction. Most chemical reactions are the equilibrium of the hydration process is largely influenced by the type of electrolyte *i.e.,* strong or weak electrolyte.

  a. Weak electrolytes upon introduction into water begin to dissociate into ions but most of the particles remain in molecular form. But the tendency of the particles in the molecular form to dissociate is same as those molecules that have already dissociated into ions.

  In fact, these molecules have also been dissociated but they immediately get associated into their molecular form. In this way, the total number of ions and molecules stays the same while each is constantly becoming the other.

  b. In the case of strong electrolytes the molecules dissociate to a great extent than weak electrolytes and the equilibrium shifts to one side of the reaction where there are more ions and no molecular particles are found in the solution.

Types of Equilbrium [2,2b,6]

Many kinds of equilibrium could be observed in solutions. The chemical and Physical equilibria are common.

i. Chemical Equilibrium

When the forward reaction rate equals the reverse reaction rate in a closed system (no energy inputs or outputs). This is called chemical equilibrium. It can be static or dynamic.

a. Static equilibrium-Amount of reactant and product are fixed and unchanging and

b. **Dynamic equilibrium-** Reactant is turning into product and product is turning into reactant, but at the same rate such that the concentrations of each are unchanged on a macroscopic scale.

ii. Physical Equilibrium

When opposing forces are balanced such that the object experiences no net acceleration.

Some examples are given below:

- ✰ Physical equilibrium can be reached between a solid salt and a saturated solution of that salt.Example is given below:

$$NaCl(s) \rightleftharpoons NaCl(Aq)$$

- ✰ Physical equilibrium between liquid water and gaseous water in a closed container. Example is given below:

$$H_2O(l) \rightleftharpoons H_2O(g)$$

- ✰ Physical equilibrium can be reached between molecular particles and ions in solution. Example is given below:

$$CH_3COOH\ (aq)+H_2O \rightleftharpoons CH_3COO^-(aq) + H_3O^+(aq)$$

- ✰ Chemical equilibrium can be reached between reactants and products. Example is given below:

$$N_2(g) + O_2(g) \rightleftharpoons 2NO(g)$$

Non Equilibrium Systems [2,2a,2b]

Such systems are typically irreversible, *e.g.* because they dissipate energy persistently pumped in from the outside, or they age towards an equilibrium state that is never reached [2a].

- ✰ Some reactions have an end point and due to the formation of precipitate these reactions are not considered equilibrium because a product formed by this reaction does not become available to react in the reverse direction, a precipitate is insoluble and will not react with ions in solution a gas escapes to the atmosphere and is no longer even physically present to react [2,2b]. This could be due to following reasons [2a]-
  - i. Products are no longer in the system *e.g.* an open system.
  - ii. Products do not react with each other due to their highly stable nature.
  - iii. Forward reaction has a negative enthalpy change and positive entropy change.

Le Chatelier's Principle [2,2b,3]

Le Chatelier's principle (also known as "Chatelier's principle" or "The Equilibrium Law") is a principle of chemistry used to predict the effect of a change in conditions on chemical equilibria.

- ☆ The principle is named after French chemist Henry Louis Le Chatelier, and sometimes also credited to Karl Ferdinand Braun. It states that when a system experiences a disturbance (such as concentration, temperature, or pressure changes), it will respond to restore a new equilibrium state.
- ☆ Reversible reactions will reach equilibrium which is a point at which no net reaction is occurring or we can say also that the reaction is progressing backward at the same time it is progressing forward. if it is left undisturbed.
- ☆ The disturbance in equilibrium of reversible reactions can be done. It is possible to disturb this equilibrium causing the reaction to progress more in one direction than another until a new equilibrium is reached. There are many conditions which will disturb this equilibrium. The details of these conditions are given below:
  - i. **Change in temperature** - Any addition or subtraction of energy from the system causes disturbance in equilibrium of reversible reaction.
  - ii. **A change in pressure** – The change in pressure of reversible reaction disturbs the equilibrium,
  - iii. **The removal or addition of a product or reactant-** The removal of product and reactant and addition of product and reactant disturb the equilibrium of reversible reaction.
- ☆ Le Chatelier said that any change in one of the variables that determines the state of system in equilibrium causes a shift in the position of equilibrium in the direction that tends to counteract the change in the variable under consideration.
- ☆ Le Chatelier's principle implies that the addition of heat to a reaction will favor the endothermic direction of a reaction as this reduces the amount of heat produced in the system.
- ☆ Increasing the concentration of reactants will drive the reaction to the right, while increasing the concentration of products will drive the reaction to the left.

Applications of Le Chatelier's Principle [2,2b,3]

It helps in biological, chemical and industrial processes.

i. **Chemical processes-**This principle helps to predict what effect a change in temperature, concentration or pressure will have on the position of the equilibrium in a chemical reaction.

ii. **Industrial processes-**This is very important, particularly in industrial applications, where yields must be accurately predicted and maximised. Le Châtelier's principle can be used to encourage formation of a desired product in chemical reactions. In the Haber process for the industrial synthesis of ammonia, nitrogen gas and hydrogen gas react to form ammonia gas in the reaction $N_2+3H_2 \rightarrow 2NH_3$; the process is exothermic, *i.e.*, one that gives off heat.

iii. **Biological processes-**Some examples of application of this principle in biological processes are given below:

- **Maintenance of blood pH-**Carbon dioxide reacts with water to form a weak acid H_2CO_3 which would cause the blood pH to fall to dangerous levels if it were not promptly removed as it is excreted by the cells. This is accomplished by combining it with carbonate ion through the reaction.
- **Oxygen transport by the blood-**Few of these are more important to warm-blooded organisms than those that relate to aerobic respiration, in which oxygen is transported to the cells where it is combined with glucose and metabolized to carbon dioxide, which then moves back to the lungs from which it is expelled.

pH Indicators [2,2b,3a]

A pH indicator is a halochromic chemical compound that is added in small amounts to a solution so that the pH (acidity or alkalinity) of the solution can be determined easily.

- A pH indicator is a chemical detector for protons (H^+) or hydronium ions (H_3O^+).Normally, the indicator causes the colour of the solution to change depending on the pH, of the solution. pH indicators are frequently employed in titrations in analytic chemistry and biology experiments to determine the extent and magnitude of a chemical reaction.
- pH indicators themselves are frequently weak acids or bases. When introduced into a solution, they may bind H^+ (Hydrogen ion) or OH^- (hydroxide) ions. The different electronic configuration of the bound indicator causes indicator's colour to change. Since determination of colour change is subjective in nature the results obtained by using pH indicators are most likely approximate and not precise.
- However, for applications requiring precise measurement of pH, a pH meter is frequently used.

☆ Indicators usually exhibit intermediate colour at pH values inside the listed transition range. For example, phenol red exhibits an orange colour between pH 6.6 and pH 8.0. The transition range may shift slightly depending on the concentration of indicator in solution and on the temperature at which it is used. The details of several common laboratory pH indicators, are given in following table-

Table 6.9: Common Laboratory pH Indicators [2,2b,3a]

Sl.No.	*Indicator*	*Transition pH Range (approximate)*	*Colour at Low pH*	*Colour at High pH*
1	Gentian violet (Methyl violet 10B)	0.0-2.0	Yellow	Blue violet
2	Thymolblue(first transition)	1.2-2.8	Red	Yellow
3	Methyl yellow	2.9-4.0	Red	Yellow
4	Bromophenol blue	3.0-4.6	Yellow	Violet
5	Congo red	3.0-5.2	Blue	Red
6	Methyl orange	3.1-4.4	Red	Yellow
7	Litmus	4.5-8.3	Red	Blue
8	Bromocresol purple	5.2-6.8	Yellow	Violet
9	Bromothymol blue (First transition)	6.0-7.6	Yellow	Blue
10	Phenol red	6.6-8.0	Yellow	Red
11	Thymol blue (second transition)	8.0-9.6	Yellow	Blue
12	Phenolphthalein(Second transition)	8.2-10.0	Colourless	Pink
13	Thymolphthalein	9.4-10.6	Colorless	Blue
14	Alizarin Yellow R	10.1-12.0	Yellow	Red
15	Indigo carmine	11.4-13.0	Blue	Yellow

Universal Indicator [2,3a]

It is a pH indicator made of a blend or mixture of different indicators that exhibits several smooth colour changes over wide range of pH values.

☆ A universal indicator is collectively a mixture of indicators which show a colour change in a solution, which interprets how acidic or basic a solution is. A universal indicator can be in paper form or present in a form of a solution [3a].Some examples of universal indicators are given below:

i. Anthocyanins occur in many different plants. They appear red in acidic solutions and blue in bases. Extracting anthocyanins from red cabbage leaves to form a crude acid-base indicator is a popular introductory chemistry demonstration.

i. If Ka(cation) > Kb(anion), the solution of the salt is acidic.

ii. If Ka(cation) = Kb(anion), the solution of the salt is neutral.

iii. If Ka(cation) < Kb(anion), the solution of the salt is base.

The summary of salt formed from and nature of salt solution is tabulated below [2,3]:

Table 6.6: Summary of Salt Formed

Sl.No.	*Salt Formed from*		*Salt Solution*
	Acid	*Base*	
1	Strong acid	Strong base	Neutral
2	Strong acid	Weak base	Acidic
3	Weak acid	Strong base	Basic

Sources

1. *K.S. Sharma, Physical Chemistry of milk, Agrimoon.com.*

1a. *Wikipedia-Salt.*

2. *Hydrolysis of salts, luman, https: //courses.lumenlearning.com.*

3. *Hydrolysis of salts, Equation, https: //chem.libretexts.org.*

Buffers [1,2,2a]

Buffer solutions are used as a means of keeping pH at a nearly constant value in a wide variety of chemical applications.

☆ In nature, there are many systems that use buffering for pH regulation. Some examples are given below [1]-

i. The bicarbonate buffering system is used to regulate the pH of blood. In human blood a mixture of carbonic acid (H_2CO_3) and bicarbonate (HCO^-_3) is present in the plasma fraction; this constitutes the major mechanism for maintaining the pH of blood between 7.35 and 7.45 [1].

ii. In industry, buffering agents are used in fermentation processes and in setting the correct conditions for dyes used in colouring fabrics [1].

iii. Buffering solutions are also used in chemical analysis and calibration of pH meters.

☆ There is often a need of having a solution whose pH does not change appreciably with a change in conditions. This will help in utilizing such buffer solutions to resist change in pH for maintaining product stability [2].

ii. A universal indicator is typically composed of water, 1-Propanol, phenol-phthalein sodium salt, sodium hydroxide, methyl red, bromothymol blue monosodium salt, and thymol blue monosodium salt [7]. The colours that indicate the pH of a solution, after adding a universal indicator, are tabulated below:

Table 6.10: Universal Indicator [3a]

Sl.No.	pH Range	Description	Colour
1	< 3	Strong acid	Red
2	3-6	Weak acid	Yellow
3	7	Neutral	Green
4	8-11	Weak base	Blue
5	>11	Strong base	Violet

The colours from yellow to red indicate an acidic solution, colours blue to violet indicate alkali and green colour indicates that a solution is neutral.

Sources

1. *Electrolyte chemistry, https: //chem.libretexts.org.*
2. *K.S. Sharma, Physical Chemistry of milk, Agrimoon.*
2a. *Non-equilibrium Systems, https: //scienceready.com.au.*
2b. *Electrolyte Equilibrium and pH Indicators, Lession-25, http: //ecoursesonline.iasri. res.in.*
3. *Wikipedia- Hydration.*
3a. *Wikipedia-pH Indicators.*
4. *Conductivity, https: //www.lehigh. edu.*
5. *Reversible reactions, https: //www.bbc.co.uk.*
6. *Types of Equilibrium, https: //rockedu.rockefeller.edu.*
7. *Universal Indicator Archived September 25, 2006, at the Wayback Machine, ISCID Encyclopedia of Science and Philosophy.*

Chapter 7

Oxidation-Reduction

Oxidation may be defined as the uptake of oxygen or the loss of hydrogen or the loss of electrons(takes up positive, or parts with negative charges) **[1,5]**.

- ☆ Likewise reduction is defined as the loss of oxygen or the uptake of hydrogen or the gain of electrons (takes up negative or parts with positive charges).
- ☆ These process of oxidation and reduction go simultaneously in a system composed of more than one substance constituting the system which involve exchange of electrons simultaneously in the system. Exchange of electrons generates an electromotive force *i.e.* emf in the system which indicates the direction in which processes proceeds. This is illustrated with the following equation-

$$Fe^{2+} - e^{-} \underset{\text{Reduction}}{\overset{\text{Oxidation}}{\rightleftharpoons}} Fe^{3+} + e^{-}$$

Oxidation-Reduction Potential (Eh) [1,1c]

Oxidation reduction potential (ORP) covers a large pH range. It is an electrochemical test and can easily measured using simple and portable instruments.

- ☆ Oxidation Reduction Potential is a measure of the tendency of a chemical species to acquire electrons from or lose electrons to an electrode and thereby be reduced or oxidised respectively.

- ✰ Oxidation reduction potential (ORP), also known as REDOX, is a measurement that reflects the ability of a molecule to oxidize or reduce another molecule. Redox potential is measured in volts (V), millivolts (mV), or Eh (1 Eh = 1 mV).
- ✰ Oxidizers have a positive ORP value, while reducers have a negative ORP value.
- ✰ Each species has its own intrinsic reduction potential; the more positive (negative) the potential is, the greater the species' affinity for electrons and tendency to be reduced.

Nernst Equation [1,1c,2,3,4]

It was named after Walther Nernst, a German physical chemist who formulated the equation [2,3]. In electrochemistry, the **Nernst equation** is an **equation** that relates the reduction potential of an electrochemical reaction (half-cell or full cell reaction) to the standard electrode potential, temperature, and activities (often approximated by concentrations) of the chemical species undergoing reduction and oxidation.

- ✰ **Nernst equation** helps in determining the cell potentials of electrochemical cells even under non-standard conditions. The calculation of the cell potential of an electrochemical cell at any given temperature, pressure, and reactant concentration is often done with the use of Nernst equation. For a general reduction reaction,

$$M^{n+}(aq) + ne^{-} \rightarrow M(s)$$

Nernst developed an equation, known as Nernst equation:

$$E = E^{o} - \frac{RT}{nF} \ln \frac{[M(s)]}{[M^{n+}_{(aq)}]}$$

We can rewrite above equation as follows

$$E = E^{o} - \frac{2.303\,RT}{nF} \log_{10} \frac{[M(s)]}{[M^{n+}_{(aq)}]}$$

where, E is the electrode potential, E^{o} is the standard electrode potential (for 1M solution of metal ions at 298 K), R is gas constant, T is temperature,n is the number of electrons accepted during the change and F is Faraday of electricity (96500 coulombs).

- ✰ The molar concentration of solids in the pure state is assumed to be unity *i.e.* [M(s)] = 1. Substituting the following values in above equation to find out the value of RT/F -

$R = 8.314\ J\ K^{-1}\ mol^{-1}$, $T = 298\ K$, $F = 96500$ coulombs/mole,

$$\frac{2.303RT}{F}$$ becomes a constant

$$\frac{2.303RT}{F} = \frac{2.303X\ (8.314\ J\ K^{-1}\ mol^{-1})\ X\ (298\ K)}{(96500\ C\ mol^{-1})} = 0.059$$

Hence, the Nernst equation for the general reduction reaction at 25°C is written as:

$$E = E^o - \frac{0.059}{n} \ln \frac{1}{[M^{n+}_{(aq)}]}$$

Nernst Equation at 25°C

For measurements carried out 298K(273+25), the Nernst equation can be expressed as follows.

$$E = E^o - 0.059/n \log_{10} Q$$

Therefore, as per the Nernst equation, the overall potential of an electrochemical cell is dependent on the reaction quotient.

- ☆ The Nernst Equation allows for cell potential determination under non - standard conditions. It relates the measured cell potential to the quotient of the reaction and allows the exact determination of constants of equilibrium (including constants of solubility).

Limitation of Nernst Equation [4a]

In dilute solutions, the Nernst equation can be expressed directly in the terms of concentrations (since activity coefficients are close to unity).

- ☆ At higher concentrations, the true activities of the ions must be used. This complicates the use of the Nernst equation, since estimation of non-ideal activities of ions generally requires experimental measurements.
- ☆ At very low concentrations of the potential-determining ions, the potential predicted by Nernst equation approaches toward $\pm\infty$. This is physically meaning-less because, under such conditions, the exchange current density becomes very low, and there may be no thermodynamic equilibrium necessary for Nernst equation to hold

Application of Nernst Equation

The Nernst equation has great utility for analytical chemistry as well as in important life processes (nerve conduction and membrane potential).

- ☆ The Nernst equation's prime application is to determine the ion concentration.
- ☆ Electrochemical cells and hence the Nernst equation is widely used in the calculation of the followings:
  - i. Solution pH.
  - ii. Solubility product,
  - iii. Constant equilibrium and other thermodynamic properties,
  - iv. Potentiometric titrations.
  - v. The equation is used in marine environment and in oxygen and vi. The calculation of cell membrane resting potentials.

Electrochemical Cell [1,1c,4]

An **electrochemical cell** is a device that can generate electrical energy from the chemical reactions taking place in it.

- ☆ It can also be explained as use the electrical energy supplied to it to facilitate chemical reactions in it.
- ☆ Electrochemical cells are capable of converting chemical energy into electrical energy, or vice versa.
- ☆ An electrochemical cell can be created by placing metallic electrodes into an electrolyte where a chemical reaction either uses or generates an electric current. The following points help in understanding the reaction between electrode and electrolytes:
  - **a.** Immersion of a metal plate in water or in a solution containing ions of that metal, a double electric layer is formed at the interfacial boundary between the metal plate and solution resulting in the establishment of a potential difference between them. The magnitude of potential difference established between the metal plates and the solution of the salts of that metal depends upon the followings:
    - i. Properties of the metal and the solution in particular.
    - ii. On the concentration of the ions of the metal in the solution,
    - iii. On the nature of the interaction between the particles of the double layer.
  - **b.** If two metal plates *i.e.*, zinc and copper are dipped in solutions of their salts separated by a porous membrane. Each of the metals emits a certain quantity of ions into the solution corresponding to its equilibrium state. Since the equilibrium potential for all the

metals is not similar, there is greater tendency for the zinc to give off its ions in to solution than copper. Due to this release of zinc ions in to the solution it acquires more negative charge.

$Zn = Zn^2 + 2e-$ at the negative electrode

$Cu = Cu^2 + 2e-$ at the positive electrode.

c. If the plates are now connected through a wire this difference in the potential between the plate and its, will allow the electrons to flow from the zinc plate to the copper plate. This would result in the disturbance of equilibrium of the double layer at both the plates.

- Consequently another proton of zinc ion will pass from the zinc plate to the solution and corresponding number of copper ions are discharged on the copper plate. This would further difference in the charge of the plates causing transition of the electrons from the zinc plate (negative electrode) to the copper (Positive electrode) resulting in the further transport of the ions.
- Consequent to this spontaneous process taking place with the dissolving of the zinc plate (Oxidation) and deposition of copper on the copper plate due to the discharge of the copper ions (reduction) there will be passage of electrons along the wire from zinc to the copper plate. This passage is responsible for the electric current.
- The electrical work that is obtained by means of redox potential is maximum when the cell is operating under close to reversible conditions. This maximum potential difference is called the Electro motive force (EMF).

Types of Electrochemical Cells [4]

There are following two types of electro-chemical cells:

i. Galvanic Cell

It is also called Voltaic cell. It is named after Luigi Galvani, or Alessandro Volta respectively. Galvanic cells derives its energy from spontaneous redox reactions. The appliances which could be designed to create electrical current by means of chemical reactions are known as **galvenic cell.**

- Galvenic cells involve in the reaction of oxidation reduction potential. Electrochemical cells which generate an electric current are called **voltaic cells or galvanic cells**, and common batteries consist of one or more such cells.

  Electrolytic Cell- Electrolytic cells involve non-spontaneous reactions and thus require an external electron source like a DC battery or an AC power source. These cells generate chemical reactions, via electrolysis.

☆ An electrolytic cell uses electrical energy to drive a non-spontaneous redox reaction. An electrolytic cell is a kind of electrochemical cell. It is often used to decompose chemical compounds, in a process called electrolysis.

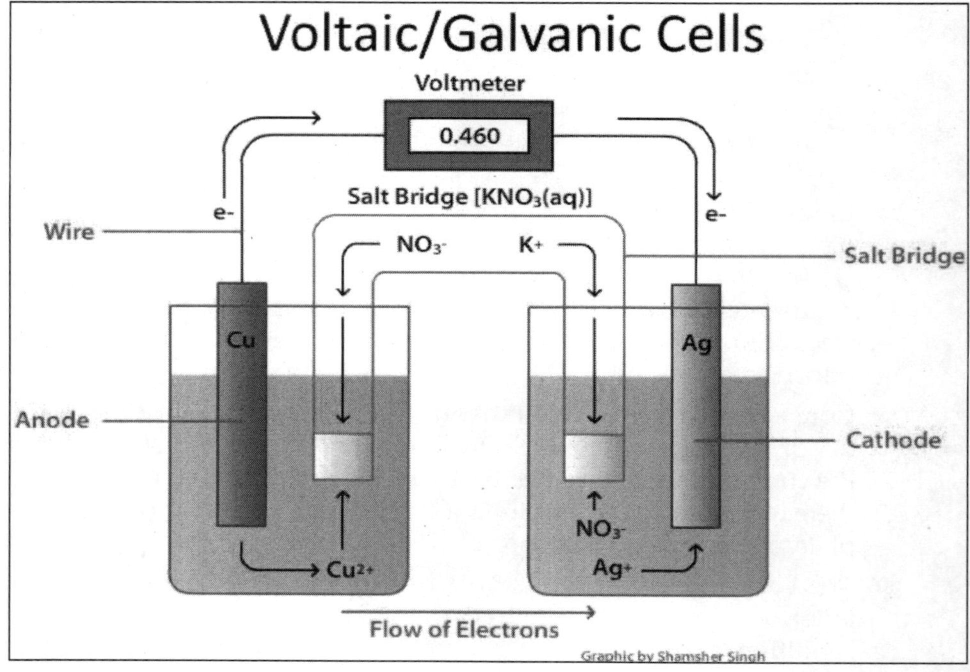

Figure 7.1: Electrochemical Cell [4b].

Electrodes [1,1c,8]

An **electrode** is an electrical conductor used to make contact with a nonmetallic part of a circuit. An electrode is a solid electric conductor that carries electric current into non-metallic solids, or liquids, or gases, or plasmas, or vacuums. Electrodes are typically good electric conductors, but they need not be metals. In an electrochemical cell, reduction and oxidation reactions take place at the electrodes. The electrode at which reduction takes places is called the cathode. Oxidation takes place at the anode. The following types of electrodes are generally used-

1. Hydrogen Electrode [1,1c,4]

Hydrogen electrode is the reference electrode universally used the standard galvanic cell. The **standard hydrogen electrode** (abbreviated **SHE**), is a redox electrode which forms the basis of the thermodynamic scale of oxidation-reduction potentials.

- The electrode potential of a given electrode is the quantity of its potential (EMF) with that of the standard hydrogen electrode. This quantity is known as electrode potential and is designated by the letter E, which is similar to the EMF of the cell.
- Standard Hydrogen electrode is a reference electrode against which the electrode potentials of all electrodes are measured. When Hydrogen gas at 1 atm-pressure is adsorbed over platinum electrode dipped in 1M HCl at 25°C, it is standard hydrogen electrode and its potential is, E0=±0 volt.

Construction of Hydrogen Electrode

This is usually consists of a platinized platinum foil immersed in a solution containing hydrogen ions and around which a current of hydrogen gas flows.

Characteristics

- In the standard hydrogen electrode the concentration of the hydrogen ions in solution corresponds to the activity $a_{H+}=1$ and a gaseous hydrogen pressure of 1 atmosphere. It functions on the basis of the reaction.

  $$\frac{1}{2} H_2 \leftrightarrow H^+ + e^-$$

  Platinum here plays only the part of an inert carrier and may be replaced by palladium, iridium, gold and certain other metals.
- The hydrogen electrode may be used at any hydrogen pressure at any hydrogen ion concentration in the solution and any temperature. It's potential depends on the operating conditions.
- Such a hydrogen electrode is taken as a reference electrode (to which 0 potential is assigned) for which the activity of the hydrogen ion in solution is unity ($a_{H+} = 1$) and the hydrogen gas pressure is one atmosphere, both the hydrogen electrode and the other electrode of the cell being at the same temperature. Hence $Eh^+ = 0$.

Sensitiveness of Hydrogen Electrode

It is very sensitive to the operating conditions as such, it is necessary to maintain high purity with respect to the hydrogen and platinum surface.

- When correctly used hydrogen electrode gives very sensitive results, reproducible to 0.00001V. The high sensitivity of this electrode to the environmental conditions greatly hampers its utilization.

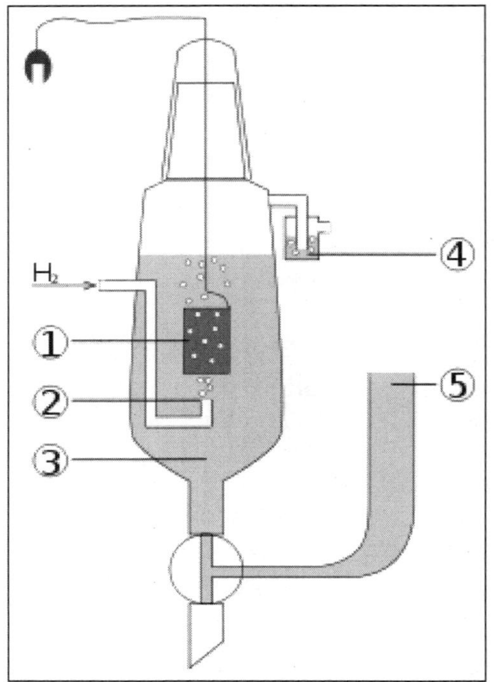

1-Platinized platinum electrode.2- Hydrogen gas.3-Acid solution with an activity of H+=1 mol/l.4-Hydroseal for prevention of oxygen interference.5-Reservoir via which the second half-element of the galvanic cell should be attached.

Figure 7.2: Standard Hydrogen Electrode Scheme [4].

2. *Reference Electrodes [1,1c,4]*

A reference electrode is an electrode which has a stable and well-known electrode potential. The high stability of the electrode potential is usually reached by employing a redox system with constant (buffered or saturated) concentrations of each participants of the redox reaction.

These electrodes are reversible, have little **hysteresis**, follow Nernst equation and have stable potential with time. The reference electrodes are Aqueous and Non aqueous.

1. **Aqueous reference electrodes [4]-** Both the Saturated calomel electrode (SCE) and saturated Ag/AgCl are aqueous electrodes based around saturated aqueous solution. While for short periods it may be possible to use such aqueous electrodes as references with nonaqueous solutions the long-term results are not trustworthy.
2. **Non aqueous reference electrodes [4]-**The standard hydrogen electrode (SHE) might seem to be a reasonable reference for nonaqueous work

as it turns out the platinum is rapidly poisoned by many solvents including acetonitrile.

The followings are the examples of reference electrodes-

i. *Saturated Calomel Electrode [1,1c,4]*

The saturated calomel electrode (SCE) is a reference electrode based on the reaction between elemental mercury and mercury(I) chloride. It has been widely replaced by the silver chloride electrode, however the calomel electrode has a reputation of being more robust.

A diagram of the calomel electrode is shown in following figure:

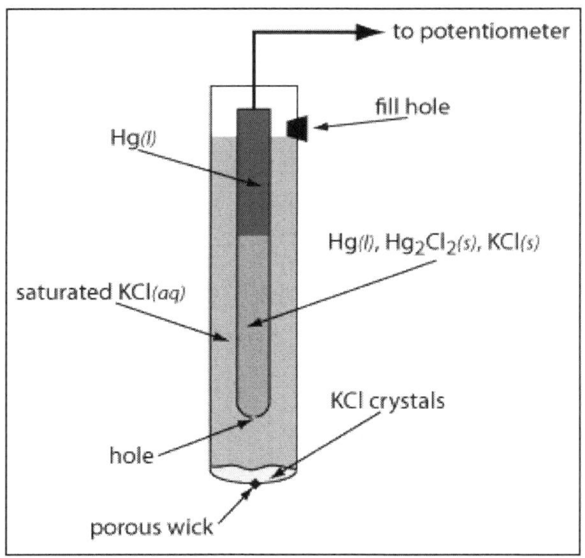

Figure 7.3: Calomel Electrode.

a. *Construction of Saturated Calomel Electrode*

SCE consists of an outer glass tube fitted with a frit at the bottom to permit electrical contact with the outside solution.

- ☆ The aqueous phase in contact with the mercury and the mercury(I) chloride (Hg_2Cl_2, "calomel") is a saturated solution of potassium chloride in water. The electrode is normally linked via a porous frit to the solution in which the other electrode is immersed. This porous frit is a salt bridge [4].
- ☆ In side there is another tube, the bottom of which is packed with glass wool to allow further electrical connection between the contents of the inner tube and the contents of the outer tube.

b. *Chemical Reaction*

The inner tube is packed with a paste of mercury and mercurous chloride dispersed in a saturated solution of potassium chloride represented by

$$Hg/Hg_2Cl_2KCl\ (xM)\ saturated$$

And the reaction for the electrode will be,

$$Hg_2Cl_2 + 2e\ Hg_2Cl_2 \rightleftharpoons 2Hg + 2Cl^-$$

c. *Electrode Potential*

The electrode potential will depend on the concentration of the potassium chloride and, thus, the electrode potential must be reported together with the potassium chloride concentration. Thus, for the saturated calomel electrode the common reference voltage is +0.244 V.

d. *Operating Temperature and Rate of Progress towards Equilibrium*

The calomel electrode operating temperature is restricted to below 80°C due to the mercurous chloride disproportioning into mercury and mercuric chloride at higher temperatures. Unfortunately due to its design and the thermal characteristics of the materials from which it is made, it's rate of progress towards equilibrium after a temperature change is rather slow.

$$Hg/Hg_2Cl_2(Sat'd),\ KCl\ (a\text{-}xM)$$

Half cell for calomel electrode -

$$Hg_2Cl_2(s) + 2e^- - \leftrightarrow 2Hg(I) + 2\ Cl^-$$

Position of equilibrium affected by a Cl from KCl, so E0 depends on a_{cl}^- for a most common saturated calomel electrode SCE [(Cl)-] ~ 4.5M)

ii. *Silver/Silver Chloride Electrode [1,1c,4]*

A silver chloride electrode is a type of reference electrode, commonly used in electrochemical measurements. The silver/silver chloride electrode is the most common and popular electrode used in electrochemistry.

- ☆ For environmental reasons it has widely replaced the saturated calomel electrode. It is usually the internal reference electrode in pH meters and it is often used as reference in reduction potential measurements [4].

Construction

The electrode consists of a silver wire coated with silver chloride and immersed in a solution of potassium chloride saturated with silver chloride. The silver/silver chloride electrode is easily constructed, can be used over a relatively wide range of temperatures and can be used in non-aqueous solutions [1].

Chemical Reaction

It involves the following chemical reaction-

$$Ag/AgCl(saturated), KCl\ (xM)$$

And the half reaction is,

$$AgCl + e\ Ag + Cl$$

The actual "redox" action occurring at the electrode is

$$Ag^{+} + e^{-} \rightarrow Ag + Cl$$

Electrode Potential

The potential developed will be a function of the (Ag+) concentration. Now as the (Ag+) concentration depends on the solubility product equilibrium of AgCl then the electrode potential will also depend on the chloride ion concentration. The standard potential of the Ag/AgCl electrode at 25°C is 0.2223 V.

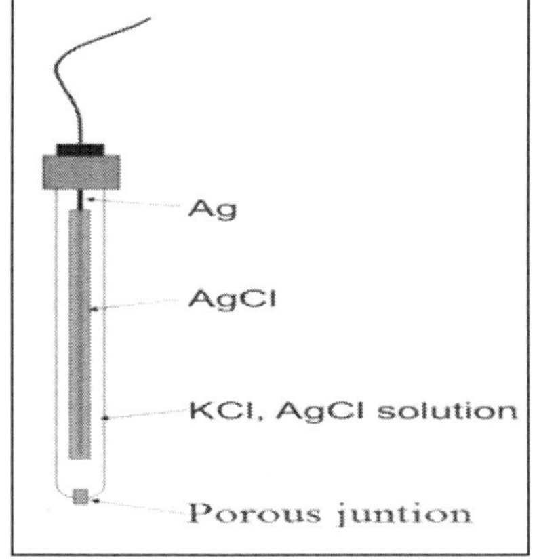

Figure 7.4: Silver Electrode

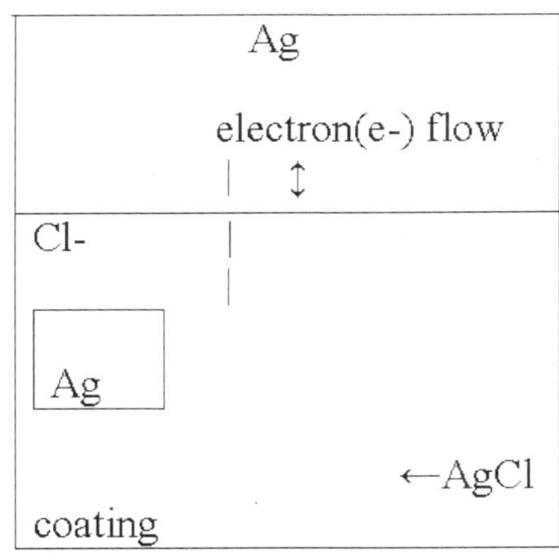

Figure 7.5: Electrode Reaction- Ag + Cl- ↔ AgCl + Electron

$$Ag/Ag2Cl(Sat'd),KCl\ ((a\text{-}xM)$$

$$AgCl(S) + e\text{-} \leftrightarrow AgCl\ (s) + Cl$$

Again depends on a_{cl-} but commonly sat (~3.5M). Ag/AgCl better for uncontrolled temperature (lower T coefficient) Ag reacts with more ions.

3. *Indicator Electrodes for Ions [1,1b,1c]*

Ideal indicator electrode responds rapidly and reproducibly to changes in the concentration of an analyte ion (or groups of analyte ions) [1b].Electrodes used with reference electrodes to measure potential of unknown solutions. Potentials are proportional to ion activity. These electrodes may be specific (one ion) or selective (Several ions).

$$E_{cell} = E_{indicator} - E_{reference}$$

Types of Indicator Electrode [1,1b,1c]

Generally there are following two types (Metallic and Membrane) of electrodes are available -

i. *Metallic Indicator Electrodes [1,1b,1c]*

The following three types of electrodes are being used-

Electrodes of the First Kind

Electrodes of the first kind respond directly to changing activity of electrode ions *e.g.* Copper indicator electrode but other ions can be reduced at Cu surface. The metals with higher +ve E0 and having better oxidizing agents than Cu such as Ag, Hg, Pd can also be used.

- ☆ In general electrodes of first kind are simple, they are not very selective, and some metals are easily oxidized (Deaerated solutions) some metals like Zn and Cd dissolve in acidic solutions.

Problems [1b]

These types of electrodes have following problems:

a. Electrode of the first kind is not very popular because:

- ☆ Metallic indicator electrodes are not very selective and respond not only to their own cations but also to other more easily reduced cations.
- ☆ Many metal electrodes can be used only in neutral or basic solutions because they dissolve in the presence of acids
- ☆ Easily oxidized, can be used only when analyte solutions are deaerated to remove oxygen
- ☆ Certain hard metals (Fe, Cr, Co, Ni) do not provide reproducible potentials.

b. Limited electrodes are-

- ☆ Ag/Ag^{+} and Hg/Hg^{2+} in neutral solutions and Cu/Cu^{2+}, Zn/Zn^{2+}, Cd/Cd^{2+}, Bi/Bi^{3+}, Tl/Tl^{+}, and Pb/Pb2+ in deaerated solutions.

Electrodes of the Second Kind [1,1b,1c]

The electrodes of the second kind respond to changes in ion activity through formation of complex. Metal electrode responds to the activities of anions that form sparingly soluble precipitates or stable complexes *e.g.* Silver wire in KCl (Sat'd) forms Ag Cl layer on the surface of the electrodes [1] and Detection of EDTA ion (Y^{4-}) with Hg Electrode [1b].

Electrodes of the Third Kind [1,1b,1c]

The electrodes of the third kind respond to changes of different ion than metal electrodes [1,1c]. Metal electrodes respond to a different cation and Linked to cation by an intermediate reaction [1b].

ii. Membrane (Or Ion Selective) Electrodes [1,1c]

Membranes are of low solubility which may be solids, semisolids and polymers, since part of membrane binds/reacts with analyte so they are selective. These electrodes have following characteristics [1b]-

- ☆ Electrodes based on determination of cations or anions by the selective adsorption of these ions to a membrane surface.
- ☆ Often called **Ion Selective Electrodes (ISE) or pIon Electrodes**, because the data obtained from them are usually presented as p-functions, (pH, pCa, pNO_3).
- ☆ The general mechanism by which an ion selective potential develops in these devices depends on the nature of the membrane and is entirely different from the source of potential in metallic indicator electrodes.
- ☆ The potential of a metallic electrode arises from the tendency of an oxidation-reduction reaction to occur at an electrode surface but **In membrane electrodes, in contrast, the observed potential is a kind of junction potential that develops across a membrane that separates the analyte solution from a reference solution.**

Desired Properties of ISE's [1b]

ISE's have following desired properties-

i. **Minimal solubility-**Membrane will not dissolve in solution during measurement. Many membranes are formed from large molecules or molecular aggregates: silica glasses, polymeric resin, low solubility inorganic compounds (AgX) can be used as membranes

ii. **Need some electrical conductivity-**Generally, this conduction takes the form of migration of singly charged ions within the membrane.

iii. **Selective reactivity with the analyte.**

Three types of binding are encountered: ion-exchange, crystallization and complexation. The former two are the more common, and we will largely focus on these types of bindings.

Types of Membrane Electrode

Generally following two types of membranes are used:

a. Non Crystalline Membranes [1,1a,1c]

The membrane is composed of a silicate glass. Glass electrodes can also be constructed that are sensitive to other cations such as sodium. These are Glass silicate glasses for H^+, Na^+ liquid ion exchanger for Ca^{2+} immobilized liquid-liquid/PVC matrix for Ca^{2+} NO_3.

b. *Crystalline Membranes [1,1a,1c]*

The membrane is composed of an insoluble inorganic salt. An ion-exchange process leads to the formation of a potential at the membrane. These could be single crystal or polycrystalline or mixed crystals - AgS for S^2 and Ag+.

4. *Biosensor Membrane Electrodes [1,1c]*

These are very important and are being developed after extensive research on this aspect.

- ☆ Immobilized enzymes bound to gas permeable membrane and the catalytic enzyme reaction produces small gaseous molecules (H^+, NH_3, CO_2) then gas sensing probe will measure the change in the gas concentration in the internal solution.

Advantages

The advantages of these electrodes are given below:

i. They are fast.

ii. Very selective

iii. Used *in vivo*

Disadvantages

The disadvantages of these electrodes include the followings-

i. They are expensive.

ii. Only few enzymes alone could be immobilized, immobilization would change the activity.

iii. The operating conditions are limited (pH, temperature, ionic strength).

Redox System in Milk [1,5-7,9]

A chemical reaction between two substances in which one substance is oxidized and the other reduced is called redox. In a redox system when half of the system is having oxidation reaction and the other half is having a reduction reaction there will be no flow of electrons either in to the system or go out of the system [1].

- ☆ In normal milk there are several complicated biological systems with varying composition and concentration.
- ☆ In addition to this microorganism gaining entry in to milk contribute certain redox systems to it depending upon the type of the organisms [1].

- Accurate measurement of Eh of milk takes considerable time as ther are several systems in milk which overlap and these reactions are either slowly reversible or even irreversible. Measuring OR potential of milk is complicated due to entry of oxygen during determination.
- Potential measured for the reversible systems is similar to the potential measured at the hydrogen electrode while determining the hydrogen ion concentration *i.e.,* pH. The quantity factor for the oxidation reduction is the overall concentration of active substances [Ox] + [Red]. The following several factors influence this potential in a system-
  i. Reversibility of the system.
  ii. Standard potential of the system(e0).
  iii. Or the position on the scale of the potential and
  iv. Concentration of the active components of this system.
- Milk has a positive oxidation reduction potential. It ranges between +0.2 and +0.3 volt for cow milk with an average value falling between +0.23 and +0.25 volt.
  i. This is primarily due to the presence of dissolved oxygen in milk.
  ii. Flushing of oxygen of milk with nitrogen decreases the oxidation - reduction potential of milk.
  iii. Individual milk samples in equilibrium with air generally have Eh's in the range of +0.25 to +0.35 V at 25°C at their normal pH of milk *i.e.* 6.6 to 6.7.
- Not much information is available on the capacity of the oxidation-reduction system in milk. Fresh raw whole milk will reduce 0.6 to 0.8 m mol of ferricyanide to ferrocyanide per litre at 50° C for 20 min.
- Heating of milk such as HTST Pasteurization milk or preheating prior to making milk powder produces reducing substances in milk. This lowers Eh values in heated milk or dried milk. Thus superior quality with higher storage capacity milk powders can be prepared by increasing reducing substances in it.
- The ascorbic acid content of fresh milk is about 0.25 meq litre-1. The milk is drawn from the udder all the ascorbate present in milk will be in its reduced form but reversible oxidation to dehydro ascorbate occurs at rates dependent on temperature. *etc.* In order to inhibit growth of microorganisms milk is held at temperature between 1 to 4°C, and the reduced form of ascorbate gradually decreases over several days as dehydro ascorbate levels are constant (0.05 mEq Litre. -1) their ratio remains large until the system disappears.
- Milk acidification by DL-starter cultures depends on the oxidation-reduction (redox) potential in milk.

☆ **Bacterial contamination** of milk affects Eh of milk. A lowering of Eh occurs due to consumption of oxygen by microorganisms. Also reducing substances are formed during the course of bacterial **metabolism.**

  i. Methylene blue reaction is based on Eh. A negative Eh is obtained when the dye becomes colorless. This dye is reduced by freshly drawn milk when it is drawn from udder anaerobically indicating a more negative potential than methylene blue system.

  ii. Exposure to oxygen will change this potential to be more positive than the mehtylene blue system.

  iii. Apart from this the chief oxidation and reduction systems present in milk are Ascorbate, lactate and riboflavin.

☆ Presence of copper in milk also influences Eh. Copper acts as an oxidizing agent with very strong electron acceptance. Presence of copper raises Eh of milk.

☆ In milk oxidation reduction systems resist change in the potential when the oxidant and reductant are near equilibrium state. This phenomenon is similar to the buffering action in acid base equilibrium. This phenomenon is known as **'poising'.** It is difficult to obtain meaningful data on the poising index though milk has considerable poising capacity.

  i. The concentration of free riboflavin is only 4μM, it plays an important role in the photo oxidation. It does not contribute significantly either to the O-R potential *i.e.,* Eh or poising action.

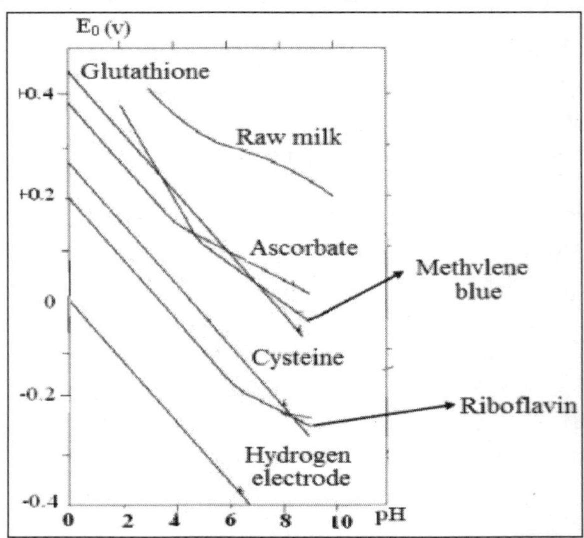

Figure 7.6: The Redox Potential (E0) of Milk and the Standard Potential (Eh) of Various Systems in Relation to pH [6a].

- ✰ Concentration of lactate-pyruvate system in milk is negligible. Enzymatic activation of this system would influence the redox system. At higher pH the aldehyde group of lactose is oxidizable to carboxyl group, but this reaction is not a reversible and hence it will not influence the Eh at pH of 6.6.
- ✰ Only low concentration of small molecular weight thiols is present in milk. Thiols of the native proteins are not active in oxidation reduction systems. Due to heat denaturation of proteins unfolding and uncoiling would occur due to which these thiols groups contribute to the O-R potential of milk. The concentration of such activated thiols in milk heated under UHT conditions may attain 0.18 mEq.liter-1.
- ✰ Redox reactions in milk systems are influenced by heat treatment, by concentration of O_2 and metal ions such as Cu^{2+}, by exposure to light and by oxido reductases of milk and or micro organisms. The redox potential of metal ions may depend closely on their binding to various ligands, including some (Metallo) proteins [1].

Determination of Oxidation-Reduction Potential [5]

Method of determining oxidation-reduction potential is based on the principle of oxidation-reduction.

- ✰ Substances giving or taking electrons in a solution creating potential difference can be measured using electrodes. Potential difference can be measured by platinum electrode (*e.g.*, donating or accepting electrons) and a reference calomel electrode in the presence of a potentiometer.
- ✰ The voltage measured under these conditions gives the oxidizing or reducing capacity of the system. This is simply called as oxidation-reduction potential and is designated by the symbol Eh.
- ✰ A positive potential involving loss of electrons from the platinum electrode is indicative of oxidizing properties, whereas a negative potential, which involves gain of electrons at the platinum electrode, gives reducing capacity.

Sources

1. *K.S. Sharma, Physical Chemistry of milk, Agrimoon.*

1a. *Ion Selective electrodes, https: //chem.libretexts.org.*

1b. *Potentiometric methods, http: //web. iyte.edu.tr.*

1c. *Oxidation Reduction Potential, Nernst Equation and Electrochemical Cell, Lesson 26. http: //ecoursesonline.iasri.res.in.*

2. *Orna, Mary Virginia; Stock, John, 1989, Electro-chemistry, past and present. Columbus, OH: American Chemical society. ISBN 978-0-8412-1572-6.*

3. *Wahl, 2005, "A Short History of Electro-chemistry".Galvanotechtnik, 96 (8), 1820-1828.*

4. *Wikipedia- Electrochemical cells.*

4a. *Wikipedia-Nernst Equation.*

4b. *Electrochemical cell, https: //www.slideserve.com.*

5. *Manual on Milk Production and Quality of Milk, School of Agriculture, Indira Gandhi National Open University.*

6. *R.Jenness and S.Patton, 1959, Principles of Dairy Chemistry, John Wiley, New York.*

6a. *P. Walstra, and R. Jenness , Dairy chemistry and physics, 1984, ISBN 9780471097792 – 467, John Wiley and Sons, New York.*

7. *Ling, E.R. (1956). Text Book of Dairy Chemistry, Vol 142 Chapman and Hall, London.*

8. *Chemicool Dictionary, https: //www.chemicool. com/definition/electrodes.*

9. *Redox system in Milk, Lesson 27, http: //ecoursesonline.iasri.res.in.*

Chapter 8

Nuclear Chemistry and Isotopes

Atomic theory in the nineteenth century presumed that nuclei had fixed compositions. In any compound the electrons around the atom's nucleus are involved in various chemical reactions **[1,1a,2]**.

- ☆ French scientist Henri Becquerel has found radiation emitted by uranium and this has expanded this field to nuclear changes.
- ☆ Marie Sklodowska Curie has initiated the study of radioactivity and contributed much to the study of nuclear changes. She was the first to coin the term "radioactivity" [1,1a].
- ☆ Ernest Rutherford (of gold foil experiment fame) has investigated and named three of the most common types of radiation [1a].

Radiation and Nuclear Reactions [1,1a,2]

As per the theory proposed by Frederic Soddy, "radioactivity is the result of a natural change of an isotope of one element into an isotope of a different element [1]."

- ☆ The spontaneous change of an unstable nuclide into another is radioactive decay. The unstable nuclide is called the **parent nuclide**, the nuclide that results from the decay is known as the **daughter nuclide [1a]**.

  i. The daughter nuclide may be stable, or it may decay itself. The radiation produced during radioactive decay is such that the daughter nuclide lies closer to the band of stability than the parent nuclide, so the location of a nuclide relative to the band of stability can serve as a guide to the kind of decay it will undergo [1a].

- ✰ Nuclear reaction is a change in the identity or characteristics of an atomic nucleus, induced by bombarding it with an energetic particle. The bombarding particle may be an alpha particle, a gamma-ray photon, a neutron, a proton, or a heavy ion.
- ✰ In any case, the bombarding particle must have enough energy to approach the positively charged nucleus to within range of the strong nuclear force.
- ✰ A typical nuclear reaction involves two reacting particles- a heavy target nucleus and a light bombarding particle and produces two new particles-a heavier product nucleus and a lighter ejected particle [2].
- ✰ In nuclear reactions, the particles in an atom's nucleus will change and thus changing the atom itself due to nuclear reactions.
- ✰ In chemical reactions the molecules are formed whereas due to the nuclear reactions there will be transmutation of one element into a different isotope or a different element altogether.
  - i. All elements heavier than bismuth (Bi) exhibit natural radioactivity and thus can "decay" into an altogether lighter elements [1].
  - ii. Normally the number of protons in an atom will determine a particular element.
  - iii. As the protons are changing in the nucleus due to radiation the atom of that element will also change [1].

Types of Radiation [1,3,4]

Radiation is energy. It can come from unstable atoms that undergo radioactive decay. Radiation travels from its source in the form of energy waves or energized particles. Radioactive decay is the emission of energy in the form of ionizing radiation. The ionizing radiation that is emitted can include alpha particles, beta particles and/or gamma rays. Radioactive decay occurs in unstable atoms called radionuclides [3].Rutherford's experiments demonstrated that there are three main forms of radioactive emissions and nuclear changes.

i. Alpha radiation (α) [1,1a,3,4]

This is symbolized by the Greek letter α. Emission of an alpha particle from an atom's nucleus is called the alpha radiation. An alpha particle is positively charged, contains 2 protons and two neutrons and is similar to He Nucleus (24 He).

- ✰ When an atom emits α particle the atom's atomic mass will decrease by four units (because two protons and two neutrons are lost) and the atomic number (z) will decrease by two units [1,1a].

  The element is said to "transmute" into another element that is two z units smaller. *e.g.* Uranium decays into Thorium by emitting alpha particle [1].

$$^{238}_{92}U \rightarrow\ ^{4}_{2}He +\ ^{234}_{90}Th$$

In nuclear chemistry, element symbols are traditionally preceded by their atomic weight (upper left) and atomic number (lower left).

- ☆ The above reaction is nuclear reaction to emphasize that the change occurs in an atomic nucleus. The **law of conservation of matter** says that matter cannot be created or destroyed. This means we must have the same number of protons and neutrons on both sides of the nuclear equation. The above reaction obeys above law [1a]
- ☆ Alpha particles come from the decay of the heaviest radioactive elements, such as uranium, radium and polonium. Even though alpha particles are very energetic, they are so heavy that they use up their energy over short distances and are unable to travel very far from the atom [3].

ii. Beta Radiation (β) [1,1a,3,4]

It is symbolized by the Greek letter β.Beta particles are small, fast-moving particles with a negative electrical charge that are emitted from an atom's nucleus during radioactive decay. Beta particles are more penetrating than alpha particles [3].

- ☆ The transmutation of a neutron into a proton and an electron followed by the emission of the electron from the atom's nucleus and hence A beta particle is an electron ejected from the nucleus (not from the shells of electrons about the nucleus) [1a].
- ☆ When an atom emits β particle the atom's mass will not change since there is no change in the total of nuclear particles.
    i. There will be an increase in the atomic number by one because of the transmutation of a neutron to an additional proton.
- ☆ Beta particles are emitted by certain unstable atoms such as hydrogen-3 (tritium), carbon-14 and strontium-90. Carbon -14 an isotope of Carbon into the element Nitrogen [1,1a].

$$^{14}_{6}C \rightarrow\ ^{0}_{-1}e +\ ^{14}_{7}N$$

- ☆ Again, the sum of the atomic numbers is the same on both sides of the equation, as is the sum of the mass numbers [1a].

iii. Gamma Radiation (γ) [1,1a,3,4]

It is symbolized by the Greek letter γ. Gamma (γ) radiation consists of photons with a wavelength less than $3x10^{-11}$ meters (greater than 1019 Hz and 41.4 keV) [4].

- Gamma rays (γ) are weightless packets of energy called photons. Unlike alpha and beta particles, which have both energy and mass, gamma rays are pure energy [3].
- It is the emission of electromagnetic energy from an atom's nucleus [1,1a].
- There will be no emission of particles during gamma radiation and gamma radiation does not itself cause the transmutation of atom [1,1a].
- It is often observed that γ radiation during the transmutation of atoms to, α or β radioactive decay. *e.g.* X-rays emitted during the beta decay of Cobalt -60 [1].

$$^{99m}_{43}\mathrm{Tc} \longrightarrow {}^{99}_{43}\mathrm{Tc} + {}^{0}_{0}\gamma$$

γ ray

Figure 8.1: Emission of γ Radiation [1a]

Penetration Power of Radiations [1a]

Alpha, beta, and gamma emissions have different abilities to penetrate matter.

- The relatively large alpha particle is easily stopped by matter (although it may impart a significant amount of energy to the matter it contacts).
- Beta particles penetrate slightly into matter, perhaps a few centimeters at most.
- Gamma rays can penetrate deeply into matter and can impart a large amount of energy into the surrounding matter.

Isotopes [1,4,4a]

The term isotope is formed from the Greek roots isos (4óïò "equal") and topos (τόπος "place"), meaning "the same place"; thus, the meaning behind the name is that different isotopes of a single element occupy the same position on the periodic table. It was coined by a Scottish doctor and writer Margaret Todd in 1913 in a suggestion to chemist Frederick Soddy [4].

- Isotopes are variants of a particular chemical element which differ in neutron number, and consequently in nucleon number. All isotopes of a given element have the same number of protons but different numbers of neutrons in each atom [4]. Isotopes are two or more types of atoms that have the same atomic number and position in the periodic table and that differ in nucleon numbers (mass numbers) due to different numbers of neutrons in their nuclei [4].

- Isotopes may be defined as the atoms having the same nuclear charge but are having different masses. Such atoms have practically same structure of their electrons shells and belong to the same chemical element of the periodic system [1].
- Consequent to the development of mass spectroscopy, it became possible to discover isotopes among the naturally occurring compounds of non radio active elements.
  - **i.** Further with the development of nuclear physics it has become possible to obtain different isotopes for different elements artificially [1].
  - **ii.** Example of Isotopes-
    - **a.** Hydrogen ($_1H$) has three naturally occurring isotopes, sometimes denoted ^1H (contains single proton), ^2H(contains one proton and one neutron) and ^3H (contains one proton and two neutron). Hydrogen is the only element whose isotopes have different names: the ^2H (or hydrogen-2) isotope is **deuterium** and the ^3H (or hydrogen-3) isotope is **tritium [4a]**.
    - **b.** Carbon has (carbon-12, carbon-13, and carbon-14)three isotopes of the element carbon with mass numbers 12, 13, and 14, respectively. The atomic number of carbon is 6, which means that every carbon atom has 6 protons, so that the neutron numbers of these isotopes are 6, 7, and 8 respectively [4b].
- Some isotopes/nuclides are radioactive, and are therefore referred to as radioisotopes or radionuclides, whereas others have never been observed to decay radioactively and are referred to as stable isotopes or stable nuclides. For example, ^{14}C is a radioactive form of carbon, whereas ^{12}C and ^{13}C are stable isotopes [4b].

Nature of Isotopes [1,4]

The composition of Isotopes of naturally occurring elements is markedly constant. Isotopes are different forms of the same element. Carbon and Hydrogen both have three naturally occurring isotopes, as discussed earlier.

- Scientists estimate that the elements that occur naturally on Earth (some only as radioisotopes) occur as 339 isotopes (nuclides) in total. Only 252 of these naturally occurring nuclides are stable in the sense of never having been observed to decay as of the present time.

☆ Isotopes of atoms that occur in nature come in two flavors- stable and unstable (radioactive). Some of the unstable isotopes are only moderately unstable and can therefore still persist in nature today. The isotope 238-U is a good example.

Sources

1. *K.S. Sharma, Physical Chemistry of milk, Agrimoon.*

1a. *Radioactivity and Nuclear Chemistry, CH103: Allied health Chemistry, Western Oregon University, https: //wou.edu.*

2. *Nuclear reaction, https: //www. britannica.com.*

3. *Radiation protection, https: //www.epa. gov/radiation.*

4. *Wikipedia-Isotopes.*

4a. *Wikipedia- Isotopes of Hydrogen.*

4b. *Wikipedia-Isotopes of Carbon.*

Half-Life [1,1a,1b,2]

The term Half Life is commonly used in nuclear physics to describe how quickly unstable atoms undergo, or how long stable atoms survive radioactive decay. The term is also used more generally to characterize any type of exponential or non-exponential decay [2].

☆ The original term, half-life period, dating to Ernest Rutherford's discovery of the principle in 1907, was shortened to half-life in the early 1950s. Rutherford applied the principle of a radioactive element's half-life to studies of age determination of rocks by measuring the decay period of radium to lead-206 [2].

☆ Half-life, in radioactivity, the interval of time required for one-half of the atomic nuclei of a radioactive sample to decay (change spontaneously into other nuclear species by emitting particles and energy), or, equivalently, the time interval required for the number of disintegrations per second of a radioactive material to decrease by one-half [3].

☆ Each radioactive nuclide has a characteristic, constant **half-life** ($t_{1/2}$). An isotope's half-life allows us to determine how long a sample of a useful isotope will be available, and how long a sample of an undesirable or dangerous isotope must be stored before it decays to a low-enough radiation level that is no longer a problem [1b].

☆ The half-life ($T_{½}$) is the amount of time necessary for one-half of the radioactive material to decay. Radioactive decay proceeds according to a principal called the half-life [1,1a].

Example

The following example of radioactive decay of Bismuth is helpful in understanding the halflife [1,1a,1b]-

a. Bismuth

The radioactive element bismuth (^{210}Bi) undergoes alpha decay to form the element thallium (^{206}Tl) with a reaction half-life equal to five days.

i. If we begin an experiment starting with 100 g of bismuth in a sealed lead container, after five days we will have 50 g of Bismuth and 50 g of Thallium in the jar.

ii. After another five days (ten from the starting point), one-half of the remaining bismuth will decay and we will be left with 25 g of Bismuth and 75 g of Thallium in the jar. The reaction proceeds in halfs, with half of whatever is left of the radioactive element decaying every half-life period [1,1a].

b. Cobalt-60

Cobalt-60, an isotope that emits gamma rays used to treat cancer, has a half-life of 5.27 years. In a given cobalt-60 source, since half of the $^{60}_{27}Co$ nuclei decay every 5.27 years, both the amount of material and the intensity of the radiation emitted is cut in half every 5.27 years. Thus, a cobalt-60 source that is used for cancer treatment must be replaced regularly to continue to be effective.

- ☆ The decay reaction and T½ of a substance are specific to the isotope of the element undergoing radioactive decay. The example given in following table [1,1a,1b]:

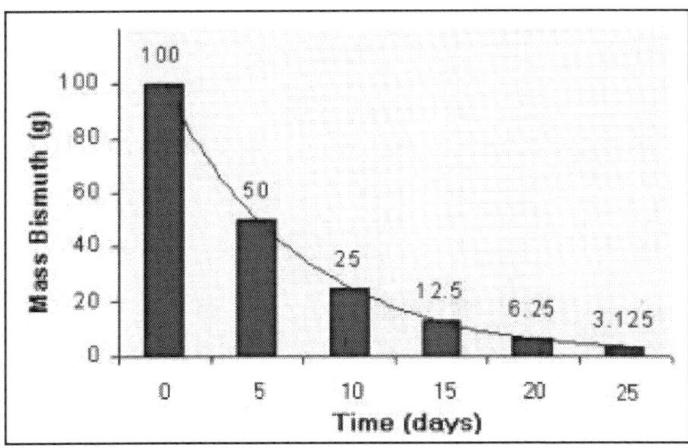

Figure 8.2: Radioactive Decay of Bismuth-210 (T 1/2=5 DAYS).

Radio Active Decay

- ☆ Bi210 can undergo decay to Tl206 with a T½ of five days.
- ☆ Bi215, by comparison, undergoes decay to Po215 with a T½ of 7.6 minutes,
- ☆ Bi208 undergoes yet another mode of radioactive decay (called electron capture) with a T½ of 368,000 years!

Radioactive Decay of different Isotopes of Bismuth.

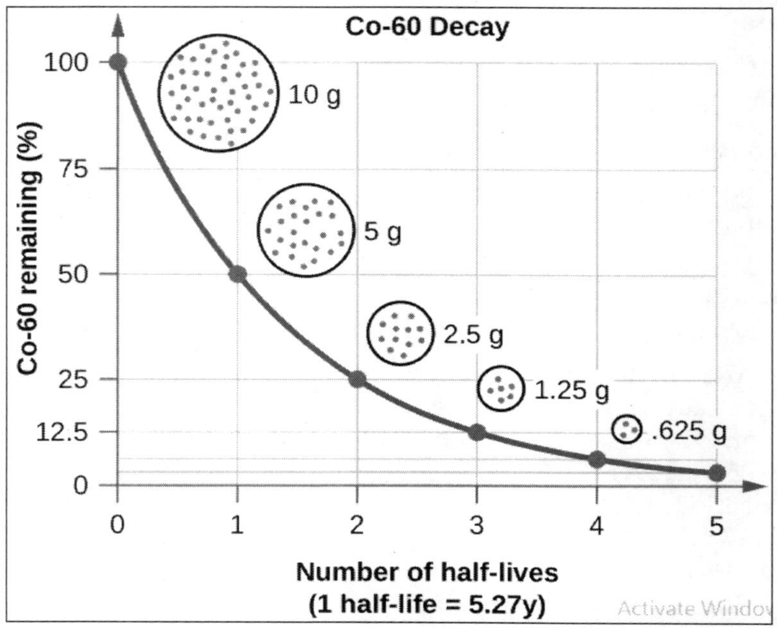

Figure 8.3: Cobalt- 60 Decay.

Calculation of Remaining Parent Material [1,1a,1b,2]

The following equation is used for calculation of fraction of parent material that remains after radioactive decay:

$$\text{Fraction remaining} = \frac{1}{2n}$$

where, n = number of half-lives elapsed.

The amount of a radioactive material that remains after a given number of half-lives is therefore:

Amount remaining = Original amount x Fraction remaining.

Half-life is constant over the lifetime of an exponentially decaying quantity, and it is a characteristic unit for the exponential decay equation. The following table shows the reduction of a quantity as a function of the number of half-lives elapsed [2]:

Table 8.1: Reduction of a Quantity as a Function of the Number of Half-Lives

Number of Half-lives Elapsed	Fraction Remaining	Percentage Remaining
0	1/1	100
1	1/2	50
2	1/4	25
3	1/8	12.5
4	1/16	6.25
5	1/32	3.125
6	1/64	1.5625
7	1/128	0.78125
n	$1/2^n$	$100/2^n$

Calculation of Half Life of Isotopes [1,1a,3,4]

The formula useful for calculating the half life of radio isotopes is

Nt = No x (0.5) number of half-lives

where, Nt = amount of radioisotope remaining, No = original amount of radioisotope, number of half-lives = time ÷ half-life.

The relationship between the half-life, T1/2, and the decay constant is given by $\mathbf{T_{1/2} = 0.693/\lambda}$. This formula is also written as follows which can be used to calculate half life of radioactive element [4]:

$$T_{1/2} = 0.693 \frac{\text{Number of atoms}}{\text{Number of decay}}$$

Table 8.2: Calculated Half Life of Radio Active Isotopes [1]

Sl.No.	Number of Half-lives	Time (Years)	Per cent Strontium -90 Remaining	Per cent Strontium -90 that has decayed
1	0	0	100	0
2	1	28	50	50
3	2	56	25	75
4	3	84	12.5	87.5
5	4	112	6.25	93.75
6	5	140	3.125	96.875
7	6	168	1.5625	98.4375

Half Life of Radioactive Elements [1]

The half life of some of the commonly used is being shown here under:

Table 8.3: Half Life of some Elements

Sl.No.	*Radioactive element*	*Isotopes*
1	Barium	Ba137m - 2.552 minutes, Ba 139 - 82.7 minutes, Ba-140 - 12.74 days, Ba-141.
2	Carbon	C-11 - 20.38 minutes, C-14 - 5730 years
3	Cesium	Cs-134 - 2.062 Years, Cs-134m - 2.90 Hours, Cs-135 - 2.3E6 Years, Cs-136 - 13.1 Days, Cs-137 - 30.0 years, Cs-138 - 32.2 minutes.
4	Iodine	I-123-13.2 hours,I-125 - 60.14 days,I-129 -1.57E7 years, I-130 - 12.36 hours I-131-8.04 days,I-132 - 2.30 hours,I-133 - 20.8 hours,I-134 - 52.6 minutes,I-135 - 6.61 hours.
5	Plutonium	Pu-238-87.74Years, Pu-239-24065 years, Pu-240-6537 years, Pu-241 - 14.4 years, Pu-242 -3.76E5 years, Pu-243 - 4.956 hours, Pu-244 - 8.26E7 years.
6	Potassium	K-40-1.27E9Years,K-42 - 12.36 Hours, K-43 - 22.6 hours
7	Radium	Ra-223-11.434 Days,Ra-224-3.66 Days,Ra-225-14.8 days, Ra-226 - 1600 years, Ra-228 - 5.75 years.
8	Rubidium	Rb-86 - 18.66 Days, Rb-87 - 4.7E10 Years, Rb-88 - 17.8 Minutes, Rb-89 - 15.2 minutes.
9	Selenium	Se-75 - 119.78 Days, Se-79 - 65000 years.
10	Sodium	Na-22 - 2.602 Years, Na-24 - 15.00 hours.
11	Strontium	Sr-85-64.84 Days,Sr-87m- 2.81 Hours, Sr-89 - 50.5 Days, Sr-90 - 29.12 years Sr-91 - 9.5 Hours, Sr-92 - 2.71 hours.
12	Sulfur	S-35 - 87.44 days.
13	Thallium	Tl-201 - 73.06 hours, Tl-207 - 4.77 minutes, Tl-208 - 3.07 minutes, Tl-209 - 2.20 minutes.
14	Thorium	Th-227-18.718 Days,Th-228- 1.913 Years, Th-229 - 7340 years, Th-230-7.7E4 years,Th-231 -25.52 hours,Th-232 - 1.41E10 years,Th-234 - 24.10 days.
15	Tin	Sn-119m- 293.1 days, Sn-123-129.2 days,Sn-125-9.64 days, Sn-126 - 1.0E5 years.
16	Tungsten	W-181-121.2 days,W-185-75.1 days, W-187 - 23.9 hours.
17	Uranium	U-232 - 72 Years, U-233 - 1.59E5 years, U-234 - 2.445E5 years, U-235 - 7.03E8 years, U-236 -2.34E7 years, U-237 - 6.75 days, U-238 - 4.47E9 years.
18	Vanadium	V-48 - 16.238 days.
19	Xenon	Xe-131m-11.9 days,Xe-133-5.245 days, Xe-133m - 2.188 days, Xe-135 - 9.09 hours, Xe-135m -15.29 minutes, Xe-138 - 14.17 minutes.
20	Zinc	Zn-65 - 243.9 days, Zn-69 - 57 minutes.

Note- Some of the longer half-lives are written in scientific notation (*i.e.,* 7.2E1 is equal to 7.2 x 10, or 72.).

Sources

1. *K.S. Sharma, Physical Chemistry of milk, Agrimoon.*

1a. *Half life period of Radio Isotopes and Measurement of radio activity, Lesson-29, http: //ecoursesonline.iasri.res.in.*

1b. *Radioactivity and Nuclear Chemistry, CH103: Allied health Chemistry, Western Oregon University, https: //wou.edu.*

2. *Wikipedia-HalfLife.*

2a. *Wikipedia-Isotopes of Hydrogen.*

3. *Half life radioactivity, Encyclopaedia Britannica.*

4. *Decay constant, Nuclear physics, https://www.britannica.com.*

Types of Nuclear Reactions [1,2]

There are following four main types of nuclear reactions:

1. Nuclear Decay (Radioactive Decay) [1,2]

This Occurs when an unstable atom loses energy by emitting ionizing radiation. Radioactive decay is a random process at the level of single atoms, in that, according to quantum theory, it is impossible to predict when a particular atom will decay. There are many types of radioactive decay [2].

The decay reaction and T½ of a substance are specific to the isotope of the element undergoing radioactive decay. The details related to the decay of Bismuth has already discussed (See earlier figure showing decay of Bismuth 210, 215 and 208) [1].

2. Stimulated Nuclear Reactions (Fusion and Fission) [1,1a]

Many elements undergo radioactive decay naturally. Nuclear reactions can also be stimulated artificially. Although these reactions also occur naturally but the most familiar are stimulated reactions.

- ☆ There are two such types of nuclear reactions *e.g.* Nuclear Fission and Nuclear Fusion. Nuclear fission and nuclear fusion are two types of nuclear reactions [1].
- ☆ Fission means to split and fusion means to combine. Therefore in fission reaction a heavy nucleus of atom is bombarded so that it splits and in fusion reaction two lighter nuclei are heated at high temperature so that they combine to form a heavy nucleus [5]. The details of Nuclear Fission and Nuclear Fusion are given below:

a. Nuclear Fission [1,1a,4,5]

It is a nuclear reaction or a radioactive decay process in which the nucleus of an atom splits into two or more smaller, lighter nuclei.

- ☆ The fission process often produces gamma photons, and releases a very large amount of energy even by the energetic standards of radioactive decay [4].
- ☆ Most commonly this is done by "firing" a neutron at the nucleus of an atom. The energy of the neutron "bullet" causes the target element to split into two (or more) elements that are lighter than the parent atom [1].
- ☆ Often, fission produces excess neutrons that will sometimes be captured by other nuclei, possibly inducing additional radioactive events [1a].

For Example

When uranium-235 atoms are bombarded with neutrons then the heavy uranium nucleus splits to produce barium-139 and krypton-94 with the emission of three neutrons [5].

- ☆ During the fission of U235, three neutrons are released in addition to the two daughter atoms. If these released neutrons collide with nearby U235 nuclei, they can stimulate the fission of these atoms and start a self-sustaining nuclear chain reaction. This chain reaction is the basis of nuclear power. As uranium atoms continue to split, a significant amount of energy is released from the reaction. There is two types of Nuclear Chain Reactions. Concept simulation – Reactions controlled and uncontrolled nuclear chain reactions [1].

Application of Nuclear Fission [1,1a,5]

Fission is the radioactive process used in nuclear power plants and one type of nuclear bomb [1a]. The following two major applications of nuclear fission are discussed here:

i. **Nuclear power plant**-The heat released during this reaction is harvested and used to generate electrical energy. Nuclear fission reactions are done to generate electricity at nuclear power plants.

ii. **Atom Bomb-** Nuclear bomb is based on the nuclear fission reaction of uranium-235 and plutonium-239.

- ☆ The fission reaction is deliberately allowed to go out of control so as to produce large amount of energy in a very short while. The atom bombs based on nuclear fission of uranium-235 ad plutonium-239 were dropped on Japanese cities of Hiroshima and Nagasaki in 1945 during the Second World War. This caused tremendous loss of human life.

b. Nuclear Fusion [1,5]

The meaning of fusion is to join or to combine. Therefore, the process in which two or more nuclei are combined to form a heavy nucleus is nuclear fusion. During the process energy is being released.

☆ The nuclei of atoms are positively charged and thus they repel each other. So in order to combine or fuse these two nuclei to form one heavy nucleus a lot of heat energy and high pressure is required. This shows that nuclear fusion is carried out by heating lighter atoms to an extremely high temperature at high pressure. Some mass is also lost in this process which gives tremendous amount of energy [5].

Example

A good example is the fusion of two "heavy" isotopes of hydrogen (deuterium: 2H and tritium: 3H) into the element helium [1]. When deuterium atoms are heated to an extremely high temperature under high pressure than two deuterium nuclei combine to form helium which has a heavy nucleus, a neutron is emitted and a lot of energy is liberated.

☆ **Nuclear fusion reaction is opposite of nuclear fission reaction. The energy produced in nuclear fusion reaction has not been controlled yet and is much more than nuclear fission reaction [5].**

Application of Nuclear Fusion [1,3,4,5]

The following applications of Nuclear Fusion are discussed here-

i. Fusion reactions release tremendous amounts of energy and are commonly referred to as **thermonuclear reactions because these reactions occur at e**xtremely high temperature. Although we think of the sun as a large fireball, the sun (and all stars) is actually enormous fusion reactors. Stars are primarily gigantic balls of hydrogen gas under tremendous pressure due to gravitational forces. Hydrogen molecules are fused into helium and heavier elements inside of stars, releasing energy that we receive as light and heat [1].

ii. **Hydrogen Bomb-** This reaction is used in producing hydrogen bomb which causes mass destruction. Isotopes of hydrogen, deuterium (2H) and tritium (3H), along with an element lithium-6 is used in making a hydrogen bomb.

☆ The explosion of hydrogen bomb is done by using an atom bomb. This is because when atom bomb is exploded then its fission reaction produced a lot of heat which raises the temperature of deuterium and tritium in a few microseconds. Thus the fusion reaction takes place and hydrogen bomb is exploded producing enormous energy. Hydrogen bomb causes destruction of life [5].

3. Transmutation [3,4]

Transmutation, conversion of one chemical element into another. A transmutation entails a change in the structure of atomic nuclei and hence may be induced by a nuclear reaction (q.v.), such as neutron capture, or occur spontaneously by radioactive decay, such as alpha decay and beta decay. An example is the natural decay of potassium-40 to argon-40, which forms most of the argon in the air [3,4].

A transmutation can be achieved either by nuclear reactions (in which an outside particle reacts with a nucleus) or by radioactive decay, where no outside cause is needed. There are following two types of transmutation [4]:

i. **Natural Transmutation-** One type of natural transmutation observable in the present occurs when certain radioactive elements present in nature spontaneously decay by a process that causes transmutation, such as alpha or beta decay. An example is the natural decay of potassium-40 to argon-40, which forms most of the argon in the air.

ii. **Artificial Transmutation-**Artificial transmutation may occur in machinery that has enough energy to cause changes in the nuclear structure of the elements. Such machines include particle accelerators and tokamak reactors. Conventional fission power reactors also cause artificial transmutation, not from the power of the machine, but by exposing elements to neutrons produced by fission from an artificially produced nuclear chain reaction.

Sources

1. *K.S. Sharma, Physical Chemistry of milk, Agrimoon.*

1a. *Radioactivity and Nuclear Chemistry, CH103: Allied health Chemistry, Western Oregon University, https: //wou.edu.*

2. *Reactor Physics, https: //www.reactor-physics.com.*

3. *Transmutation, https: //www. britannica.com.*

4. *Wikipedia.*

5. *Nuclear Fission and Fusion, https: //www. kruss-scientific.com.*

Measurement of Radioactivity [1,1a,2,3]

Radioactive material will lose its radio activity by emitting the radiation into the atmosphere. Consequently the ability to emit radiation by a given radioactive material progressively gets reduced. The time taken for half of the radionuclide's atoms to decay is commonly referred as half life. This property is related to decay constant. The radioactivity of any radionuclide is based on its half-life.

- ✰ The radioactive materials are of use in several research activities especially to monitor the movement of substances across the body and as a diagnostic tool for some of the abnormalities in the body.
- ✰ These radio isotopes are also of some use in the dairy field, because of the fact that they are often found in milk in traces.

The nuclides are high radioactive or weak radioactive. **A radio nuclide is said to be highly radioactive if the half life is short while a radionuclide having longer half-life is considered to radioactively weak.** There is wide variation in half-life of known radionuclides which ranges from 10^{19} years to 10^{-23} seconds [1].

Unit of Measurement of Radioactivity [2,3]

The units of measure for radioactivity are the curie (Ci) and becquerel (Bq).

- ☆ Radioactivity is measured in disintegrations per second and its unit of measure is the Becquerel (Bq), in honour of the physicist Henry Becquerel. Exposure describes the amount of radiation traveling through the air.
- ☆ Many radiation monitors measure exposure. The units for exposure are the roentgen (R) and coulomb/kilogram (C/kg). The famous **Geiger-Mueller tube, commonly called a Geiger counter,** is designed to measure the electrical response produced by the newly formed ions [2,3].

Radiation Detection [1,1a,4]

The different principles used for detection of radiation are given below:

1. The Oldest Principle

This is the oldest principle for detection of radiation. This principle for detecting radiation is by darkening photographic emulsion and it is used in the personnel dosimetry.

- ☆ The film badge consists of a radiation sensitive film held in a plastic holder. Filters of copper and lead are attached to the holder to differentiate exposure from different types and energies of radiation.
- ☆ It is cost effective and very popular for personnel monitoring and provides reasonably accurate readings of exposures from beta, gamma and x-ray radiations.

2. Thermoluminiscence

It is another principle for detecting the radiation. Thermoluminiscent dosimeters (TLD) work on this principle and are mostly used as finger dosimeters.

- ☆ Several inorganic crystals (*e.g.* LiF) can accumulate radiation energy and hold it. If the crystal is heated from 300 to 400°C, it emits light in amounts proportional to the absorbed energy. These inorganic crystals are held in a plastic holders and plastic rings in TLD.

  i. It gives an accurate exposure reading and can be reused.

3. Converting the Energy of Radiation to Electric Current [1,4]

Converting the energy of radiation to electric current is also used for detecting the radiation. There are two basic principles which are based on ionization and excitation.

i. **Principle based on Ionisation:** First principle is based on ionization of gas molecules.

ii. **Principle based on excitation**: The second is based on excitation and ionization of solid, liquid or plastic material in a Scintillator, which emits photons of light after absorbing radiation. Light is then converted to the electric current by means of photomultiplier tube, and electric current is measured.

The ionisation and Excitation are well described in following figure [4]-

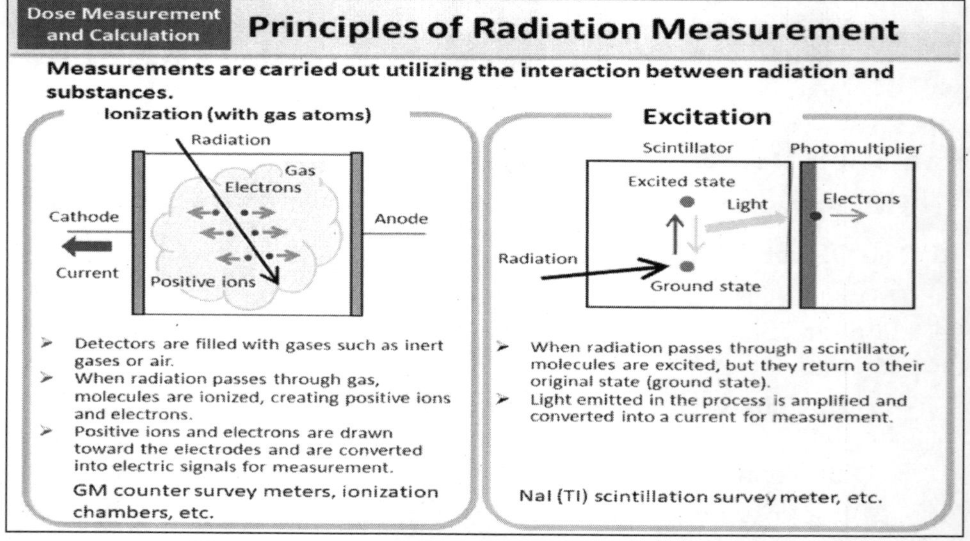

Figure 8.4: Principles of Radiation Measurement.

A. Gas-Filled Detectors [1,1a,4]

In gas-filled detectors voltage is applied between two electrodes and the ion pairs formed due to excitation are collected as a current. The measured current is proportional to the applied voltage and the amount of radiation.

☆ Gas-filled detectors are cylindrical chambers with a central wire filled with air or different gases. These detectors are primarily used for measuring high intensity radiation.

i. Application of Lower Voltage

In ionization chambers lower voltage ranging from 50 to 300 V is applied resulting in the formation of primary ion pairs by the initial radiation.

- **Dose calibrators and pocket dosimeters** are the common ionization chambers used in nuclear medicine. The dose calibrator is one of the most essential instruments in nuclear medicine for measuring the activity of radionuclides and radiopharmaceuticals.
- It must be regularly checked for constancy, accuracy, linearity and geometry.

ii. Application of High Voltage

At higher voltages from 1000 to 1200 V, the current becomes identical regardless of how many ion pairs are produced by the incident radiation. **Geiger-Müller counters** operate in this region.

- Geiger-Müller counters are used to monitor the radiation level in different work areas and they are called area monitors or survey meters. Geiger-Müller counters are more sensitive than ionization chambers but cannot discriminate between energies. They are almost 100 per cent efficient for counting alpha and beta particles but have only 1 to 2 per cent efficiency for counting gamma and x rays [1].
- Geiger Muller (GM) counter survey meters and ionization chambers utilize the ionization between radiation and gas atoms. Ionization effect refers to the process in which radiation ejects electrons from nuclei in a substance.
- Detectors of GM counter survey meters and ionization chambers are filled with gases. When radiation passes inside a detector, it causes ionization of gas atoms, separating atoms into positive ions and electrons. Separated electrons and positive ions are attracted to the electrodes, causing a current to flow. This is converted into electric signals, which are then measured as the amount of radiation [4].

B. Scintillation Detectors [1,1a,4]

This types of detectors consist of scintilator emitting flashes of light after absorbing gamma or x radiation [1].

- Organic scintillators such as anthracene, stilbene, and many of the commercially available liquid and plastic scintillators are being used.
- NaI (Tl) scintillation survey meters utilize excitation with substances. Radiation gives energy to electrons of nuclei, and when an electron jumps to an outer orbit, this phenomenon is called excitation. An atom in that state is unstable (excited), and when it returns to a stable state (ground state), it gives off energy in the form of light. This is called the excitation effect [4].

Process

i. A photomultiplier tube converts the emitted light photons to an electrical pulse. A linear amplifier is used for amplification of electric pulses.

ii. The amplified pulses are sorted by a pulse-height analyzer and then registered as a count.

- ✰ Aside from NaI (Tl) scintillation survey meters, germanium semiconductor detectors also utilize the excitation effect for radiation measurement [4].
- ✰ Different solid or liquid scintillators are used for different types of radiation.
- ✰ In nuclear medicine, sodium iodide solid crystals with a trace of Thallium sodium iodide (NaI(Tl)) are used for gamma and x ray detection [1].

Sources

1. *K.S. Sharma, Physical Chemistry of milk, www.Agrimoon.com.*

1a. *Half life period of Radio Isotopes and Measurement of radio activity, Lesson-29, http: //ecoursesonline.iasri.res.in.*

2. *Radioactivity measurement, http: //www. eniscuola.net.*

3. *Measuring radiation, https: //www.nrc.gov.*

4. *Principle of radiation measurement, https: //www.env.go.jp/en/chemi.*

Occurrence of Radio Nuclides in Milk and Milk Products [1,1a,2,3]

Radio nuclides are often secreted into the milk by the consumption of contaminated feed and water and are ultimately found in milk and milk products [1,1a].

- ✰ Measurements of radioactivity in environment and in foodstuffs are extremely important for controlling radiation levels to which mankind is direct or indirectly exposed.
- ✰ Besides natural radionuclides, due to several nuclear weapon tests and numerous nuclear reactor accidents, various artificial radioactive elements were introduced in the biosphere [3].
- ✰ Another important fact is that, importation of contaminated food from any region that suffered a nuclear accident, can be indirectly affect people health around the world [3].

Source of Radio Nuclides in Milk [1,1a,2]

There are different pathways through which radionuclides enter the human body. They travel through the air, in water (both ground water and surface water), and through the food chain.

- Radionuclides may enter the human body by eating or drinking, by inhalation, or by absorption through the skin [1,1a,2].
- A general path of such contamination could be diagrammatically represented as below [2]-

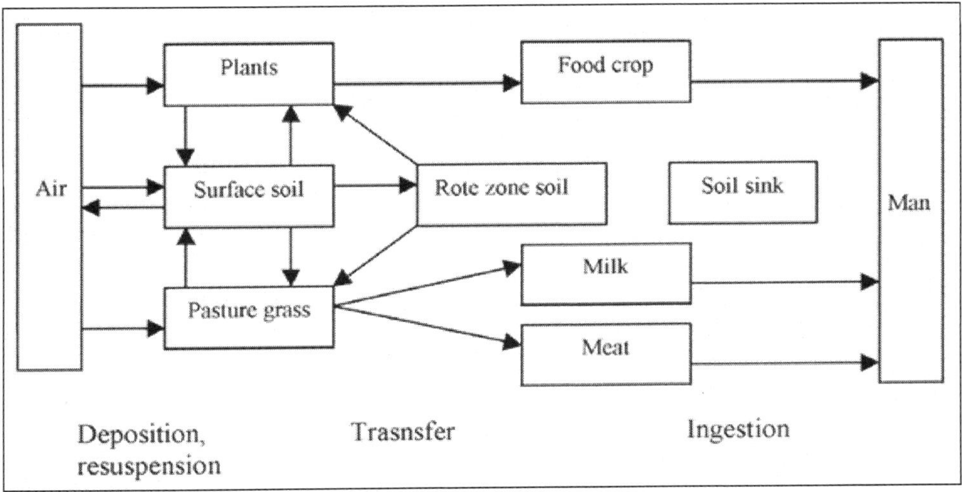

Figure 8.5: A Typical Set of Pathways through the Environment from a Radiation Source Exposed to the Air.

Radionuclides sources on human diet are come mainly from milk, meat and other products more frequently consumed.

- Powdered milk, is an important ingredient used to make cookies, ice cream, yogurt, chocolate, powdered chocolate and it was found highly contaminated with radionuclides. In a global scale, radio-nuclide elements coming form the fallout incorporates in biosphere [2].
- Environmental radionuclide contamination is primarily man made. Nuclear explosions, manufacture and processing of nuclear fuel, accidents and or operation of nuclear reactors, increasing use of nuclear and disposal of radioactive waste will discharge radionuclide elements into the biosphere [1,1a].
- These elements present in atmosphere contaminate plants, soil and water and by different ways contaminate the environment [2].When cows graze ingesting grass from contaminated soil radionuclides enters milch animals. Even if milch animals are kept indoors, contamination of milk may occur by inhalation of radionuclides or ingesting them through contaminated drinking water and feed. Milk from goat and

sheep, should be checked periodically over a longer period because of their ranch system of grazing [1,1a].

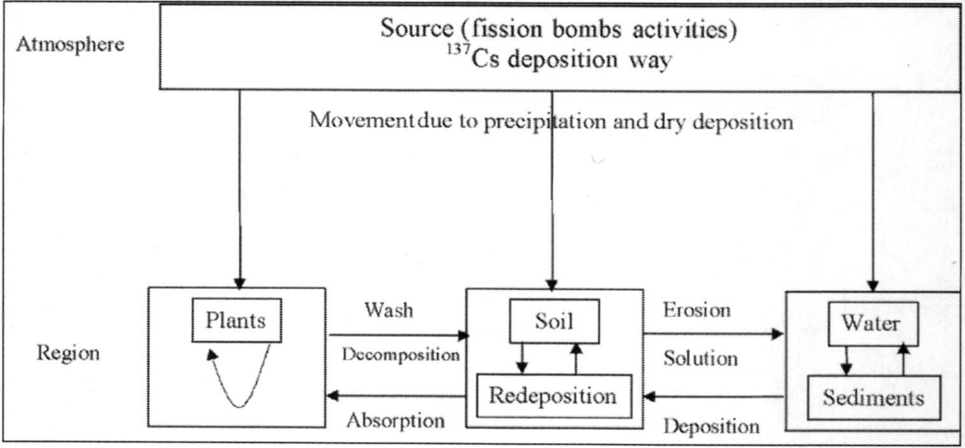

Figure 8.6: 137 Cs Incorporation Cycle Diagram in the Nature [2].

Vegetation when decomposed will result in the entry of radionuclides into the soil. Erosion of soil by rain, rivers *etc.*, would result in the contamination of water. Sediments from contaminated water will in turn contaminate the soil. The plants growing in such a soil absorb the radionuclides.

Sources

1. *K.S. Sharma, Physical Chemistry of milk, Agrimoon.*

1a. *Occurrence of radio nuclides in Milk products, Lesson-30, http: //ecoursesonline. iasri.res.in.*

2. *Fabio Luiz Melquiades and Carlos Appoloni, Natural radiation level in powdered milk samples, Ciência e Tecnologia de Alimentos, 24(4), 2004.*

3. *F.L.Melquiades, and C.R. Appoloni, 40K, 137Cs and 232Th activities in brazilian milk samples measured by gamma ray spectrometry. Indian Journal of Pure and Applied Physics. v. 40, p. 5-11, 2002.*

Common Radionuclides found in Milk [1, 2,4]

Since the end of Second World War in 1948, research works on radionuclide contamination of food in the environment, and its transfer mechanism and pathway to animals and human population have been reported with vigor (ICPR, 1993 and 2000). milk has been one of the staple foodstuff products that was featured prominently in the food items studied.

A **radionuclide** (radioactive nuclide, radioisotope or radioactive isotope) is a nuclide that has excess nuclear energy, making it unstable. This excess energy can be used in one of three ways [1a]:

i. Emitted from the nucleus as gamma radiation.
ii. Transferred to one of its electrons to release it as a conversion electron.
iii. Or used to create and emit a new particle (alpha particle or beta particle) from the nucleus.

- ^{134}Cs, ^{137}Cs, ^{137m}Ba, ^{131}I, ^{89}Sr, ^{90}Sr, ^{238}Pu, ^{239}Pu, ^{241}Am, ^{242}Cm and tritium are the most important radionuclides associated to internal radiation exposure (due to ingestion of contaminated water and food by the animal) and the contamination of the environment.
- With regard to milk, radionuclides of specific interest are ^{134}Cs, ^{137}Cs, ^{131}I, ^{89}Sr, ^{90}Sr. As they decay at different rates, the doses from these elements and other radionuclides are delivered in different manners. The distribution of Radionuclides in milk and milk products is given in following table:

Table 8.4: Distribution of Radio Nuclides in Milk and their Half Life [3]

Sl.No.	*Radio Nuclide*	*Physical Half Life**	*Biological Half Life***	*Distribution*
1	^{89}Sr	52 Days	1-50 years	80 per cent in casein micelles
2	^{90}Sr	28 years	1-50 years	80 per cent in casein micelles
3	^{131}I	8 days	100 days	Serum
4	^{133}I	21 hrs	-	-
5	^{137}Cs	33 years	230 days	Serum
6	^{140}Ba	13 days	< 3 years	plasma

* Physical half life is the period needed to reduce the concentration/amount of radioactive emission by an isotope to half of its original level, outside the biological system (*i.e.* in-vitro). ** Biological half life is the time taken for a radioactive isotope to be reduced to half of the amount in the body of after ingestion by an animal/person.

Atomic Energy Regulation Board (AERB) Standards for Radio Activity [1,2,4]

AERB standards for radioactivity are based on an annual exposure of 10 mrem when all items of food are contaminated to the recommended levels and intake is throughout the year [1,2].

- From milk and milk products alone exposure would be 1.6MR/year which is quite small compared to the natural and other radiation exposure that is received during the day to day activities [1,2].
- Milk and milk products are principal pathways for radiation contamination in the event of a radiological emergency [4]. Radio nuclides in milk and milk products observed in different countries are invariably lower than that of recommended limits in those countries. Presence of radio nuclides in milk and milk products varies from country to country.

The details related to the presence of radionuclides in milk and milk products are given below [1,2,4]-

a. Radionuclides in Milk

Strontium

It accumulates in bone and remains there for a long time.

i. ^{90}Sr uptake causes a prolonged exposure to harmful radiation as it has a long physical half life. It is distributed in milk in much similar way to calcium.

ii. However, strontium phosphate in milk is less soluble than calcium phosphate and a greater proportion of it associated with the casein micelles.

iii. The mean value for 90Sr is reported to be 3.9pCi/l.

Iodine

It accumulates in the thyroid gland where it attains high concentration and radioactive isotopes of ^{131}I are considered hazardous as they produce intense radiation.

i. Small portions of ^{131}I are being strongly bound to protein while about 2 per cent of it is found in the fat globules.

ii. ^{131}I present in milk is to an extent of 20 Bq/l.

Cesium^{137}Cs

It chemically behaves like Na^+ and K^+ ions. In India ^{137}Cs has been reported to be 0.02 Bq/l in milk.

i. The ratio of $^{90}Sr/^{137}Cs$ in European milk is 1.8 while it is 0.5 -1.58 in Indian milk.

Barium

It behaves more nearly like calcium.

Potassium^{40}K

It is some what more hazardous due to higher energy of radiation compared to ^{137} Cs.

i. Natural radioactivity due to ^{40}K present in milk is in the range of 30-40 Bq/l in Indian milk and 50-80 Bq/l in foreign milk.

ii. It has also been reported that cow milk contains 44 Bq/l and buffalo milk contains 38Bq/l.

b. Radionuclides in Milk Products

Potassium^{40}K

Milk powder-Natural radioactivity due to ^{40}K present in milk powder is in the range of 600 - 800 Bq/kg.

Butter-Since potassium salts are very soluble 40K being retained in liquid phase in the preparation of butter and it contains only 1.3Bq/kg.

Cheddar cheese- It is reported to have 41Bq/kg of 40K.

Cesium^{137}Cs

Milk powder-In India ^{137}Cs has been reported to be 0.6 Bq/l kg in milk powder which is far below than the prescribed levels of 30Bq/l in milk and 210 Bq/l kg in milk powder.

Dairy Products-Most of the dairy products in general contain ^{137}Cs to the extent of 40Bq/1 kg.

☆ The problem of radioactive pollution is extremely urgent in Russia. 363 samples of milk, dry milk, butter, cheese and yogurt from Novosibirsk region were examined. Cs-137 level was 3.7. 9.2 times higher than Sr-90 one in milk, cheese and yogurt. At the same time the level of these radio nuclides in butter was identical (8.03 Bk/kg) [4].

Sources

1. *K.S. Sharma, Physical Chemistry of milk, Agrimoon.*

1a. *Wikipedia-Radionuclide.*

2. *Occurrence of Radio nuclides in milk products, http: //ecoursesonline.iasri.res.in.*

3. *P. Walstra and R. Jenness, Dairy chemistry and Physics, John Wiley and Sons, New York- ISBN 9780471097792 - 467, 1984.*

4. *Marmuleva, N. I.; Barinov, E. Ya.; Petukhov, V. L., Radionuclides accumulation in milk and its products, Journal de Physique IV (Proceedings), Volume 107, Issue 1, May 2003, pp.827-829.*

Chapter 9

Molecular Spectroscopy

Molecular spectroscopy is the measurement of interactions between electromagnetic waves and matter.

- ☆ The scattering of the sunlight produces a colorful spectrum when a narrow beam of light is passed through a triangular glass prism.
- ☆ This visible light is simply a part of the whole spectrum of electromagnetic radiation, extending from radio waves to cosmic rays.

Spectrum of Electromagnetic Radiation [1,5]

The energy from the Sun that reaches the Earth arrives as solar radiation, part of a large collection of energy called the electromagnetic radiation spectrum. Solar radiation includes visible light, ultraviolet light, infrared, radio waves, X-rays, and gamma rays [2].

- ☆ A spectrum comprising the different electromagnetic radiations according to energy, frequency, or wavelength is called an electromagnetic spectrum.
- ☆ Our eye perceives the difference in the colour of various compounds and things for example chlorophyll is green, quinone is yellow, 2,4-dinitrophenylhydrazone derivatives of aldehydes and ketones range in color from bright yellow to deep red and aspirin is colourless.
- ☆ Our eye is working like a spectrophotometer analysing the light reflected from the surface of solid or passing through a liquid.
- ☆ The most used electromagnetic radiation by humans is visible light. Our eye can see objects by the light rays that are reflected by these objects. Sunlight looks like as single white colour but in fact it is

composed of a wide range of radiation having different wavelengths in the ultraviolet, visible and infrared portions of the spectrum [1,5].

- ☆ Electromagnetic radiation (EMR) consists of waves of the electromagnetic (EM) field, propagating through space, carrying electromagnetic radiant energy [2a].
- ☆ Electromagnetic radiation is energy which is transmitted at the speed of light through oscillating electric and magnetic fields. The electromagnetic spectrum of an object is the characteristic distribution of electromagnetic radiation emitted or absorbed by that particular object. In principle this spectrum is infinite and is continuous [1,5].

Classification of Electromagnetic Radiation [1,5,2a]

The sunrays bend in accordance to their wave lengths when these are passed through a prism.

- ☆ Electromagnetic radiation (EM radiation or EMR) refers to the waves (or their quanta, photons) of the electromagnetic field, propagating (radiating) through space, carrying electromagnetic radiant energy [2a,3]. It includes radio waves, microwaves, infrared, (visible) light, ultraviolet, X-rays, and gamma rays [2a,4].

  The electromagnetic spectrum comprises the span of all electromagnetic radiation and consists of many subranges, commonly referred to as portions, such as visible light or ultraviolet radiation.
- ☆ The various portions bear different names based on differences in behaviour in the emission, transmission, and absorption of the corresponding waves and also based on their different practical applications. There are no precise accepted boundaries between any of these contiguous portions, so the ranges tend to overlap [2b].
- ☆ Electromagnetic radiation such as visible light is commonly treated as a wave phenomenon, characterized by a wavelength or frequency.
  - **i.** Wavelength is defined as distance between adjacent peaks (or troughs), and may be designated in meters, centimeters or nanometers (10^{-9} meters).
  - **ii.** Frequency is the number of wave cycles that travel past a fixed point per unit of time, and is usually given in cycles per second, or hertz (Hz) [1,5].

Generally wave length of electromagnetic radiation (EMR) is the characteristic on which EMR is classified. Visible wavelengths range from approximately 400 to 800 nm. The spectrum along with the wave length for the visible light is shown in following figure [2a]-

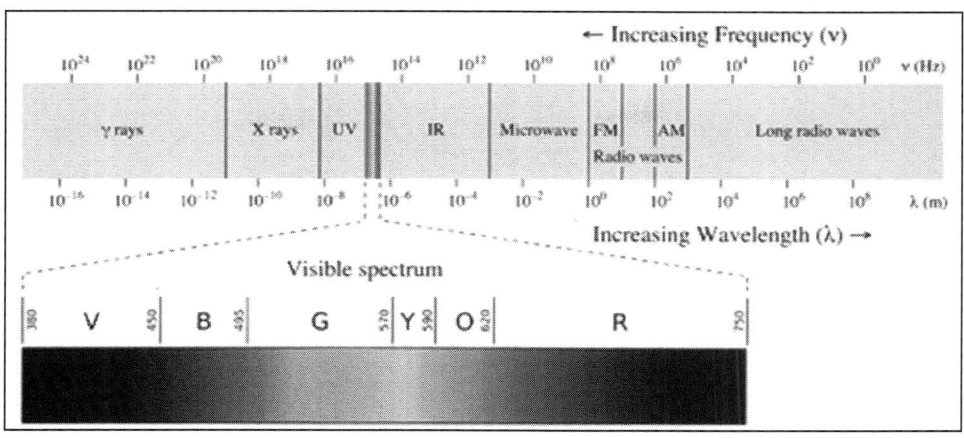

Figure 9.1: Wavelength in nm-visible Spectrum [2a].

It is revealed from the aforesaid spectrum [2a] that red colour has the longest wavelength while violet has the shortest wavelength and the remaining colours are in decreasing order of wavelengths.

This radiation is perceived as the visible light, ultraviolet (UV) rays, x-rays and gamma rays. The behaviour of electromagnetic radiation depends on its wavelength and on the amount of energy it carries with when it interacts with single atoms and molecules. This energy is measured per quantum (photon).

Applications of Electromagnetic Waves [1a,2a]

The application of aforesaid electromagnetic waves are summarised in following table [1a]:

Table 9.1: Applications of Electromagnetic Waves

Sl.No.	*Types of Vaves*	*Application*
1	Visible light portion	Helps in all visual aids and see all objects, colour. Photosynthesis, Human vision.
2	Radio waves	Useful for wireless television, radio and mobile. Communications and remote controls
3	Microwaves	Communications, ovens, radar
4	Infrared	Thermal imaging, heating.
5	X-rays	Medical Security. Proved to be useful in medical field for detecting ailments or deformities in bones.
6	Gamma rays	Nuclear medicine, Security. Useful for ionization purposes and for nuclear medicine.
7	High Ultraviolet radiation	Provide energies to ionize the atoms causing chemical reactions.

Electromagnetic Spectrum

Spectrum of electromagnetic radiation is shown in following figure [5] and table [5a]-

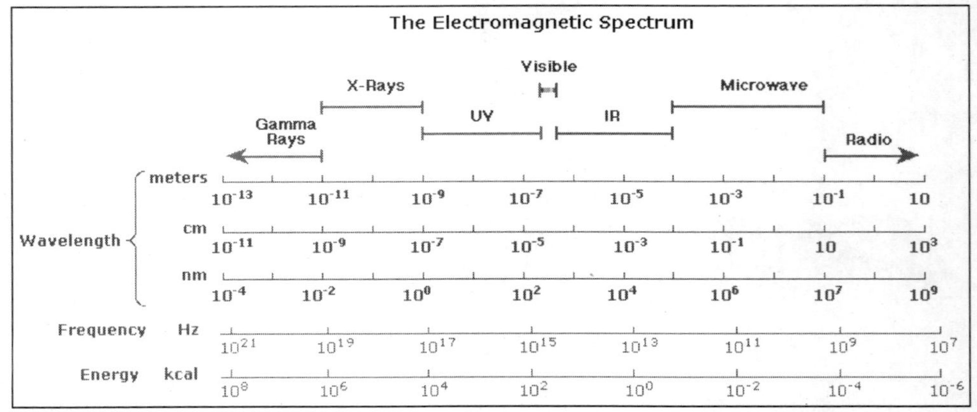

Figure 9.2: Spectrum of Electromagnetic Radiation [5].

Electromagnetic spectrum can broadly be grouped into the followings [1,2c,5]-

i. Non-ionizing

This refers to any type of electromagnetic radiation that does not carry enough energy per quantum (photon energy) to ionize atoms or molecules.

- ☆ The non ionizing radiation when obtained by low induced current is non thermal while high induced current generates heat which is effectively used in the microwave ovens. Photo electric excitation is used for producing photochemical effect.
- ☆ In non-ionizing radiation, the extremely low frequency rays and very low frequency are included. These are generally referred as radio waves.
- ☆ The next range of frequency is categorized as microwave, which is followed by infra red rays.
- ☆ The visible range is forming the next group of rays.

ii. Ionizing Radiation

This consists of electromagnetic waves that have sufficient energy to ionize atoms or molecules by detaching electrons from them.

- ☆ The ionizing radiation is more useful in medical field. Rays generated during ionization produce several effects such as damage to the DNA and also breaking of the bonds which are of immense help in treatment of cancer, correcting birth defects and also causing mutation.

☆ The ultra violet rays, X rays and gamma rays are considered under the category of ionizing radiation.

Table 9.2: Regions of the Electromagnetic Spectrum [5a]

Sl.No.	*Wave Type*	*Wavelength (m)*	*Frequency (Hz)*	*Energy (J)*
1	Radio waves	> 0.1	< 3×10^{9}	< 2×10^{-24}
2	Microwaves	10^{-3} - 0.1	3×10^{9} - 3×10^{11}	2×10^{-24} -2×10^{-22}
3	Terahertz waves	10^{-3} -10^{-4}	3×10^{11} - 3×10^{12}	2×10^{-22} -2×10^{-21}
4	Infrared	7×10^{-7} -10^{-3}	3×10^{11} -4×10^{14}	2×10^{-22} -3×10^{-19}
5	Optical(visible light)	4×10^{-7} - 7×10^{-7}	4×10^{14} -7.5×10^{14}	3×10^{-19} - 5×10^{-19}
6	Ultraviolet	10^{-8} - 4×10^{-7}	7.5×10^{14} -3×10^{16}	5×10^{-19} - 2×10^{-17}
7	X-rays	10^{-11} -10^{-8}	3×10^{16} - 3×10^{19}	2×10^{-17} - 2×10^{-14}
8	Gamma rays	< 10^{-11}	> 3×10^{19}	> 2×10^{-14}

Sources

1. *K.S. Sharma, Physical Chemistry of milk, Agrimoon.*

1a. *The Electromagnetic Spectrum, https: //courses.lumenlearning.com.*

2. *EMR, Radiation. www.esrl.noaa.gov.*

2a. *Wikipedia- Electromagnetic radiation.*

2b. *Electro- magnetic spectrum, https://www. britannica.com.*

2c. *Wikipedia- Ionising and Non ionising radiation.*

3. *Purcell and Morin, Harvard University,2013, Electricity and Magnetism, 820p (3rd ed.). Cambridge University Press, New York. ISBN 978-1-107-01402-2. p 430.*

4. *Browne, Michael,2013, Physics for Engineering and Science, p427 (2nd ed.), McGraw Hill/Schaum, New York. ISBN 978-0-07-161399-6, p319.*

5. *Spectrum of Electromagnetic Radiation, Lesson-31, ecoursesonline.iasri.res.in.*

5a. *http://csep10.phys.utk.edu/ astr162/lect/light/spectrum. html.*

Lambert–Beer Law [1, 1a]

It is also known as Beer's law, Beer-Lambert law, or the Beer-Lambert-Bouguer law and this law relates to the absorption (attenuation) of light and to the properties of material through which the light travels [1,1a,2].

☆ The law is commonly applied to chemical analysis measurements and used in understanding attenuation in physical optics, for photons, neutrons, or rarefied gases [2].

Transmittance [1,1a]

It is the term used to the amount of light penetrating a solution. It is expressed as the ratio between the intensity of transmitted light (It) to the intensity of initial light of light beam (lo).

$$T = I_t/I_o$$

where, T -Transmittance, I_t - Intensity of the transmitted light and I_0 - Intensity of the initial light beam.

The Absorbance [3]

The absorbance, A, of the solution is related to the transmittance and incident and transmitted intensities through the following relations:

$$A = \log_{10} \frac{I_0}{I}$$
$$A = -\log_{10} T$$

The absorbance has a logarithmic relationship to the transmittance, with an absorbance of 0 corresponding to a transmittance of 100 per cent and an absorbance of 1 corresponding to 10 per cent transmittance [3,4].

- ✰ The law states that there is a logarithmic dependence between transmission (or transmissivity) of light through a substance, T, and the product of absorption coefficient of substance, α, and the distance the light travels through the material (*i.e.* the path length), ℓ.

  The absorption coefficient can, in turn, be written as a product of either a molar absorptivity of the absorber, ε, and the concentration c of absorbing species in the material, or an absorption cross section, σ, and the (number) density N′ of absorbers.

  For liquids, these relations are usually written as:

$$T = \frac{I}{I_0} = 10^{-\alpha l} = 10- \varepsilon^{lc}$$

  The transmission (or transmissivity) is expressed in terms of an absorbance which, for liquids, is defined as-

$$A = -\ln \left(\frac{I}{I_0}\right)$$

  Thus, if path length and molar absorptivity (or the absorption cross section) is known and if the absorbance is measured, then the concentration of the substance (or the number density of absorbers) can be deduced [1,1a].

- ☆ The Beer-Lambert law states that there is a linear relationship between the concentration and the absorbance of the solution, which enables the concentration of a solution to be calculated by measuring its absorbance [3,4].
  - i. If the concentration is expressed as a mole fraction *i.e.* a dimensionless fraction, the molar absorptivity (ε) takes the same dimension as the absorption coefficient, *i.e.* reciprocal length (*e.g.* m^{-1}).
  - ii. However, if the concentration is expressed in moles per unit volume, the molar absorptivity (ε) is used in $L \cdot mol^{-1} \cdot cm^{-1}$, or sometimes in converted SI units of $m^2 \cdot mol^{-1}$. The absorption coefficient α' is one of many ways to describe the absorption of electromagnetic waves [1,1a].

Lambert-Beer Law Limitations [2]

The law tends to break down at very high concentrations.

- ☆ With this law it becomes easier to study the absorptivity coefficient of the sample when the concentration is low ie; <10mM.
- ☆ As the concentration becomes high ie; >10mM there is a deviation as the electrostatic interactions become more.

Application of Lambert-Beer Law [2]

This law is widely used in analytical chemistry and it is also applicable for atmosphere.

- **i. For Atmosphere-** This law is also applied to describe the attenuation of solar or stellar radiation as it travels through the atmosphere. In this case, there is scattering of radiation as well as absorption.
- **ii. Analytical Chemistry-** This law can be applied to the analysis of a mixture of compounds by spectrophotometry, without the need for extensive pre-processing of the sample.

Mass Spectrometry [1,1a,5,7]

Mass spectrometry is an analytical tool useful for measuring the mass-to-charge ratio (m/z) of one or more molecules present in a sample.

- ☆ These measurements can often be used to calculate the exact molecular weight of the sample components as well [6].
- ☆ In mass spectrometry, a compound is ionized and the resulting ions are separated according to their mass-to-charge ratio, and the ion abundances are measured.

Principle

The basic principle of mass spectrometry (MS) is to generate ions from either inorganic or organic compounds by any suitable method (or ionizing chemical compounds to generate charged molecules or molecule fragments), to separate these ions by their mass-to-charge ratio (m/z) and to detect them qualitatively and quantitatively by their respective m/z and abundance [5].

- ✰ Although there are many different kinds of mass spectrometers, all of them make use of electric or magnetic fields to manipulate the motion of ions produced from an analyte of interest and determine their m/z [7].
- ✰ The mass spectrum can be used to determine the mass of the analytes, their elemental and isotopic composition, or to elucidate the chemical structure of the sample [4].
- ✰ MS is an experiment that must take place in gas phase and under vacuum (1.33 * 10–2 to 1.33 * 10–6 pascal). Therefore, the development of devices facilitating the transition from samples at higher pressure and in condensed phase (solid or liquid) into a vacuum system has been essential to develop MS as a potent tool for identification and quantification of organic compounds and peptides [7].

Instrumentation [1,1a,6]

A mass spectrometer can measure the mass of a molecule only after it converts the molecule to a gas-phase ion. To do so, it imparts an electrical charge to molecules and converts the resultant flux of electrically charged ions into a proportional electrical current that a data system then reads.

- ✰ The mass spectrometer is an instrument which can measure the masses and relative concentrations of atoms and molecules. It makes use of the basic magnetic force on a moving charged particle. This is an analytical technique that measures the mass-to-charge ratio of charged particles.
- ✰ In addition to the determination of masses, this technique is also useful for the determination of elemental composition of a sample or a molecule, in finding out the chemical structure of molecules such as peptides and other chemical compounds.
- ✰ Mass spectrometer is the instrument used for this analytical technique. Mass spectrophotometer has following three associated components-
  - i. **The ion source [1,1a,6,7]**- This component of the instrument will help in ionization of the sample in which an electron is removed from cations [1].

    Molecules are converted to gas-phase ions so that they can be moved about and manipulated by external electric and magnetic fields [6].

1. For liquid and solid biological samples, Electrospray ionization (invented by John Fenn) and Matrix-assisted laser desorption/ ionization(MALDI, developed by K. Tanaka),are used.
2. Others include glow Discharge, field desorption (FD), Fast atom bombardment (FAB), Thermo-spray, Desorption/ ionization on silicon (DIOS), Direct Analysis in Real Time (DART), Atmospheric pressure chemical ionization (APCI), Secondary ion mass spectrometry (SIMS), Spark ionization and Thermal ionization (TIMS) [7].

ii. **The Mass analyser [1,1a,6,7]**-This component of the instrument will sort out ions and separate them according to their mass and charge [1].

- ✰ Once ionized, the ions are sorted and separated according to mass-to-charge (m/z) ratios.
- ✰ There are a number of mass analyzers currently available, each of which has trade-offs relating to speed of operation, resolution of separation, and other operational requirements. The mass analyzer often works in concert with the ion detection system [6]. Mass analyzers are nothing but sieve.
- ✰ There are many types of mass analyzers, using either static or dynamic fields, and magnetic or electric fields, but all operate according to the above differential equation. Each analyzer type has its strengths and weaknesses. Many mass spectrometers use two or more mass analyzers for tandem mass spectrometry (MS/MS) [7,8].

iii. **The detector [1,1a,6,7]** - The sorted and separated ions are then measured and the results are recorded.

- ✰ The separated ions are then measured and sent to a data system where the m/z ratios are stored together along with their relative abundance.
- ✰ A mass spectrum is simply the m/z ratios of the ions present in a sample plotted against their intensities.Each peak in a mass spectrum shows a component of unique m/z in the sample, and heights of the peaks connote the relative abundance of the various components in the sample [6].
- ✰ The detector allows to monitor the ion current and, after its amplification, to record the data in the form of mass spectra. The ions separated by mass analysers for their individual m/z values, are detected by Ion Detectors.
- ✰ The most common types of MS detectors include the electron multiplier, the photon multiplier, and the microchannel plate (MCP) detector [7].

Working of Mass Spectrophotometer [7,8]

Initially, the analyzed sample has to be introduced into the ion source of the instrument, where the sample molecules undergo the ionization process. These ions are then extracted into the analyzer where they are separated according to their m/z. The separated ions are detected and the signal containing information about the m/z values is stored together with their relative abundance (intensity) as a mass spectrum. To allow the ions traveling from one end of the instrument to the other part without any hindrance originated from air molecules, both the analyzer and detector (and often the ion sourceas well) are kept under high vacuum.

Applications [1,1a,7]

MS techniques have proved to be an excellent tool for qualitative characterization and quantitative determination of various food components because of their high sensitivity and specificity [7,8].

☆ MS is now in very common use in analytical laboratories that study physical, chemical, or biological properties of a great variety of compounds/Dairy products/Food [7]. The mass spectroscopy is useful in the following fields:

a. In Physical and Chemical Fields [1,1a,7]

1. Measuring the size of nano particles in the absence of sophisticated transmission electron microscopy or x-ray diffraction.
2. Characterisation of Proteins and Aroma/Flavour [7].
3. Characterisation of Carbohydrates [7].
4. Determination of fat and Identification of Vitamins [7].

b. In Food Analysis and Food Safety [1,1a,7]

1. MS and proteomic techniques can be used in the field of food safety. In addition to characterisation of proteins, mass spectroscopy is also successfully being used for three areas related to food safety [7]-
   i. the detection of microorganisms.
   ii. the determination of modifications induced by food processing.
   iii. the identification of protein components responsible for food allergy and intolerance.
2. Detecting toxins in pharmaceuticals, biotechnology, and other foods.
3. Can identify multiple compounds of pesticides at once that enter the food supply.

Nuclear Magnetic Resonance (NMR) [1,1a,2,7]

Nuclear magnetic resonance spectroscopy, most commonly known as NMR spectroscopy or magnetic resonance spectroscopy (MRS), is a spectroscopic technique to observe local magnetic fields around atomic nuclei [2]. There are two types of NMR spectrometers:

i. Continuous-wave (cw).

ii. Pulsed or Fourier-Transform (FT-NMR).

- ✰ Nuclear magnetic resonance (NMR) is a physical phenomenon in which magnetic nuclei in a magnetic field absorb and re-emit electromagnetic radiation. The technique is based on the nuclear magnetic resonance of the atomic nuclei in the molecules [1].
- ✰ Nuclear magnetic resonance is defined as a condition when the frequency of the rotating magnetic field becomes equal to the frequency of the processing nucleus [7].
- ✰ The sample is placed in a magnetic field and the NMR signal is produced by excitation of the nuclei sample with radio waves into nuclear magnetic resonance, which is detected with sensitive radio receivers. The intramolecular magnetic field around an atom in a molecule changes the resonance frequency, thus giving access to details of the electronic structure of a molecule and its individual functional groups [2].
- ✰ In the scientific field this technique is being used to investigate properties of the organic molecules. Samples for which this technique is used range from small compounds with one dimensional protons or carbon–13 to large protein or nucleic acids using 3 or 4 dimensional techniques [1]. The most common types of NMR are proton and carbon-13 NMR spectroscopy, but it is applicable to any kind of sample that contains nuclei possessing spin [2].

Principle [7]

The principle behind NMR is that many nuclei have spin and all nuclei are electrically charged.

i. If an external magnetic field is applied, an energy transfer is possible between the base energy to a higher energy. When molecules are placed in a strong magnetic field, the nuclei of some atoms will begin to behave like small magnets.

ii. The energy transfer takes place at a wavelength that corresponds to radio frequencies and when the spin returns to its base level, energy is emitted at the same frequency. The resonant frequencies of the nuclei are then measured.

iii. The signal that matches this transfer is measured in many ways and processed in order to yield an NMR spectrum(displays all of the right frequencies as peaks on a graph) for the nucleus concerned.

Instrumentation [7]

The basic elements of a typical NMR spectrometer consist of the following main parts:

i. **Magnet** - It is with strong, stable homogeneous field. The field must be constant over the area of the sample.

ii. **Radio frequency oscillator (transmitter)-**It is connected to coil which transmits energy to the sample in a direction perpendicular to the magnetic field.

iii. **Sample container**-It is usually a glass tube spun by an air driven turbine to average the magnetic field over the sample dimensions.

iv. **Radio frequency receiver**- It is connected to a coil encircling the sample. The two coils are perpendicular to each other and to the magnetic field.

v. **A read out system-** The other supporting parts are - consisting of an amplifier, recorder and additional components for increasing sensitivity, accuracy or convenience.

vi. **Sweep generator**- It supplies a variable d-c current to a secondary magnet so that the total applied magnetic field can be varied (swept) over a limited range.

Application of NMR

Nuclear Magnetic Resonance (NMR) spectroscopy is an analytical chemistry technique used in quality control and research for determining the content and purity of a sample as well as its molecular structure [7].NMR has following characteristics and applications [1,2,7]:

a. In Chemical Field [1,1a,2]

☆ Nuclear magnetic resonance spectroscopy is a research technique which is used for determining physical and chemical properties of atoms or the molecules of a compound.

☆ It also gives information about the structure, dynamics, reaction state, and chemical environment of the molecules.

☆ This technique provides a wide range of information for a large number of samples including solutions and solids.

☆ NMR spectra are unique, well-resolved, analytically tractable and often highly predictable for small molecules.

b. In Food and Food Safety Field [1,1a,2,7]

- ☆ It comprises the use of high-resolution NMR spectroscopy for the molecular study of dairy compounds such as casein and whey protein [7].
- ☆ Characterization of cow and buffalo milk and pressure treated SMP, analysis of Ricotta cheese and High resolution analysis of milk and Determination of milk ii
- ☆ NMR spectroscopy is employed in a wide range of food safety areas broadly aimed at averting significant chemical, biological or microbiological threats to the food chain. This technique is more useful about new or immerging threats. Under such circumstances the analysis of specific compounds using traditional methods is time-consuming, expensive and often unsuccessful in reaching the goal of controlling risk.
- ☆ Biochemists use NMR to identify proteins and other complex molecules. Some examples of Biomolecular NMR Spectroscopy are given in following table [2]

Table 9.3: Examples of Biomolecular NMR Spectroscopy

Sl.No.	*Compounds*	*Observations*
1	Proteins/Nucleic acid	i. A common goal of this spectroscopy is to obtain high resolution 3-dimensional structures of the protein. ii. Used to obtain information on the dynamics and conformational flexibility of different regions of a protein. iii. Used to obtain information about the structure and dynamics of polynucleic acids, such as DNA or RNA.
2	Carbohydrates	i. The analysis of carbohydrates by 1H NMR is challenging due to the limited variation in functional groups. ii. For smaller carbohydrates, the dispersion of the anomeric proton resonances facilitates the use of 1D TOCSY experiments to investigate the entire spin systems of individual carbohydrate residues.
3	Drugs	NMR can be used to guide drug design hypotheses,
4	New compounds from complex mixtures	The first examples for using NMR spectroscopy to identify new compounds from complex mixtures originated from arthropod natural products studies

- ☆ Besides identification, NMR spectroscopy provides detailed information about the structure, dynamics, reaction state, and chemical environment of molecules.
- ☆ NMR is focussed on the development of non-targeted methods that are able to rapidly determine the presence of unspecified hazards.

- NMR spectroscopy has also been used to study the effect of microbes on food composition.
  - i. Applications include monitoring maturation processes, shelf-life determination.
  - ii. The measurement of freshness like storage history of a food product may be determined by considering the biochemical composition.
- This spectroscopy is playing a role in understanding a range of activities, for example, assessing ingredient purity, characterising functional ingredients and investigating the effect of novel processes on final product composition.
- NMR methodologies can be used for food authentication *i.e.* for the verification of geographical and botanical origin, and for the verification of labelling claims such as "organic" and those relating to more direct health implications including the characterisation of pronutrients.

Sources

1. *K.S. Sharma, Physical Chemistry of milk, Agrimoon.*

1a. *Spectrum of Electromagnetic Radiation, Lesson-31, ecoursesonline.iasri. res.in.*

2. *Wikipedia- NMR Spectroscopy.*

3. *Lambert Beer Law, Edinburgh Instruments, https: //www.edinst.com.*

4. *IUPAC, Compendium of Chemical Terminology, 2nd ed. (the "Gold Book"); Compiled by A. D. McNaught and A. Wilkinson. Blackwell Scientific Publications, 1997.*

5. *Basic principle of Mass Spectrometry, DGMS, https: //dgms.eu/en.*

6. *What is mass spectrometry, Broad Institute, https: //www. broadinstitute.org*

7. *M.K. Srivastava, Instrumentation in analysis of foods and milk products, 2019, ISBN: 9789388173520, Astral International Pvt Limited, New Delhi.*

8. *Mass Spectrometry and Hyphenated Instruments in Food Analysis by Tomas Cajka, Jana Hajslova, and Katerina Mastovska.*

Visible and Ultraviolet Spectrophotometers [1,1a,2]

The electromagnetic radiation and its spectra are of immense value in developing instruments useful form the analysis of various compounds. Development of spectrophotometer is an outcome of the study of this spectrum.

- The absorption or reflectance tested using the visible light or near Ultra violet (UV) and Near Infrared (NIR) is referred as absorption spectroscopy or reflectance spectroscopy. The absorption or reflectance in the visible range directly affects the perceived color of the chemicals involved [1,2].

- ☆ White light covers the entire visible spectrum 400-780.For Colorimetry and spectrophotometry, the Visible region and the adjacent Ultraviolet region are of major importance.
- ☆ The following table helps in understanding the composition of Ultraviolet and visible light [3,4]-

Table 9.4: Composition of Ultraviolet and Visible Light

A: UV Region		
Less than 400 nm	Ultra Violet Region	
B: Visible Region		
Wavelength of Max. Absorption (nm)	*Colour Absorbed*	*Colour Observed*
400-420	Violet	Green- Yellow
420-440	Violet –Blue	Yellow
440-470	Blue	Orange
470-500	Blue –Green	Red
500-520	Green	Purple
520-550	Yellow- Green	Violet
550-580	Yellow	Violet -Blue
580-620	Orange	Blue
620-680	Red	Blue -Green
680-780	Red	Green

- ☆ In visible region of the electromagnetic spectrum, molecules undergo electronic transitions. This technique is complementary to fluorescence spectroscopy.
  - **i.** In fluorescence transition from excited state to ground state is measured.
  - **ii.** whereas in absorption transition from the ground state to the excited state is measured [1,2].
- ☆ This spectroscopy is routinely used in analytical chemistry for the quantitative determination of different analytes. Determination is usually carried out in solutions [3,4].

Principle of Spectrophotometer [1,1a,3,4]

UV radiations have enough energy to induce electronic transitions. Molecules containing π-electrons or non-bonding electrons (n-electrons) can absorb energy in the form of ultraviolet or visible light to excite these electrons to higher anti-bonding molecular orbitals.

- The energetically most favourable $\pi \geq \pi^*$ excitation occurs from the highest energy bonding pi-orbital (HOMO) to the lowest energy antibonding pi-orbital (LUMO). The more easily the electrons excited the higher wavelength of light it can absorb [1,2].
- There are four possible types of transitions (π-π^*, n-π^*, σ-σ^*, and n-σ^*), and their can be ordered as follows: σ-σ^* > n-σ^* > π-π^* > n-π [1,3,4].
- The visible region of the spectrum comprises of photon energies of 36 to 72 kcal/mole. Ultraviolet region has photon energy 143 kcal/mole and Ultraviolet radiation having wavelengths less than 200 nm is difficult to handle.
- UV radiation is rarely used as a routine tool for structural analysis. The energies noted above are sufficient to promote or excite a molecular electron to a higher energy orbital. Consequently, absorption spectroscopy carried out in this region is sometimes called **"electronic spectroscopy".**
- Accurate measurements of light absorption at different wavelengths in and near the visible part of the spectrum are made to understand why some compounds are coloured and others are not and also to determine the relationship of conjugation in the compounds to colour [1].
- The Beer-Lambert law states that the absorbance of a solution is directly proportional to the concentration of absorbing species in the solution and the path length. Thus, for a fixed path length, UV/Vis spectroscopy can be used to determine the concentration of absorber in a solution. It is necessary to know how quickly the absorbance changes with concentration. This can be taken from references (tables of molar extinction coefficients), or more accurately, determined from a calibration curve [1,2,3].
- Different compounds may have different absorption maxima and absorbance. Intensely absorbing compounds must be examined in dilute solutions, so that significant light energy is received by the detector, and this requires the use of completely transparent (non-absorbing) solvents. Because the absorbance of a sample will be proportional to its molar concentration in the cuvette, a corrected absorption value, known as the molar absorptivity, is used when comparing the spectra of different compounds [1,3].

Procedure

When sample molecules are exposed to light having an energy that matches a possible electronic transition within the molecule, some of the light energy will be absorbed as the electron is promoted to a higher energy orbital.

i. An optical spectrometer records the wavelengths at which absorption occurs, together with the degree of absorption at each wavelength.

ii. The resulting spectrum is presented as a graph of absorbance (A) versus wavelength.

Example

The isoprene spectrum is shown below. Since isoprene is colorless, it does not absorb light in the visible part of the spectrum and this region is not displayed on the graph.

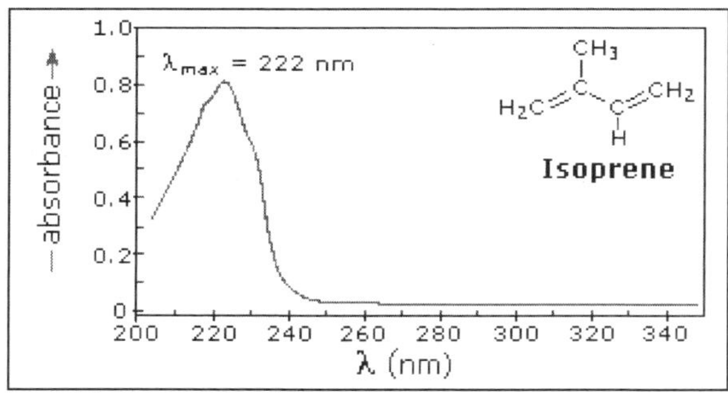

Figure 9.3: UV-visible Spectrum of Isoprene Showing Maximum Absorption at 222 nm [5].

i. Absorbance usually ranges from 0 (no absorption) to 1 (99 per cent absorption) and is precisely defined in context with spectrometer operation [1,5].

ii. The UV visible spectrum for isoprene (as shown in above Figure) shows maximum absorptivity (λ max) of 222nm.

iii. Absorption of light in 200 to 800 nm region by molecular moieties is due to the function of pi.-electron and hetero atoms having non-bonding valence-shell electron pairs. Such light absorbing groups are referred as chromophores [1,5].

iv. Conjugation will help in moving the absorption maxima to longer wavelength as observed in the isoprene spectra.

✰ Though UV-Visible spectroscopy can detect the presence of these chromophores in a molecule, most instruments cannot provide absorption data for wavelengths below 200nm.

Solvents [1,2]

The oxygen non-bonding electrons in alcohols and ethers do not give rise to absorption above 160 nm. Consequently, pure alcohol and ether solvents may be used for spectroscopic studies. The most commonly used solvents are water, ethanol, hexane and cyclohexane. Solvents having double or triple bonds, or heavy atoms (*e.g.* S, Br and I) are generally avoided.

Molar Absorptivity [1,1a,2]

It is important characteristic. As the absorbance of a sample will be proportional to the number of absorbing molecules in spectrometer light beam (*e.g.* their molar concentration in the sample tube), it is necessary to correct the absorbance value and other operational factors if the spectra of different compounds are to be compared in a meaningful way. The corrected absorption value is called **"molar absorptivity"**. Molar absorptivity (ε) is defined as-

$$\text{Molar absorptivity} = \varepsilon = A/cl$$

where, A- absorbance, c - sample concentration in moles/liter and l - length of light path through the sample in cm.

- ☆ Molar absorptivity is particularly useful when comparing the spectra of different compounds and determining the relative strength of light absorbing functions (chromophores). By UV-visible spectra it is possible to identify these structural features.
- ☆ Molar absorptivity may be very large for strongly absorbing chromophores (>10,000) and very small if absorption is weak (10 to 100). The magnitude of ε reflects both the size of the chromophore and probability that light of a given wavelength will be absorbed when it strikes the chromophore. The terms used to define shifts in absorption are shown in Table given below:

Table 9.5: Terminology for Absorption Shifts

Sl.No.	Nature of Shift	Descriptive Term
1	To longer wavelength	Bathochromic
2	To shorter wavelength	Hypsochromic
3	To greater absorbance	Hyperchromic
4	To lower absorbance	Hypochromic

- ☆ Thus extending conjugation generally results in bathochromic and hyperchromic shifts in absorption.
- ☆ The appearance of several absorption peaks or shoulders for a given chromophore is common for highly conjugated systems, and is often solvent dependent.
  - **i.** This fine structure reflects not only different conformations that conjugated systems may assume, but also electronic transitions between different vibrational energy levels possible for each electronic state.

ii. Vibrational fine structure of this kind is most pronounced in vapor phase spectra, and is increasingly broadened and obscured in solution as the solvent is changed from hexane to methanol [1].

Instrumentation [2,3,4]

The instrument used in ultraviolet-visible spectroscopy is called a UV/Vis spectrophotometer. It measures the intensity of light after passing through a sample, and compares it to the intensity of light before it passes through the sample.

- ☆ A spectrophotometer consists of two instruments, namely a spectrometer for producing light of any selected color (wavelength), and a photometer for measuring the intensity of transmitted light/radiation.
- ☆ The instruments are arranged so that liquid in a cuvette can be placed between the spectrometer beam and the photometer. The amount of light passing through the tube is measured by the photometer. The photometer delivers a voltage signal to a display device, normally a galvanometer. The signal changes as the amount of light absorbed by the liquid changes.
- ☆ UV/VIS spectro-photometers have the following components-
  - i. Radiation source.
  - ii. Sample container and optics.
  - iii. Wavelength selector (Filters),
  - iv. Radiation detector and.
  - v. Signal processor and read out.

Types of UV-Visible Spectrophotometers [1,2,3]

Single beam and Double beam UV/VIS spectrophotometers are available.

i. Single Beam-In a single beam instrument,all of the light passes through the sample cell must be measured by removing the sample. This was the earliest design and is still in common use in both teaching and industrial labs.

ii. Double Beam- In a double-beam instrument, the light is split into two beams before it reaches the sample. One beam is used as the reference; the other beam passes through the sample.

Application of different Spectroscopic Methods for Analysis of Milk and Dairy Products [1,3]

Routine analytical methods used for measuring various milk components are sample destructive, expensive and time and labour intensive. The increasing awareness among consumers about the quality and safety of milk and milk products compels the manufacturers to determine the concentrations of constituents such as water, protein, fat and carbohydrate in milk by using classical absorption spectroscopy. The following spectroscopic methods are being used for analysis of milk and milk products:

i. UV/VIS Spectroscopy [3]

a. Chemspec 150(Bentley Instrument), an automatic, fast (upto 150 samples/hour) and compact instrument being used by dairy for determination of Urea (or MUN) in milk [3].

b. The methods based on this spectroscopy are used for analysis of proteins, Melamine, Uric acid in cereals, Synthetic colours in biscuits/cakes, Non milk fat solids in breads, Titanium dioxide in chewing gums/bubble gums, higher alcohols in alcoholic beverages, Nitrite in meat *etc.* [3].

i. IR Spectroscopy [1,1a,3]

Most commonly used instrument for milk analysis is Milko-scan which is based on the infra red spectroscopy. Milkoscan Minor, Lactoscope and Foodscan are working on IR where as Milk screen, Milkoscan FT and Milkoscan FT120 are working on FTIR. IR Spectroscopy is used to determine nearly all major constituents in all dairy products. Some examples are given in following table [3]:

Table 9.6: Examples of IR Spectroscopy

Products	*Constituents*	*Products*	*Constituents*
Cream	Fat, Dry matter	Milk powder	Moisture, Fat, Protein, Lactose
Creamer	Moisture(water),Fat	Ice cream mix	Fat, Dry matter
Coffee creamer	Moisture, Fat, Protein	Raw/Market milk	Fat, Protein, Lactose, Dry matter
Evaporated milk	Fat, Dry matter	Skim Milk	Drymatter, Protein, Casein
Skim milk powder	Moisture, Fat	Feeding stuffs	Various constituents

ii. Near IR Spectroscopy [1,3]

The greatest advantage of NIR spectroscopy is that it is rapid and accurate. This technique can be used for analyzing various dairy products like milk powders, whey, cheese *etc.*

a. InfraXact (compositional analysis of Food and Feed) and Process Analytics™ Dairy(for compositional analysis of Food and Feed) are available [3].

b The analysis of water by near infrared spectroscopy (NIRS) was the first successful application of this rapid technology.

c. NIR spectroscopy is used routinely for the compositional, functional and sensory analysis of food ingredients, process intermediates and final products. Studies carried on NIR spectroscopy have made it possible to analyze milk for adulteration and for several constituents like fat, and lactose *etc.* [1].

d. An additional advantage of NIR spectroscopy is that it could be used online in the process of milking for effective farm management [1].

iii. Atomic Absorptive Spectroscopy [1,3]

It is helpful in the study of essential micronutrients such as iodine content in infant milk formulae and milk powder samples [1]. The characteristics of different commonly used instruments based on atomic absorption spectroscopy *e.g.* FAAS, ETAAS, ICPMS *etc.*) for analysis of trace and major elements (Nutritionally important- Fe, Ca, P, Mg, Cu, Zn, I *etc.* or toxic - Pb, Hg, As *etc.*) in milk and milk products are given below [3]-

Table 9.7: Applications of AAS

Sl.No.	*Absorption/Emission Techniques*	*Sample Volume*	*Limit of Detection*	*Multi-element*
1	FAAS	5-10	1-1000	Possible
2	FAES	5-10	1-100	Yes
3	ETAAS,	0.01-0.1	0.01-0.1	Possible
4	ICPAES	1-10	0.05-10	Yes
5	ICPMS	1-10	0.001-0.01	Yes

The applications of different Atomic absorption and Atomic emission techniques in analysis of milk and milk products are tabulated below [3]:

Table 9.8: Applications of AAS and AES in Analysis of Milk and Milk Products

Sl.No.	*Atomic Absorption and Atomic Emission*	*Applications*
1	Flame Atomic Absorption Spectrometry (FAAS)	Detection/determination of Cu, Zn,Ca, Fe, Mg and Mn in dairy products. It is good for liquid samples *e.g.* milk *etc.* and does not need sample preparation.
2	Flame Atomic Emission Spectrometry (FAES)	Determination of relatively high concentrations of Sodium (Na) and Potassium(K) in milk products.
3	Electro thermal Atomic Absorption Spectrometry (ETAAS)	Capable of in situ sample preparation and hence it can determine several elements in solid sample directly.
4	Inductively Coupled Plasma Mass Spectrometry (ICPMS)	Detection of Non metal *e.g.* P and I (Iodine) in milk and milk products.

Sources

1. *K.S. Sharma, Physical Chemistry of milk, Agrimoon.*

1a. *Mathur, M. P., Datta, R. D. and Dinakar, P. 2005. Textbook Of Dairy Chemistry. Indian Council Agricultural Research Publ., New Delhi.*

2. *Wikipedia-UV-Visible spectroscopy.*

3. *M.K. Srivastava, Instrumentation in analysis of foods and milk products, 2019, ISBN: 9789388173520, Astral International Pvt Limited, New Delhi.*

4. *Vogel's Text Book on Quantitative chemical analysis revised by G.H. Jeffery, J. Bassett, J. Mendham and R.C. Denney*

5. *UV-visible spectrum of isoprene, www. Pharmax change.info.*

Part 2

Chemistry of Milk

Chapter 10

Definition and Structure of Milk and Factors affecting the Composition of Milk

Milk is a complex fluid containing many components in different states of dispersion. The role of milk in nature is to nourish and provide immunological protection for the mammalian young. All species of mammals secrete milk to provide nutrients required for the optimum growth of the new born(neonate). As the development of the young one in all species of mammals is not uniform and hence the composition of the milk secreted by these mammals will also vary depending up on the nutritional needs of the young one. Milk and honey are the only articles of diet whose sole function in nature is food. It is not surprising, therefore, that the nutritional value of milk is high. Milk is also a very complex food with over 100,000 different molecular species found [1,1e].

Milk Definition [1,2]

Chemical and Physical definition along with definition under regulatory agencies (PFA, FSS Regulation-2011) are discussed here.

Chemical Definition

Milk is complex chemical substance and heterogeneous colloidal product in which fat and protein are dispersed in water and other substances likes Sugar, Minerals are held in solution form where the water is dispersion medium for dispersed phase and solvent for other soluble constituents [2,3]. Fat is present in the form of an emulsion, protein and some mineral matter in the colloidal state and lactose with some minerals and soluble proteins in the form of true solution [1].

Physical Definition [2,3]

Milk is white coloured colloidal substance in which fat and protein are dispersed in water and is produced by mammary glands of mammalian [2,3].

Definition under P.F.A. and FSS Regulation-2011 [1,2]

Milk is normal mammary glands secretion derived from completely milking of healthy milch animal without either addition there to or extraction there from. It shall be free from colostrums [2].

Under PFA-1954

Milk in legal terms may be defined as whole, fresh, clean, lacteal secretions, obtained by complete milking of one or more healthy milch animals, excluding that obtained 15 days before or 5 days after or such periods as may be necessary to render the milk practically colostrum free and containing legally prescribed minimum percentage of fat and Solids-not-fat (SNF) [1].

Under Food Safety and Standards (Food Products Standards and Food Additives) Regulations- 2011

"Milk" means the normal mammary secretion derived from complete milking of healthy milch animal, without either addition thereto or extraction there from, unless otherwise provided in these regulations and it shall be free from colostrum. The legal standards for different types of milks (Buffalo, Cow, Goat, Sheep, Camel, Mixed, Toned, Double toned, Skim, Standardised, Full cream milk) are given in following table [2a]:

Table 10.1: Legal Standards for different Milk

Sl. No.	*Class of Milk*	*Locality or State or Area*	*Min. Milk Fat (per cent, m/m).*	*Min. Milk Solids Not-Fat (SNF) (per cent, m/m).*
1	Buffalo Milk	Assam, Bihar, Chandigarh, Delhi Gujarat, Haryana, Jharkhand, Maharashtra Meghalaya Punjab Sikkim, Uttar Pradesh, Uttarakhand West Bengal.	6.0	9.0
		Andaman and Nicobar Islands, Andhra Pradesh, Arunachal Pradesh, Chhatisgarh, Dadra and Nagar Haveli, Goa, Daman and Diu, Himachal Pradesh, Jammu and Kashmir, Karnataka, Kerala, Lakshadweep, Madhya Pradesh, Manipur, Mizoram, Nagaland, Odisha, Puducherry, Rajasthan, Tamil Nadu, Telangana, Tripura.	5.0	9.0

Sl. No.	Class of Milk	Locality or State or Area	Min. Milk Fat (per cent, m/m).	Min. Milk Solids Not-Fat (SNF) (per cent, m/m).
2	Cow Milk	All India	3.2	8.3
3	Goat or Sheep Milk*	Chandigarh, Haryana, Kerala, Madhya Pradesh, Maharashtra.	3.5	9.0
		Punjab, Uttar Pradesh, Uttarakhand, Andaman and Nicobar Islands, Andhra Pradesh, Arunachal Pradesh, Assam, Bihar, Chhattisgarh, Dadra and Nagar Haveli, Delhi, Goa, Daman and Diu, Gujarat, Himachal Pradesh, Jammu and Kashmir, Jharkhand, Karnataka, Lakshad-weep, Manipur, Meghalaya, Mizoram, Nagaland, Odisha, Puducherry, Rajasthan, Sikkim, Tamil Nadu, Telangana, Tripura, West Bengal.	3.00	9.0
4	Camel Milk	All India	2.0	6.0
5	Mixed Milk	All India	4.5	8.5
6	Standardized Milk	All India	4.5	8.5
7	Toned Milk	All India	3.0	8.5
8	Double Toned Milk	All India	1.5	9.0
9	Skimmed Milk	All India	Not more than 0.5	8.7
10	Full Cream Milk	All India	6.0	9.0

Note:* As Food Safety and Standards (Food Products Standards and Food Additives) Regulations-2011(Version-XIV -25.03.2021), the milk of sheep and goat shall have Min. fat-3.0 per cent and Min. SNF-8.0 per cent for all India.

Structure of Milk [1,1d,1e,2]

Milk can be defined as a dispersion of milk fat globules (fat particles) and casein micelles (protein particles) in a continuous phase of water, sugar (lactose), whey proteins, and minerals [1e]. Milk is a secretion of the epithelial cells of the mammary gland and because of this, it has specific interactive forces which are working between its components and help in maintaining the integrity of the whole system under different conditions and thereby maintaining its specific structure. The interactive forces between the components are important and they determine the integrity of the whole system under various conditions [1,1d]. The main structural elements are shown schematically in following figure [1,1e].

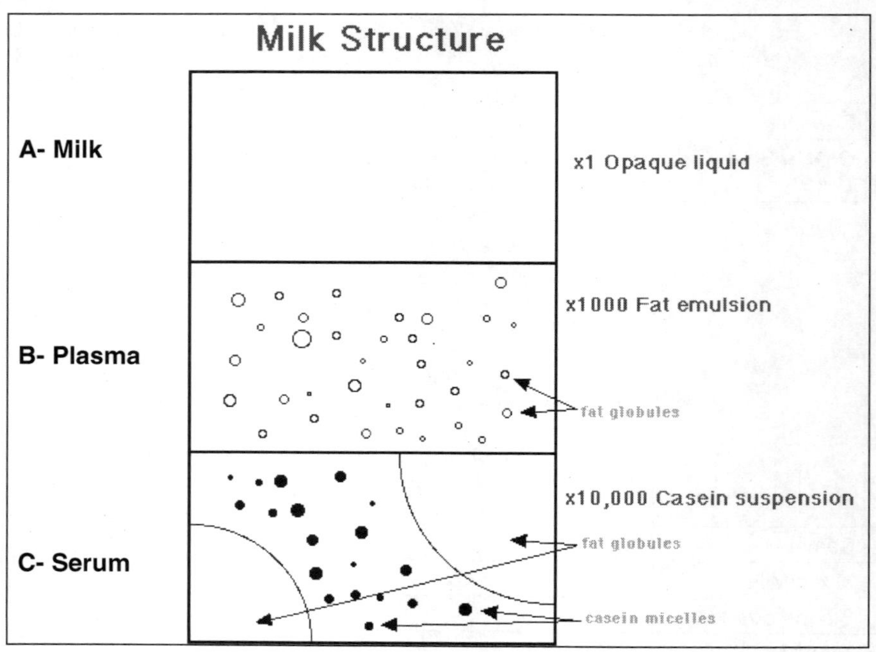

Figure 10.1: Milk Viewed at different Magnifications [1c,1e].

The structural elements are microscopic or sub-microscopic in scale. The structure of milk is very simple and extensive studies are being conducted on this aspect. Since milk constituents exist in different physical state influencing milk properties in addition to the chemical nature of the constituents.

Physical Structure of Milk [1,1e]

The details related to the views of milk drop under different level of magnification are given below:

i. **View under five Times (5X) magnification**-If a drop of milk is viewed with only 5 times magnification it appears as uniform liquid, however, it cannot be homogeneous since it is turbid in nature.

ii. **View under another 100 Times (500X) magnification** -With an increase in the magnification by another 100 times one can observes spherical droplets of fat floating in plasma.

iii. **View under another 100 Times magnification more than step ii (50,000X)-** A further increase of magnification by another 100 times the proteinaceous particles (casein micelles) become visible.

Fat, casein micelles and other proteins form the major structural elements for milk, and these elements are microscopic or sub microscopic in their size.

- ☆ Although fat is considered to be present in the form of an emulsion but it is not just a simple emulsion, because of the fact that the fat globules are being covered with the fat globule membrane which is derived during its biosynthesis.
- ☆ In addition to the fat globules some of the fat in the milk is found outside the globules while the membrane material includes several enzymes and proteins and polar lipids. As such fat and fat globules are not identical.

Structural Elements [1,1d,1e]

Milk is a dispersion of milk fat globules (fat particles) and casein micelles (protein particles) in a continuous phase of water, sugar (lactose), whey proteins, and minerals. Milk can be described as [1e]-

- ☆ An **oil-in-water emulsion** with the fat globules dispersed in the continuous serum phase
- ☆ A **colloid suspension** of casein micelles, globular proteins and lipoprotein partilcles
- ☆ A **solution** of lactose, soluble proteins, minerals, vitamins other components.

The properties of various structural elements of milk are shown in the following table:

Table 10.2: Properties of Main Structural Elements [1d,7a]

Sl. No.	*Structural Element*	*Fat Globules*	*Plasma*		
			Casein Micelles	*Serum*	
				Globular Protein	*Lipo- protein Particles*
1	Main Components	Lipids	Casein, water, salt	Serum protein	Lipoprotein particles
2	Considered as	Emulsion	Fine dispersion	Colloidal solution	Colloidal dispersion
3	Content (per cent DM)	3.8	2.8	0.6	~ 0.01
4	Volume fraction	0.042	~0.06	0.006	~ 0.0001
5	Particle diameter	0.1-10 µm	10-300nm	3-6 nm	~ 10nm
6	Density(20°C) g/ml	0.92	1.11	1.34	1.10
7	Visible with	Microscope	Ultra Microscope	Electron Microscope	Electron Microscope
8	Light Scattering	Very turbid, white	Turbid bluish	Light haze	Negligible

Sl. No.	*Structural Element*	*Fat Globules*	*Plasma*		
			Casein Micelles	*Serum*	
				Globular Protein	*Lipo- protein Particles*
9	Separable with	Dairy separator	High speed centrifuge	Gel Filtration	Gel Filtration
10	Iso electric pH	~3.8	~4.6	~4.5-5.5	~4?
11	Flocculation	By agglutininin cold	By acid or by rennet	By heat	–
12	Filterable with	Fine sintered glass	Chamberland filter	Cellophane	Cellophane

Milk Fat [1,1g]

Fat is the costliest component of milk. The unique feature of milk fat is its fatty acid composition. Milk fat is rich in saturated fatty acids and mono unsaturated fatty acids and also acts as a carrier for the fat soluble vitamins. About 98 per cent of milk fat is a mixture of triacylglycerols, with much smaller amounts of free fatty acids, mono-and diacylglycerols, phospholipids, sterols, and hydrocarbons. Milk fat exists in the form of fat globules. The unique feature of these fat globules is that milk fat globules are surrounded with a membrane which is derived from the apical membrane of the mammary secretory cell. Milk fat also contains pigments (*e.g.* carotene, which gives butter its yellow colour), and waxes. Milk fat acts as a solvent for the fat-soluble vitamins A, D, E and K and also supplies essential fatty acids (linoleic, linolenic and arachidonic).

Proteins [1,1d]

Milk is rich in protein content and has a unique protein namely the casein. It is present in the form of colloidal dispersion in milk and is responsible for several physical properties of milk. Milk contains Casein and whey proteins. Casein is the major protein of bovine milk and exists mainly as micellar form, where as in human milk major proteins are whey proteins.

- ☆ **Casein**- The principal group of milk proteins, the caseins, are not soluble in water and exist in milk as small particles (<300 nm) called micelles.

  The Casein micelle (structural element of milk) consists of water, casein, salts and some minor components including lipase and proteinase. Casein micelles are built of smaller particles called sub-units or sub-micelles.

- ☆ Milk plasma free from casein micelles is known as milk serum.
- ☆ The liquid that is obtained on clotting of milk either with rennet or by acidification to pH 4.6 (isoelectric point of casein) is known as whey.

This whey differs in its composition from that of serum in having some of the polypeptides cleaved from casein by the action of rennet.

- Casein micelles do not completely account for the casein content since some of the casein is also present in a solution form. As casein binds cations especially the Ca^{++} and Mg^{++} ions at the normal pH of milk *i.e.* ~6.6, it is referred as caseinate. Amorphous calcium phosphate and small amounts of citrates are the constituents associated with the casein micelle [1d].
- Calcium caseinate phosphate complex is the term applicable to describe the casein micelles. Phosphate being a part of the colloidal particles, the term colloidal phosphate is also used [1d].
- **Whey proteins [1,1d]**-The proteins present in whey are termed as whey proteins of milk. The whey proteins are mainly globular proteins.
- The non micellar proteins constitute the whey proteins, but it is often the globular proteins present as such are considered to be the whey proteins [1d]
- Lipoprotein particles sometimes called microsomes vary in their quantity, composition, and shape. They consists remnants of cell membranes, microvilli, *etc.* [1].

Somatic Cells

The somatic cells consist of mainly leukocytes of various types is considered as extraneous particles although they are always present. They are about 10 μ m in diameter, number is about 100,000 per milliliter and accounts for about 0.005 per cent of the volume of milk. They contain all cytoplasmic components notable nucleic acids and enzymes. They are rich in catalase. In mastitic milk the cell counts are much higher.

Milk Plasma [1,1d,1e]

Milk gets its structure with the structural elements. The main structural element being the fat, all that portion of milk which is free from the fat is considered as milk plasma. Milk Plasma is left after the fat globules is separated and it is equivalent to skim milk for practical purposes. It is often a practice to use cream separators for getting the skim milk, but all the fat present in milk will not be separated as such Milk plasma can not be considered as synonymous to skim milk. Here we consider fat free -milk to be plasma but not skim milk. Although the words serum and plasma are used for fractionation of blood here in milk these words are denoting the physical fractions of milk.

Milk Serum [1a,1d,1e]

Milk Serum is left after both fat globules and casein micelles are removed from milk and equivalent to cheese whey for most practical purposes. Milk plasma free from casein micelles is known as milk serum. All that portion of milk excluding the casein micelles is referred as milk serum. By using either rennet or dilute acids the casein micelles could be separated from the plasma to obtain milk serum. The globular proteins consists mainly the whey proteins. They also bind some cations and a little water. Lipoprotein particles present in milk are also referred as the milk microsomes.

Factors (Genetic, Physiological and Environmental) affecting the Composition of Milk [1,1a,2,2a,2c]

The composition of milk is not uniform in all the species. Depending upon the needs of the new born there will be variation in the proportion of the major constituents in milk. In addition to aforesaid differences several other factors also influence the proportion of various milk constituents. The composition of milk will show variation due to following several factors:

1. Animal factors,
2. Environmental Factors
3. Miscellaneous factors

1. Animal Factor [1,1a,2,2a,2c]

a. Genetic

The composition of milk is an important trait of milch animals and is basically a inherited character influenced by genes. The genetic factor can be modified by proper planning of breeding practices and It plays role in the quality and quantity of milk produced [1,2a,2c].

- ☆ The potential fat content of milk from an individual cow is determined genetically. Thus, selective breeding can be used to upgrade milk quality.
- ☆ Heredity also determines the potential milk production of the animal [1,2a].

b. Species

The milk composition varies between species and is designed to provide different amount of nutrients to new born and is an inherited trait for that species. The faster the rate of growth the more concentrated are the milk components needed for this growth. The following examples help us in understanding the aforesaid facts-

- ☆ Cat milk contains about 7.0 per cent protein since the new born has to double its birth weight in 9.5 days,
- ☆ The lamb will double its birth weight in about 15 days as such the sheep milk has only 4.8 per cent protein.

☆ Human infants take substantially longer time (180 days) to double their birth weight, since the human milk has least protein content (~ 1.0 per cent).

c. Breed

There are several breeds in domestic animals which are reared exclusively for milk production. Milk fat being the costliest component of milk, there is considerable selection among the breeds to ensure milk with a considerably higher fat content. Between and within breeds, fat varies the most and lactose the least. No much variation could be observed in the remaining constituents. Composition of the milk for some exotic and zebu cattle are presented in following table for understanding this influence of breed on the milk composition:

Table 10.3: Average Composition (per cent w/v) of Milk from different Breeds of Exotic Cow [4]

Sl.No.	*Breed*	*Water*	*Fat*	*Protein*	*Lactose*	*Ash*	*SNF*
1	Jersey	85.27	5.14	3.8	5.04	0.75	9.59
2	Guernsey	85.47	4.96	3.84	4.98	0.75	9.57
3	Ayrshir	87.1	3.85	3.34	5.02	0.69	9.05
4	Short Horn	87.43	3.03	3.32	4.89	0.73	8.94
5	Friesian	88.01	3.45	3.15	4.65	0.68	8.48

Zebu cattle also show variation in milk composition between the different breeds. The data of the milk composition due to breed of the cow is given in following table:

Table 10.4: Average Composition (per cent w/v) of Milk from different Breeds of Indigenous Cow [5]

Sl.No.	*Breed*	*Water*	*Fat*	*Protein*	*Lactose*	*Ash*
1	Sindhi	86.07	4.9	3.42	4.91	0.7
2	Gir	86.44	4.73	3.32	4.85	0.66
3	Tharparker	86.58	4.55	3.36	4.83	0.68
4	Sahiwal	86.42	4.55	3.33	5.04	0.66

The variation among the milk constituents contributing to the SNF is very marginal. The influence of breed on the fat and SNF composition is presented in the following table [5]:

Table 10.5: Average composition (per cent w/v) of Milk from different Breeds of Buffalo [5]

Sl.No.	Breed	Fat per cent	SNF per cent
1	Murrah	6.8	10.1
2	Jaffarabadi	7.3	10.1
3	Surti	8.4	10.3

d. Individual Cow

Although individual animals maintain the breed characteristic of milk produced by them. However, there will be a variation with the individual animals and this also depends on the genetic characteristics of the individual animals.

- ☆ By following good breeding practices it is possible to get higher fat per cent in the individual or milch animal.
- ☆ A similar variation could be observed in the protein and lactose levels in individual animals within a given breed.

Thus, selective breeding can be used to upgrade milk quality apart from the inherited characteristics, the environment and various physiological factors influence the composition of milk produced by individual animal.

e. Lactation Period [1,1a,1b,2a]

Similar variation in the fat and SNF percentage during a lactation period is also observed as noticed in the case of breed and individual animals. The fat, lactose and protein contents of milk vary according to stage of lactation. Some important variations are given below:

a. Variation in Fat Content [1,1a,1f]

- ☆ Fat content is higher after parturition, which decreases to a low during the first and second months of lactation and then gradually increases during the remaining period of the lactation [1,1f]. Fat content is high immediately after calving but soon begins to fall and continues to do so for 10 to 12 weeks, after which it tends to rise again until the end of the lactation [1,2a]. The concentration of milk fat and protein is highest in early and late lactation and lowest during peak milk production through mid lactation. During the first half, the proportions of short- and intermediate-chain fatty acids increase, and the proportion of long-chain fatty acids decreases. No further changes occur during the last half of lactation [1f]. Normally, an increase in milk yield is followed by a decrease in the percentages of milk fat and protein [1b]. The composition of the fat also found to be changing during the lactation [1a].

In cow, Milk production peaks at about 60 days of lactation and declines there after, Fat and protein are minimum at 60 days and increase thereafter. Protein/fat ratio peaks at sixty days and decreases thereafter [1g].

b. *Variation in SNF Content*

- ✰ Solids-not-fat content is usually highest during the first 2 to 3 weeks, after which it decreases slightly.
- ✰ A slight decrease in Lactose percentage could be observed during the last stage of lactation.
- ✰ An increase in the chloride content was observed with the advancement of lactation.
- ✰ Calcium concentration reduces to a minimum and further increase whereas the total phosphorous declines throughout the lactation period.
- ✰ Natural acidity was observed to be at higher level at the beginning of lactation and will decrease to normal value during the first three months and remains fairly steady during the remaining lactation period till the last month it declines sharply.

2. Environmental Factors [1,1b,2,2a,2c]

a. Interval between Milking

The fat content of milk varies considerably between the morning and evening milking as there is shorter interval of time between the morning and evening milking than between the evening and morning milking. If cows and buffaloes are milked at 12-hour intervals, there would be negligible variation in fat content between milking. However this is not practicable on most farms. Normally, SNF content varies a little even if the intervals between milking vary considerably [1,2a].

b. Age

Because of the several physiological changes occurred in the body of the older individual cow/buffalo, the composition of milk produced by these older animals too shows variation in their composition. As cows grow older, the fat content of their milk decreases by about 0.02 percentage units per lactation. The fall in SNF content is much greater [1,1a,2a]. Milk fat content remains relatively constant, milk protein content gradually decreases with advancing age [1b].

c. Feeding Regime

All the precursors of milk constituents are derived from blood of milch animal. The suitable and well balanced feeding to milking animal, plays an important role in maintaining requisite levels of constituents in the milk produced.

Underfeeding a lactating animal will only reduces the milk yield but also level of fat and SNF. However changes in SNF content are more sensitive to the feeding level than the fat content. Fat content and its composition are influenced more by the roughage (fiber) intake. The solids-not-fat content can fall if the cow/buffalo is fed a low-energy diet, but is not greatly influenced by protein deficiency, unless the deficiency is acute [1,1a,2a].

d. Disease

Milk production is a physiological function of a lactating animal and hence general health condition of individual cow plays a very important role in maintaining good milk composition. Both fat and SNF contents reduce when animal suffers from any specific disease. Mastitis, being a disease specific to the udder, has substantial effect on milk composition; it reduces lactose content with a corresponding increase in salt content of milk [1,2,2a].

Mastitis results in a reduction in fat and casein content and an increase in whey content of milk. The decrease in fat percentage, however, is less (about 10 per cent) than that observed for lactose or casein (about 15 per cent) [1,1f]. These changes in the milk proteins, in conjunction with alterations in lactose, mineral content and milk pH, result in lower cheese yields and altered manufacturing properties. Milk from cows with elevated somatic cell counts (greater than 500,000 somatic cells/ml) has longer coagulation time and forms weaker curds than milk from cows with lower somatic cell counts [1b].

e. Completeness of Milking

The fat percentage of milk increases continuously during the milking process, with the lowest fat milk drawn first and the highest fat milk drawn last [1f]. During the interval between the two milking, milk will be held in gland cistern and teat cistern, consequently there is possibility that fat being the lightest component of milk would be moving to the surface, resulting in the variation of fat percentage in different portions of milk during the process of milking. It is advised to draw the milk completely from the udder during each milking. This will not only ensure that all the milk produced by the animal is drawn but also the fat is not left in the udder. The last part of the milking (Stripping) has a higher fat content as compared to the fore milk (first milk). The completeness of milking ensures uniform quality of milk [1,1a,2a].

If cows/Buffaloes are not milked out completely, fat percentage will be lower than normal, but, at the next milking, fat content will be higher than normal. Furthermore, when milking intervals are unequal, the highest fat percentage is obtained after the shortest interval. Milk fatty acid composition is not affected by milking interval or time of day milking [1f].

f. Season [1,1b,1f]

In India, there is wide variation in climate and also type of vegetation available in different seasons of the year. Due to seasonal variation in the availability of fodder in India, there are following changes in the milk composition-

- ☆ The butter fat content in milk is maximum in the month of May and minimum in the month of November.
- ☆ The SNF is highest in October, and minimum in July and September.
- ☆ The variation in the milk production is dependent on several other factors *viz.*, temperature, hours of sun shine, length of the day, exercise, rain fall, drought, stall feeding or pasture feeding, *etc.* that are associated with the season.
- ☆ Lush spring pastures low in fiber depress milk fat. Hot weather and high humidity decrease dry matter intake and increase feed sorting, resulting in lower forage and fiber intake [1b].

Seasonal variations in milk fat percentages are well recognized, with summer months averaging 0.4 percentage units less than winter months. The higher environmental temperatures during the summer also affect milk fatty acid composition. Milk fat in the summer tends to be lower in palmitic acid relative to stearic and octadecanoic acids than milk fat from the same cows during the winter [1f].

g. Weather

In dry months the total yields of milk tend to decrease along with a decrease in SNF and increase in fat percentage. In wet months there may be a decrease in both SNF and fat percentage along with or without the increase in the yield of milk which depends on the level of nutrition.

3. Miscellaneous Factors [1,2c,8]

a. Exercise

The regular exercise by the lactating animal will ensure good blood circulation through out the body and keeps the animal in good condition. For sufficient reserve of energy such animals will eat more which would indirectly increase the milk yield. It was observed that the feed consumption of the animal will increase resulting in an increase in milk yield along with an increase in the fat percentage when compared with the period of rest.

Slight exercise apparently increases the fat content from 0.2to 0.3 per cent without reducing the quantity of milk secreted. Moderate to high exercise in high producing cows will result in reduction milk secretion and increases in fat content. Exercise causes no apparent changes in the SNF content of milk [8].

b. Excitement [1,8]

Hormones play an important role in the animal, especially during the milking time. Excitement effects the milk yield as well as milk composition. The factors related to excitement cannot be generalized as they may be specific to certain individual animal at a given geographic location.

- ☆ Favourable conditions during milking of animals result in good milk yield.
- ☆ Harassment and frightening conditions such as barking of dog, presence of strangers, bursting of crackers, *etc.* disturbs the animal and reduces milk yield.
- ☆ A change in the milk composition may also be expected by aforesaid conditions.
- ☆ Animal in oestrus will also reduce the milk yield.

The summary of different factors affecting the composition of milk are tabulated below [6,7]:

Table 10.6: Different Factors affecting the Composition of Milk

Sl.No.	*Category*	*Factors*	*Affects*
1	**Inheritance**	**I- Species**	Each species yields milk of a characteristic composition.
		II-Breeds	Breeds producing the largest amounts milk yield milk of lower fat.
		III-Individuality	Each animal tends to yields milk of a composition that is characteristic of the individual.
2	**Environmental**	**I-Feed**	Tends to give temporary effect only. Under and Over feeding alter the composition.
		II-Season	Both fat and SNF per cent have slight and well defined variation during the course of the year.
		III-Temperature	Affects the fat and SNF per cent. High temperature lowers the yield and hence increases fat per cent and decreases SNF. Low Temp. (< 40 C) increases both fat SNF.
		IV-Place	Affects the fat and SNF.
		V-Stage of Lactation	The changes occur at the beginning and at the end of lactation. First secretion after calving (colostrum) differs from milk in composition.
03	**Physiological**	**I-Age**	Fat declines slightly as animal grows older.
		II-Health	Poor health effects the composition of milk.
		III-Udder infection	Tends to alter the composition especially when fall in yield. Lowers the fat, SNF, lactose and casein and increases the serum and chloride.

Sl.No.	Category	Factors	Affects
		IV-Udder quarter Variation	The quarters of animal udder are anatomically separate. The fat content in milk of the udder milked last alters.
		V- Gestation Period	Affects the fat and SNF per cent.
		VI-Yield	Increased yield accompanied by lower fat per cent.
		VII-Drugs and Hormones	Injection or feeding of hormones results in increase in both yield and fat.
04	**Miscellaneous**	**I-Fear/Excitement**	Both yield and composition are liable to transient fluctuation.
		II-Efficiency of Milker	Affects the fat and SNF per cent.
		III-No. of Offsprings	Affects the Fat and SNF per cent.
05	**Procedures of Milking**	**I-Frequency of Milking**	No great effect on fat if animal milked one.two or three times.
		II-Interval of Milking	Longer interval associated with more milk with a lower fat.
		III-Completeness of Milking	If animal is not completely milked then it gives milk with lower solids.
		IV-Irregularity in Milking	Frequent changes in time and interval of milking results in lower tests.
		V- Day to day Milking	May show variations for the individual.
		VI- Portion of Milking	Fore milk is low in fat content while strippings are highest. Other constituents are slightly affected on a fat free basis.

Sources

1. *K.S. Sharma, Chemistry of milk, Agrimoon.com.*

1a. *R.C.Laben, Factors responsible for variation in milk composition, Journal of Dairy Science, vol-46, issue-11, 1963, p-1293-1301.*

1b. *Michael Looper, Factors Affecting Milk Composition of Lactating Cows, Dairy cattle, 2012, https: //en.engormix.com.*

1c. *H. Mulder and P. Walstra, The milk fat globule, 1974, Emulsion Science as Applied to Milk Products and Comparable Foods, Commonwealth Agricultural Bureaux, Farnham Royal and Centre for Agricultural Publishing and Documentation, Wageningen.*

1d. *Structural elements of milk-fat globules, casein micelles, globular proteins, lipoproteins particles and their properties and grading of milk, Lesson-3, http: // ecoursesonline.iasri.res.in.*

1e. *Milk structure, University of Guleph, https: //www.uoguelph.ca.*

1f. *J.G.Linn, Factors Affecting the composition of Milk from Dairy Cows, https: // www.ncbi.nlm.nih.gov.*

1g. *Art Hill, Milk Composition and Properties, University of Guelph, www.foodscienceca.*

2. *M.K. Srivastava, Handbook on Analysis of Milk, CBS Publications, New Delhi, 2010. ISBN 13: 9788123926681.*

2a. *Food Safety and Standards (Food Products Standards and Food Additives) Regulations- 2011.*

2b. *Factors affecting milk composition, https: //www.groupe-esa.com.*

2c. *Definition of milk and factors affecting its composition, http: //courseware.cutm.ac.in.*

3. *D. Ramaswamy, A.V. Shibu and H. Gopi, Dairy Technologists Handbook.*

4. *M.M. Rai, Dairy chemistry and animal nutrition, 1964.*

5. *K, S.Rangappa and K. T. Achaya, Indian Dairy Products, Asia Publ. House, Bombay, 1974.*

6. *Sukumar DE, Outlines of dairy technology 1990.*

7. *Robert Jenness and Stuart Patton, Principles of Dairy Chemistry, John Wiley and Sons, Inc., NY, 1995.*

7a. *P. Walstra and Robert Jenness, Dairy Chemistry and Physics, 1984, John Wiley and Sons, New York, ISBN 9780471097792-467.*

8. *G.C. Banerjee, A text book of animal husbandry, ISBN 81-204-1260-5, 2010.*

Chapter 11

Milk Proteins (Casein and Whey Proteins)

Proteins and peptides present in milk have important nutritional, functional, biological and technological properties. Milk Protein is widely consumed human food proteins for people of all ages, particularly infants and children. About two thirds of all cells consist of proteins and they have functional and structural roles in human body such as catalyst (enzymes), regulator (hormones), protection (Immunoglobulins), carrier proteins (haemoglobulin, lactoferrin, *etc.*) and as structural proteins (collagens) [1].

The milk proteins present in colloidal form. These are Casein, Albumin, Globulin, Proteose-peptones and enzymes and account for 95 per cent of nitrogen. Milk proteins are **92-93 per cent digestible** [3]. There are 2 major categories of milk protein that are broadly defined by their chemical composition and physical properties [1a].

i. **The casein family**: It contains phosphorus and will coagulate or precipitate at pH 4.6. Caseins are highly digestible than whey proteins and are important for growth and development of infants.

ii. **The serum (whey) proteins**: These proteins do not contain phosphorus, and these proteins remain in solution in milk at pH 4.6 [1a]. Whey proteins contain four major proteins *i.e.* α-lactalbumin (α-lac), β-lactoglobulin (β-lg), blood serum albumin (BSA) and immunoglobulins (Ig). Besides these, the whey fraction contains proteoses and peptones (PP), lactotransferin, serotransferin, osteopontins, vitamin binding proteins, lactoferrin and about 60 indigenous enzymes [1c]. When substantial whey protein is not digested fully in the intestine, some of the intact protein may stimulate a localized intestinal or a systemic immune response. This is sometimes referred to as milk protein allergy and is most often thought to be caused by β-Lg. Milk protein allergy is only one type of food protein allergy. Since camel milk is devoid of β-Lg or contains a very small amount of it, and could be interesting as a new raw material for nutritional infant formula in countries where these animals thrive [1b].

Caseins and whey proteins are the main proteins present in various milk species in different proportion. The ratio of casein and whey proteins is 40:60 in human milk while in milk of cow, sheep, goat and buffalo; it is 80:20.

Casein is unique due to their characteristic physical properties which are different from globular proteins. These proteins influence the behaviour and properties of other milk constituents. The protein content in a given milk products influences the structure of that product [1].

Milk proteins are synthesized in the mammary gland, but 60 per cent of the amino acids used to build the proteins are obtained from the diet of cow and buffalo. Total milk protein content and amino acid composition varies with cow breed and individual animal genetics [1a].

Definition and Structure [1,1d,2,3]

a. Definition

Proteins are defined as high molecular weight polymers of α-amino acids that are formed by living organisms. All these amino acids have L-configuration except glycine [1]. In proteins,the chains of amino acid molecules are connected by peptide bonds. The amino acids within protein chains can bond across the chain and fold to form 3-dimensional structures.

b. Protein Structure

Proteins differ from each other not only in their length of polypeptide chain but also in the proportion and sequence of the amino acids residues in them. Proteins of various species are said to be homologous if they are considered to have originated from a common ancestor. The most definitive criterion of homology is similarity of the amino acid sequence. The homologous proteins usually exhibit a common biological function but coincidence of function also can arise by convergence from different ancestral lines [1,2]. It is convenient to describe protein structure in terms of 4 different aspects of covalent structure and folding patterns. The different levels of protein structure are known as primary, secondary, tertiary, and quaternary structure. The conformation or three-dimensional organization of proteins involves the following secondary, tertiary, and quaternary structures [3,3a]. All four types of structures of proteins are discussed below:

i. **The primary structure** of proteins consists of a polypeptide chain of amino acids residues joined through peptide bonds [2]. The milk proteins consists of a polypeptide chain of amino acids joined together by peptide linkages, which may also be cross-linked by disulphide bridges. Amino acids contain both a weakly basic amino group, and a weakly acid carboxyl group both connected to a hydrocarbon chain, which is unique to different amino acids [3].

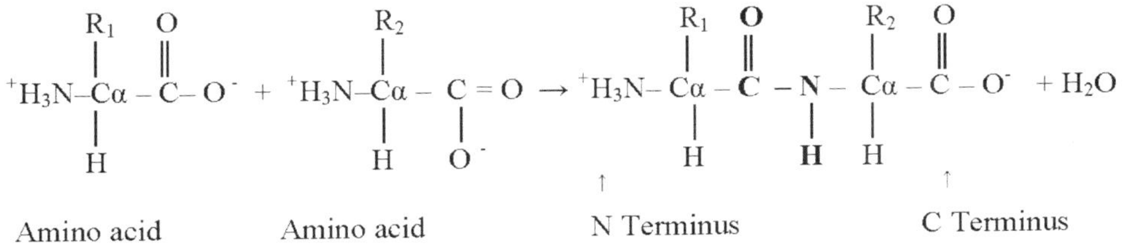

$$^{+}H_3N{-}\underset{H}{\overset{R_1}{C\alpha}}{-}\overset{O}{\overset{\|}{C}}{-}O^{-} + {}^{+}H_3N{-}\underset{H}{\overset{R_2}{C\alpha}}{-}\underset{O^{-}}{C}{=}O \rightarrow {}^{+}H_3N{-}\underset{H}{\overset{R_1}{C\alpha}}{-}\overset{O}{\overset{\|}{C}}{-}\underset{H}{N}{-}\underset{H}{\overset{R_2}{C\alpha}}{-}\overset{O}{\overset{\|}{C}}{-}O^{-} + H_2O$$

Amino acid Amino acid N Terminus C Terminus

Figure 11.1: Structure of Protein [2].

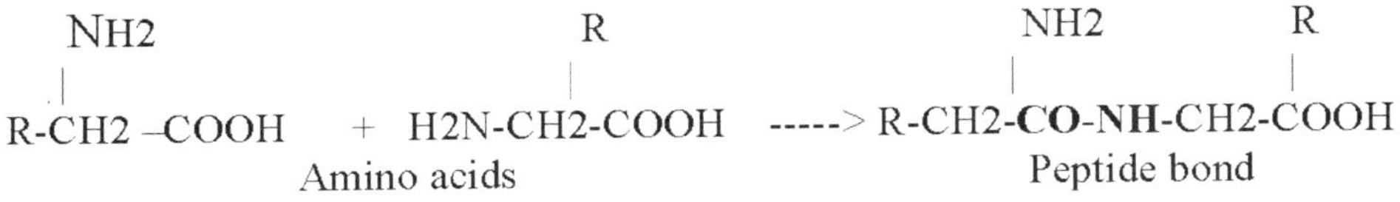

$$R{-}\overset{NH2}{\overset{|}{CH2}}{-}COOH + H2N{-}CH2{-}\overset{R}{\overset{|}{COOH}} \dashrightarrow R{-}\overset{NH2}{\overset{|}{CH2}}{-}\mathbf{CO{-}NH}{-}CH2{-}\overset{R}{\overset{|}{COOH}}$$

Amino acids Peptide bond

Figure 11.2: Structure of Protein [3].

ii. **The secondary structure** [3,3a] has spatial arrangement of amino acid residues that are near one another in the linear sequence- example - alpha-helix and β -pleated sheet.

iii. **The tertiary structure [3,3a]** has spatial arrangement of amino acid residues that are far apart in the linear sequence, giving rise to further coiling and folding. The tightly coiled and folded into a somewhat spherical shape protein is called a **globular protein** where as the protein consists of long polypeptide chains (intermolecularly linked) are called **fibrous proteins**.

iv. **Quaternary structure [3,3a]]** is derived when proteins with two or more polypeptide chain subunits are associated.

The proteins contain amino acids and other materials such as Sulphur and phosphorous in small amounts. A vide variety of these acid is found in each of the proteins of milk. Out of the nine essential amino acids, eight amino acids are present in milk. Buffalo milk is a relatively abundant source of Leucine, Lysine, Isoleucine, Valine, Threonine and phenylalanine. The concentration of amino acids in 100 gm milk of different milch animals are given in following table:

Table 11.1: The Concentrations of Amino Acids in 100 gm of different Type of Fresh Milk [4]

Sl.No.	Amino Acids	Units	Cow	Goat	Buffalo	Sheep	Human	Soy
1	Tryptophan	gm	0.046	0.044	0.053	0.084	0.017	0.043
2	Threonine	gm	0.149	0.163	0.182	0.268	0.046	0.113
3	Isoleucine	gm	0.199	0.207	0.203	0.338	0.056	0.144
4	Luecine	gm	0.322	0.314	0.366	0.587	0.095	0.241
5	Lysin	gm	0.261	0.29	0.28	0.513	0.068	0.179
6	Methionine	gm	0.083	0.08	0.097	0.155	0.021	0.040
7	Cystine	gm	0.03	0.049	0.048	0.035	0.019	0.047
8	Phenylalamine	gm	0.159	0.155	0.162	0.284	0.046	0.151
9	Tyrosine	gm	0.159	0.179	0.183	0.281	0.053	0.112
10	Valine	gm	0.22	0.24	0.219	0.448	0.063	0.141
11	Arginine	gm	0.119	0.119	0.114	0.198	0.043	0.214
12	Histidine	gm	0.089	0.089	0.078	0.167	0.023	0.071
13	Alanine	gm	0.113	0.118	0.132	0.269	0.036	0.122
14	Aspartic acid	gm	0.25	0.21	0.309	0.328	0.082	0.341
15	Glutamic acid	gm	0.689	0.626	0.477	1.019	0.168	0.550
16	Glycine	gm	0.07	0.05	0.08	0.041	0.026	0.120
17	Proline	gm	0.319	0.368	0.364	0.58	0.082	0.162
18	Serine	gm	0.179	0.181	0.227	0.492	0.043	0.144

Nomenclature of Milk Proteins [1-5]

The Committee on Milk Protein Nomenclature, Classification and Methodology of the Manufacturing Section of American Dairy Science Association (ADSA) has made efforts to suggest suitable nomenclature to the milk proteins.

- ☆ Several research workers made attempts to suggest nomenclature basing on the resolution of this protein on the free boundary electrophoresis. The proteins which are having similar electrophoretic characteristics at pH 8.6 were actually heterogeneous mixtures of proteins [1,2].
- ☆ Scientists [3] have shown that α-casein could be separated into calcium sensitive α s-casein and calcium insensitive crude k-casein based upon its dissociation and differential solubility in the presence of calcium ions.
- ☆ A new terminology was suggested by Scientists [3]for the action of rennin on α-casein fraction k casein for the primary reaction product of rennin on k casein and αs-casein for the clots which are formed.

But the role of other α-casein components and of β and or γ-casein in this transformation was not explained.

The committee felt that in view of the advancements in the knowledge of the casein systems the committee preferred not recommend any nomenclature for the α-casein complex. The American Dairy Science Association's committee on milk protein nomenclature, classification and methodology has given a classification system for the known proteins in milk [4,5].

According to this system of classification a Greek letter with or without a numerical precedes the class name when necessary to identify the family of proteins. The genetic variant of the protein is indicated by an upper case Arabic letter with or without a numerical super script immediately following the class name. Post translational modifications are added in sequence: β-casein B 5P (f1 to 105) indicates that

- **i.** Class of the protein is caseins.
- **ii.** The family is β,
- **iii.** The genetic variant is B,
- **iv.** There are 5 post translational phosphorylation and.
- **v.** The amino acid sequence from N-terminal amino acid through residue 1 through residue 105.

Sources

1. *K.S. Sharma, Chemistry of milk, Agrimoon.com.*

1a. *Milk Protein, Milk Facts, http: //milkfacts.info.*

1b. *L.C.Laleye, B. Jobe, and A.A.H. Wasesa, 2008. J. Dairy Sci., 91: 4527-4534.*

1c. *P.F.Fox, 2003, Advanced dairy chemistry, 3rd ed., Part A. Plenum Publishers, New York, London, pp. 5-30.*

1d. *Introduction, definition ans nomenclature of milk proteins, Lesson 4, http: // ecoursesonline. iasri.res.in.*

2. *http: //www. web-books.com/MoBio/Free/Ch2B.htm .*

3. *David F. Waugh and Peter H. von Hippel, κ-Casein and the Stabilization of Casein Micelles, J. Am. Chem. Soc. 1956, 78, 18, 4576-4582.*

3a. *Milk Proteins, University of Guleph, https: //www.uoguelph.ca.*

4. *W.N.Eigel, J.E. Butler, C.A. Ernstrom, H.M. Farrell, V.R. Harwalkar, R.Jenness and R. McL.Whitney, Nomenclature of Proteins of Cow's Milk: Fifth Revision, Journal of Dairy Science, Volume 67, Issue 8, 1984, Pages 1599-1631.*

5. *H. M. Farrell, Jr., R. Jimenez-Flores, G. T. Bleck, E. M. Brown, 1 J. E. Butler, L. K. Creamer, C. L. Hicks, C. M. Hollar, K. F. Ng-Kwai. Hang, and H. E. Swaisgood, Nomenclature of the Proteins of Cows' Milk-Sixth Revision, 2004, J. Dairy Sci. 87: 1641-1674.*

Classification and Distribution of Milk Proteins [1,1a]

The details related to chronological development in classification of milk proteins are given below:

- ☆ Earlier proteins were classified largely on the basis of rather empirical fractionation procedures.
- ☆ In most cases the proteins were classified partly on the basis of their solubility and partly on the basis their composition.
- ☆ With the availability of better analytical techniques it is established that the proteins which were earlier considered to be individual proteins are actually a mixtures of several other fractions.

Common Method for Classification [1,1a,1b,2,3]

At present the proteins in milk are classified on the basis of their fractionations and their behaviour during electrophoresis, difference in their solubility in various solutions, difference in their sedimentation rate *etc.* A common method for the classification and distribution of milk proteins is represented as follows:

Table 11.2: Classification and Distribution of Milk Proteins

Sl. No.	*Types of Milk Proteins (Genus Bos (30-35 g/L)*	*Fractions of Milk Proteins*	
		Sub class	*Fractions/Genetic Variants*
1	**Caseins (24-28 g/L)**	**α -Caseins**	**αs1 -Caseins (12-15 g/L)** **i.** αs1-Casein X-8P (X = Genetic variants-A, B, C, D-9P, and E). **ii.** αs1-Casein X-9P (X = Genetic variants-A, B,C, D-10P, and E). **iii.** αs1-Casein fragments.
			αs2 caseins (3-4 g/L) **i.** αs2 -Casein X-l0P (X =Genetic variants-A, B, C-9P, and D-7P). **ii.** αs2 -Casein X-l1P (X =Genetic variants-A,B, C-10P, and D-8P). **iii.** αs2-Casein X-12P (X =Genetic variants-A, B, C-11P, and D-9P). **iv.** αs2 Casein X-13P (X =Genetic variants-A,B, C-12P, and D-10P).
		β - Casein (9–11 g/L)	☆ βCasein X-5P (X =Genetic variants-A1, A2, A3,B, C-4P, D-4P, and E). ☆ β-Casein X-IP (f 29-209) (X=Genetic variants-A1, A2, A3,and B). ☆ β-Casein X-(f 106-209) (X =Genetic variants-A2, A3, and B).

Sl. No.	Types of Milk Proteins (Genus Bos (30-35 g/L)	Fractions of Milk Proteins	
		Sub class	Fractions/Genetic Variants
			✰ β-Casein X-(f 108-209) (X=Genetic variants-(Fragments from A1, A2,A3) and B). ✰ β - Casein X-4P (f 1-28) *. ✰ β -Casein X-5P (f 1-105) *. ✰ β -Casein X-5P (f 1-107) *. ✰ β -Casein X-IP (f 29-105) *. ✰ β Casein X-IP (f 29-107) *.
		κ -Caseins (2-4 g/L)	**i.** κ -Casein X-IP (X = Genetic variants-A and B). **ii.** Minor κ -caseins X -l, -2, -3, *etc.* (x = Genetic variants-A and B).
2	**Whey proteins (5-7 g/L)**	**β Lactoglobulins (2-4 g/L)**	**i.** β -Lactoglobulins X (X = genetic variants-A,B, C, D, Dr, E, F, and G).
		α-Lactalbumins (0.6-.7g/L)	**i.** α-Lactalbumin X (X = Genetic variants-A and B). **ii.** Minor α-Lactalbumins.
		Bovine serum albumin (0.2-0.4 g/L).	—
		Immunoglobulins (0.5-1.8 g/L)	**i. IgGimmunoglobulins** ✰ IgG1, Immunoglobulins. ✰ IgG2, Immunoglobulins. ✰ IgG fragments. **ii. Ig Mimmunoglobulins**. **iii. IgA immunoglobulins** ✰ IgA Immunoglobulins. ✰ Secretory IgA Immunoglobulins. **iv. IgE Immunoglobulins.** **v. J-chaincomponent.** **vi. Free secretory component.**
3	**Milk Fat Globule Membrane (MFGM) Proteins [2,3]**	**MFGM-A15 -127, C, S**	Where MFGM indicates the protein satisfied the operational definition for a MFGM protein, A15 designates the zone in a 15 per cent Laemmliacrylamide gel, 127 is the "apparent molecular" weight in k-daltons, and C, S designates that the protein is both Coomassie brilliant blue and periodic acid/Schiff positive (PAS). **i.** Six major proteins readily stained with Coomassie brilliant blue are XDH/XD; CD36, BTN,ADPH, PAS 6/7 And FABP. **ii.** Two stained by Periodic Acid/Schiff (PAS) or silver stain are Muc-1 and PAS III.

Sl. No.	Types of Milk Proteins (Genus Bos (30-35 g/L)	Fractions of Milk Proteins	
		Sub class	Fractions/Genetic Variants
4	**Minor proteins [4]**	**Serum transferrin ***	Genetic variants of these fragments have not been specifically identified.

The following figure also helps in understanding the fraction of milk proteins:

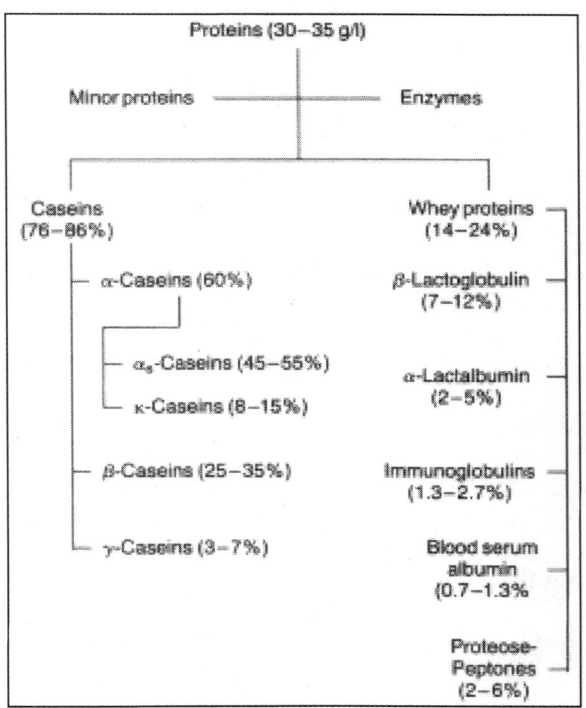

Figure 11.1: Milk Protein Fractions.

Sources

1. *K.S. Sharma, Chemistry of milk, Agrimoon.com.*

1a. *Milk proteins-classification, distribution, genetic polymorphism and its importance, Lesson-5, http://ecoursesonline.iasri.res.in.*

1b. *Milk Protein-Casein,http://courseware.cutm.ac.in.*

2. *W.N.Eigel, J.E. Butler, C.A. Ernstrom, H.M. FarrellJr. V.R. Harwalkar, R. Jenness, R. McL. Whitney, Nomenclature of Proteins of Cow's Milk: Fifth Revision, Journal of Dairy Science, Volume 67, Issue 8, 1984, Pages 1599-1631.*

3. *H. M. Farrell, Jr., R. Jimenez-Flores, G. T. Bleck, E. M. Brown,1 J. E. Butler, L. K. Creamer, C. L. Hicks, C. M. Hollar, K. F. Ng-Kwai Hang, and H. E. Swaisgood, Nomenclature of the Proteins of Cows' Milk-Sixth Revision, 2004, J. Dairy Sci. 87: 1641-1674.*

4. *Noble P.Wong, Robert Jenness, Mark Keeney and Elmer H. Marth,Fundamentals of Dairy chemistry, Chapman and Hall, New York,1988, ISBN 0-442-20489-2.*

Polymorphism in Milk Proteins [1,1a,1b]

Polymorphism is a term for a condition where there are at least 2 genetic variants (alleles) of some sign in the population. At the same time its frequency of occurrence in the population exceed 1 per cent. Milk protein polymorphism has been studied in depth since Aschaffenburg's discovery of genetically distinct forms of β-lactoglobulin in 1955.

- In 1956, it was discovered by Aschaffenburg and Drewry that the β lactoglobulin exists in two forms A and B and these forms differ from each other by only a few amino acids.
- The milk of any few individual animal may contain β lg A or β lg B or both and the milk is indicated as AA, BB or AB with respect to β lg. **This phenomenon is referred as genetic polymorphism** and has since been shown to occur in all the milk proteins.
- A total of about 30 variants have been demonstrated by Polyacrylamide Gel Electrophoresis (PAGE). Since PAGE differentiates on the basis of charge and molecular weight, only polymorphs which differ in charge alone are detected *i.e.* a charged residue is replaced by a non charged or vice versa.

  Therefore it is likely that more than 30 polymorphs exist. The genetic variant present is indicated by a latin letter eg: α s1 -CN A-8P,α s1- CN B – 8P and α s1 CN B – 9P *etc.*
- The frequency with which certain genetic variant occur is breed specific and hence genetic polymorphism has been useful in phylogenic classification of cattle and other species.
- Various technologically important properties of the milk proteins are related to the genes that occur in the primary polypeptide chain of milk proteins, which will result in variation of the behaviour of that protein either on electrophoretic behaviour or its gel filtration pattern. This is usually due to the changes occurring in genetic code for that protein. **This deviation or change is referred as polymorphism**. There are several such variations observed in the milk proteins.

The casein genotype significantly influences the milk yield, fat yield and protein yield with highest yields obtained for the genotype BB. Cheese yield on a fixed amount of milk and fat yield were significantly related to κ polymorphism observed to be β lg genotype with highest estimates obtained for BB. Protein percentage was influenced by αs1– casein and κ casein with the genotype BC and BB respectively having the highest percentages.

Distribution of Milk Proteins [1,1b,2]

- ☆ Milk contains proteins which are chiefly the caseins, whey proteins and some minor proteins. Caseins(major fraction) are present in micellar form while the whey proteins (considered as soluble proteins) exists in colloidal form.
- ☆ Since enzymes are protein in nature, they are also grouped along with other proteins.
- ☆ Enzymes play a significant role in the raw milk storage and processing of the milk and also found to have industrial importance.
- ☆ Minor proteins include vitamin binding protein, lactoferrin, metallo-protein, MFGM protein, Ceruloplasmin, *etc.*

The distribution of the proteins in milk is presented in following figure:

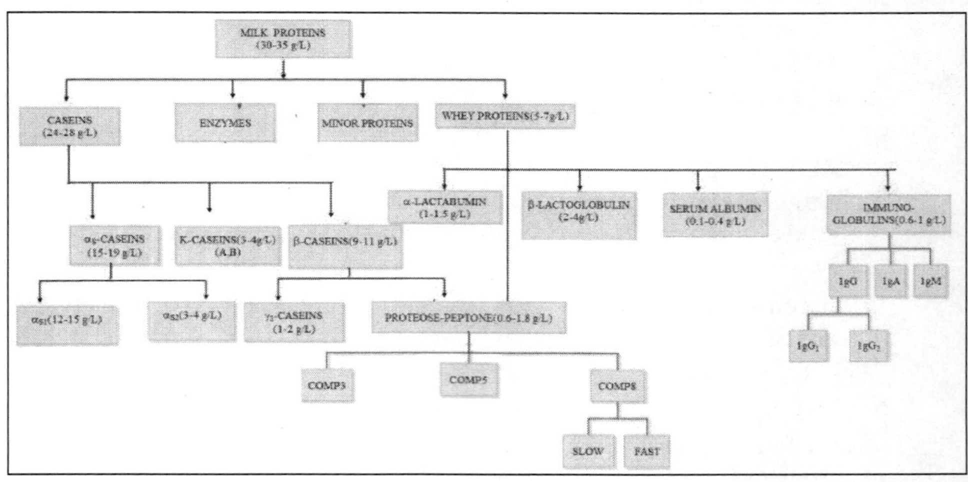

Figure 11.2: Distribution of Protein Fractions in Bovine Milk [1,3].

Functional and Nutritive Properties of Milk Proteins [2,2a]

Milk is only source of animal protein for vegetarians. It is a building block of body (1gm = 4 cal.) of growing creatures,easily digestible and contains 20 essential amino acids. Its biological value is 85 and it has immunological property. Milk proteins are helpful in biological activities. Milk proteins concentrate may relieve joint pains and enhances Iron adsorption due to

Iron binding property of Lactoferrin. Proteins for binding Vit.B_{12}, Folic acid, Riboflavin assist in enhancing their bio-availability from milk.

- ☆ Casein micelle is minerals carrier and protects against calcification in mammary gland.
- ☆ Whey proteins offer protection against tumor incidence.
  - **i.** Immunoglobulins play a role in the animal's immune system. Immuno globulins *e.g.* Gi, lg, G2, lgA, lgM provides passun immunity.
  - **ii.** The function of β-lactoglobulin is thought to be a carrier of vitamin A. It is interesting to note that β-lactoglobulin is not present in human milk. α-Lactalbumin plays a critical role in the synthesis of lactose in the mammary gland and enhances Calcium adsorption. α-Lactalbumin and Lactoferrin promote growth of Bifido bacteria. Lactoferrin and transferrin play an important role in iron absorption.
- ☆ Butyrophilin (Fat globule membrane protein) is also a part of immune system. Milk protein hydrolyzate (PRODIET -F200) shows anti stress effect.

Sources

1. *K.S. Sharma, Chemistry of milk, Agrimoon.com.*

1a. *R.Aschaffenburg and J, Drewry, Genetics of the β-lactoglobulins in cow's milk, Nature 180, 1957, 376-378.*

1b. *Milk Protein-Casein, http: //courseware.cutm.ac.in.*

2. *M.K. Srivastava, Handbook on Analysis of Milk, CBS Publications, New Delhi, 2010. ISBN 13: 9788123926681.*

2a. *Milk Proteins, Milk Facts, http: //milkfacts.info.*

3. *P. F. Fox, Developments in Dairy Chemistry, Volume 4, Applied Science Publishers, 1989 - Dairy products.*

Types of Milk Proteins [1,3]

The milk protein mainly contains Caseins and Whey proteins. The nitrogen content of milk is distributed among **caseins (76 per cent), whey proteins (18 per cent), and non-protein nitrogen (NPN) (6 per cent)**. This does not include the minor proteins that are associated with the FGM. The milk proteins can be separated first by precipitation at pH 4.6 to separates caseins from whey nitrogen and then by Precipitation with sodium acetate and acetic acid (pH 5.0) to separates total proteins from whey NPN. The details related to the presence of different types of proteins in milk are given in following table:

Table 11.3: The Concentration of Proteins in Milk [5]

Different Proteins	*Grams/litre*	*Per cent of Total Protein*
Total Protein	33	100
Total Caseins	26	79.5
αs1	10	30.6
αs2	2.6	8.0
β	9.3	28.4
Kappa	3.3	10.1
Total Whey Proteins	6.3	19.3
α lactalbumin	1.2	3.7
β lactoglobulin	3.2	9.8
BSA	0.4	1.2
Immunoglobulins	0.7	2.1
Proteose peptone	0.8	2.4

1. Caseins [1,3,3a]

Casein is considered the specific protein of milk and exits in milk with calcium. The casein content of milk represents about 80 per cent of milk proteins. The caseins have a relatively random, open structure due to the amino acid composition (high proline content). It is present in milk as colloidal dispersion. Casein is present partly in soluble form and partly in the form of suspended micelles.

Casein is dispersed in milk in the form of micelles. The micelles are stabilised by the K-casein. Caseins are hydrophobic but K-casein contains a hydrophilic portion known as the **glycomacropeptide** and it is this that stabilises the micelles.

The distinguishing property of all caseins is their low solubility at pH 4.6 [3]. Their common property of low solubility at pH 4.6 serves as the basis for a rather convenient operational definition (at least for bovine milk). At this pH, all of the caseins except some of the proteolytic derivatives precipitate. As the solubility of caseins is much less and the whey proteins are having better solubility than casein the separation of the casein has become possible by lowering the pH value to 4.6 [1,2,3a].

The common compositional factor is that caseins are **conjugated proteins**, most with phosphate group(s) esterified to serine residues' [3]. The unique feature of the caseins is their ester-bound phosphate. All of the casein polypeptide chains have at least one such group per molecule; whereas, none of the whey proteins have any ester bound phosphate [1,3a].

These phosphate groups are important to the structure of the casein micelle. Calcium binding by the individual caseins is proportional to the phosphate content. The high phosphate content of the casein family allows it to associate with calcium and form calcium phosphate salts. The abundance of phosphate allows milk to contain much more calcium than would be possible if all the calcium were dissolved in solution, thus casein proteins provide a good source of calcium for milk consumers [3].

The conformation of caseins is similar to that of denatured globular proteins. The high number of **proline residues** in caseins causes particular bending of the protein chain and inhibits the formation of close-packed, ordered secondary structures.

The casein family of protein consists of several types of caseins (α-s1, α-s2, β, and 6) and each has its own amino acid composition, genetic variations, and functional properties.

- ✰ The αs1 and β-caseins contain no cysteine residues, while αs2 and κ-caseins each have two cysteine residues.
- ✰ Proline contents of caseins are rather high (αs1-, αs2-, β -,and κ- caseins contain 17,5, 17 and 12 mol per cent, respectively).

There is no organized secondary structure for the various fractions of casein. The various analytical studies revealed that they have short lengths of α -helix or β -sheet structure in them. Their ionizable groups of the amino acids are accessible to titration and are also involved in several other side chains to reaction. Denaturing agents and heating seems to have no effect on the secondary structure of these proteins. Thus their conformation appears to be much like that of denatured globular proteins.

As the protein is having large proportion of proline residues, closely packed orderly secondary conformation is being prevented in this protein molecule. The four caseins differ greatly from each other in charge distribution and the tendency to aggregate in the absence and presence of Ca^{2+} ions [1,3a].Caseins contain no disulfide bonds and tertiary structure (More stabile against heat denaturation because there is very little structure to unfold).

Types of Caseins [1,1a,3,3a]

There are four types of caseins in milk, each having appropriate amino acids that make the milk essential for growth and development of the young. They are also phosphorylated to various degrees, which plays an important role in the stabilization of the casein itself. The principal casein fractions are **alpha [(s1)** and **(s2)] caseins, β -casein, γ –casein** and **kappa-casein**. The proportions of these fractions depend upon species, breed and stage of lactation.

1. The molecular weight, structure and Ca precipitation of aforesaid caseins are summarized in following table [3,4]:

Table 11.4: Summary of Molecular Weight, Structure and Ca Precipitation of Caseins

Properties	*Alpha casein -*	*β-casein*	*kappa-casein*
Molecular weight	(s1)- 23,000; 199 residues, 17 proline residues. (s2)- 25,000; 207 residues, 10 prolines	24,000; 209 residues, 35 prolines	19,000; 169 residues, 20 prolines
Structure	(s1)-Two hydrophobic regions, containing all the proline residues, separated by a polar region, which contains all but one of eight phosphate groups. (s2)- Concentrated negative charges near N-terminus and positive charges near C-terminus.	Highly charged N-terminal region and a hydrophobic C-terminal region. Very amphiphilic protein acts like a detergent molecule. Self association is temperature dependant; will form a large polymer at 20° C but not at 4° C.	Rennet cleavage at the Phe105-Met106 bond eliminates the stabilizing ability, leaving a hydrophobic portion, para-kappa-casein, and a hydrophilic portion called kappa-casein glycomacropeptide (GMP), or more accurately, caseino-macropeptide (CMP).
Ca Precipitation	Both can be precipitated at very low levels of calcium.	Less sensitive to calcium precipitation.	Very resistant to calcium precipitation stabilizing other caseins

The details of aforesaid caseins are discussed below [1,1a,3]-

A. α-Caseins [1,1a,3a]

There are following two types of α-Caseins. The details related to these α-Caseins are given below:

i. αs1-Caseins

The polypeptide chain of αs1-casein consists of two predominantly hydrophobic regions (residues 1-44, and 90-199) and a highly charged polar zone (residues 45-89).

- ☆ All but one of the phosphate groups is in the 45-89 residues segment, and the prolines are distributed at intervals in the hydrophobic segments. Thus, this protein can be visualized as a rather loose flexible polypeptide chain.
- ☆ Self-association of αs1-casein depends markedly on its concentration and on the pH, ionic strength, and kind of ion in the medium, but it is relatively independent of temperature.

Physical measurements (lights scattering, sedimentation, and viscosity) indicate about the dissociation and association of αs1- casein and the details areb given below:

a. αs1- casein is completely dissociated to a flexible random chain monomer of MW = 24,000 at pH 12, in 3 M guanidinium chloride.

b. At neutral pH and 0.01 ionic strength, it associates at neutral pH and higher ionic strength, the extent of association depending on protein concentration, it binds about 8 moles Ca^{2+} per mole near pH 7, probably to the ester phosphate groups. It aggregates and precipitates at very low concentrations of Ca^{2+} (7 mM Ca^{2+},28 mM NaCl). A small amount of peptides, sometimes called A-casein, is present in milk, these appear to originate from proteolysis of αs1-casein.

ii. αs2-Casein

αs2-casein has a remarkable dipolar structure with a concentration of negative charges near the N-terminus and positive charges near the C-terminus.

- ☆ It binds Ca strongly and is even more sensitive to precipitation by Ca^{2+} than αs1-casein.
- ☆ It self-associates at neutral pH in the absence by Ca^{2+}, and the association depends markedly on ionic strength and is at a maximum at an ionic strength of about 0.2.

B. β-Casein [1,1a,3a]

β-casein has a strong negatively charged N-terminal portion. The net charge of the 21- residue N-terminal sequence is 12 at pH 6.6, and the rest of the chain has virtually no net charge.

- ☆ The large number of Pro residues effectively precludes extended helix formation. The β casein molecule is somewhat like that of an anionic detergent with a negatively charged head and an uncharged essentially hydrophobic tail.
- ☆ The outstanding characteristics of the association of β-casein in both the absence and the presence of Ca^{2+} are its strong dependence on temperature. In the absence of Ca^{2+}, only monomer is present at 4°C, but large polymers (20-24 monomers) are formed at room temperature.
- ☆ Removal of the 20- residue C-terminal sequence destroys the ability of β-casein to associate, suggesting that specific hydrophobic interactions maybe involved.
- ☆ β -casein tightly binds about 5 Ca^{2+} per mole, consistent with its ester phosphate content. The revised nomenclature for the fragments of β-caseins in milk is given in following table [1]:

Table 11.5: Revised Nomenclature for Fraction of β-casein [4a]

Sl.No.	Fragment	Recommended Nomenclature	Former Nomenclature
1	29-209	β -casein X* (f29-209)	γ 1-Casein
2	106-209	β -casein X (f106-209)	γ2-Casein
3	108-209	β -casein X (f108-209)	γ3-Casein

Sl.No.	Fragment	Recommended Nomenclature	Former Nomenclature
4	1-105	β -casein X (f1-105)	PP *Component 5
5	1-107	β -casein X (f1-107)	PP Component 5
6	29-105	β -casein X (f29-105)	PP Component 8- slow
7	29-107	β -casein X (f29-107)	PP Component 8- slow
8	1-28	β -casein X (f1-28)	PP Component 8- Fast

Note- X*- Genetic variants of Casein Fractions, **PP** * Proteose – Peptone.

C. γ-Caseins [1,1a]

Group of caseins designated as γ-caseins have been known for some time to correspond to C terminal portions of the β -casein sequence. The γ-caseins are formed by cleavage of β-casein at positions 28/29, 105/106, and107/108 by the enzyme plasmin.

- ☆ The fragments 29-209, 106- 209, and 108-209 constitute the y-caseins. The smaller fragments resulting from the cleavage appear in the whey when casein is precipitated by acid and constitute part of what has long been designated as the proteose-peptone fraction of the whey. Thus, fragments 1-105 and 1-107 were called as whey component 5, fragment 1-28 was whey component 8-fast, and fragments 29-105 and 29-107 were named as whey component 8-slow.

D. κ-Caseins [1,1a]

About one-third of the κ casein molecules are carbohydrate-free and contain only one phosphate group (SerP-149).

They have varying numbers of N-acetyl neuraminic acid (NANA) residues and one, at least, appears to have a second phosphate (SerP-127). Three different glycosyl oligomers linked to Thr-133 have been identified.

- ☆ The N-terminal residue of κ-casein is glutamic acid. In the isolated protein it is present as the cyclic derivative pyroglutamic acid.
- ☆ κ-casein as isolated from milk consists of a mixture of polymers probably held together by intermolecular disulfide bonds, these polymers range in molecular weight from about 60,000 (trimers) to more than 150,000.
- ☆ κ-casein is rapidly hydrolyzed at the Phe (l05)-Met (l06) bond by the enzyme chymosin (EC 3.4.23.4), and by other proteases, yielding N-terminal fragment called para- κ casein, which contains the two cysteine residues, a C-terminal fragment of 64 residues called the macropeptide, containing all of the carbohydrate and phosphate groups as well as the genetic substitutions. κ-casein binds about 2 moles Ca^{2+} per mole of protein at neutral pH but differs markedly from the other caseins in its solubility over a wide range of Ca^{2++} concentration.

- ☆ The naturally occurring mixture of bovine κ-casein variants polymerizes, as previously mentioned, through -S-S- linkages to subunits containing three to eight monomers. These further polymerize by no covalent association to particles of about 6,50,000 daltons. This polymerization is insensitive to concentration of Ca^{2+} and to temperature.

Sources

1. *K.S. Sharma, Chemistry of milk, Agrimoon.com.*

1a. *Milk Protein-Casein, http: //courseware.cutm.ac.in.*

2. *http: //www. web-books.com/MoBio/Free/Ch2B.htm).*

3. *M.K. Srivastava, Handbook on Analysis of Milk, CBS Publications, New Delhi, 2010. ISBN 13: 9788123926681.*

3a. *R. Aschaffenburg and J. Drewry, 1959, New procedure for the routine determination of the various non- casein proteins of milk. Proceedings of the 15th International Dairy Congress, 29 June-3 July 1959, London, 1959, Vol.3, pp. 1631-1637.*

4. *Fred Peaquod and Department of U.S. Agriculture.*

4a. *P.Walstra and R. Jenness, 1984, Dairy chemistry and physics. Wiley, New York.*

5. *Milk Proteins, Manual of University of Guelph, https: //www.uoguelph.ca.*

Proteins Fractionation [1,1a,2,3]

The behaviour of individual milk proteins and their concentration will be known only when they are fractionated from milk. This can be done by chemical or by several other means.

- ☆ A classical scheme for fractionation of milk proteins by chemical method was developed by Rowland in 1938.
- ☆ This was further modified by Aschaffenburg and Drewry in 1959.

The following methods for fractionation of proteins, are in application-

1. Rowland Method of Fractionation [1,1a]

In this method, the following steps will be taken for protein fractionation [1]:

i. Total protein in milk will be fractioned initially into protein nitrogen and non protein nitrogen using the insoluble property of NPN compounds in trichloro acetic acid (TCA).

ii. The casein components are precipitated by the utilizing solubility of non-casein proteins in 10 per cent acetic acid.

iii. The NPN components are further fractioned by utilizing the insolubility of the components with variation.

- Rowland observed that when milk was heated to 95°C for 10 min, 80 per cent of the nitrogenous compounds in whey were denatured and co-precipitated with the casein when the pH of the heated milk was adjusted subsequently to 4.6 [1a].
- He considered that the heat-denaturable whey proteins represented the lactoglobulin and lactalbumin fractions and designated the remaining 20 per cent 'proteose-peptone' [1a].
- The proteose peptone fraction, which is quite heterogenous is precipitated by 12 per cent trichloroacetic acid (TCA) but some nitrogenous compounds remain soluble in 12 per cent TCA and are designated as non-protein nitrogen [1a].

2. Aschaffenburg and Drewry Method [1,1a]

This is a modified method [3] suggested by Aschaffenburg and Drewry in 1959

- The solubility of some fractions of the proteins in sodium sulphate solutions is utilized where the total albumin along with NPN compounds will be soluble.
- In saturated sodium sulphate the albumins get precipitated.

The following figures help in understanding the aforesaid methods.

3. Differential Solubility Methods [1,1a,4,5]

Several methods have been developed to obtain one or more of the various caseins from whole casein or directly from skim milk based on their differential solubility.

a. Scientists [4] developed following two procedures which have been used extensively or partly incorporated into other methods.
   - The first one is based on the differential solubility of casein in 50 per cent alcohol in the presence of ammonium acetate by carrying the pH temperature and ionic strength.
   - The second procedure involves the dispersion of whole casein in 6.6M urea and separation of casein fractions by dilution, pH adjustment and finally addition of $(NH_4)_2 SO_4$.

   The order of precipitation of the casein in both methods is α-casein, β-casein and then γ- casein.

b. Warner s casein fractions [5] are now identified as follows:
   - α- casein is a mixture of αs and κ - caseins.
   - The αs caseins and κ - caseins have been prepared largely from either whole casein or the α casein fraction of Hipp and co workers [4].

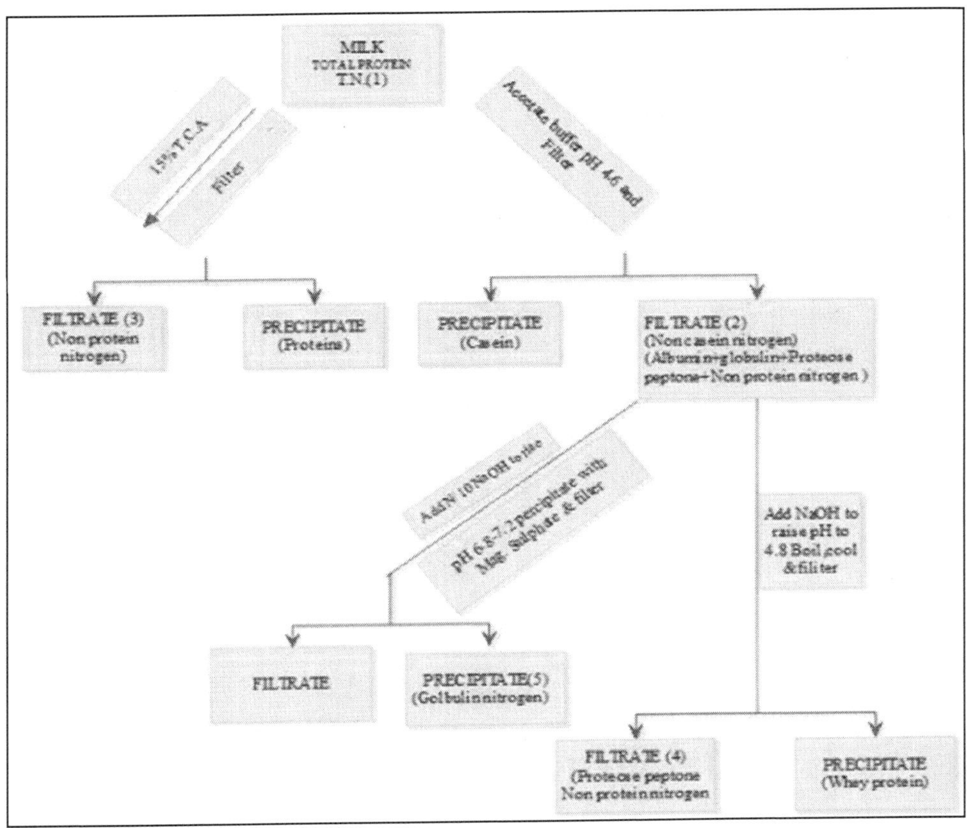

Figure 11.3: Rowland Fractionation [1,1a,2].

- Crude αs -casein is prepared from casein obtained by $CaCl_2$ precipitation at 37°C by removing calcium with oxalate to solubilize the casein and reprecipitation with 0.25M $CaCl_2$ at 37°C and pH 7.0.
- The κ-caseins are removed from the supernatant by precipitation with Na_2SO_4, followed by re-precipitation from 50 per cent ethanol with ammonium acetate or by using calcium oxalate as a precipitate to enhance the removal of the other caseins. Adjustment of the pH of whole casein dispersions in urea also has been used to precipitate the αs -casein either by adjusting 6.6 M urea dispersion to pH 1.3 to 1.5 with H_2SO_4 or by adjusting a 3.3M urea to pH 4.5.
- The crude k caseins can be obtained from supernatant of the H_2SO_4 method, by precipitating with $(NH_4)_2SO_4$ and purified by reprecipitation from aqueous ethanol.
- More than 90 per cent of αs1 -caseins have been recovered from acid casein by precipitation with 75mM $CaCl_2$ at 5°C.

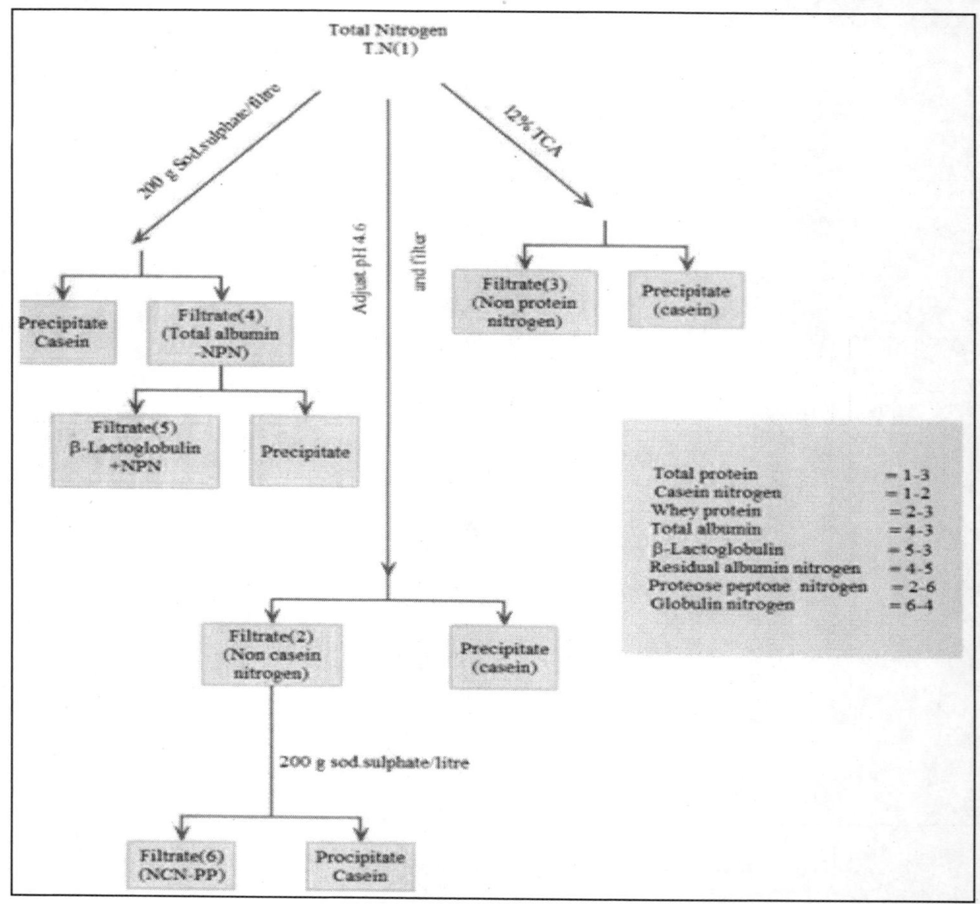

Figure 11.4: Protein Fractionation by Asch-affenburg and Drewry (1959) [1,1a,3].

c. Starting with α -casein fraction of Hipp and co workers [4] the αs -caseins can be precipitated by $CaCl_2$ treatment and the κ-caseins can be removed from the supernatant by pH adjustment to 4.7.

- ☆ Some research workers have added 12 per cent tricholoro acetic acid (TCA) to a 6.6 M urea dispersion of the same fraction at 30°C and precipitated αs-fraction. After removal of the urea and TCA from the supernatant they adjusted the pH to 7.0 added $CaCl_2$ to 0.25 M and removed the precipitate. κ-casein was finally obtained from the supernatant at pH 4.4.
- ☆ Wake in 1959, prepared κ-casein from the supernatant remaining from the β- casein precipitation by the first procedure of Hipp and his co-workers [4]by adjusting the pH to 5.7. A κ-casein concentrate has been prepared from commercial casein based on the differential solubility s of the caseins in $CaCl_2$ solutions.

d. Preparation of β-casein component-The major β-casein component can be prepared by a simplification of the urea method of Hipp and his co workers [4]. Whole casein is dispersed in 3.3M urea at pH 7.5 and adjusted to pH 4.6 which precipitates the bulk of αs1 and κ-caseins. The supernatant is adjusted to pH 4.9 diluted to 1.0M urea and warmed to 30°C precipitating the major β- casein.

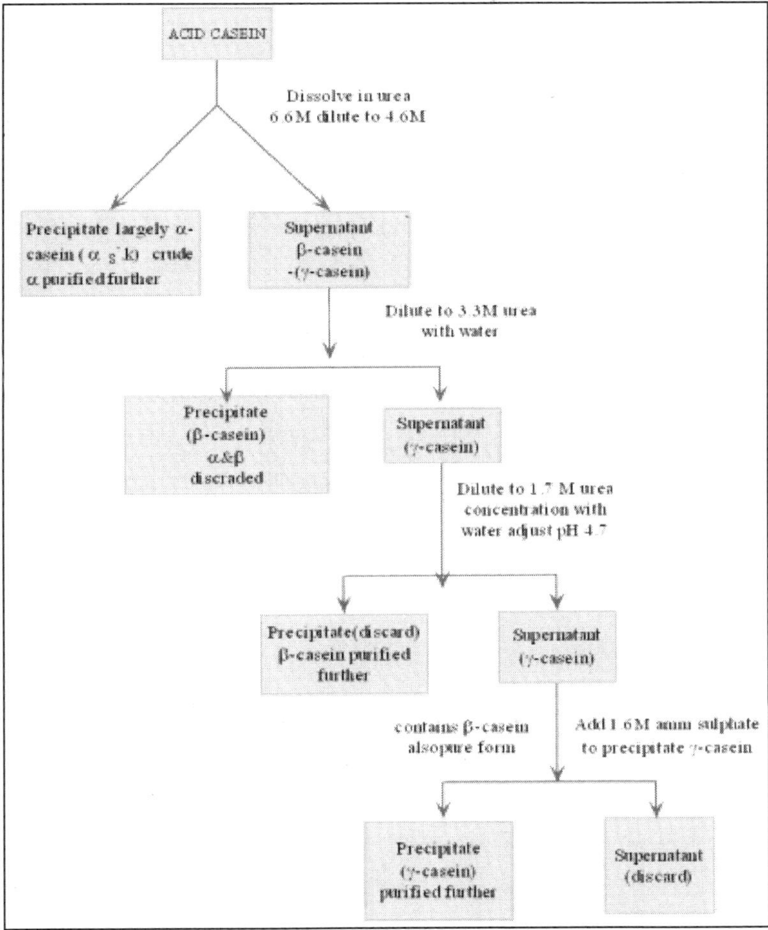

Figure 11.5: Fractionation of Casein by Hipps *et al.*, Base on Urea Solubility [4].

Sources

1. *K.S. Sharma, Chemistry of milk, Agrimoon.com.*

1a. *Milk Protein-Casein, http: //courseware.cutm.ac.in.*

2. *S.J. Rowland, The determination of the nitrogen distribution in milk, J. Dairy Res., 9, 1938, 30.*

3. *R. Aschaffenburg and J. Drewry, 1959, New procedure for the routine determination of the various non- casein proteins of milk. Proceedings of the 15th International Dairy Congress, 29 June-3 July 1959, London, 1959, Vol.3, pp. 1631-1637.*

4. *N.J. Hipp, M.L.Groves, J. H. Custer and T. L.Mcmeerin, Separation of α-, β, and γ-Casein. J. Dairy Sci., 35: 272, 1952.*

5. *Robert C. Warner, Separation of α- and β-Casein, J. Am. Chem. Soc. 1944, 66, 10, 1725-1731.*

Physico-chemical Characteristics of Caseins [1,1a2]

The physico-chemical parameters for the casein fractions are shown in following table. Number of these parameters is derived from the composition based on determination of the primary structure.

Table 11.6: Physico-chemical Characteristics of Caseins [2]

Protein	Charge at pH 6.6	Isotonic pH	Partial specific volume	Absorptivity (cm^2/g) 280nm	Molecular weight
α_{s1}-casein					
A	-20.0	4.97	0.722	1.13	22066
B	-20.9	4.96	0.725	1.05	23612
C	-20.0	5.00	0.725	1.06	23540
D	-22.6	4.91	0.723	1.05	23722
α_{s2} - casein A					
P_{10} (α_{s6})	-13.2	5.39	0.722	1.11	25148
P_{11} (α_{s4})	-14.8	5.32	0.721	1.10	25228
P_{12} (α_{s3})	-16.4	5.25	0.720	1.10	25308
P_{13} (α_{s2})	-18.0	5.19	0.718	1.10	25388
β-casein					
A^3	-12.8	5.11	0.742	0.46	23.971
A^2	-12.3	5.19	0.742	0.46	23980
A^1	-11.8	5.27	0.742	0.46	24020
B	-10.8	5.35	0.742	0.46	24089
C	-8.2	5.53	0.742	0.46	24939
κ-casein					
A	-3.9	5.43	0.734	0.95	19037
B	-3.0	5.64	0.734	0.95	19005

Sources

1. *K.S. Sharma, Chemistry of milk, Agrimoon.com.*

1a. *Milk Protein-Casein, http: //courseware.cutm.ac.in.*

2. *P.F.Fox, Development of Dairy Chemistry, Volume - 4, Applied Science Publishers, 1989.*

Structure of Casein Micelle [1,1b,1c,2]

Water, casein and salts along with some minor components such as lipases and proteinases constitute this structural element of milk [1].Casein micelles does not completely account for the casein content since some of the casein is also present in a solution form. As casein binds cations especially the Ca^{++} and Mg^{++} ions at the normal pH of milk *i.e.* ~6.6, it is referred as caseinate. To provide proteins along with its calcium and phosphate to offspring, the casein in cow milk forms intricate particles which are recognized as casein micelles. Highly insoluble material (calcium phosphate) has to be carried without disturbing either stability or increase in its size [1].

Most of the casein proteins are present in a colloidal particle known as the **casein micelle**. It carries large amounts of highly insoluble CaP to mammalian young in liquid form and to form a clot in the stomach for more efficient nutrition. The micelle in addition to casein protein, calcium and phosphate also contains citrate, minor ions, lipase and plasmin enzymes, and entrapped milk serum. These micelles are rather porous structures and occupying about 4 ml/g and 6-12 per cent of the total volume fraction of milk [3].

Amorphous calcium phosphate and small amounts of citrates are the constituents associated with the casein micelle. Calcium caseinate phosphate complex is the term applicable to describe the casein micelles. Phosphate being a part of the colloidal particles, the term colloidal phosphate is also used [1].

- ☆ The whole casein will form aggregate when it is in solution at a concentration, pH, and ionic strength as in milk and low Ca^{2+} activity. The micelles contain about ten to one hundred casein molecules.
- ☆ The aggregates like globular proteins will have a dense hydrophobic core in which most hydrophobic parts of the casein molecules are buried and a more loosely packed, hydrophilic outer layer containing most of the acidic (carboxylic and phosphoric) and some of the basic groups.
- ☆ Each of these small aggregates of the whole casein, usually called submicelles contains different casein molecules. The relative proportion of α s1: α s2: β and κ in casein micelle is 3: 0.8: 3:1 respectively.
- ☆ More over κ-casein probably exists in milk as an oligomer containing several molecules and held together by covalent bonds (S-S linkages). Consequently there may be essentially two types of submicelles with and without κ-casein.

Casein Micelle Models [1,1c,2]

There have been many models developed over the years to explain the structure of the casein micelle, based on all of the information available about its composition and reactivity. Earlier following casein micelle models were described:

i. Core-coat model,
ii. Internal structure model and.
iii. Sub-unit model

The summary of Core Coat models and Internal Structure model is tabulated below [2a]-

Table 11.7: Summary of Core Coat Models and Internal Structure Model

Sl.No.	*Name of Models*	*Descriptions*
1	Core Coat Models	i. The first model fit in this class was proposed by Waugh and Nobel in 1965. This model is based originally on their studies of the casein solubility in Ca2+ solutions. ii. The second model falls in this category was proposed in 1966 by Payens based on his experimental data on the association of caseins. iii. The last model in this category was proposed by Paquin and co-workers in 1987. This model describes the micelle core as a framework of αs1-caseins and CCP, where β-caseins are bound by hydrophobic interactions.
2	Internal structure models	i. The first internal structure model was proposed by Rose in 1969. He used the known endothermic polymerization of β-casein as the foundation for his micelle structure. ii. Two alternative models, which fall into internal structure category, have been proposed by Holt in 1992 and by Horne (1998).

The Encyclopaedia of Dairy Sciences (2004) states that there is no universally accepted model for the structure of the casein micelle. The sub –unit model, the Holt model and the dual-binding model were introduced to overcome criticism levelled at the first two models (Core-coat model and Internal structure model) [1].

The following models for structure of casein micelle are discussed in details-

a. Casein Sub-micelle/Sub unit Model [2,2a,4]

- ☆ The first sub-micelle model was proposed by Morr in 1967. This model was based on the results obtained from a study of the influence of urea and oxalate treatment on the disruption of casein micelles [2a].

- ✰ The second model falls in this category was proposed by Slattery and Evard in 1973. The model was based on results of experiments on the influence of calcium on the sedimentation behavior of the particles formed in mixtures of caseins' [2a].
- ✰ The most commonly accepted model in the sub-micelle model category was proposed by Walstra in 1984 [2a].

As per this model, prominent for the last several years with out universal acceptance, there are small aggregates of whole casein, containing 10 to 100 casein molecules, called **submicelles**. There are two different kinds of submicelle; with and without kappa-casein and these submicelles contain a hydrophobic core and are covered by a hydrophilic coat which is at least partly comprised of the polar moieties of kappa-casein. The hydrophilic **caseino-macropeptide** (CMP) of the kappa-casein exists as a flexible hair.

The casein sub-micelle model was prominent for many years, but there is sufficient evidence now to conclude that there is not a defined sub-micellar structure to the micelle at all.

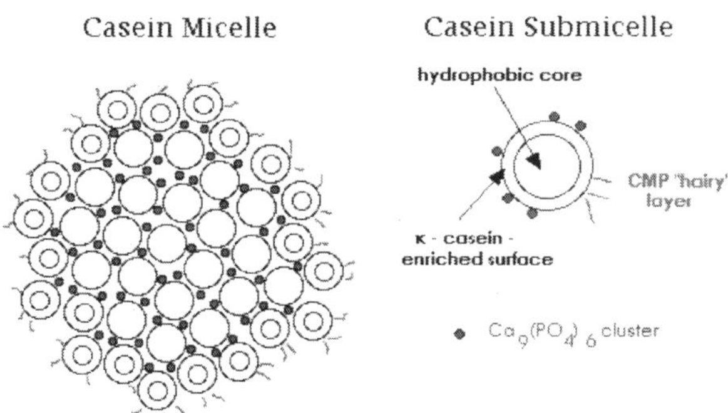

Figure 11.6: Casein Micelle Structure [4].

b. Open Model [2,4]

Another open model of a more **open structure** is also defined. This model also suggests that there are more dense and less dense regions within the micelle, but there is less of a well-defined structure. Calcium phosphate nanoclusters bind caseins and provide for the differences in density within the casein micelle. Colloidal calcium phosphate (CCP) acts as a cement between the hundreds or even thousands of submicelles that form the casein micelle. Binding may be covalent or electrostatic. Submicelles rich in kappa-casein occupy a surface position, whereas those with less are buried in the interior. The hairy layer acts to prohibit further aggregation of submicelles by steric repulsion. The casein micelles are not static and three dynamic equilibria between the micelle and its surroundings *i.e.*

i. The free casein molecules and submicelles.
ii. The free submicelles and micelles
iii. The dissolved colloidal calcium and phosphate, are responsible for that.

c. Dual Binding Model for Micelle Assembly and Structure [1,1c,2a]

Recently, the dual bonding model of Horne (Horne, 1998) was proposed. According to this model micellar assembly and growth take place by a polymerization process involving two distinct forms of bonding:cross linking through hydrophobic regions of the casein or bridging across calcium phosphate nanoclusters.

- ✰ The micellar integrity and stability is maintained by a localised excess of hyrophobic attractions over electro static repulsion.
- ✰ Each casein micelle effectively functions as a block co polymer, with each bloc possessing different different and possibly multiple functionality for the cross linking paths.
- ✰ κ-casein is the most important of the caseins in this model of micellar assembly and structure. It can link in to the growing chain through its hydrophobic N-terminal block but its C-terminal block has no phophoserine cluster and therefore cannot extend the polymer chain through an anocluster link. Final is a chain and network growth of casein micelle with a surface layer of κ -casein molecules.

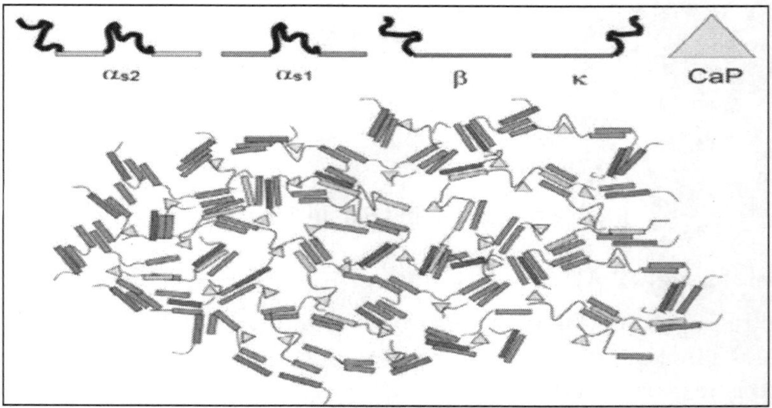

Figure 11.7: Dual Binding Model of Casein Micelle [2a,3].

- ✰ There will be considerable increase in the association of the casein micelle with increase in the Ca^{2+} activity.
- ✰ The calcium phosphate acts as cementing agent. The aggregation will go on until a gel or a precipitate was formed if there were no κ -casein, which thus acts as a **"protective colloid"**.

- ☆ The C-terminal part of κ -casein is very hydrophilic particularly those molecules that contain carbohydrate and it also has a considerable negative charge. This part of the molecule sticks partly out in to the surrounding medium as a flexible "hair" more or less behaving as a random coil polymer chain.

Some areas which having steric repulsion causing hinderance for aggregation of the sub micelles. In this way aggregation of sub micelles would go on until the surface of the micelle was more or less covered with κ-casein. The κ- casein will be inters paced with other caseins, instead of covering the entire surface of the micelles.

Most of the k- casein is at the outside and the protruding chains of its C- terminal end give the micelles a hairy surface which are flexible and show perpetual Brownian motion. The effective thickness of the hairy layer is at least 5nm. A small part of κ- casein is in the interior.

- ☆ The casein micelle is fairly voluminous. About half of the voluminosity of casein micelle is due to the sub micelles while remaining volume is serum between sub micelles. The serum inside the micelles however is depleted of part of the large solute molecules. The pore width in the micelles is probably a few nanometers, Permitting easy access by many solute molecules but restricting the access of globular proteins.
- ☆ Taking the hairy outer layer in to consideration micelles voluminosity is higher on average about 4 ml per gram of casein.
- ☆ The hydrodynamic voluminosity increases with the decreasing micelles size.
- ☆ The casein micelles are not static and have three dynamic equilibrium with its surroundings. The equilibrium is between the free casein molecules and sub micelles(a tiny part of the casein is in true solution) the equilibrium strongly depends on temperature, between the dissolved and colloidal Ca and phosphate, between free sub-micelles and micelles.
- ☆ The casein micelles are rather variable particularly in milk of individual cows. Casein composition varies though not very much.
- ☆ The proportion of κ-casein varies by about 9-15 per cent which will be reflected in the average casein micelle size. The voluminosity varies by a factor of 2 variations in the size and voluminosity usually from 6-9 g per 100g dry casein. There will also be a variation in the composition of inorganic matter.
- ☆ The quantity of calcium phosphate in the micelles increases on heating and very much decreasing when lowering the pH.

- The casein micelles will be dispersed into smaller units by any treatment that dissolves a considerable proportion of the calcium phosphate at constant pH such as adding Na Cl.
- Micelles disperse in to submicelles when milk is subjected to high pressure(100MPa for 1 hr) this disintegration is irreversibly and it may be caused by the formation of hydroxyl apatite from the calcium phosphate. Casein micelles also dissolve when submicelles dissociate. In the sub micelles the molecules are held together mostly by hydrophobic interaction and by "H" bonds in the hydrophobic interior of the submicelles.
- Consequently addition of large quantities of urea or guanidinium chloride and sodium dodocyl sulfate in small quantities will dissolves the micelles.
- Electro static interactions (*i.e.* internal salt bridges) also participate in keeping the submicelles together, but they cannot be broken without dissolving the calcium phosphate. Reagents that break S-S linkage do not disintegrate the micelles fully. Lowering the temperature (eg: to 5 °C) considerably effects the casein micelles.
- Hydrophobic interactions become much weaker and part of the casein particularly of the beta casein dissociate from the micelles. The voluminosity of the micelles increases, probably in part from increased "hairiness" as β-casein chains may protrude from the micelle surface. A small part of the calcium phosphate dissolves. These changes may be the cause of the slight disintegration of the micelles despite the increase in voluminosity their average size decreases to some extent (by 15 per cent).

Properties of Casein Micelle [1, 1a,1b,1c]

A casein molecule is composed of several similar proteins forming a multi molecular granular structure [1a].

- Almost all casein in un-cooled milk is present in roughly spherical particles mostly 0.02 to 0.30 μ m in diameter, comprised of some 20 to 150000 casein molecules.
- These micelles also contain inorganic matter mainly calcium phosphate, about 8 g per 100g casein.
- The casein micelles are voluminous and hold more water than pure casein. They also contain small quantities of other proteins such as the milk enzymes lipase and plasmin and part of proteose-peptone [1,1b,1c].

Importance of Casein Micelle [1,1b,1c]

A study of the properties and structure of casein micelles will help in proper understanding of various phenomena occurring in milk and milk products. The existence and properties of casein micelles have several consequences in the use of milk.

- ✰ The stability of milk products during heating, concentrating, and holding is largely determined by these micelles.
- ✰ The rheological properties of sour milk and concentrated milk are largely determined by the changes occurring in these micelles.
- ✰ The interaction of these micelles with air- water and oil- water interfaces is important in homogenization.
- ✰ The casein micelles also influence some of the properties of the homogenized products.

Sources

1. *K.S. Sharma, Chemistry of milk, Agrimoon.com.*

1a. *Structural properties of casein micelles in milk, The effect of salt, temperature and pH, Master Thesis by: Elsaid Younes Supervisor: Torbjorn Akesson and Marie Skepo, Theoretical Chemistry, Lund University, Sweden, February 5, 2014.*

1b. *Milk Protein-Casein, http: //courseware.cutm.ac.in.*

1c. *Casein micelle: structure, properties and its importance, Lesson-8, http: // ecoursesonline.iasri.res.in.*

2. *M.K. Srivastava, Handbook on Analysis of Milk, CBS Publications, New Delhi, 2010. ISBN 13: 9788123926681.*

2a. *C.Phadungath, Casein micelle structure: a concise review Songklanakarin J. Sci. Technol., 2005, 27(1): 201-212.*

3. *Horne D.S. Structure, self-assembly and gelation, Curr. Opin. Colloid Interface, Sci. 7, 2002, 456-461*

4. *Milk Proteins, Manual of University of Guelph. https: //www. uoguelph.ca.*

2. Whey Proteins [1,3]

Whey protein is a mixture of proteins isolated from whey, the liquid material created as a by-product of cheese production. Whey or serum proteins are soluble milk proteins and these are available as an ingredient in the form of whey powder, whey protein concentrates, and whey protein isolates, mainly manufactured from whey.

The non micellar proteins constitute the whey proteins, but it is often the globular proteins present as such are considered to be the whey proteins [1]. Proteins present in the supernatant of milk after precipitation at pH 4.6 are collectively called whey proteins. Like the other major milk components, each whey protein has its own characteristic composition and variations.

- The unique characteristics of whey proteins such as solubility over a wide range of pH, favorable nutritional profile with respect to essential amino acids, diverse functionality, and relatively low cost make them an ideal ingredient in formulating wide array of food products.
- The serum (whey) protein family consists of approximately 50 per cent β-lactoglobulin, 20 per cent α-lactalbumin, blood serum albumin, immunoglobulins, lactoferrin, transferrin, and many minor proteins and enzymes.
- Albumin and globulin are found in comparatively large quantities in colostrum (12 or more time greater than normal milk). Mastitis milk also contains more concentration of Albumin and globulin. Their proportion rises during extended lactation.

  Whey proteins do not contain phosphorus, by definition, but do contain a large amount of sulfur-containing amino acids. These form disulfide bonds within the protein causing the chain to form a compact spherical shape. The disulfide bonds can be broken, leading to loss of compact structure, a process called denaturing.
- These globular whey proteins are more water soluble than caseins and are subject to heat denaturation. Denaturation increases their water holding capacity.
- The whey proteins serve the body in the same way as the casein does The whey proteins have good gelling and whipping properties.

The whey or serum proteins are not precipitated or coagulated by acid,or by enzyme (Rennin) so they are found in whey [3]. It forms a complex with K-casein when milk is heated to more than 75°C, and this complex affects the functional properties of milk. Denaturation of β-lactoglobulin causes the cooked flavour of heated milk.

Isolation and Fractionation of Whey Proteins [1]

The information related to the separation of whey proteins and the properties of these proteins are helpful for proper understanding of chemistry of milk and could be used to develop suitable procedure in the manufacture of several dairy products. The following procedures have been suggested for isolating the whey proteins from other whey components [1]-

- **By complex formation methods**-The study conducted by Hansen and co- workers in 1971 has revealed that the whey proteins could be separated by complexing the whey proteins with carboxymethyl cellulose. Similarly ferripolyphosphate, hexa meta phosphate and polyacrylic acid were also used to isolate whey proteins by forming a complex.

☆ **Other Instrumental methods**-Gel filtration with Sephadex G-25, ultrafiltration using cellulose acetate membranes, reverse osmosis and electro dialysis are some of the methods available for isolation of whey proteins from whey.

Fractionation of Whey Proteins by Gel Filtration [1,3]

Separation of individual components of whey protein mixture was originally accomplished by fractionations based on differences in solubility and by crystallization.

☆ The gel filtration (which fractionates on the basis of molecular size) and ion exchange(which separates on the basis of net charge on protein) are the preferred methods of fractionation for both analytical and preparative purposes.

☆ The use of Gel filtration technique is particularly useful for fractionation of bovine whey proteins because of their great variation in their molecular size. The fact that β-lactoglobulin (MW = 18,277) forms a noncovalent dimer makes it readily separable from α-lactalbumin (MW = 14,175), as the latter does not aggregate.

The Types of Whey Proteins [1,3]

The principle portions are β -lactoglobulin, alpha-lactalbumin, bovine serum albumin (BSA), and immunoglobulins (Ig). The details of the some of the whey proteins are given below:

1. α-Lactalbumins [1,3]

This whey protein have Molecular weight 14,000; 123 residues. α-Lactalbumin has a highly ordered secondary structure, and a compact, spherical tertiary structure [1].

☆ Thermal denaturation and pH <4.0 results in the release of bound calcium. These proteins contain eight cysteine groups, all involved in internal disulfide bonds, and four tryptophan residues [3].

☆ This family of proteins consists of a major component and several minor components. Three genetic variants of α-lactalbumin have been identified [1].

1. Two genetic variants, A and B, of this protein exist. They differ by a single substitution, A having Gln and B having Arg at position 10.
2. In the milk of European breeds and yaks only B variant is observed while Both (a) and (b) variants occur in the milk of Indian cattle. The complete primary structure of the major α-lactalbumin has been determined.

- ☆ The B variant consists of 123 amino acid residues with a calculated molecular weight of 14,178 and the A variants differ from it only in having Gln instead of Arg at position 10 [1].
- ☆ Some minor forms of bovine α-lactalbumin are revealed by electrophoresis. Some of these contain covalently bound carbohydrate groups; the major component of bovine α-lactalbumin is devoid of carbohydrate. Other minor components seem to have fewer amide groups than the major ones, and one minor α-lactalbumin containing three instead of four disulfides has been reported. In total, the major components do not account for more than 10 per cent of the αlactalbumin [1].
- ☆ The amino acid sequence of α-lactalbumin is similar to that of lysozyme. Indeed, bovine α-lactalbumin B and chicken egg white lysozyme have identical amino acid residues at 49 positions, and the four disulfide groups are located identically (positions between 6and 120; 28 and 111; 61 and 77; and 73 and 91, respectively) in α-lactalbumin. The two proteins have different biological activities without mutual interference.
- ☆ The biological activity of α-lactalbumin is its interaction with galactosyltransferase to promote the transfer of galactose from uridine diphosphate galactose (UDP - galactose) to glucose to form lactose.
- ☆ It has been shown that α-lactalbumin binds two atoms of Ca^{2+} very tenaciously. In fact it is probable that all preparations of this protein have carried this Ca undetected. Removal of the bound Ca with ethylenediamine tetraacetate renders α-lactalbumin more susceptible to denaturation by heat or by addition of guanidine [1].

2. β-Lactoglobulins [1,3]

This whey protein have Molecular weight 18,000; 162 residues. This group, including eight genetic variants, comprises approximately half the total whey proteins. The conformation includes considerable secondary structure and exists naturally as a noncovalent linked dimer.

- ☆ At the isoelectric point (pH 3.5 to 5.2), the dimers are further associated to octamers but at pH below 3.4, they are dissociated to monomers. β –Lactoglobulin has two internal disulfide bonds and one free thiol group [3].
- ☆ Aschaffenburg and Drewry in 1957 demonstrated that there are two components of β-lactoglobulin in the electrophoretic pattern of this protein in the western cattle. However, two more variants have also been identified by other workers. These genetic variants differ in their electrophoretic mobilities in starch or poly acrylamide gel in the ascending order as A > B > C > D.

- Bovine β-lactoglobulin 'B' consists of 162 amino acid residues. Their calculated molecular weight for monomer is 18,227 and dimer is 36000 respectively. The dimer contains five cysteine residues per mole, of which four are involved in disulfide linkages. Location of one disulphide bond always occurs between Cys residues at 66 to 160 positions and the other link is between 106 and119 or 121.
- The single free thiol appears to be equally distributed between Cys119 and Cys 121. The existence of this thiol group is of great importance for changes occurring in milk during heating, as it is involved in reactions with other proteins, notably κ-casein and α-lactalbumin.
- It is likely that considerable portions of the sequence of β-lactoglobulin exist in the α-helix and β-sheet structures.
    1. **α-helix** -Regions that are most likely helical are residues 21-37, 51-63, 127-143, and 154-159 respectively.
    2. **β-sheet**- These structures are likely in 2-19, 39-43,76-88, 91-99, and 101-107, while the structure of the sequence 114-125 is not clearly established.

  β-Lactoglobulin exists naturally as a dimer of two monomeric subunits which is covalently linked. When more than one genetic variant is present, hybrid dimers are formed.
- Dissociation to the monomer occurs below pH 3.4. β-Lactoglobulin A associates to form an octamer at pH 4.5 and low temperature. The B variant (predominant in Western cattle) octamerizes to a smaller extent [1].

3. Bovine Serum Albumin [1,6]

Whey is an important by product from Paneer/cheese industry and it has a large number of important nutrients which must be utilized effectively and efficiently [1]. Bovine serum albumin **(BSA or "Fraction V")** is a serum albumin protein derived from cows. It is often used as a protein concentration standard in lab experiments.

- The nickname "Fraction V" refers to albumin being the fifth fraction of the original Edwin Cohn purification methodology that made use of differential solubility characteristics of plasma proteins. [6].
- Bovine Serum Albumin (BSA) is a major component of blood serum and is synthesized in the liver.BSA gains entrance to milk through the secretary cells and it comprises only about 1.2 per cent of the total milk protein [1].
- The protein as isolated from bovine milk could not be differentiated from that isolated from bovine blood by methods available in 1950.

1. As no work done on the serum albumin isolated from milk and because of this it has been assumed to be identical to that in blood.
2. The protein isolated from blood consists of a single polypeptide chain of 582 amino acid residues. Its tertiary structure reveals three equal-sized globular domains. It has one free thiol and 17 disulfide linkages, which neatly hold the protein in a multi loop structure.

☆ BSA has following important properties [6]-

Table 11.8: Important Properties of BSA

Sl.No.	Properties	Value
1	Molecular weight	66,463 Da (= 66.5 kDa)
2	Isoelectric point in water at 25 °C	4.7
3	pH of 1 per cent Solution	5.2-7
4	Intrinsic viscosity, η	0.0413

☆ There is no specific role of this protein in the function of mammary gland. The behaviour of this protein in milk and milk products and its possible influence on their properties is not known [1].

☆ BSA has numerous biochemical applications including ELISAs, immunoblots, and immunohistochemistry. BSA is also the main constituent of fetal bovine serum, a common cell culture medium. [6].

4. Immunoglobulins [1,6]

Immunoglobulins, also known as antibodies, are glycoprotein molecules produced by plasma cells (white blood cells). It is heterodimeric proteins.

☆ They act as a critical part of the immune response by specifically recognizing and binding to particular antigens, such as bacteria or viruses, and aiding in their destruction [6a].

☆ An immunoglobulin (Ig) is a large, Y-shaped protein used by the immune system to identify and neutralize foreign objects such as pathogenic bacteria and viruses. The antibody recognizes a unique molecule of the pathogen, called an antigen [6]. Immunoglobulins are antibodies synthesized in response to stimulation by macromolecular antigens foreign to the animal. They are polymers with following two kinds of polypeptide chains [1,6]-

i. **Light chain** (L) of MW 22,400,

ii. **Heavy chain** (H)-These are of several types, including γ (MW 52,000), α (MW 52,000-56,000) and μ (MW 69,000).

Each of the Light and Heavy chains consists of a relatively constant and highly variable sequence and appears to be coded for by two genes.

Immunoglobulin Classes (Isotypes) [1,6a]

The nomenclature of the classes of chains and of antibodies was developed for human antibodies and was extended to other species, including cattle, on the basis of homology of the heavy chains as revealed by immunological cross reaction with those of humans [1].

The various antibodies produced by plasma cells are classified by isotype, each of which differs in function and antigen responses primarily due to structure variability. This classification is based on differences in amino acid sequence in the constant region (Fc) of the antibody heavy chains. The following Five major antibody classes have been identified in placental mammals [6a]-

IgA, IgD, IgE, IgG and IgM.

The various immunoglobulin classes and subclasses (isotypes) differ in their biological features, structure, target specificity and distribution. Hence, the assessment of the immunoglobulin isotype can provide useful insight into complex humoral immune response. Assessment and knowledge of immunoglobulin structure and classes is also important for selection and preparation of antibodies as tools for immunoassays and other detection applications [6a].

IgG [1,6b,6c]

It is a monomer with an approximate molecular weight of 146 Kd and a serum concentration of 9.0 mg/mL. IgG is said to be divalent *i.e.,* it has two identical antigen-binding sites that comprise 2 L chains and 2 H chains joined by disulfide bonds.

- ✰ IgG is synthesized mostly in the secondary immune response to pathogens. IgG can activate the classical pathway of the complement system, and it also is highly protective.
- ✰ The four subclasses of IgG include IgG1, IgG2, IgG3, and IgG4. IgG1 is around 65 per cent of the total IgG. IgG2 forms an important host defense against bacteria that are encapsulated.

IgG1and IgG2

IgG and IgA are further grouped into subclasses (*e.g.,* in human IgG1, IgG2, IgG3, IgG4, IgA1 and IgA2) based on additional small differences in the amino acid heavy chain sequences [6a].

- ✰ These are each polymers of two light chains and two heavy chains of the γ type (γ1and γ2). The chains are joined by disulfide linkages to form two antibody sites, each consisting of the variable portion of an H and L chain.
- ✰ IgGI and IgG2 have about 2.9 per cent bound carbohydrate and MW of about 1,50,000. They differ slightly in electrophoretic mobility [1].

IgA and IgM [1,6b,6c]

These immunoglobulin likewise have the basic structure of two Heavy and two Light chains joined by disulfide bridges.

- **IgA**-It appears in 2 different molecular structures: monomeric (serum) and dimeric structure (secretory).
- The serum IgA has a molecular weight of 160 Kd and a serum concentration of 3 mg/mL.
- Secretory IgA (sIgA) has a molecular weight of 385 Kd and a mean serum concentration of 0.05 mg/mL [6b,6c].

In IgA, the H chains are of α type.IgA is secreted as a dimer of two of the basic four-chain units joined by a polypeptide of MW about 25,000, called J-component, and associated with another called secretory component, SC. This complex is called secretory IgA (SIgA) and has a MW of about 3,85,000. The secretory component is a protein of MW about 75,000, consisting of a single polypeptide chain with number of internal disulphide salt bridges. The carbohydrate content is high consists of N-acetyl galactose amine, D-glucose, D-mannose, L-fucose and N-acetyl neuraminic acid. These sugars are bound to the SIgA [1].

IgM

It has a molecular weight of 970 Kd and an average serum concentration of 1.5 mg/ml. In IgM the H chains are of μ type. It is mainly produced in the primary immune response to infectious agents or antigens. It is a pentamer and activates the classical pathway of the complement system. IgM is regarded as a potent agglutinin (*e.g.*, anti-A and anti-B isoagglutinin present in type B and type

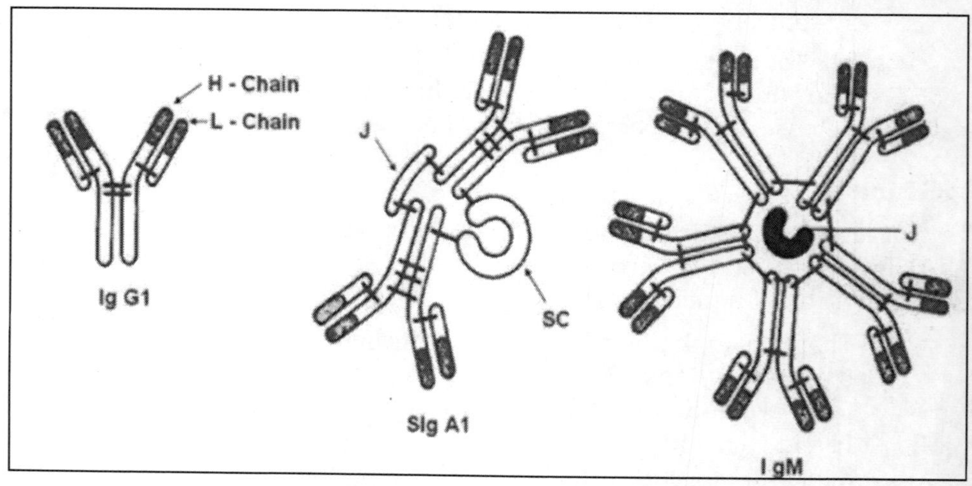

Figure 11.8: Schematic Conformation of Immunoglobulins (IgA,SIgA, IgM---- = di-sulfides linkage. Hatched = Variable portion) [7].

A blood respectively) and a monomer of IgM is used as a B cell receptor (BCR) [6b,6c]. IgM consists of pentamer of the basic four chain units joined by J component and has molecular weight of 9,00,000 and carbohydrate content of 11-12 per cent [1].

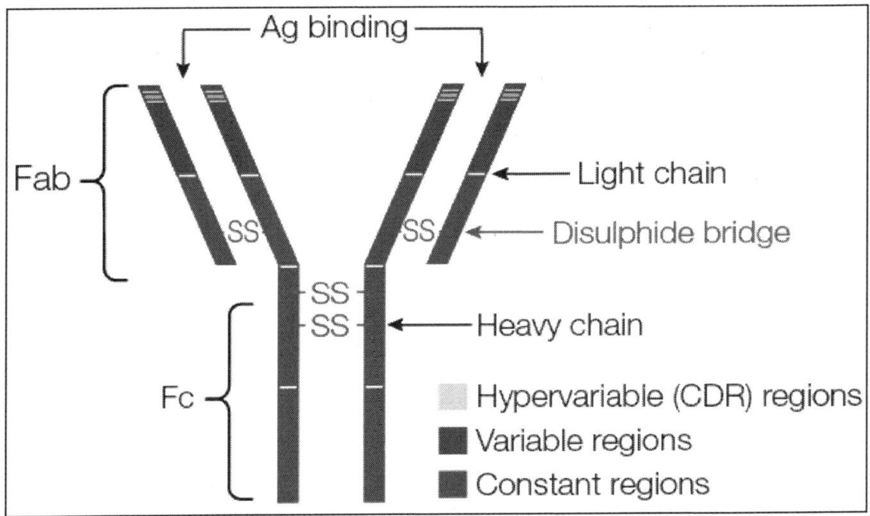

Figure 11.9: Immune System [8]

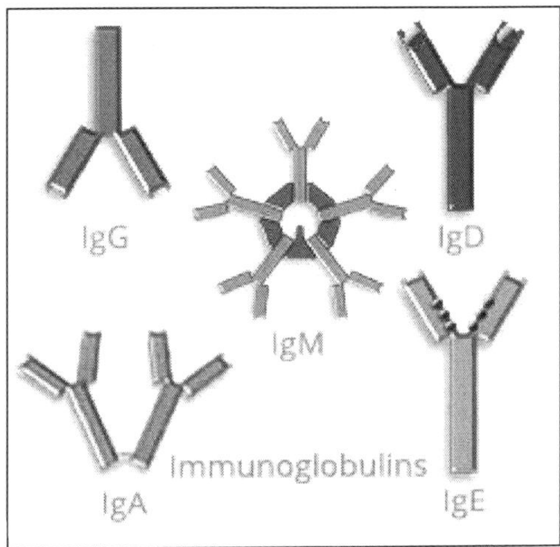

Figure 11.10: Five Classes of Immunoglobulins [9].

Sources

1. *K.S. Sharma, Chemistry of milk, Agrimoon.com.*
2. *http: //www. web-books.com/MoBio/Free/Ch2B.htm).*

3. *M.K. Srivastava, Handbook on Analysis of Milk, CBS Publications, New Delhi, 2010. ISBN 13: 9788123926681.*
4. *Fred Peaquod and Department of U.S. Agriculture.*
5. *Manual of University of Guelph.*
6. *Wikipedia.*

6a. *Introduction to Immunoglobulins, https: //www.thermofisher.com.*

6b. *Vasilev N, Smales CM, Schillberg S, Fischer R, Schiermeyer A. Developments in the production of mucosal antibodies in plants. Biotechnol Adv. 2016, Mar-Apr; 34(2): 77-87.*

6c. *Angel A. Justiz Vaillant; Zohaib Jamal; Kamleshun Ramphul, Immunoglobulin, https: //www.ncbi.nlm.nih.gov.*

7. *P.Walstra and R. Jenness, 1984, Dairy chemistry and physics. Wiley, New York.*
8. *D. Fix., Professor at Southern Illinois University, Immuno-globulin and B cell development, https: //oncology pro.esmo.org.*
9. *Five classes of Immuno globulins, https: //oncology pro. esmo.org.*

Proteose Peptone [1,2,3]

Proteose and peptones are the polymers of amino acids which are of lower molecular weight than proteins. They often are formed by the partial hydrolytic degradation of proteins. A proteose is any of various water-soluble compounds that are produced during digestion by the hydrolytic breakdown of proteins short of the amino acid stage. It forms after breaking down of polypeptides by gastric pepsin. In addition to proteoses, peptones are also formed at this stage.

- ☆ Peptones are different from proteoses in that proteoses are precipitated from solution by half saturation with ammonium sulfate, while peptones are not, even with fully saturated ammonium sulfate [2,3].
- ☆ The difference between peptone and proteose is that peptone is any water-soluble mixture of polypeptides and amino acids formed by the partial hydrolysis of protein while proteose is a mixture of peptides produced by the hydrolysis of proteins.

Rowland has defined the conditions and procedures for obtaining the various fractions of protein in milk and to quantify them. It was observed that –

- **i.** if milk is heated, about 80 per cent of the whey proteins consisting mainly of α lactalbumin and β-lactoglobulins which precipitate with the casein when it is precipitated by acidification to pH 4.6.
- **ii.** The remaining 20 per cent is a separate protein to which he applied the name proteose-peptone.

Proteose and peptones are usually not heat denaturable and hence it was easy for Rowland to reason by analogy that the proteins of milk which remain soluble in acid after heating are proteoses and peptones. The nature of proteins in this fraction has not been clearly established. The issue is complicated by the fact that the fraction may consist in part of native proteins and in part of breakdown products resulting from heat treatment [1].

Sources

1. *K.S. Sharma, Chemistry of milk, Agrimoon.com.*
2. *Wikipedia.*
3. *Miall, Stephen, 1940, A New Dictionary of Chemistry. Longmans. pp. 420–421.*

Non-Protein-Nitrogenous Compounds in Milk [1,2]

Non-protein nitrogen accounts for 5-6.8 per cent, 5 per cent, around 20-30 per cent and 5-6 per cent of the total nitrogen in sheep milk [1],buffalo milk [2], human milk [3] and cow milk [4] respectively. Milk contains a significant and variable (3-8 per cent) NPN fraction. NPN corresponds to the fraction soluble in 12 per cent trichloroacetic acid (TCA), and whey proteins N corresponds to the difference between non-casein N and NPN [5]. Milk serum also contains a large number of nitrogenous substances, each in relatively small concentration. They are collectively called non-protein nitrogen (NPN).

Milk on addition of 12 per cent Trichloro acetic acid (TCA),would result in the precipitation of all the caseins, α lactalbumin and β lactoglobulins leaving the non protein nitrogenous compounds in the filtrate. The major compounds identified from this filtrate are given below [6]-

i. Uric acid.
ii. Creatine
iii. Creatinine
iv. Orotic acid.
v. α-aminonitrogen
vi. Hippuric acid
vii. Indicans
viii. Phosphoglyceroethanolamine
ix. o-phospho-ethanolamine
x. phenylacetyl glutamine.

The compounds present in the urine of dairy animals have a remarkable similarity between those present in non protein nitrogen fractions of milk. As a matter of fact the compounds or substances present in the urine of dairy animals are the resultant waste metabolites of dairy animal's body. It is apparent that the bulk of these waste metabolites in the urine of dairy animals originate from the blood and hence this entry or appearance and levels in milk or urine are due to the protein metabolism of the animals.

Effect on NPN of Milk [6]

Non protein nitrogen in the mixed milk varies within narrow limits with an average of 5.7 per cent under the influence of environmental factors and season. The point wise details are given below:

i. Intake of feed by the animals is directly proportional to the presence of these compounds or substances in milk.

ii. NPN in milk varies from season to season and has no biological value as protein. It cannot be utilized by the body as a substitute of protein nor can it increase the cheese yield.

iii. Pasteurization by itself has no effect. However pasteurization with homogenization causes an increase in the non protein and amino nitrogen content.

iv. Increase in NPN was observed when concentrated milk is stored.

Sources

1. *M.Ramos, M. Juarez, Milk/Sheep milk, in Encyclopedia of Dairy Sciences (Second Edition), 2011.*
2. *J.S. Sindhu, S. Arora, Milk/Buffalo milk, in Encyclopedia of Dairy Sciences (Second Edition), 2011.*
3. *L.Kalbe, B.Reusens, C.Remacle, Diet and the prevention of degenerative disease, in Functional Foods, Ageing and Degenerative Disease, 2004.*
4. *Johnson, A. H., 1974, The composition of milk. Page 1 in Fundamentals of Dairy Chemistry. B. H. Webb, A. H. Johnson, and J.A. Alford, ed. AVI Publ. Co., Inc., Westport, cr.*
5. *D. Dupont, D. Lefier, Milk proteins, in Encyclopedia of Dairy Sciences (Second Edition), 2011.*
6. *K.S. Sharma, Chemistry of milk, Agrimoon.com.*
7. *M. Chrif, A. El Hourch, S. Chouni, S. Aitbenyouf, A.El abidi, The variation of the non-protein nitrogen content by region and season and their impact on the analysis of milk proteins.*

Other Whey Proteins

The following other whey proteins are also found in milk:

A. Lactoferrin and Transferrin [1,2]

The following two iron-binding proteins are found in milk [1]-

i. Transferrin (Tf)

One of two iron binding proteins is transferring (Tf). It is a common blood plasmaprotein [1]. Transferrins are glycoproteins found in vertebrates which bind to and consequently mediate the transport of Iron (Fe) through blood plasma [2,2a]. It is produced in the liver and contains binding sites for two Fe3+ atoms [2,2b]. Transferrin has a molecular weight of around 80 kDa and contains two specific high-affinity Fe(III) binding sites [2].

ii. Lactoferrin (Lf)

The other iron binding protein is lactoferrin (Lf) and it is secreted not only by mammary glands but also by lacrymal, bronchial, and salivary glands and by kidney and endometrial mucosa [1]. Lactoferrin (LF), also known as lactotransferrin (LTF), is a multifunctional protein of the transferrin family. Lactoferrin is a globular glycoprotein with a molecular mass of about 80 kDa that is widely represented in various secretory fluids, such as milk, saliva, tears, and nasal secretions [2]. It also is found in specific granules in heterophilic leukocytes. Lactoferrin can be purified from milk or produced recombinantly. Lactoferrin is one of the components of the immune system of the body; it has antimicrobial activity (bacteriocide, fungicide) and is part of the innate defense, mainly at mucoses [2,3]. Lactoferrin is one of the transferrin proteins that transfer iron to the cells and control the level of free iron in the blood and external secretions. It is present in the milk of humans and other mammals [4]. Lactoferrin has antibacterial, antiviral and antifungal activities [2]. Lactoferrin is an inhibitor of bacteria because it deprives them of iron.

Structure and Properties of Tf and Lf [1,2]

i. Both Tf and Lf appear to be large single chain polypeptides of 600-700 amino acid residues. The recent work favours 75,000 to 77, 000 for Tf, but values for Lf are not so consistent, 77, 000 or 93,000 being reported. In both proteins about 4 mol per cent of the residues are Cys, and both have covalently linked carbohydrate consisting of N-acetylglucosamine, mannose, galactose, and N-acetylneuraminic acid.

ii. All transferrins and lactoferrins appear to bind 2 mol of Fe^{3+} per mole.

iii. Tf and Lf differ markedly from each other in amino acid composition and in electrophoretic mobility. They can be detected readily in electrophoretic patterns by auto radiography with 59Fe. Electrophoretic patterns of milk and blood preparations from individual animals reveal the occurrence of genetic variants of both proteins.

iv. No immunological cross-reaction between Tf and Lf has been demonstrated even when both are from a single species. Amino acid analyses and partial sequences of human Lf and Tf indicate some degree of homology between the two and some internal homology of peptide segments with in each; sequencing is far from complete, however.

v. Determination-Both Tf and Lf can be determined quantitatively in a biological fluid by immuno diffusion method using a specific antiserum. The concentrations and ratios of Tf and Lf in milk vary greatly among species and with stage of lactation. The concentration of Lf in colostrums is about 1250 mg. kg-I in mid-lactation, the concentration falls to less than 100 mg kg-I. Concentrations of Tf in milk have not been determined accurately but may be similar to those of Lf. The concentration of Tf in milk have not been determined accurately but may be similar to those of Lf. Lactoferrin is an inhibitor of bacteria because it deprives them of iron. The concentration of Lf in bovine milk is so low, however, that it does not exert any significant antibacterial effect.

Sources

1. *K.S. Sharma, Chemistry of milk, Agrimoon.com.*

2. *Wikipedia.*

2a. *Crichton RR, Charloteaux-Wauters M, May 1987, "Iron transport and storage". European Journal of Biochemistry. 164 (3): 485-506.*

2b. *Hall DR, Hadden JM, Leonard GA, Bailey S, Neu M, Winn M, Lindley PF, January 2002, "The crystal and molecular structures of diferric porcine and rabbit serum transferrins at resolutions of 2.15 and 2.60 A, respectively". Acta Crystallographica. Section D, Biological Crystallography. 58 (Pt 1): 70-80.*

3. *Sánchez L, Calvo M, Brock JH, May 1992, "Biological role of lactoferrin". Archives of Disease in Childhood. 67 (5): 657-61.*

4. *Johansson B, Virtanen AI, Tweit RC, Dodson RM, 1960, "Isolation of an iron-containing red protein from human milk" (PDF). Acta Chem. Scand. 14 (2): 510–512.*

B. Milk Fat Globules Membrane Proteins (MFGM) [1,2]

The details are given separately under milk Fat Globules Membrane Proteins.

C. Acid Glycoproteins [1,2]

It is Formerly called as orosomucoid. Orosomucoid (ORM) or alpha-1-acid glycoprotein (α1AGp, [3] AGP or AAG) is an acute phase protein found in plasma [2]. It belongs to the structurally unrelated group of acute phase proteins. It is modulated by two polymorphic genes.

- ☆ It has been isolated from human serum, colostrum, and milk, and from bovine serum and colostrum, but it has not been detected in bovine milk.
- ☆ It consists of a polypeptide chain of 181 residues to which five heteropolysaccharide groups are linked to asparagine residues. The

carbohydrate constitutes about 45 per cent of the total molecule. In any event, α1-acid glycoprotein comprises only a small portion of the acid glycoproteins obtainable from fractionation of colostrum or milk on DEAE-cellulose.

- ☆ Five other fractions have been obtained in varying states of homogeneity [1,4]. All contain carbohydrate and phosphate and promote the growth of Bifido bacteriumbifidum var. pennsylvanicus (formerly Lactobacillus bifidus). The possibility that some of these glycoproteins represent partial degradation products of caseins or membrane materials has not been elucidated [1].

Milk-AGP content is made up of at least two isoform groups, a low MW group (44 kDa) that is produced in the mammary gland (MG-AGP) and a higher MW group (55–70 kDa), that is produced by somatic cells (SC-AGP). Identical SC-AGP isoforms can be found both in milk and blood PMN cells. Immunohistochemistry showed that AGP is found in all tissues that make up the mammary gland, but that it is most likely produced for the main part by the alveoli. [4].

The function of this protein is not known but available literatures now establish its some functions [4]. AGP, on one hand is an immuno-modulatory protein and on the other hand, AGP is one of the most important binding proteins in plasma. This protein regulates every single event related to inflammation, including binding of pathogens and modulating white blood cells activity throughout the entire leukocyte attacking sequence. AGP is a lipocalin with immunomodulatory and anti-inflammatory activity and plays an important role in the regulation of local inflammatory reactions [5].

Sources

1. *K.S. Sharma, Chemistry of milk, Agrimoon.com.*
2. *Wikipedia.*
3. *Logan, Carolynn M., Rice, M. Katherine, 1987, Logan's Medical and Scientific Abbreviations. Philadelphia, J. B. Lippincott Company. P-3, ISBN 0-397-54589-4.*
4. *Fabrizio Ceciliani, Vanessa Pocacqua, Cristina Lecchi and Paola Sartorelli, Differential expression and secretion of a1-acid glycoprotein in bovine milk, Journal of Dairy Research, 2007, 74 374-380.*
5. *Fabrizio Ceciliani and Cristina Lecchi, The Immune Functions of α1 Acid Glycoprotein, Current Protein and Peptide Science, Volume 20, Issue 6, 2019.*

D. Folate Binding Protein [1,2,4,5]

Folate binding proteins (FBP) have been separated and identified in both bovine and human milk [1].

- ☆ FBP is another important human milk protein and it has been found in both particulate and soluble form in human milk. The soluble FBP is glycosylated to about 22 per cent, which may help it to survive proteolytic digestion [2].
- ☆ A specific protein that binds folate (FBP) has been isolated from milk. Affinity chromatography on Sepharose to which folate has been attached is especially effective in isolating this protein. Its concentration in normal milk is about8 mg/100 ml [3].

Uses [4]-

- **i.** FBP can refer to protein(s) (*e.g.* extracted from cow's milk) used to do folic acid assays. [4,5]
- **ii.** FBP can be measured to help in tumour classification to aid treatment, as it is over-expressed (20-80 fold) in more than 90 per cent of ovarian and endometrial cancers, as well as 20-50 per cent of breast, lung, colorectal, and renal cell carcinomas. FBP has very limited tissue distribution and expression in non-malignant tissue, making it a good immunotherapy target. [4]

Sources

1. *Mark A. Fenelon and Eoin G. Murphy, Whey proteins in Infant formula, in Whey Proteins, 2019.*
2. *G.M. Hendricks, M. Guo, Bioactive components in human milk in Human Milk Biochemistry and Infant Formula Manufacturing Technology, 2014.*
3. *K.S. Sharma, Chemistry of milk, Agrimoon.com.*
4. *Wikipedia.*
5. *Sobczyñska-Malefora A, Harrington DJ (November 2018). "Laboratory assessment of folate (vitamin B9) status". Journal of Clinical Pathology. 71 (11): 949-956.*

E. β2-Microglobulin [1,2]

β2 microglobulin also known as B2M is a component of MHC class I molecules, MHC class I molecules have α1, α2, and α3 proteins which are present on all nucleated cells (excludes red blood cells) [3]. In humans, the β2 microglobulin protein is encoded by the B2M gene [3,4].

β-2 microglobulin is a light chain of the major histocompatibility complex. It is a LMW (17 kDa) ubiquitously expressed protein [1]. This protein consists of a single polypeptide chain of about 100 amino acid residues and MW of 11,800 daltons. It is present in several body fluids and in membranes of various types of cells.

- ☆ Its amino acid sequence indicates homology with the constant regions of immunoglobulin light and heavy chains.

- ✰ A protein that had long previously been crystallized from bovine milk and designated lactollin was shown in the year 1977 to be bovine β2-microglobulin.
- ✰ It is a polypeptide of 98 amino acid residues, two of which are Cys.
- ✰ Direct quantification of microglobulin concentration have not been made in bovine milk; the amounts of lactollin that have been isolated from colostrum and milk are about 6 and 2 mg per liter, respectively [1a].

Sources

1. *J. Tarrant, Experimental ADME and Toxicology, in Comprehensive Medicinal Chemistry III, 2017.*

1a. *K.S. Sharma, Chemistry of milk, Agrimoon.com.*

2. *Wikipedia.*

3. *Güssow D, Rein R, Ginjaar I, Hochstenbach F, Seemann G, Kottman A, Ploegh HL, 1987, "The human beta 2-microglobulin gene. Primary structure and definition of the transcriptional unit". Journal of Immunology. 139 (9): 3132-8.*

4. *Suggs SV, Wallace RB, Hirose T, Kawashima EH, Itakura K (November 1981). "Use of synthetic oligonucleotides as hybridization probes: isolation of cloned cDNA sequences for human beta 2-microglobulin". Proceedings of the National Academy of Sciences of the United States of America. 78 (11): 6613-7.*

Factors affecting the Stability of Casein Micelle and Casein Micelle System [1,2]

The **Concentration of Ca^{++}** (The removal of Ca^{++} leads to reversible dissociation of β -casein where as the addition of Ca^{++} leads to aggregation), **Hydrophobic Interactions** (Caseins, hydrophobic proteins play a role in the stability of the micelle. Hydrophobic interactions are very temperature sensitive), **Electrostatic Interactions** (The micellar structure is very loose and open and some subunit interactions may be due to the ionic bonding), **H Bonding** (occurs between the individual caseins in the micelle), **Disulfide Bonds** (α (s1) and β-caseins having no cysteine residues and if any S-S bonds present within the micelle, they are not the driving force for stabilization), **Steric stabilization-** (The hairy layer interferes with interparticle approach) and **van der Waals Forces** are the factors having considerable importance while assessing the stability of the casein micelle where as **Salt content** (Affects the calcium activity in the serum and calcium phosphate content of the micelles), **Temperature** (beta-casein begins to dissociate from the micelle at 4°C and 0°C, no micellar aggregation at 0°C), **Heat Treatment** (Whey proteins become adsorbed, altering the behaviour of the micelle), **Dehydration-**By ethanol leads to aggregation of the micelles) and **pH** (Lowering the pH leads to dissolution of calcium phosphate until, at the isoelectric point -pH 4.6), and are the factors having considerable importance while assessing the stability of the casein micelle system.

Casein Micelle Aggregation [1,2]

Caseins are able to aggregate if the surface of the micelle is reactive. Although the casein micelle is fairly stable, there are four major ways in which aggregation can be induced-

i. **Heat** - The casein micelles will irreversibly aggregate at temperatures above the boiling point. On heating, there is change in buffer capacity of milk salts, carbon dioxide is released, organic acids are produced which causes precipitation of tricalcium phosphate and casein phosphate with the release of hydrogen ions.

ii. **Acid Coagulation** – The Addition of acid to milk causes the casein micelles to destabilize or aggregate by decreasing their electric charge to that of the isoelectric point and at the same time, the acidity increases the solubility of minerals so that organic calcium and phosphorus contained in the micelle gradually become soluble in the aqueous phase. Casein micelles disintegrate and casein precipitates. Aggregation results because of the entropically driven hydrophobic interactions.

iii. **Enzyme Coagulation -Rennet** or Chymosin, is generally used for enzyme coagulation. During the **primary stage**, rennet cleaves the Phe (105)-Met (106) linkage of kappa-casein resulting in the formation of the soluble CMP which diffuses away from the micelle and para-kappa-casein. During the **secondary stage**, the micelles aggregate. This is due to the loss of steric repulsion of the kappa-casein as well as the loss of electrostatic repulsion due to the decrease in pH.

   When the pH of milk is changed, the acidic or basic groups of the proteins will be neutralised. At the pH at which the positive charge on a protein equals exactly the negative charge, the net total charge of the protein is zero. **This pH is called the isoelectric point of the protein (pH 4.6 for casein).** As the pH approaches its isoelectric point (pH 4.6), the caseins aggregate. The casein micelles aggregate because of hydrophobic interactions. Calcium assists coagulation by creating isoelectric conditions and by acting as a bridge between micelles. The increase in temperature up to 40° C increases the rate of the rennet reaction and during the secondary stage, increased temperatures increase the hydrophobic reaction. In the **tertiary stage** of coagulation,there is rearrangement of micelles after a gel has formed.

iv. **Age Gelation [1,2]-** Age gelation is an aggregation phenomenon which affects the quality of shelf-stable, sterilized dairy products, such as concentrated milk and UHT milk. The weeks to months storage of aforesaid products some time have a sudden sharp increase in viscosity accompanied by visible gelation and irreversible aggregation of the micelles into long chains forming a three-dimensional network. This phenomenon may be carried either due to the Proteolytic breakdown of the casein or Chemical reactions (polymerization of casein and whey proteins due to Maillard type or other chemical reactions). The formation of kappa-casein-β -lactoglobulin complexes may also be referred as a cause of age gelation.

Casein coagulates with acid, enzymes and alcohol. Acid coagulation neutralizes the charges on the casein particles and separates the casein from the calcium or phosphate naturally attached to the casein. Casein is also liable to enzyme coagulation but in this case, the mineral components with which it is associated in milk remain with the casein hence **Rennet casein has higher minerals than acid casein.** The casein of normal fresh milk is not readily coagulated by alcohol. If milk has 0.06-0.08 per cent developed acidity or 0.20 per cent total titratable acidity then it will coagulate when equal quantity of such milk and 68-70 per cent alcohol are mixed. Colostrum and Mastitis milk coagulate with alcohol. This reaction of the casein with alcohol is primarily a result of improper salt or mineral balance in the milk.

Sources

1. *M.K. Srivastava, Handbook on Analysis of Milk, CBS Publications, New Delhi, 2010. ISBN 13: 9788123926681.*
2. *Milk composition, University of Guleph, https: //www. uoguelph.ca.*

Denaturation and Coagulation of the Proteins [1,2]

The handling and processing of the milk will change the physico-chemical environment which would result in the denaturation of the protein molecule [1] which ultimately reduces the nutritional value of milk proteins. Milk Proteins play an important role in the human nutrition. The definition of denaturation of milk protein is given below:

Denaturation is a phenomenon that involves transformation of a well-defined, folded structure of a protein, formed under physiological conditions, to an unfolded state under non-physiological conditions [1].

Protein denaturation is the net effect of alterations in the biological, chemical, and physical properties of the protein by mild disruption of its structure.

- Denaturation involves the breaking of many of the weak linkages, or bonds (*e.g.*, hydrogen bonds), within a protein molecule that are responsible for the highly ordered structure of the protein in its natural (native) state. Denatured proteins have a looser, more random structure; most are insoluble [1a].
- Denaturation is a process in which proteins or nucleic acids lose the quaternary structure, tertiary structure, and secondary structure which is present in their native state, by application of some external stress

or compound such as a strong acid or base, a concentrated inorganic salt, an organic solvent (*e.g.*, alcohol or chloroform), radiation or heat. [1d,3,4]

- Denaturation of the proteins and coagulation or clotting characterizes the early stage of protein change. Heating fresh milk precipitates soluble calcium salts naturally present. This increases the time required for coagulation. The addition of soluble calcium reduces the coagulation time of milk both on addition of rennet or on heating at 120°C. Sodium and Potassium cause similar changes but phosphate and citrate ions increase heat coagulation time. Heating of milk beyond 75°C can activate the "Sulphydryl groups " making proteins more liable to oxidation [2].
- Denaturation causes a change in the physical structure of proteins, but generally does not affect the amino acid composition and thus the nutritional properties. Denatured proteins lose their biological activity (*e.g.* as enzymes), but not their nutritional value. Indeed, their digestibility is improved compared with the native structures, which are relatively resistant to enzymatic hydrolysis [1,2].

Reversibility and Irreversibility in Denaturation of Proteins [1d,2,3]

Protein denaturation can be reversible or irreversible.

i. In many cases, denaturation is reversible (the proteins can regain their native state when the denaturing influence is removed). This process can be called **renaturation**. [4] This understanding has led to the notion that all the information needed for proteins to assume their native state was encoded in the primary structure of the protein, and hence in the DNA that codes for the protein, the so-called "Anfinsen's thermodynamic hypothesis" [5].

ii. Denaturation can also be irreversible. This irreversibility is typically a kinetic, not thermodynamic irreversibility, as generally when a protein is folded it has lower free energy. Through kinetic irreversibility, the fact that the protein is stuck in a local minimum can stop it from ever refolding after it has been irreversibly denatured [6].

Factors/Agents Responsible for Protein Denaturation [1,2]

A variety of factors are responsible for change in the structure of proteins. Some factors are encountered frequently while others are less and more of theoretical interests. The chemical breakdown of milk proteins can be caused by heat, acid, enzymes, alcohol *etc.*

- A change in the structure of protein by heat, acid, alkali, or other agents such as sound waves, surface forces, pressure, UV radiation and ionizing radiations, results in loss of solubility and coagulation which is otherwise known as denaturation.

- Treatment with organic solvents such as alcohol, acetone, and solutes like urea, guanidine and ionic detergents would also result in protein denaturation. It is normally irreversible.

The pointwise details related to the factors affecting the denaturation of protein, are discussed below:

1. Thermal Denaturation [1,2a]

Heat treatment of milk and milk products is an essential operation in commercial dairy processes in order to provide acceptable safety and shelf life. The most important reactions that occur during milk heat treatment are the whey proteins denaturation, its interactions with the casein micelles and aggregation/dissociation of the casein micelles.

- Denaturation alters several important properties of proteins from the viewpoint of food technology. The denatured protein is generally less soluble or even insoluble, promotes an increase in feed viscosity and the reactivity of their side groups are intensified. Among the milk proteins, whey proteins are the most thermolabile.
- Pasteurization or UHT processing causes partial, irreversible unfolding of β-lactoglobulin, thereby exposing hydrophobic surface and the sulfhydryl group. The subsequent interaction with κ-casein, stabilized by a sulfhydryl-disulfide interchange, alters the surface properties of casein micelles [2a].
  - **i.** The caseins are stable to heat treatment. Typical high temperature short time (HTST) pasteurization conditions will not affect the functional and nutritional properties of the casein proteins. High temperature treatments can cause interactions between casein and whey proteins that affect the functional but not the nutritional properties. For example, at high temperatures, β-lactoglobulin can form a layer over the casein micelle that prevents curd formation in cheese.
  - **ii.** The whey proteins are more sensitive to heat than the caseins. HTST pasteurization will not affect the nutritional and functional properties of the whey proteins. Higher heat treatments may cause denaturation of β-lactoglobulin. Severe heat treatments such as ultra high pasteurization may cause some damage to heat sensitive amino acids and slightly decrease the nutritional content of the milk. The whey protein α-lactalbumin, however, is very heat stable
- On heating (exposed to increasing temperature) of milk proteins, there are losses of solubility or enzymatic activity over a fairly narrow range. A number of bonds in the protein molecule are weakened when

heating temperature is increased. Depending on the nature of proteins, the thermal denaturation of proteins is reversible and irreversible as discussed earlier. The weakening of bonds takes place in the following manners [1]-

a. The first affected are the long range interactions that are necessary for the presence of tertiary structure. As these bonds are first weakened and are broken, the protein obtains a more flexible structure and the groups are exposed to solvent. If heating ceases at this stage the protein should be able to readily refold to the native structure.

b. As heating continues, some of the cooperative hydrogen bonds that stabilize helical structure will begin to break.

   i. As these bonds are broken, water can interact with and form new hydrogen bonds with the amide nitrogen and carbonyl oxygen of the peptide bonds.

   ii. The presence of water further weakens nearby hydrogen bonds by causing an increase in the effective dielectric constant near them. As the helical structure is broken, hydrophobic groups are exposed to the solvent.

   iii. The effect of exposure of new hydrogen bonding groups and of hydrophobic groups is to increase the amount of water bound by the protein molecules.

c. Heating at higher temperature-The unfolding that occurs increase the hydrodynamic radius of the molecule causing the viscosity of the solution to increase. The net result will be an attempt by the protein to minimize its free energy by burying as many hydrophobic groups while exposing as many polar groups as possible to the solvent. While this is analogous to what occurred when the protein folded originally, it is happening at a much higher temperature. This greatly weakens the short range interaction that initially direct protein folding and the structures that occur will often be vastly different from the native protein.

d. Cooling-Upon cooling, the structures obtained by the aggregated proteins may not be those of lowest possible free energy, but kinetic barriers will prevent them from returning to the native format.

   ☆ Any attempt to obtain the native structure would first require that the hydrophobic bonds that caused the aggregation be broken. This would be energetically unfavourable and highly unlikely.

   ☆ Only when all the intermolecular hydrophobic bonds were broken, could the protein begin to refold as directed by the energy of short range interactions.

- ☆ The exposure of this large number of hydrophobic groups to the solvent, however, presents a large energy barrier that make such a refolding kinetically unlikely. Exposure of most proteins to high temperatures results in irreversible denaturation.
- ☆ Some proteins, like caseins, however, contain little if any secondary structure and have managed to remove their hydrophobic groups from contact with the solvent without the need for extensive structure. This lack of secondary structure causes these proteins to be extremely resistant to thermal denaturation.
- ☆ The increased water binding noted in the early stages of denaturation may be retained following hydrophobic aggregations. The loss of solubility that occurs will greatly reduce the viscosity to a level below that of the native proteins.

2. Effect of pH on Protein Denaturation [1,2a,3]

The pH plays very important role in protein denaturation. Besides time and heating temperature, several other factors influence the extent of association of denatured whey proteins with the casein micelles, which include the pH of the milk prior to heating, the soluble calcium and phosphate concentrations and the concentration of solids in milk [2a].

- ☆ Denaturation can also be caused by changes in the pH which can affect the chemistry of the amino acids and their residues. The ionizable groups in amino acids are able to become ionized when changes in pH occur. A pH change to more acidic or more basic conditions can induce unfolding [3]. Acid-induced unfolding often occurs between pH 2 and 5, base-induced unfolding usually requires pH 10 or higher [3].
- ☆ Most proteins at physiological pH are above their isoelectric points and have a net negative charge. When the pH is adjusted to the isoelectric point of the protein, its net charge will be zero. Charge repulsions of similar molecules will be at minimum and many proteins will precipitate. Even for proteins that remain in solution at their isoelectric points, this is usually the pH of minimum solubility.
- ☆ If the pH is lowered far below the isoelectric point, the protein will lose its negative and contain only positive charges. The like charges will repel each other and prevent the protein from readily aggregating. In areas of large charge density, the intramolecular repulsion may be great enough to cause unfolding of the protein. This will have an effect similar to that of mild heat treatment on the protein structure [1].
- ☆ **Reversible and irreversible denaturation**- In some cases the unfolding may be extensive enough to expose hydrophobic groups and cause

irreversible aggregation. Until this occurs such unfolding will be largely reversible.

Some proteins contain acid labile groups and even relatively mild acid treatment may cause irreversible loss of function. This generally results from the breaking of specific covalent bonds and thus should be considered separately from denaturation.

- ☆ **Effect at low pH**-Exposure to strong enough acid at elevated temperatures will first release amide nitrogen from glutamine and asparagine groups and eventually lead to hydrolysis of peptide bonds.
- ☆ **Effect at high pH**- The effects of high pH are analogous to those of low pH. The proteins obtain a large negative charge which can cause unfolding and even aggregation.
  - **i.** The use of high pH to solubilize and alter protein structure is very important to the formation of fibers from proteins of plant origin. A number of reactions can cause chemical modification of proteins at alkaline pH's that are commonly encountered in protein processing. Many of these involve cysteine residues. Perhaps the most important are the base catalyzed beta eliminations of sulfur to yield dehydroalanine which can react with lysine to form lysinoalanine. This result in a **loss of nutritive value of the protein and the products of there action may be toxic.**
  - **ii.** Exposure of protein molecules to high pH should be minimized as much as is possible. Exposure to very high pH at elevated temperatures results in alkaline hydrolysis of peptide bonds.

3. Effect of Changes in Dielectric Constant [1,1a]

Dielectric properties determine the response of a material to an electromagnetic field. The dielectric constant is an indication of the polarizability of the molecules and their ability to store electrical energy. The dielectric loss factor is related to the energy absorption and dissipation of electromagnetic energy from the field. When proteins denature, they change their water and salt binding capacity, expelling or binding water, and thus should change the dielectric properties of the protein-containing food item. [1a].

The dielectric constant of the system is changed by addition of less and more polar solvent. The details are given below [1]:

- ☆ The addition of a less polar solvent (miscible with water) will lower the dielectric constant of the system. This will tend to increase the strength of all electrostatic interactions between molecules that were in contact with water. Many of the protein hydrogen bonds are effectively removed from the solvent and will not be affected.
- ☆ The presence of the less polar solvent will also have the effect of weakening the hydrophobic bonds of the proteins. These bonds

depend upon an increase in the order of water when they are broken for their existence. As there is less water in the system, this becomes less important and at some level of replacement, these groups are at a lower energy level when in contact with the solvent. The structure of the protein will be changed and hence, it will be denatured.

☆ The reversibility of the process depends to a large extent on the nature of the non-polar solvent, the extent of unfolding, the temperature of the system and the rate of solvent removal.

i. When large amounts of the solvent are present, the protein will be largely unfolded with extensive exposure of the hydrophobic groups. If the protein could be instantaneously transferred to pure water at room temperature, the protein would most likely aggregate and precipitate. The sudden exposure of the hydrophobic groups to water would cause them to try to remove themselves from the aqueous phase as soon as possible. Even before the short range interactions could redirect the folding of the protein aggregation would occur.

ii. If the solvent exchange were slow, there would be a better chance that the hydrophobic groups would be able to return to the interior of the molecule and prevent aggregation.

iii. If the exchange occurred at low temperatures, the chances of regaining the native structure would be even better. At low temperatures, the hydrophobic groups may in part be stable in the aqueous phase or at least not as unstable. In this case, the removal of the solvent has little affect.

iv. When the temperature is subsequently increased, the normal course of protein refolding can occur. Solvent precipitation is often utilized as a means of purifying and concentrating enzymes. It is extremely important that both the solvent and the protein solution be cold when they are mixed and that the subsequent removal of the solvent be performed at reduced temperature. This helps to insure the recovery of enzyme activity.

4. Denaturation at Interfaces [1,1c]

The denaturation of proteins can occur when these are exposed to either liquid-air or liquid-liquid interfaces. Like many other substances with both hydrophilic and hydrophobic groups, soluble proteins tend to migrate into the interface between air and water or oil and water. The motion of protein molecules at the air–water interface has been used to determine the molecular weight of proteins [1c].

i. **A liquid-liquid interface [1,1c]**- As a liquid-liquid interface the protein comes into contact with a hydrophobic environment. If allowed to remain at this interface for a period of time, proteins will tend to unfold and place as many of their hydrophobic groups as possible in the non-aqueous layer while maintaining as much charge as possible in the water layer.

- ☆ The tertiary structure of a protein is not rigid and because of this protein unfolds at hydrophobic interfaces. There are continued fluctuations about an average configuration. Any change in conformation that yields a higher energy state will spontaneously go back to the state of lowest energy. As a part of this process, hydrophobic groups will occasionally be positioned so that they have increased contact with the aqueous phase. When this occurs, these groups will assume the configuration of lowest free energy and will be removed from the water.
- ☆ If a hydrophobic group is exposed while a protein is in contact with a polar solvent, these groups will find a state of lower energy exists if they enter into the solvent phase. This will continue to occur until random fluctuations in protein structure can no longer yield a configuration of lower free energy.
- ☆ The amount of unfolding that occurs at such an interface will depend on how rigid the three dimensional protein structure is on the number and location of hydrophobic groups in the molecule.
  - **a.** A flexible, non-cross linked protein will be able to unfold easier than will a highly structured and cross linked one.
  - **b.** If energy is applied to cause shear, the process will be accelerated. The shear can cause the protein to unfold, thus exposing its hydrophobic groups to the non - aqueous phase. It can also increase the interfacial area between the two phases and allow more proteins to come into contact with the non-aqueous phase. This unfolding is essentially non-reversible because of the large energy barriers. Even if the phases should separate and the protein is forced into the aqueous phase the protein will not regain its original structure. Rather an association of hydrophobic groups will cause the protein to aggregate.

ii. A liquid-air interface [1,1c] - When proteins are exposed to liquid-air interface, denaturation can takes place. The same forces are in operation when a protein migrates to a liquid-air interface. Hydrophobic groups tend to associate in the air and the protein unfolds. The presence of shear helps to unfold the protein and to introduce more air into the solution.

- **a.** Both of these effects can be minimized by keeping the temperature low (to weaken hydrophobic bonds) and by minimizing the interfacial area.

b. If the interface is limited, then only a small amount of protein will be able to denature.

c. The presence of this denatured protein will serve as a barrier to further denaturation. Proteins are often utilized in food products to stabilize emulsions or to incorporate air.

5. Ionic Strength [1]

Ionic strength also has significant effect on protein denaturation.

i. **More solubility in dilute salt solution-**Proteins are usually more soluble in dilute salt solutions than in pure water. The salts are thought to associate with oppositely charge groups in the protein. This combination of charged groups binds more water than do the charged groups alone and protein hydration is increased.

ii. **Less solubility in concentrated salt solution-**With most proteins there is little change in solubility as more salt is added until some very high salt content is reached. At very high levels of salt there is a competition between the ions and the proteins for water of hydration. When the salt concentration is high enough, the proteins will be sufficiently dehydrated to lose solubility. Removal of the salt or dilution to a low enough concentration will usually result in the recovery of native structure.

6. The Effect of Protein Cross Linkers [1,1e,1f]

Crosslinking is the process of chemically joining two or more molecules by a covalent bond. The technique, often called bioconjugation when referring to its use with proteins and other biomolecules, is an essential component of many proteomics methods [1f].

- ☆ Cross linking reagents (or crosslinkers) are molecules that contain two or more reactive ends capable or chemically attaching to specific functional groups (primary amines, sulfhydryls, *etc.*) on proteins or other molecules. This article describes the chemistry and variety of crosslinkers that exist [1f]. **The use of crosslinking reagents for the study of proteins is widespread. Crosslinking reagents have also been used extensively for studying enzyme topology [1e]**
- ☆ The technique of chemical crosslinking has been used to enhance the stability of proteins and enzymes. In this procedure, the molecule is braced with chemical crosslinks either intramolecularly or intermolecularly to another species to reinforce its active structure. Various chemicals have been used for this purpose [1e].
- ☆ The presence of groups that cross link protein molecules will tend to lower the extent of protein denaturation. There are following two main reasons for this type of behaviour [1].

i. First, when proteins are cross linked it is more difficult for them to unfold. As energy is added to the system and secondary bonds are weakened, the presence of cross linkers will tend to maintain structure. This is especially true if the cross links are covalent as in the case of disulfide bonds. The more compact the molecule is and the greater the number of disulfide linkages present, the greater the stability of the protein. While secondary forces may be weakened and some bonds can be broken, the cross linkers will tend to keep these groups in fairly close proximity. They also tend to prevent the exposure of large numbers of hydrophobic groups to the solvent. When conditions are returned to the native state, there is now a much greater chance for the proper secondary interaction to occur and for the protein to assume the native configuration [1].

ii. A second effect has to do with the differences in entropy between the native and unfolded states. If a protein can be caused to assume a completely random coil conformation, there will be a large increase in entropy compared to the native structure. This entropy must be overcome if the protein is to refold into a native conformation. When cross linking groups are present, a completely random coil conformation can not be assumed. These groups introduce order into the structure and there is a considerable loss in the amount of disorder that can be achieved in the most denatured state. Because of this, the entropy change between the native and denatured state is not nearly as great and there will be less of a driving force for denaturation. If the cross linking groups are broken before denaturation and thus allowed to randomly form after denaturation, no stability will be added to the protein by the pressure of these groups [1].

No reactions involving primary covalent bonds (such as peptide linkages) occur during denaturation the unfolding of the molecules often exposes groups which may undergo chemical reactions (oxidation of sulfhydryl groups by the atmospheric oxygen).

Hydrolysis of Protein [1,1g,2,7]

Peptides and proteins consist of many amino acids that are bound together via peptide bonds. Proteins are polymers of amino acids which are held by the polypeptide bonds between them. When these bonds are broken the protein splits in to smaller peptide and may even proceed to that extent of release for amino acids or smaller peptides [1,1g].

- ☆ The solution containing these smaller peptides and even amino acids is called hydrolysate solution. This process will help the digestion of the proteins and also study of the various amino acids present in the protein [1].

Protein hydrolysis is carried out by chemical and enzymatic methods.

i. Acid Hydrolysis [1g]

Acid hydrolysis is the most common method for hydrolyzing a protein and the method can be performed in either vapor or liquid phase. Although a range of different acids can be used for this reaction, the most common is 6 M HCl. Because HCl is evaporative, it can also be used to recover the hydrolysate in smaller amounts of buffer, a feature that is particularly useful for small amounts of sample.

The acid-hydrolysis reaction with 6 M HCl results in the addition of water to each covalent peptide bond, yielding the desired individual amino acids.

- However, not all amino acids are completely recovered under hydrolysis by HCl. Some amino acids are hydrolyzed to their acid forms, such as asparagine and glutamine, which form aspartic acid and glutamic acid, respectively.
- In addition, other amino acids cannot be reliably measured. For example, tryptophan is destroyed during the reaction.
- Sulfur-containing amino acids (*e.g.*, cysteine, methionine) cannot be reliably measured due to partial destruction of the amino acids.
- Furthermore, amino acids such as tyrosine, serine, and threonine may have lower recoveries due to the nature of acid hydrolysis.

ii. Enzymatic Hydrolysis

Proteolytic enzymes hydrolyze proteins at the optimum temperature and pH and usually target specific peptide cleavage bonds, resulting in digestion consisting of amino acids and peptides of varying size. Protein hydrolysis is the breakdown of protein into smaller peptides and free amino acids. The sequence starts from proteins which is broken down to peptones and then to polypeptides and finally to amino acids [7].

Benefit of Enzymatic Hydrolysis of Protein [7]

Compared to acid and alkaline hydrolysis of proteins, the main advantages of enzyme hydrolysis of proteins are that: (a) the hydrolysis conditions (*e.g.*, like temperature and pH) are mild and do not result in any loss of AAs; (b) proteases are more specific and precise to control the degree of peptide-bond hydrolysis.

Sources

1. *K.S. Sharma, Chemistry of milk, AgriMoon.com.*

1a. *Denaturation, https: //www.britannica.com.*

1b. *S.A.Barringer and C. Bircan, Use of the Dielectric Properties to detect Protein Denaturation, Advances in Microwave and Radio Frequency Processing pp 107-118, 2015.*

1c. *Protein denaturation, https: //www.britannica.com/science/protein/Protein-denaturation.*

1d. *Denaturation-Wikipedia.*

1e. *Michael P. Byrne and Wesley E. Stites, Chemically crosslinked protein dimers-Stability and denaturation effects, Protein Science, 1995, 4: 2545-2558.*

1f. *Chemistry of Crosslinking, Protein Biology Resource Library, Thermo Fisher Scientific, https: //www.thermofisher.com.*

1g. *Introduction to Hydrolysis, Types of hydrolysis, Water, https: //www.waters.com.*

2. *M.K. Srivastava, Handbook on Analysis of Milk, CBS Publications, New Delhi, 2010. ISBN 13: 9788123926681.*

2a. *Verruck S, Sartor S, Marenda FB, et al., Influence of heat treatment and microfiltration on the milk proteins properties. Adv Food Technol Nutr Sci Open J., 2019, 5(2): 54-66.*

3. *Wikipedia.*

4. *Mosby's Medical Dictionary (8th ed.). Elsevier. 2009. Retrieved 1 October 2013.*

5. *Anfinsen CB., 1973, "Principles that govern the folding of protein chains", Science, 181, 4096, 223-30.*

6. *Wetlaufer, D.B., 1988, "Reversible and irreversible denaturation of proteins in chromatographic systems". Makromo-lekulare Chemie. Macromolecular Symposia. 17(1): 17-28. ISSN0258-0322.*

7. *Protein hydrolysis, https: //www. sciencedirect.com/topics/food-science/protein-hydrolysis.*

Fat Globule Membrane Proteins, their Properties and Role [1,1a2]

Membrane material surrounding the lipid globules of milk is called the milk lipid or milk fat globule membrane (MFGM). The milk fat globule membrane (MFGM) is the membrane surrounding lipid droplets during their secretion in the alveolar lumen of the lactating mammary gland.

- ☆ Milk-fat globules originate near the basal region of the secretory cells as small droplets of fat. They migrate through the cytoplasm, gradually increasing in size, as the synthesis of triacylglycerol proceeds. The milk fat globules are secreted from the apical surface of the cell, surrounded by a membrane thin bilayer, the Milk Fat Globule Membrane (MFGM). MFGM are formed by a unique and quantitatively small subcategory of milk proteins (approximately 2-4 per cent of total protein in human milk), the content of which is still largely unknown.
- ☆ The fat globule membrane contains approximately 50 per cent protein and accounts for about 1 per cent of the total protein of the milk. There

is also wide variation in the nature of these proteins several enzymes have been reported to be associated with this membrane.

- Some of the protein constituents of the membrane are enzyme, but it is not possible at present to estimate the ratio of enzymatic and non-enzymatic components.
- The fat globule membrane proteins are difficult to resolve analytically and to separate preparatively because they interact strongly with one another and with lipids, and most of them are insoluble. Part is released from the globules by cooling the milk. Some success has been achieved in solubilizing the membrane proteins with detergents such as sodium dodecyl sulfate coupled with reduction of disulfides [1,1a].
- The MFGM proteins are having the nutraceutical and biological importance. Realizing the importance of these proteins there is an increase in the studies on MFGM. The MFGM is a rich source of membrane proteins, and applied proteomic analysis of these membrane proteins, has highlighted some of the possible signalling and secretory pathways used by the mammary gland [1,1a].

Structure and Composition of MFGM [1,1a,2,3]

a. Structure

MFGM is a complex and unique structure composed primarily of lipids and proteins that surrounds milk fat globule secreted from the milk producing cells of humans and other mammals. The MFGM comprises a variety of phospholipids, glycolipids, proteins, and glycoproteins, along with cholesterol and other lipids. Specific lipids and proteins are localized to different layers of the membrane, with carbohydrate chains of glycoproteins and glycolipids directed toward the outer surface of the milk fat globule; the lipid-to-protein weight ratio in MFGM is approximately 1:1 [2,3]. Milk fat globule membrane is composed of proteins and lipids in a 1:1 weight ratio [1,1a].

b. Composition of Fat Globule Membrane Proteins [1,1a]

The membrane is predominantly proteinaceous in nature. The composition of the fat globule varies considerably depending on different factors.

- At least 25 different enzymes have been reported to be isolated from the fat globule membrane which includes the 5 nucleotidase, alkaline phosphatase, acid phosphatase, aldolase, xanthine oxidase.
- Although a major portion of the membrane material contains higher proportion of enzymes, presently it is not possible to estimate the ratio of enzymatic and non-enzymatic components.
- A part from the enzymes the proteins of the fat globule membrane protein exists as polypeptides of varying molecular weight. These micro lipid droplets are encircled by a special membrane composed

of lipid bilayer and proteins. This membrane has been designated the milk fat/lipid globule membrane.

- ☆ The MFGM originates from the plasma membrane of the mammary gland secretory cells by extrusion of lipid droplet from the cell. Because each lipid droplet in milk has its own membrane envelope, *i.e.*, the MFGM, the lipid droplets are present in milk in a dispersed form and, therefore, do not aggregate with each other. Only after destruction of the structure of the MFGM through, for example, a mechanical force like churning, do the lipid droplets aggregate and subsequently form large fat clumps known as butter.
- ☆ Many of these proteins are present in the MFGM as glycoproteins.
  - **a.** The main protein of the MFGM is the glycoprotein butyrophilin (about 40 per cent of the total proteins of the MFGM),
  - **b.** The second representative protein of the MFGM is xanthine oxidase, which comprises 12 to 13 per cent of the total proteins.
  - **c.** Other proteins are present in MFGM each at 5 per cent or less. The physiological role of the MFGM proteins is not completely clear despite numerous research studies.

However, there are some reasonable suggestions about the physiological function of these proteins in the cell because these proteins are an integral part of the plasma membrane of secretory mammary gland cell.

Nomenclature [2,3]

The current nomenclature for milk fat globule membrane (MFGM) proteins recommended by the Milk Protein Nomenclature Committee of the American Dairy Science Association, the major proteins of MFGM in order of increasing mobility upon SDS-PAGE separation,are the mucin MUC-1, the redox enzyme, xanthine dehydrogenase/oxidase (XDH/XO),glycoprotein (PASIII), cluster of differentiation 36 (CD36), butyrophilin (BTN),adipose differentiation related protein (ADRP), and glycosylated variants of some polypeptide backbone designated PAS6/7 and fatty acid binding protein (FABP).

Role and Health Benefits of Fat Globule Membrane Proteins [1,1a,2]

A great deal of knowledge has been accumulated on health beneficial factors, protein and non-protein, of bovine milk fat globule membrane (MFGM).

- ☆ Among the health beneficial components of the MFGM are cholesterolemia lowering factor, inhibitors of cancer cell growth, vitamin binders, inhibitor of Helicobacter pylori, inhibitor of beta glucuronidase of the intestinal Escherichia coli, xanthine oxidase as a bactericidal agent, butyrophilin as a possible suppressor of multiple sclerosis, and phospholipids as agents against colon cancer, gastrointestinal pathogens, Alzheimer s disease, depression, and stress. Because of the aforesaid factors, bovine MFGM is considered as a potential nutraceutical [1,1a].

- The work done by researchers, has indicated that MFGM, or components thereof, may potentially play roles in brain development and cognitive function, immunity and gut health, and cardiovascular health [2].
- MFGM may also play a role in supporting cardiovascular health by modulating cholesterol and fat uptake. [2].
- MFGM glycoproteins seem to contribute to the prevention of pathogenic organisms infections, being able to act as specific bacterial and viral ligands in the stomach of new borns, to prevent the attack of the intestinal mucosa [1,1a].
- The diversity of the glycans found in MFGM is thought to enable the glycoproteins to perform this function in the acidic environment of the stomach. It has been noted that some forms of gastric diseases such as peptic ulcer, chronic type B gastritis and gastric cancer can be attributed to the colonization of gastric mucosa by Helicobacter pylori Non-defatted and defatted MFGM preparations, given orally, caused equal healing effect on H. pylori infection of gastric mucosa in BALB/ CA mice, leading to the conclusion that the major role in inhibition of H. pylori infection is played by the protein [1,1a].

Sources

1. *K.S. Sharma, Chemistry of milk, Agrimoon.com.*

1a. *Fat globule membrane proteins their properties and role, Lesson-12, http: // ecoursesonline. iasri.res.in.*

2. *Wikipedia.*

3. *Kanno, C., 1990, "Secretory membranes of the lactating mammary gland". Protoplasma. 159 (2-3): 184-208.*

Quantification of Proteins in Milk [1,1a,2,2a,3]

Milk protein content is an important component of milk, especially from a nutritional point of view and also for payment purposes. Several analytical reference and routine methods are reported for milk protein determination and quantification

- **The Reference methods** perform accurate analyses comply according to the International Standard ISO.
- **Routine methods** perform analyses using routine instrumental techniques for faster and cheaper results with acceptable accuracy and a large number of processed samples.

- ✰ The performance of reference method results is usually more reliable and also more expensive than routine method results.

In most of cases, using of routine indirect methods for milk protein analysis requires their specific calibrations according to biological kind of measured milk (cow's, goat's or sheep's milk) or specific conditions of milk technology treatment [2a]. The determination of the protein content of materials in which proteins occur in mixtures with other biological materials is not simple.

Methods for the determination of proteins in milk can be divided into three broad categories [1a]-

a. Determination of Total Nitrogen

For more than a century, food analysts have determined the protein content of foods by determining the total nitrogen content and then calculating the protein content by using a suitable "nitrogen conversion factor". Kjeldahl method and Dumas method are the examples of such types of determination. The advantage of these methods is that they have high reliability and accuracy. A disadvantage is that they require dedicated laboratory equipment and skilled staff which makes them expensive and time-consuming to carry out.

b. Direct Protein Determination

The quantification of specific protein is also possible. A variety of methods are available for protein component determination. Dye binding assay, Chromatographic methods,Electrophoretic method and Immunological method are the examples of direct protein determination.

c. Indirect Protein Determination

Spectroscopic methods can be carried out rapidly at low costs in a production environment, but rely on calibration to a chemical or reference method to ensure consistency. Near and mid infrared spectroscopy are in widespread use for both liquid and solid dairy products in production laboratories and independent testing laboratories alike. The following well validated ISO/IDF standard methods are also being used:

i. ISO 8968-1/2 | IDF 020-1/2- Milk - Determination of nitrogen content - Part 1/2: Determination of nitrogen content using the Kjeldahl method.

ii. IDF Standard 185: 2002(E), ISO 14891:2002(E). Milk and Milk Products. Determination of Nitrogen Content Routine Method by Combustion According to the Dumas Principle.

The proteins in milk can be estimated by using the following chemical and physical methods [1]:

i. By using their component amino acids or,

ii. By their chemical reactions or,

iii. By their physical forms.

Comparison of Individual Methods [2a]

The following table shows the comparison of different methods for proteins estimation in milk and milk products:

Table 11.9: Comparison of different Methods for Proteins Estimation

Sl.No.	Method	Principle	Use
1	Dumas	A burning of sample in pure oxygen (a releasing of CO_2, water and NO_2 and measuring of the chemiluminescence reaction by TCD)	Milk and dairy products
2	Formol titration	Free amino acids and protein bound amino acids and peptides react with formaldehyde, producing methylene amino acid derivatives and changing the pKa of these amino groups	Milk and dairy products
3	Coulometry	A measuring of quantity of electricity (applying Faraday 's laws) after previous neutralization, oxidation, reduction, eventually precipitation (organic material has to be mineralized in mixture with concentrated sulphuric acid and catalyst under conditions of high temperature for longer time, similarly to Kjeldahl method)	Raw milk
4	Dye binding	A forming an insoluble dye protein complex (a stoichiometrically connection of negatively charged sulfonic groups in azo dyes with positively charged constituents in protein→separation of the insoluble complex and determination of the unbound dye by spectrophotometer)	Milk and dairy products
5	Ultrasound procedure	A measurement of high frequency ultrasound radiation passing through the sample material milk	Milk
6	IR spectroscopy	A measurement of infrared irradiation (in near, mid *etc.* spectrum) absorbed or reflected by sample; it occurs changes in the rotational vibrational energy states of the molecule depending on variations of the dipole moment of the molecule	Milk and Dairy products

1. Chemical Methods [1,2,2a,2b,4]

A. Determination of Nitrogen

The following methods are used for determination of nitrogen in foods-

i. Dumas Method [4]

This method was described by Jean Baptiste Dumas in the early 19th century. In principle, the sample is burned (approximately 900 °C) in pure oxygen. This leads to release of carbon dioxide, water, and nitrogen oxides. The gases pass through special sorption columns that absorb carbon dioxide and water. Nitrogen is measured of the chemiluminescence reaction of the NO_2 or in the elemental form by thermal conductivity detection.

Enhanced Dumas Method [2b]

Recently, an automated instrumental technique has been developed which is capable of rapidly measuring the protein concentration of food samples. This technique is based on a method first described by a scientist called Dumas over a century and a half ago. It is beginning to compete with the Kjeldahl method as the standard method of analysis for proteins for some foodstuffs due to its rapidness [2b].

The instrument is calibrated by analyzing a material that is pure and has a known nitrogen concentration, such as EDTA (= 9.59 per cent N). Thus the signal from the thermal conductivity detector can be converted into a nitrogen content.

Advantages and Disadvantages [2b]

This method has following advantages and disadvantages-

Advantages- This method is much faster than the Kjeldahl method (under 4 minutes per measurement, compared to 1-2 hours for Kjeldahl). It doesn't need toxic chemicals or catalysts. Many samples can be measured automatically. It is easy to use.

Disadvantages-High initial cost. It does not give a measure of the true protein, since all nitrogen in foods is not in the form of protein. Different proteins need different correction factors because they have different amino acid sequences. The small sample size makes it difficult to obtain a representative sample.

ii. Formol Titration [1,2,2a]

This is the next fast, practical and technological method in the dairy system to determine milk protein. This method was described by Steinegger (1905) and modified by Pyne (1932), Cole (1969) and Peeples and Heath (1979) [2a]. Considerable use has been made of titration with formaldehyde as a means of determining protein content in milk and milk fractions. Such titrations have been of particular interest as rapid methods of determining the casein content of milk for cheese making. As per Bureau of Indian Standards, for the estimation of total milk protein a factor of 1.7 is used and while for casein the factor 1.38 is used [1,2].

Principle [2a]

In principle, free amino acids and protein bound amino acids and peptides react with formaldehyde, producing methylene amino acid derivatives and changing the pKa of these amino groups. The value pKa = – log 10 Ka is the co logarithm of acid dissociation constant (acid dissociation constant Ka is a quantitative measure of the strength of an acid in solution),which measures the tendency for a group to give up a proton. The method can be applied to raw milk, drinking milk (whole, semi skimmed and skimmed milk), milk powder or ice cream.

Method [1,2a]

All such methods involve titration of a sample of milk to the end point of an indicator such as phenolphthalein, or adding a solution of formaldehyde, and titrating the acid liberated to the same end point. The amount of alkali used in the second titration is a measure of the amino groups that were originally present and combined with the formaldehyde.

iii. Kjeldahl Method [1,2,2a.2b]

The most widely used method for determining protein content is the Kjeldahl procedure for nitrogen. The Kjeldahl method was developed in 1883 by a brewer called Johann Kjeldahl. Because the Kjeldahl method does not measure the protein content directly a conversion factor (F) is needed to convert the measured nitrogen concentration to a protein concentration. A conversion factor of 6.25 (equivalent to 0.16 g nitrogen per gram of protein) is used for many applications, however, this is only an average value, and each protein has a different conversion factor depending on its amino-acid composition. The Kjeldahl method involves a three-step approach to the quantification of protein-

a. **Digestion**
b. **Distillation**
c. **Titration**.

This method involves the oxidation of the sample with sulfuric acid and a catalyst (copper sulphate and potassium sulphate, 1: 10). Carbon and hydrogen are oxidized to CO_2 and H_2O, and reduced forms of nitrogen (such as $-NH_2$ and >NH) are retained in the digest as ammonium ions. The digest may be made alkaline and the ammonia distilled, and titrated, or it may be determined colorimetrically directly in the digest by means of Nessler s reagent.

Difficulties

This is widely accepted and used method but in spite of this, the Kjeldahl method suffers from following serious difficulties [1,2]-

a. There is the problem of separating protein from other nitrogenous materials. In milk about 5 per cent of the total nitrogen is in the form of low molecular weight non-protein nitrogenous materials. The protein can be separated from these by precipitation with trichloroacetic acid. This precipitation, however, does not separate protein from nitrogen-containing lipids. Since the amount of lipid nitrogen is small, comprising only about 0.270 of the total milk nitrogen, it is usually neglected (*i.e.*, included as protein nitrogen).

b. A second difficulty with nitrogen methods for determining protein arises from variation in the nitrogen content of various proteins. Proteins vary in nitrogen content from 14 to 19 per cent and thus a single universal conversion factor cannot be used. Fortunately the principal milk proteins exhibit much less variation ranging from 15.3 to 16.0 per cent. An average factor of 6.38 (corresponding to 15.65 per cent nitrogen) is commonly used for milk proteins to convert nitrogen to protein. In precise work on an individual milk protein the correct factor for that protein should be used.

c. A third group of difficulties concerned with the Kjeldahl method is the problem of proper digestion and no loss of nitrogen. A great many different modifications of the original Kjeldahl procedure have been suggested to accelerate the digestion while still attaining complete digestion. Copper, mercury, or selenium are used as catalysts and frequently Na_2SO_4 or K_2SO_4 is added to elevate the boiling point during digestion.

Principle and Method [5]

Kjeldahl method starts with the digestion of samples with sulfuric acid. The principle is to convert quantitatively all the organic and inorganic (NH_3/ NH_4^+ but not NO_3) nitrogen into ammonium sulphate by digestion at 400-420°C in a mixture of K_2SO_4/H_2SO_4 with a catalyst. This solution is neutralized under alkali conditions and then distilled into a boric acid. The final step is the titration of borate anions with hydrochloric acid. The nitrogen concentration is then estimated by acid titration of ammonia released by the distillation of ammonium sulfate in the presence of sodium hydroxide.

It is clear that Kjeldhal method is the most reliable method so far, however there are three different steps which are both time and chemical consuming. The use of boric acid to receive the ammonia as it is distilled off is an advantage in that only one reagent, the standard acid for titrating, need to be standardized and measured accurately.

The detailed method is discussed in author's another book "Handbook on analysis of milk" [2c].Several instrument manufacturers like Foss(Kjeltech), Pelican(Kelplus), Borosil(Kjeldatherm) have developed instruments working on the principle of Kjeldahl method for rapid analysis of protein.

Advantages and Disadvantages [2b]

This method has following advantages and disadvantages:

Advantages- The Kjeldahl method is widely used internationally and is still the standard method for comparison against all other methods. Its universality, high precision and good reproducibility have made it the major method for the estimation of protein in foods.

Disadvantages- It does not give a measure of the true protein, since all nitrogen in foods is not in the form of protein. Different proteins need different correction factors because they have different amino acid sequences. The use of concentrated sulfuric acid at high temperatures poses a considerable hazard, as does the use of some of the possible catalysts The technique is time consuming to carry-out.

B. Dye Binding Method [2a,2b,5,6]

These are used to determine milk protein as indirect, routine procedures. Several authors described dye binding methods. Dye-binding methods are based on the quantitative reaction between ionized groups of proteins and ionized groups of the dye, *i.e.* sulphonic acid. The dyes belong to a group so called "azo dyes" which have sulfonic groups in their molecules. In acid buffered solutions(pH 2.4), negatively charged sulfonic groups in azo dyes (must be in excess, to form an insoluble complex) are stoichiometrically connected with positively charged constituents in proteins (positive charge on terminal amino group and the basic amino acid residues of proteins) to form an insoluble dye protein complex. The unbound dye after removal of complex,is determined by measuring the abosrbance. The amount of protein present in the original solution is proportional to the amount of dye that bound to it: $dye_{bound} = dye_{initial} - dye_{free.}$

The most commonly used dyes are Orange G (no. CAS 1936 15 8), Acid Orange 12 (no. CAS 1934 20 9) and Amido Black 10B (no. CAS 1064 48 8). The amido black method which uses CI Acid Black I, is routinely used in different countries.

Procedure

a. The dye and protein react and form insoluble complex. Subsequently, the insoluble complex is separated and the unbound dye can be determined by spectrophotometer (*e.g.* Pro Milk with using of Amido Black 10B by Foss Electric, Denmark). In addition, hydrophobic interactions occur between free and bound dye. After centrifugation, the absorbance of the supernatant is measured at about 630 nm. As the quantity of dye bound per unit weight of protein varies with the amino acid composition of the proteins, the proportions of the different proteins in a mixture influence the trueness of the method. Whey proteins bind approximately 27 per cent more dye than caseins, and the NPN fraction does not bind dye [5].
b. Certified 92 per cent Orange G dye was used, but it was adjusted to a concentration of 1 g of 100 per cent dye content per liter. A 2-ml milk sample and 20 ml of buffered Orange G dye were placed in a 1-oz-capacity plastic bottle and thoroughly shaken for complete mixing. Next, the contents were centrifuged at 820 rpm for 5 rain in a Babcock centrifuge. The solutions were then filtered through No. 595 S and S filter paper to get a clear filtrate, free from fatty material. A short-light-

path absorption cell was used to read the percentage of transmittance in an Evelyn colorimeter at 490 mμ. Refrigeration was used for holding either the dye or the dye-plus-milk samples [6].

2. Physical Methods [1]

The following physical methods are being used for quantification of proteins in milk-

i. *Direct Weighing [1]*

Since the ultimate objective of protein analyses is to determine the weight of protein in a given quantity of material, it might seem that separation of the protein and direct weighing would be the simplest method to use.

- ☆ Since removing lipids, salts, other solutes, and water completely is difficult it is not generally employed for routine analyses.
- ☆ Obviously, however, direct weighing is the ultimate standard for all other methods. For example, the factor for converting nitrogen to protein was arrived at originally by preparing a sample of pure protein and determining the ratio between nitrogen and dry weight.

ii. *Volume and Density Measurements [1,7]*

a. *Volume Measurement [1]*

Methods based on precipitating protein, centrifuging in a calibrated tube, and measuring its volume, have often been suggested. The Hart method for casein is of this type. Fat is extracted with chloroform and the casein precipitated with acetic acid, centrifuged, and measured.

- ☆ Methods based on volume measurements of proteins are not used to any extent in the dairy industry at present except for assessing the amount of insoluble material in milk powders (solubility index).

b. *Density Measurement [2b,7]*

The density of a protein is greater than that of most other food components, and so there is an increase in density of a food as its protein content increases. Thus the protein content of foods can be determined by measuring their density.

iii. *Turbidimetric Methods [1,1b,1c]*

In analytical chemistry,Turbidimetry is method for determining the amount of cloudiness, or turbidity, in a solution based upon measurement of the effect of this turbidity upon the transmission and scattering of light. Turbidimetric techniques are rapid and convenient, but they yield different values with different proteins [1b].

Protein molecules which are normally soluble in solution can be made to precipitate by the addition of certain chemicals, *e.g.*, trichloroacetic acid. Protein precipitation causes the solution to become turbid. Thus the concentration of protein can be determined by measuring the degree of turbidity [2b].

It is sometimes convenient to form a suspension of insoluble protein and to estimate the protein content by optical methods. Either light transmittance or light scattering can be measured [1].

- ✰ Methods employing the former principle are called turbidimetric, those using the latter, nephelometric.
- ✰ In recent years a turbidimetric method devised by Harland and Ashworth has been used to a considerable extent for determining the concentration of undenatured serum proteins in heated and dry skim milks.
  - **i.** In this method the casein and denatured serum proteins are precipitated by saturation with sodium chloride. The resulting filtrate is then diluted and acidified and its light transmittance determined. The protein content is determined from a calibration curve in which per cent transmittance is plotted against protein nitrogen content of the filtrate (determined by Kjeldahl). [1]
  - **ii.** Based on the turbidimetric response of protein with 50 per cent -trichloroacetic acid (TCA), an assay method for protein in solution, is introduced. The standard procedure consists of mixing equal volume of sample solution (standard or unknown) with 50 per cent -TCA solution and measuring the absorbance at 450 nm after 20 min. This assay method is simple, reproducible, and tolerant to many interfering substances. It can detect less amount than 10 μg/ml of bovine serum albumin [1c].
- ✰ Methods of this kind are rapid, require little sample, and are conveniently adaptable to photometers and colorimeters found in most laboratories.

Limitations [1,1b]

Nevertheless, turbidimetric methods are not of the highest accuracy and precision because light absorption in turbid systems depends not only on the amount of dispersed material present but also on its degree of dispersion.

- ✰ The degree of dispersion depends on time of standing after development of turbidity, pH, salt concentration, method and rate of precipitation, and concentration of material.
- ✰ Turbidimetric methods do not permit differentiation between protein and acid-insoluble compounds such as nucleic acids [1b].

iv. Refractive Index Measurements [1,2b,7]

The refractive index of an aqueous solution increases as the protein concentration increases and therefore RI measurements can be used to determine the protein content [2b,7].

The most proteins have a refractive index increment of about 0.0018, meaning that one g. of protein dissolved in 100 ml. of solution increases the refractive index by that amount.

- ☆ With an accurate refractometer such as the Zeiss dipping refractometer, which has a sensitivity of ±0.00003, it is possible to determine protein concentrations satisfactorily. The casein of milk is isolated by acid precipitation, washed and dissolved in alkali, and the refractive index is determined.
- ☆ The refractive index of the solvent must be determined as well as that of the protein solution.
- ☆ Temperature must be carefully controlled since most proteins exhibit a considerable temperature coefficient for refractive index increment.

v. Absorption of Ultraviolet/Visible Radiation [1,2b,7]

a. Absorption of Utraviolet radiation [1,2b,7]

Tryptophan, tyrosine and pheny-lalanine absorb ultraviolet light strongly at 280 nm. The tryptophan and tyrosine content of many proteins remains fairly constant, and so the absorbance of protein solutions at 280nm can be used to determine their concentration.

- ☆ This method is particularly valuable for a known pure protein whose extinction coefficient (E11 per cent cm) is known. It may also be useful as an approximate measure of the total protein content of a mixture if an average extinction coefficient can be assumed to apply. Obviously the extinction coefficients of proteins vary with their contents of tyrosine, tryptophan, and phenylalanine [1,7].

Advantages and Disadvantages [1,2b,7]

- **i.** The advantages of this method are that the procedure is simple to carry out, it is non destructive, and no special reagents are required [2b,7].
- **ii.** The major disadvantage is that nucleic acids also absorb strongly at 280 nm and could therefore interfere with the measurement of the protein if they are present in sufficient concentrations. The presence of materials other than proteins that absorb at this wavelength would seriously limit the method, but the common salts and other solutes do not absorb radiation at 280 nm. Even so, methods have been developed to overcome this problem, *e.g.*, by measuring the absorbance at two different wavelengths [1,2b,7].
- **iii.** Methods based on this principle have not been used to any extent for routine determinations of milk proteins.

b. *Absorption of UV-Visible Radiation [1,2b,7]*

For fundamental studies in the laboratory, where pure proteins are often analyzed, UV-visible spectroscopic techniques are often preferred because they give rapid and reliable measurements, and are sensitive to low concentrations of protein [1].

A number of methods have been devised to measure protein concentration, which are based on UV-visible spectroscopy. These methods use either the natural ability of proteins to absorb (or scatter) light in the UV-visible region of the electromagnetic spectrum, or they chemically or physically modify proteins to make them absorb (or scatter) light in this region. A number of the most commonly used UV-visible methods for determining the protein content of Milk/foods are discussed below:

i. **Biuret Method [2b,7]-**A violet-purplish colour is produced when cupric ions (Cu^{2+}) interact with peptide bonds under alkaline conditions. The biuret reagent, which contains all the chemicals required to carry out the analysis, can be purchased commercially. It is mixed with a protein solution and then allowed to stand for 15-30 minutes before the absorbance is read at 540 nm.

   - ☆ The major advantage of this technique is that there is no interference from materials that adsorb at lower wavelengths, and the technique is less sensitive to protein type because it utilizes absorption involving peptide bonds that are common to all proteins, rather than specific side groups.
   - ☆ However, it has a relatively low sensitivity compared to other UV-visible methods.

ii. **Lowry Method [2b,7]-**The Lowry method combines the biuret reagent with another reagent (the Folin-Ciocalteau phenol reagent) which reacts with tyrosine and tryptophan residues in proteins. This gives a bluish colour which can be read somewhere between 500 - 750 nm depending on the sensitivity required. There is a small peak around 500 nm that can be used to determine high protein concentrations and a large peak around 750 nm that can be used to determine low protein concentrations.

   - ☆ This method is more sensitive to low concentrations of proteins than the biuret method.

Advantages and Disadvantages [2b]

These methods have following advantages and disadvantages-

Advantages- UV-visible techniques are fairly rapid and simple to carry out, and are sensitive to low concentrations of proteins.

Disadvantages- For most UV-visible techniques it is necessary to use dilute and transparent solutions, which contain no contaminating substances which absorb or scatter light at the same wavelength as the protein being analyzed. The need for transparent solutions means that most foods must undergo significant amounts of sample preparation before they can be analyzed, *e.g.*, homogenization, solvent extraction, centrifugation, filtration, which can be time consuming and laborious.

- ☆ In addition, it is sometimes difficult to quantitatively extract proteins from certain types of foods, especially after they have been processed so that the proteins become aggregated or covalently bound with other substances. In addition the absorbance depends on the type of protein analyzed (different proteins have different amino acid sequences).

3. Instrumental Methods [1,2b,7]

There are a wide variety of different instrumental methods available for determining the total protein content of food materials. A major advantages of these instrumental methods over the other techniques mentioned earlier are that these are non-destructive, require little or no sample preparation, and measurements are rapid and precise. These can be divided into following three different categories according to their physicochemical principles [2b,7]-

i. Measurement of bulk physical properties.
ii. Measurement of adsorption of radiation, and.
iii. Measurement of scattering of radiation.

Each instrumental methods has its own advantages and disadvantages, and range of foods/milk to which it can be applied. Several instrumental methods have been developed for the estimation of protein content in milk and milk products.

i. **Measurement of Bulk Physical Properties [2b,7]-** The details related to Direct weighing, measurement of volume, measurement of RI,etc are discussed above.

ii. **Measurement of Adsorption of Radiation [2b,7]**

UV-visible

Discussed above.

Infrared [1,2b,7]

Infrared techniques can be used to determine the concentration of proteins in food and milk samples. Proteins absorb IR naturally due to characteristic vibrations (stretching and bending) of certain chemical groups along the polypeptide backbone. Measurements of the absorbance of radiation at certain wavelengths can thus be used to quantify the concentration of protein in the sample. IR is particularly useful for rapid on-line analysis of protein content [7].

For quality control purposes, it is often more useful to have rapid and simple measurements of protein content and therefore IR techniques are most suitable [1].

Dairy industries are using IR (Milkoscan) and FTIR (Milkoscreen, Milkoscan FT) based instruments for rapid and accurate analysis of milk proteins.

Advantages and Disadvantages-It also requires little sample preparation and is non destructive. Its major disadvantages are its high initial cost and the need for extensive calibration.

Nuclear Magnetic Resonance (NMR) [7]

NMR spectroscopy can be used to determine the total protein concentration of foods. The protein content is determined by measuring the area under a peak in an NMR chemical shift spectra that corresponds to the protein fraction.

iii. Measurement of Scattering of Radiation [7]-

Light scattering- The concentration of protein aggregates in aqueous solution can be determined using light scattering techniques because the turbidity of a solution is directly proportional to the concentration of aggregates present.

Ultrasonic scattering-The concentration of protein aggregates can also be determined using ultrasonic scattering techniques because the ultrasonic velocity and absorption of ultrasound are related to the concentration of protein aggregates present.

The details related to the aforesaid instrumental analysis are very well discussed in author's another book Instrumentation in analysis of Milk/Milk Products and Foods [8].

Sources

1. *K.S. Sharma, Chemistry of milk, Agrimoon.com.*

1a. *Protein Determination, IDF Factsheet, May 2013.*

1b. *Ennis Layne, Spectrophotometric and turbidimetric methods for measuring proteins, Methods in Enzymology, Volume 3, 1957, Pages 447-454.*

1c. *Wahn Soo Choi, Kae Jong Chung and Sung Youl Hong, A turbidimetric determination of protein by trichloroacetic acid, Archives of Pharmacal Research volume 16, 57–61, 1993.*

2. *Quantification of proteins in milk, lesson-13, http: //ecoursesonline.iasri.res. In.*

2a. *Robert Kala, Eva Samkova, Oto Hanus and Lenka pekova, Milk Protein Analysis: An Overview of the Methods -Development and Application, 2019, Acta Universitatis Agriculturae et Silviculturae Mendelianae Brunensis 67(1): 345-359.*

2b. *Analysis of Proteins, https: //people.umass.edu.*

2c. *M.K. Srivastava, Handbook on Analysis of Milk, CBS Publications, New Delhi, 2010. ISBN 13: 9788123926681.*

3. *Wikipedia.*

4. *Jakob, E., Sievert, C., Sommer, S. et al., 1995, Automated Determination of Total Nitrogen in Milk according to the Dummas Method [in German: Automatisierte Bestimmung des Gesamt Stickstoffs in Milch Nach der Dummas-Methode]. Z. Lebensm. Unters. Forsch., 200: 239-243.*

5. *Milk Proteins - Analytical Methods, https: //hal. archives-ouvertes.fr/hal-02267859/ document.*

6. *Udy, D. C. A Rapid Method for Estimating Total Protein in Milk. Nature, 178: 314.1956.*

7. *Analysis of Proteins, file: ///C: /Users/HP/Desktop/6. per cent 20 ANALYSIS per cent 20OF per cent 20 PROTEINS.html*

8. *M.K. Srivastava, Instrumentation in analysis of foods and milk products, 2019, ISBN: 9789388173520, Astral International Pvt Limited, New Delhi.*

Chapter 12

Enzymes

Enzymes **[1,2]** are delicate organic compounds/catalysts produced by living cells (either animal or plant), but work independently from the cells that produce them. They do not necessarily remain at the site of production but are often transported to the site of action through the circulatory systems. Enzymes are proteins or combination of proteins and coenzymes and they function within a narrow pH and temperature range. In general, they are fairly sensitive to heat [1].

Enzymes are colloidal and a group of proteins that have the ability to catalyze chemical reactions and the speed of such reactions. The action of enzymes is very specific. Enzymes initiate chemical reactions but remain unchanged. They bring about most complex chemical reactions but they themselves do not enter into chemical change. The enzymes are relatively unstable and high temperature, light, unfavourable pH and certain chemicals inactivate the enzymes. Enzymes are susceptible to heat, light and pH changes [1a].

Classification of Enzymes [1,1a]

Enzymes/Milk enzymes are classified on the basis of the following activities or on chemical changes they bring about:

- **i.** Enzymes are classified on the type of substrate they attack or are named after the class of substrates they act upon- *e.g.*,
  - **a.** Proteinase or protease(protein splitting enzymes), which works on protein.
  - **b.** Lactase catalyzes the hydrolysis of lactose.
  - **c.** Lipase (fat splitting enzymes)which causes hydrolysis of fats.
  - **d.** Amylase (starch splitting enzymes).

ii. Still others are named after the function they perform- *e.g.*, Oxidases (oxidizing enzymes), Reductase (reducing enzymes) and hydrolase (hydrolysing enzymes) which catalyze the oxidation,reduction and hydrolysis of the substrate respectively.

iii. However, some of the earlier-discovered enzymes were not named in the above descriptive manner, and some still retain their original names- *e.g.*, pepsin and trypsin.

iv. Their names, in general, carry the suffix -ase, the prefix referring to the function or the substrate of the enzyme.

v. Certain enzymes are produced in an inactive form and are called zymogens. Such zymogens are then activated before they can function- *e.g.*, Trypsinogen produced by the pancreas needs to be converted to trypsin either by previously formed trypsin or by the enterokinase.

Enzymic Activity [1,2,2a]

It is determined by measuring either the disappearance of the substrate or the production of the end-products [1]. Enzyme activity is frequently expressed in units. The quantity or concentration of an enzyme can be expressed in molar amounts, as with any other chemical, or in terms of activity in enzyme units.

☆ The Enzyme Commission has defined a unit of enzyme as the quantity that will catalyze the transformation of one micromole of substrate to product(s) per minute under standard conditions.

i. Enzyme activity is a measure of the quantity of active enzyme present and is thus dependent on conditions, which should be specified. The SI unit is the katal, 1 katal = 1 mol s^{-1}, but this is an excessively large unit. A more practical and commonly used value is **enzyme unit (U) = 1 μmol min^{-1}**. 1U corresponds to 16.67 nanokatals [2a].

ii. The enzyme unit, **or international unit for enzyme (symbol U, sometimes also IU)** is a unit of enzyme's catalytic activity. 1 U (μmol/min) is defined as the amount of the enzyme that catalyzes the conversion of one micromole of substrate per minute under the specified conditions of the assay method.

For comparative purposes and quantization, initial velocities are preferred. In order to employ the relation between reaction velocity and enzyme concentration to quantitate the latter, three other parameters *viz.* substrate concentration, temperature, and pH-must be recognized and controlled. In general a plot of initial velocity against substrate concentration is a section of a rectangular hyperbola. Such a relation is rationalized and explained on the basis of the well-known Michaelis-Menten equation [2]-

$$vi = V\ max\ [S]/(Kmt + [S])$$

where, vi =initial velocity of reaction, Vmax = maximum velocity (or activity), Km = Michael is constant, and [S] = substrate concentration.

Specific Activity [2a]

The specific activity of an enzyme is another common unit. This is the activity of an enzyme per milligram of total protein (expressed in μmol min^{-1} mg^{-1}). Specific activity gives a measurement of enzyme purity in the mixture. Specific activity is equal to the rate of reaction multiplied by the volume of reaction divided by the mass of total protein. The SI unit is katal/kg, but a more practical unit is μmol/mgmin.

- ✰ Generally, enzymes are active only over a limited range of pH, and usually a distinct pH optimum is observed. This may result from an effect of pH on the V max, on affinity of enzyme for substrate, or on the stability of the enzyme.
- ✰ The activity of many enzymes depends on particular ionized groups in the active site. Obviously, these will be influenced by even small variations in pH and Ionic strength. More extreme high or low pH may denature and hence inactivate the enzyme [2].
- ✰ The overall effect of temperature on enzyme-catalyzed reactions is a resultant of the accelerating influence of temperature on the reaction itself and the thermal denaturation and inactivation of the enzyme. Thus, the rate of the catalyzed reaction passes through a maximum as it is examined at a series of increasing temperatures. The optimal temperature therefore, is generally higher when the time during which the reaction is measured is shorter [2].
- ✰ The optimum differs greatly among individual enzymes because they differ in susceptibility to thermal denaturation. It should be apparent from this brief Introduction: that measurement of enzyme activity must be performed under rigidly controlled conditions in which observed activity can be translated into activity with 100 per cent of the enzyme in the ES form. Therefore, enzyme activity is a value equivalent to V max for the concentration of enzyme available, that is, [E] k_{cat} (The values given in represent such values) [2].
- ✰ Enzymatic activities detected in bovine milk are listed in table given below. They are named and classified according to the recommendations of the Enzyme Commission of the International Union of Biochemistry. All classes of enzymes except ligases have been detected in milk [2].
- ✰ Some of the milk enzymes (*e.g.*, catalase) are constituents of leukocytes, and some (*e.g.*,plasmin) may gain entrance to milk from blood.
    - **i.** Most, however, are constituents or products of the mammary cells that enter milk as rather benign accidental constituents during the secretory processes.

ii. Some (*e.g.* the galactosyl transferase component of lactose synthase) are constituents of Golgi membranes, others (*e.g.*, alkaline phosphatase) of the cell membrane.

iii. Individual milk enzymes are associated with casein micelles, fat globules, or leukocytes, or are dispersed in the serum. Locations are specified in the appendix, for cases in which they are known.

Enzymes in Milk [1,1a,2,3]

Milk produced under normal conditions, and from normal healthy cows/ buffaloes, contains a wide variety of enzymes. While normal milk contains substrates for most of these enzymes, very little is known concerning the significance of these enzymes in milk. It is believed that these enzymes are normally present in tissues or cells and are spilled into milk during the milking process [1]. It is believed that all milk enzymes originate in the secretory epithelial cells of the mammary gland and the nature has provided these enzymes in milk for the benefit of the young [1].

Milk contains both indigenous and exogenous enzymes. The details of these types of enzymes are given below:

i. Exogenous enzymes – These mainly consist of heat-stable enzymes produced by psychrotrophic bacteria: lipases (fat splitting), and proteinases (Protein splitting), Analase (Starch splitting) and Peroxidase and Catalase.

ii. Indigenous enzymes- There are many indigenous enzymes that have been isolated from milk.

Enzymes in milk occur in following four phases-

i. Water soluble.

ii. Casein bound.

iii. Lipid bound and.

iv. Present in microsomal particles.

These enzymes are in a state of equilibrium in milk and can change their state due to agitation, homogenization, heating, *etc.* There are substances, which enhance the activity of various enzymes and are known as co-enzymes and those, which inhibit the activity, are called as anti-enzymes [1a].

Milk is a biological secretion of epithelial cells of the mammary gland and because of this, it has to undergo several biochemical reactions during its secretion [2,3].

- ☆ Consequently some of the enzymes will be entering the milk which have not been utilized during the biosynthesis of the milk.

- Similarly some enzymes get incorporated directly as a measure to protect the constituents after the production of milk.
- As such the enzymes naturally found in milk play a very significant role not only during the processing but also during their storage [2,3].

Approximately 50 enzymatic activities have been detected in bovine milk. Milk chemists are interested in quantitation of these enzymes, in deleterious or beneficial reactions that they catalyze in milk and dairy products, and in their inactivation.

Catalysis of the enzyme is not similar for all enzyme,some enzymes have a very fast activity *e.g.* conversion up to 1,000,000 mols of reactant in one minute where as some enzymes catalyse only few hundred molecules of reactant per min in to the product. The rate of catalysis per molecule of enzyme is expressed as k_{cat} sometimes called the turnover number for the enzyme. The k_{cat} is a function of the efficiency of the enzyme and the chemistry of the reaction.

Activities of some Enzymes in Bovine Milk [2,2B]

The enzymes in milk play a significant role by having an activity specific to that enzyme [2]. Milk contains about thirty enzymes which originate principally from the mammary cell cytoplasm or membranes as a result of peculiarities of or deficiencies in the secretory mechanism. Only about ten indigenous enzymes are technologically significant, from following five viewpoints [2b]-

i. Deterioration or preservation of milk quality- lipase (commercially probably the most significant enzyme in milk), proteinase, acid phosphatase, xanthine oxidase, sulphydryl oxidase, superoxide dismutase.
ii. Indices of thermal history of milk- alkaline phosphatase, lactoperoxidase.
iii. Indices of mastitic infection- catalase, acid phosphatase; the concentrations of several other enzymes increase on mastitic infection and hence could be exploited.
iv. Antimicrobial activity-lysozyme, lactoperoxidase (which is itself a natural inhibitor and can also be exploited as a component of the peroxidase-H_2O-thiocyanate systemfor the cold sterilisation of milk).
v. Commercial source of enzymes- ribonuclease.

The important enzymes along with their EC No, pH of optimum activity and temperature and the ratio of their activity in relation to the activity in bovine and human are presented in following Table.

Table 12.1: Activities of some Enzymes in Bovine Milk [2]

Enzyme	EC Number	Optimum		Activity in Bovine Milk (μmole.min^{-1} .liter^{-1})	Ratio of Activities Human/Bovine
		pH	Temp (^{0}C)		
Lactate dehydrogenase	1.1.1.27			40.	3.5
Malate dehydrogenase	1.1.1.37			40.	1.8
Xanthine oxidase	1.2.3.2	6.-9	37	175.	-
Catalase	1.11.1.6	7.0		300.	10.
Lactoperoxidase	1.11.1.7	6.-7	20	22,000.	0.03
Galactosyl transferase	2.4.1.38	7.2	37	50.	-
Lipoprotein lipase	3.1.1.34			600.	-
Alkaline phosphatase	3.1.3.1	9.8	37	500.	0.02
Phosphoprotein Phosphatase	3.1.3.16	4.0-5.5	14-50	13.	1.2
Ribonuclease	3.1.27.5	7.5	37	11.0-25.0 mg. liter^{-1}	0.3
Lysozyme	3.2.1.17	7.9	37	130.μg liter^{-1}	3000.
Glucose-6-phosphate isomerase	5.3.19			155.	2.

a. Variation in Concentration

Concentrations of milk enzymes vary greatly among species, and within a species large variations occur among individual animals and during the course of lactation. Data on such variations for most of the milk enzymes are not very complete or definitive.

b. Activities/Functions

- ☆ A few of the enzymes found in milk (*e.g.*, lactose synthetase) have known functions in the biosynthetic processes of the mammary cells.
- ☆ Some milk enzymes act on substrates present as normal constituents of milk and may under suitable conditions play either beneficial or deleterious roles in dairy processes and products.
- ☆ Hydrolases, such as lipase and proteinase, may facilitate resorption of milk constituents if and when milking is stopped. There is no well documented case of an enzyme in bovine milk being of direct benefit to the primary consumer-the calf. It has been suggested, however, that a lipase in human milk facilitates digestion of fat by the human infant.

Source and Significance of Milk Enzymes [1,1a,2,2c]

The activities of important enzymes(Approx. 14) in bovine milk are listed in by way of emphasizing the differences among species, ratios of activities in bovine and human milks are included. Xanthine oxidase, lactoperoxidase, and ribonuclease are especially prominent in bovine milk; lysozyme predominates quantitatively in human milk.

A. Hydrolases

This group includes the enzymes which are responsible for the hydrolysis of various milk constituents. These enzymes play a significant role in dairy industry especially the lipases which are responsible for the development of off flavours in milk and milk products.

- ✰ Similarly some of these hydrolases are useful in the detection of efficiency of milk processing.
- ✰ Enzymes which have the bacteriostatic action are also included in this group of enzymes.

a. Lipase (Lipolytic lipase- LPL) (EC 3.1.1.34) [2,2c]

A lipoprotein is the principal lipolytic enzyme of bovine milk. The normal function of such enzymes is to liberate fatty acids from lipoproteins and chylomicrons of the blood, the fatty acids are then resorbed by the secretory cells of the mammary gland. A low molecular weight apoprotein cofactor is necessary for the enzyme to attack its acylglycerol substrate when the latter is present as or covered with lipoproteins. Blood serum contains such cofactors, and the lipoprotein lipase is said to be serum-stimulated [2].

i. It is a hydrolytic enzymes and splits fats into glycerol and free fatty acids. This enzyme is found mainly in the plasma in association with casein micelles. It causes butyric or hydrolytic rancidity in milk, cream, butter, ghee and enhances acidity [1a]. In bovine milk the lipoprotein lipase (about 2 mg per liter) is bound largely to casein micelles. It does not attack the glycerides of the fat globules unless the membranes of the latter are damaged (*e.g.* homogenization) or if lipoproteins containing the apoprotein cofactor are added. The latter effect may arise from leakage of blood serum into milk if the tight junctions between secretory cells are impaired [2].

   - ✰ The milk fat is protected from its action by the FGM. If the FGM has been damaged, or if certain cofactors (blood serum lipoproteins) are present, the LPL is able to attack the lipoproteins of the FGM. Lipolysis may be caused in this way [1.1a].
   - ✰ Compared to fresh normal milk,late lactation milk contains higher lipase content. Homogenization, repeated heating and cooling, the presence of Cu and Fe ions and exposure to air or direct sunlight

will initiate lipolysis causing irreversible flavour and odour changes which adversely affect the quality of milk [1,1a].

- ✰ Homogenization of milk increases the surface area of fat globules and hence before or immediately after homogenization, pasteurisation is required. Raw homogenized milk will rapidly become rancid because of the activity of lipase on the greater fat surface [1,1a].
- ✰ It is destroyed on heating at 63°C for 20 minutes [1a]. This enzyme is inactivated at pH 4.6 for one hour [2].

ii. Lipoprotein lipase has been isolated from bovine skim milk using affinity chromatography on Sepharose to which heparin has been linked covalently. Its molecular weight is about 50,000, and it exists as a non- covalently linked dimer under physiological conditions; it contains about 8 per cent carbohydrate. Preparations with activities capable of releasing 300-700 µmol of fatty acids (from a triglyceride emulsion at pH 8.5) per min per mg protein have been achieved [2].

iii. Cow's colostrum contains little of the lipoprotein lipase but has a different lipase that is not bound to casein, does not bind to heparin-Sepharose, is not activated by blood serum, and is stable at pH 4.6 for 1 h. It disappears after the first few milkings following calving. The two lipases do not exhibit immunological cross-reaction. The colostral lipase probably falls under the classification of a triacylglycerol lipase (EC 3.1.1.3) but does not appear to be homologous to the bile-salt stimulated lipase of human milk.

- ✰ Lipolytic activities of skim milk and colostrum are respectively about 600 and 200 µmol·min-1·liter-1(measured with tributyrin emulsion at pH 8.75, 37°C).

b. Phosphatase Enzymes [1,1a,2,2c]

Phosphatases catalyze the hydrolysis of phosphoric acid esters. A great variety of phosphatases exists in nature-phosphomonoesterases, phosphodiesterases, phosphorylases, pyrophosphatases, nucleotidases, and phytases. Several types of phosphatases reportedly exist in normal milk but alkaline (pH 9.6) and acidic (pH 4.0) Phosphatase enzymes are discussed more. Both the alkaline and the acid phosphatase can release inorganic phosphatase from caseins and from soluble esters, and this may occur in milk or fractions thereof under appropriate conditions. The acid phosphatase is the more active of the two at the pH of milk. These two enzymes differ greatly in susceptibility to inactivation by heat [1,1a].

i. Milk Alkaline Phosphatase (EC3.1.3.1) [2,2c]

The scientific name for this enzyme is ortho phosphomonoester phosphohydrolase. This is a phosphomonoesterase. Two major isozymes have been identified, α and β phosphatase, mainly located in the milk plasma and fat globule membrane, respectively. The latter, more abundant, isozyme has been highly purified from bovine milk and found to be a dimer of two identical or very similar subunits each of MW ~ 85,000.

- ☆ It contains about five atoms of zinc per dimeric molecule. Its optimal pH for hydrolysis varies from one phosphate ester to another and with the composition of the medium.
- ☆ Monoesters, such as phosphoserine and β-glycerophosphate are hydrolyzed maximally near pH 9.0.
- ☆ Alkaline Phosphatase enzymes are able to split specific phosphoric acid esters into phosphoric acid and the related alcohols. Unlike most milk enzymes, it has a pH and temperature optima differing from physiological values; pH of 9.8.
- ☆ Milk alkaline phosphatase is used as the method of preference for determining whether the milk has been pasteurized adequately because this enzyme is destroyed by minimum pasteurization temperatures. Batch pasteurization or high-temperature, short-time pasteurization inactivates the alkaline phosphatase and, therefore, the presence of phosphatase activity in pasteurized milk indicates either improper pasteurization or post-pasteurization contamination with raw milk.
- ☆ In dairy, phosphatase test can be done to ensure proper pasteurization. Alkaline phosphatase is deactivated by heat but reversibly reactivated when milk is chilled. Inactivation of alkaline phosphatase by pasteurization is an index of destruction of Mycobacterium tuberculosis. It is possible to determine the inactivation of phosphatase enzyme by easy chemical methods.
- ☆ Concentration of the alkaline enzyme is highly variable in milk, and its concentration in milk is inversely correlated to the milk yield. It is associated with the microsomal particulates, which appear to be concentrated in the fat globule membrane. However, its concentration reportedly is not related to the fat content of the milk [1,4].

ii. Acid Phosphomonoesterase (EC 3.1.3.2) [2,2c]

A second phosphatase present in milk has a pH optimum at about 4.0. It has been assumed to be more similar to the phosphoprotein phosphatase (EC 3.1.3.16) of bovine spleen. It is primarily in the milk plasma. Its concentration is quite low (compared to alkaline phosphatase), though higher in colostrum [2,2c]

The acid phosphatase is perhaps the most heat-stable enzyme of milk. It is a highly basic protein and has been purified, 40,000-fold, using repeated chromatography on Amberlite IRC-50 resin and acetone precipitation [1].

Both the alkaline and the acid phosphatase can release inorganic phosphatase from caseins and from soluble esters, and this may occur in milk or fractions thereof under appropriate conditions. The acid phosphatase is the more active of the two at the pH of milk. These two enzymes differ greatly in susceptibility to inactivation by heat [2,2c].

c. *Ribonuclease (RNASE, EC 3.1.27.5) [1,2,2c]*

Another enzyme present in significant quantities in normal milk is ribonuclease, which catalyzes the hydrolysis of ribonucleic acid [1].

- ☆ Its content of mixed milk was found in one study to be 11 mg. liter^{-1} and 25 mg. liter^{-1} in another. RNase has been isolated from bovine milk by various methods.
- ☆ On the basis of the amino acid composition, immunological characteristics, gel electrophoretic mobility, and specific activity, ribonuclease A of milk appears to be similar to ribonuclease A of pancreas. Bovine pancreatic RNase has been well characterized. It consists of a single polypeptide chain of 124 residues, with MW 13,690 [1,2,2c].
- ☆ In spite of the low content of RNase in human milk (3mg·liter^{-1}), it has been isolated from that source by adsorption on a cation exchanger. Both RNase and lysozyme are so adsorbed, they can be differentially eluted [1,2,2c].

d. *Lysozyme (EC 3.2.1.24) [1,1a,2,2c]*

It is quantitatively an important fraction of the proteins of human milk (400 mg.liter^{-1}). It is a powerful bactericide as it attacks polysaccharides of the bacterial cell wall, causing lysis of the bacteria [2,2c].

- ☆ Lysozyme is an enzyme which lyses certain bacteria by hydroyzing the β-linkage between muramic acid and glucosamine of mucopolysaccharide of the bacterial cell wall [1,2].
- ☆ Bovine milk contains an average of 13 μg of lysozyme per 100 ml. Human milk contains 39 mg/100 ml, or nearly 3,000 times as much as in bovine milk [1]. High activity in human milk and possibly in horse milk, but very low activity in cow milk [1a].

Although the lysozyme content in bovine milk contains only about 0.1 mg. liter^{-1} it could be isolated from it. The lysozyme of human milk appears to be identical to that found in other human secretions, it is a polypeptide of 129 residues, with MW 14,602. α-lactalbumin and lysozyme are considered to be descendants of a common ancestor. Human lysozyme and α-lactalbumin, both obtained from milk, have different residues at 81 of 129 positions, Both have four disulfide bridges identically placed [1,2].

The bovine milk lysozyme has a pH optimum of 7.90 and human milk 6.35. Both are highly basic proteins. Their isoelectric points lie near pH 9.5 and 11.0, respectively [1].

e. *Plasmin (EC 3.4.21.7) [1a,2,2c]*

Plasmin is a proteolytic enzyme; it splits proteins. Plasmin attacks both β -casein and alpha(s2)-casein. It is very heat stable and responsible for the development of bitterness in pasteurized milk and UHT processed milk. The action of plasmin may produce serious defects *e.g.* bitter flavour (caused by hydrophobic-peptides of low molecular weight) and changes in viscosity and appearance, in UHT milk products. It may also play a role in the ripening and flavour development of certain cheeses, such as Swiss cheese. Plasmin fully survives pasteurization and partially resists UHT treatments. Increased activity has even been observed after heating milk at 72°C for 15 s [1,2].

i. The principal milk proteinase belongs to the alkaline serine proteinase class, it is probably identical to the plasmin of blood. Blood plasminogen (human) is a polypeptide of 790 residues; activation involves proteolytic cleavage of the C-terminal 230 residues and sometimes the N terminal 76 residues as well.

- ☆ Apparently, this enzyme enters milk from blood mostly in the form of its zymogen, plasminogen. In fresh milk only a small proportion is in the active form, milk may contain a factor that slowly activates the plasminogen.
- ☆ The enzyme is associated with the casein micelles. It attacks peptide bonds at the C-terminal side of Arg and Lys residues and thus is trypsin-like. Cleavage of Lys-X is faster than that of Arg-X. Optimal activity occurs at slightly alkaline pH and 37°C. The milk proteins most susceptible to plasmin are β- and αs2-caseins. αs1-casein also is attacked while κ-casein is relatively resistant, and the whey proteins α-lactalbumin and β-lactoglobulin are not affected.
- ☆ Plasmin action on β-casein is responsible for production of γ-caseins and the proteose-peptone fragments.

ii. A second proteinase, with maximal activity at pH 4.0,also occurs in milk. Its molecular weight is 36,000, it is heat-labile (inactivated at 70°C for 10 min), and it is partly inhibited by SH-blocking agents.

- ☆ It cleaves αs1-casein faster than β- or γ-caseins (at pH 5,37°C). Its action on these proteins produces peptides that in electrophoretic behavior resemble those produced by chymosin. It is not yet clear how this enzyme should be classified; it may well be an aspartate proteinase.

Proteases [1a] are proteolytic enzymes which hydrolyse protein into simpler compounds such as proteose, peptone and amino acids. These enzymes are involved in the hydrolysis of milk proteins *i.e.* casein and whey proteins. Proteases enzymes are destroyed at a heating temperature of 80°C.

f. *Amylase (diastase) [1a]*

It is a starch splitting enzyme (hydrolyses starch to dextrin) and considered to be least variable in quantity. This enzyme is present in milk in small amount. It is destroyed at 60-65°C.

g. *Lactases*

Lactases hydrolyse lactose into glucose and galactose. The principle enzyme is beta glycosidase. The above enzyme hydrolyse lactose, which is finally converted into lactic acid.

B. Oxidases [1,1a,2,2d]

Oxidases play a very important role especially the reactions involving oxidation and reduction. These enzymes are a group of enzymes which catalyzes an oxidation-reduction reaction involving molecular oxygen (O_2) as the electron acceptor. In these reactions, oxygen is reduced to water (H_2O) or hydrogen peroxide (H_2O_2).

- ☆ Some of the enzymes which are included in this group are helpful in increasing the shelf life of the raw milk utilizing the activity of an oxidase for the oxidation of thiocynate.
- ☆ Presence of some of the oxidases like catalase will help in the detection of diseased condition of the udder.

a. *Xanthine Oxidase (XO, EC 1.2.3.2) [1,2,2d]*

This is a rather nonspecific enzyme, in that it catalyzes the oxidation of several compounds, such as purines, pyrimidines, and aldehydes. This enzyme is very prominent in bovine milk. Most of it is associated with the fat globule membrane. Because of its relatively high XO content, milk is a source of choice for isolating this enzyme for investigation.

- ☆ Xanthineoxidase is much less prominent in milks of most other species; goat's milk has only about one-tenth the activity of cow's milk.
- ☆ The activity of XO in cow's milk depends to some extent on the Molybdenum content of the feed consumed.
- ☆ Bovine milk xanthine oxidase is a metallo flavo-protein, containing molybdenum and iron as its metal moieties. Molybdenum forms an integral part of the enzyme, in that there is a relationship between the molybdenum content and its activity [1].
- ☆ The molecular weight of xanthine oxidase is about 2,83,000; it is a dimer of two identical subunits, each carrying one flavin-adenine dinucleotide, one Mo, four Fe, and four acid-labile Satoms.

- ✰ The specific activity of the pure enzyme is 5 μmol.min-1 ·mg^{-I}. It can oxidize 12 moles of xanthine per mole per second (assuming two active centers per dimer).
- ✰ Although hypoxanthine and xanthine are its normal substrates in the pathway of purine metabolism, it catalyzes oxidation of many other substrates, including various aldehydes, Under some conditions XO catalyzes a two-electron reduction of O_2 to H_2O_2; but at high pH, high O_2^- is favoured concentration, and low xanthine concentration, the one-electron reduction to superoxide ion, O [2,2d].
- ✰ As isolated from milk, XO utilizes molecular O_2 as an electron acceptor, but if some of its disulfides are reduced with dithiothreitol it becomes a NAD^+ -dependent dehydrogenase. Reoxidizing the thiols to disulfides with the milk enzyme sulfhydryl oxidaserestores the specificity of the enzyme for O_2 as electron acceptor [2,2d].
- ✰ This enzyme is particularly important because it is fairly heat stable, being more stable than lipase and alkaline phosphatase. Contrary to the behavior of various other milk enzymes, the xanthine oxidase activity in milk increases with several heat treatments, homogenization, and by the protease and]ipase action. Such treatments presumably release xanthine oxidase from its inhibitors or complexing compounds in milk [1].
- ✰ The activity of XO in fresh milk is increased about fourfold by storing at 4°C, by heating at 70°C for 5 min, by homogenization, or by incubation with commercial proteinase or lipase preparations. Such treatments also transfer much of the enzyme from fat globules to plasma [2].

b. *Sulfhydryl Oxidase SOX (EC number not assigned) [2,2d,10]*

Sulfhydryl oxidase (SOX) is present in human milk and in milk from all species that have been studied. In human milk SOX is stable in an acid environment: 50 per cent of its activity remains after 1 h at pH 2.5. Acid stability is also characteristic of 7-glutamyltranspeptidase, another membrane- bound enzyme in skim milk [10].

Sulfhydryl oxidases (SOXs) are enzymes able to oxidise free sulfhydryl groups. Molecular oxygen is used as an electron acceptor and is reduced to hydrogen peroxid [11]. This enzyme is present in bovine milk, catalyzes oxidation of thiols to disulfides using molecular O_2 as the electron acceptor. It is an aerobic oxidase, reducing the O_2 to H_2O_2 rather than to H_2O.

$$2\ RSH + O_2 \rightarrow RSSR + H_2O_2$$

Sulfhydryl oxidase catalyzes oxidation of thiols in both small compounds and proteins. It is a large aggregate of subunits of MW 89,000. It contains about1.1 per cent carbohydrate residues by weight and 0.5 atom Fe per subunit.

At optimal pH (7.0) and temperature (35°C) its K_m for glutathione as substrate is 90 μM. Preparations with specific activities of more than 50 μmol· min-1mg $^{-I}$ have been obtained [2,2d].

- Milk may contain about 3 mg of this enzyme per liter. Sulfhydryl oxidase is closely associated with γ-glutamyl transferase (EC 2.3.2.2) in skim milk, and it has been suggested that the two activities reside in the same molecule. Such suggestions have been proved wrong, however, because the two activities can be separated. Sulfhydryl oxidaseim mobilized on glass beads has been used to oxidize thiols in UHT milk in an attempt to reduce cooked flavour [2,2d].

Application of SOX in Dairy [11]

The first application of SOX was reported in 1977 for the dairy industry.

- The SOX from milk was proved successful in removing the 'burnt-off' flavour of milk caused by ultrahigh-temperature treatment (UHT), and the process has been tested in a pilot scale with the enzyme immobilized on glass beads.
- Moreover, the addition of A. sojae SOX to UHT milk at a 9-U/l dosage was also able to completely remove the burnt off flavour.

c. *Catalase (EC1.11.1.6) [1,1a,2,2d]*

This enzyme splits the hydrogen peroxide (H_2O_2) into water(H_2O) and molecular oxygen. It is present in small amount in milk. However, on contamination of milk its concentration increases in milk. It is destroyed when heated above 90°C for 20- 25 minutes.

Catalase has been found to be present in several body organs and fluids, including milk. It is found in many tissues and is particularly prominent in liver, erythrocytes, and kidneys of animals.

- In milk its activity parallels leukocyte count and is higher in mastitic milk and colostrum than in normal milk.
- It increases with the multiplication of bacteria.
- Cells associate with the fat globules: therefore, catalase accumulates in the cream layer.
- Crystalline catalases have been prepared from several sources, but because little is present in milk it has not been fully purified therefrom. Milk catalase has a molecular weight of about 210,000, its isoelectric pH ≈ 5.5, and it contains heme iron.
- Catalase activities on the order of 300 μmol per min per liter have been reported for milk.

d. *Peroxidase/Lactoperoxidase (LP, EC 1.11.1.7) [1,1a,2,2d]*

Peroxidase is another enzyme associated with the oxidation: reduction system. It is an enzyme that breaks down hydrogen peroxide and liberates oxygen. It has antibacterial properties. It is the most abundant enzyme present in milk (more in buffalo milk than cow milk).

- ☆ Lactoperoxidase is the principle enzyme present in fresh milk. Activation of lactoperoxidase result in enhancing the keeping quality of raw milk for longer period. Heat treatments of milk such as 80° C for three and one half min.,73.5° C for 28 min. or 70° C for 150 min. destroy the this enzyme [1a].
- ☆ Compared to several other milk enzymes, peroxidase is fairly heat-stable. Technologically, therefore, it could be used as an indicator of high-beat treatments of milk [1].
- ☆ This enzyme catalyzes oxidation by H_2O_2 of a long list of electron-donor compounds, including aromatic amines, phenols, aromatic acids, leuko dyes, tyrosine and tryptophan, ascorbate, iodide, nitrite, and thiocyanate. LP may amount to as much as 1 per cent of the total serum proteins of milk (*i.e.,* 60 mg·kg^{-1}). Its activity in milk increases with advancing lactation to a maximum about 40 days postpartum and thereafter declines somewhat [2.2d].
- ☆ It is a glycoprotein (MW 77,500) containing 20-26 hexosamine residues (nosialic acid) and one heme per mole. Whether it consists of one or two polypeptide chains has not been settled [2,2d].
- ☆ LP catalyzes oxidation of thiocyanate (SCN-) to a product hypothiocyanite (OSCN-) that inhibits certain bacteria. Thiocyanate is a natural constituent of milk and H_2O_2 is produced by some bacteria themselves. Thus, in milk such bacteria exhibit self inhibition [2,2d].

e. *Superoxide Dismutase (SOD,EC1.15.1.1) [2,2d,8,9]*

Superoxide dismutases (SOD) are a group of enzymes that catalyze the dismutation of superoxide radicals (O_2–) to molecular oxygen (O_2) and hydrogen peroxide (H_2O_2), providing cellular defense against reactive oxygen species [8].

- ☆ There are three types of SOD which contain either manganese, copper/zinc or iron and in bovine milk Cu/Zn-SOD is found. The enzyme in milk serum has the same specific activity, molecular weight and electrophoretic properties as bovine erythrocyte Cu/Zn-SOD. In human milk the activity of Mn-SOD has also been identified [9].
- ☆ The enzyme consists of two identical subunits of MW 16,000, each containing one Cu and one Zn per mole. The primary structure of SOD from bovine erythrocytes is known completely. It has one free thiol and one disulfide bond. It functions as a non covalently linked dimer, and its specific activity is3300 µmol/min/mg [2,2d].

- ☆ Bovine milk serum contains a superoxide dismutase that is similar if indeed not identical to that of the erythrocytes. Estimated content in milk range from 0.15-2.4 mg/kg. It is a rather heat-stable enzyme; the activity is not decreased by heating milk at 63°C for 30 min. SOD can inhibit lipid peroxidation [2,2d].
- ☆ The concentration of SOD in cow's milk is probably not affected by stage of lactation or age of the cow, and does not vary between morning and night milking. The concentration of SOD in milk is approximately 100 times lower than that in bovine blood. The SOD concentration in human milk is 2.0 to 2.3 times higher than that of bovine milk. In cow milk SOD is exclusively present in skim milk fraction, with a concentration of 0.15 mg to 2.4 mg/L [9]. Human milk has 2.0 to 2.3 time higher concentration of SOD than cow milk.
- ☆ Cu/Zn-SOD is very resistant to various types of denaturing stress including heating. In fact commercial pasteurised milk retains the enzymatic activity at a similar level to unpasteurised milk. A minimal pasteurisation (71.7°C for 15 s) does not cause heat inactivation of SOD but heat treatments at >75°C have been shown to inactivate more than 20 per cent of the SOD. Purified SOD fractionated from bovine milk is more sensitive to thermal processing than the SOD in bovine milk serum [9].
- ☆ It may well be an important antioxidant protecting milk constituents from oxidation by superoxide ion generated by oxidations catalyzed by xanthine oxidase and lactoperoxidase as well as by riboflavin-sensitized photore actions [2,2d].

3. Transferases [2,2d,5,7]

A transferase is any one of a class of enzymes that enact the transfer of specific functional groups from one molecule (called the donor) to another (called the acceptor) **[5,6].**

The Galactosyl Transferase (EC2.4.1.74) Component of Lactose Synthase (EC 2.4.1.22)

Galactosyl transferases catalzye the synthesis of the hetero-polysaccharide moieties of complex carbohydrates. One of the best known galactosyl transferases is UDP-galactose: N-acetylglucosamine galactosyl transferase or A-protein of the lactose synthetase system. This enzyme has been purified from various animal and human body fluids, the amino acid and carbohydrate content of the bovine milk enzyme are known [7].

- ☆ In the absence of α-lactalbumin this enzyme transfers a galactosyl residue from UDP-galactose to an N-acetylglucosamine residue either free or in a protein-bound oligomer. In performing this function the enzyme is sometimes designated glycoprotein β-D-galactosyl-transferase (EC 2.4.1.38) [2,2d].

- ☆ α-lactalbumin modifies the transferase so that the Km for transfer of a galactosyl group to D-glucose is much reduced, and a β-1,4-glycosidic linkage is formed [2,2d].
- ☆ Galactosyl transferases are found in the Golgi membranes of many tissues. The transferase is found in membranes of Golgi vesicles and there performs the synthesis of lactose as α-lactalbumin transiently combines with it.
- ☆ It enters milk with golgi membrane fragments. Its concentration in bovine milk is approximately 3-4mg/liter. The specific activity of the purified enzyme is 14 μmol/min/mg proteins at 37°C and neutral pH. As isolated from milk, the transferase has a molecular weight of about 42,000. The transferase contains 12-13 per cent carbohydrate consisting of about 8 per cent neutral sugars, 1 per cent glucosamine,1 per cent galactosamine, and 2 per cent sialic acid [2,2d].

C. Other Enzymes [1a]

i. **Galactase**–This enzyme slowly reduces the milk proteins and it helps in ripening of cheese.

ii. **Reductases**- Reductases help in reduction of certain organic compounds. They reduce methylene blue added to milk and make it colourless.

Milk has numerous other enzymes including: protease activators, nucleases, glycosidases, and others.

Sources

1. *K.M.Shahani, Milk Enzymes "Their Role and Significance", Department of Dairy Science, University of Nebraska, Lincoln.*

1a. *M.K. Srivastava, Handbook on Analysis of Milk, CBS Publications, New Delhi, 2010. ISBN 13: 9788123926681.*

2. *K.S. Sharma, Chemistry of milk, Agrimoon.com.*

2a. *Enzyme assay, Wikipedia.*

2b. *P. F. Fox and P. A. Morrissey, Indigenous Enzymes of Bovine Milk, Department of Food Chemistry, University College, Cork, Ireland, Enzymes and Food Processing pp 213-23.*

2c. *Milk enzymes its sources and significance-part I, Lesson-15, http: //ecoursesonline.iasri.res.in*

2d. *Milk enzymes its sources and significance-part II, Lesson-16, http: //ecoursesonline.iasri.res.in.*

3. *Dixon and Webb, Enzymes, 1979.*
4. *Haab, W., and Smith, L. M. 1956. Variations in Alkaline Phosphatase Activity of Milk. J. Dairy Sci., 39: 1644.*
5. *Wikipedia.*
6. *Transferase, Genetics Home Reference. National Institute of Health. Retrieved 4 November 2013.*
7. *Margit Hamosh, Enzymes in Human Milk, Handbook of Milk Composition, 1995.*
8. *Emery and Rimoin's Principles and Practice of Medical Genetics (Sixth Edition), 2013, Pages 1-22.*
9. *Helena Lindmark-MaÊnsson, and B. AE kesson, Antioxidative factors in milk, British Journal of Nutrition 2000, 84, Suppl. 1, S103±S110.*
10. *Sulfhydryl oxidase in Human milk, www.nature.com.*
11. *Greta Faccio, Outi Nivala, Kristiina Kruus, Johanna Buchert and Markku Saloheimo, Sulfhydryl oxidases: sources, properties, production and applications, Appl Microbiol Biotechnol, 2011), 91: 957-966.*

Chapter 13

Milk Carbohydrates

Lactose [1,1a,1b,4] is the major carbohydrate in the milk of most species. The characteristic carbohydrate of milk is lactose. It is commonly referred as milk sugar(Latin, lac, milk). Lactose is also known as Lactobiose. Milk of mammals is the sole source of lactose. It is found only in milk. In addition to lactose, fresh milk contains other carbohydrates in small amounts, including glucose, galactose, and oligosaccharides. Some important aspects of lactose are given below:

- ✰ Human milk is a rich source of lactose and oligosaccharides. Human milk contains about 6.3-7.0 per cent lactose and between 0.3-0.6 per cent oligo-saccharides.
- ✰ Lactose is essentially unique to milk, although it has been identified in the fruit of certain plants.
- ✰ Of the mammalian species where information is available, only some marsupials have an alternative sugar other than lactose, and those sugars are generally trisaccharides of glucose and galactose.
- ✰ Of all the mammalian milk human milk contains highest lactose content (7.0 per cent) while the average lactose content of normal bovine milk is 4.8 per cent which accounts for about 50 to 52 per cent of the total solids in skim milk. Lactose comprises 4.8 to 5.2 per cent of milk, 52 per cent of milk SNF, and 70 per cent of whey solids.
- ✰ The amount of lactose present in milk depends upon the health of the udder and the breed of the milch animals. The lactose content of cows milk varies with the breed of cow, individual animals, udder infection (mastitis depresses secretion of lactose) and stage of lactation.

- ✰ Lactose is a major, readily digestible source of glucose which provides energy for the neonate.
- ✰ Lactose intolerance can occur in adult animals or animals who do not have lactase activity in their intestines.
- ✰ The lactose content in milk is increased slightly by the overfeeding of carbohydrates and decreased by mastitis infection in udder.
- ✰ Lactose plays a major role in milk synthesis. It is the major osmole in milk and the process of synthesis of lactose is responsible for drawing water into the milk as it is being formed in the mammary epithelial cells. Because of the close relationship between lactose synthesis and the amount of water drawn into milk, lactose content is the least variable component of milk.

Milk also contain traces of bound carbohydrates. Besides lactose small amount of other carbohydrates are found in milk partly in a free form and partly bound to proteins, lipid or phosphate. Cow's milk contains monosaccharide glucose and galactose in concentrations of about 10mg/10ml. The amount of oligosaccharides is 100 mg/litre. This disaccharide would act as a controlling factor in fermented and ripened dairy products. It contributes to the nutritive value of milk and milk products and plays essential role in the body and texture and solubility of certain stored products. It has an essential role in the colour and flavour of heated products. Essential role in the colour and flavour of the high heated dairy products [1,1b].

Important Characteristics of Lactose [1a

Lactose has following characteristic:

i. Is the major carbohydrate in milk.
ii. Is present in milk of most mammals,
iii. Is a readily digestible source of glucose (energy) for the neonate,
iv. Is digested by lactase enzyme in the neonate,
v. Is unique to the mammary gland and
vi. Is the major osmole of milk (the primary component that draws water into the gland).

Lactose-Nomenclature and Structure [1,1a,1b,2,4]

Nomenclature

Lactose is a disaccharide found in the milk of many mammals, including humans and cows. It consists of a d-glucose and a d-galactose molecule joined in a β-1,4-glycoside linkage. The IUPAC name of lactose is 4-O-(β-d-galactopyranosyl)-d-glucopyranose. Its systemic name is β-*O*-d-galacto-pyranosyl-(1-4)-α-d-glucopyranose (α-lactose) or β-*O*-d-galactopyranosyl-(1-4)-

β-d-glucopyranose (β-lactose). It occurs in both alpha and beta isomeric forms which differ in the steric configuration of the hydroxyl group of C-1 moiety of glucose [2a]. Lactose yields one molecule of D-glucose and one molecule of D-galactose on hydrolysis. The glucose can be in either the α-pyranose form or the β-pyranose form, whereas the galactose can only have the β-pyranose form: hence α-lactose and β-lactose refer to the anomeric form of the glucopyranose ring alone.

- ☆ Lactose is hydrolysed to glucose and galactose, isomerised in alkaline solution to lactulose, and catalytically hydrogenated to the corresponding polyhydric alcohol, lactitol. Lactulose is a commercial product, used for treatment of constipation. [4].
- ☆ The predominant carbohydrates encountered in the body are structurally related to the aldotriose glyceraldehyde and to the ketotriose dihydroxyacetone.

All carbohydrates are optically active because they contain at least one asymmetrical (chiral) carbon. In addition, carbohydrates can exist in either of the two conformations, as determined by the orientation of the hydroxyl group about the asymmetric carbon atom. Farthest from the carbonyl. With a few exceptions, those carbohydrates that are of physiological significance exist in the D-conformation. The mirror-image conformations, called enantiomers, are in the L-conformation.

Structure of Lactose [1,1b,4]

Lactose is derived from the condensation of galactose and glucose, which form a β-1→4 glycosidic linkage. The two monosaccharides of lactose are linked through the aldehyde group of D-galactose. Thus the aldehyde portion of lactose is on glucose residue. The configuration of the D-galactoseresidue is of the beta form, which was shown by using the enzyme β- D-galactosidase that hydrolyses both lactose and a methyl β - D- galactopyranoside but not the α anomer.

- ☆ The D-galactose in the lactose molecule is in the beta form was also shown by its synthesis from D-glucose and D-galactose. The point of union of the two monosaccharides was established through products of hydrolysis or methylated lactose.

The aldehyde and ketone moieties of the carbohydrates with five and six carbons will spontaneously react with alcohol groups present in neighboring carbons to produce intra molecular hemiacetals or hemiketals, respectively. This results in the formation of five- or six-membered rings [1,1b].

Furanose

Because the five-membered ring structure resembles the organic molecule furan, derivatives with this structure are termed furanoses. If reaction is between the C-4 hydroxyl and the aldehyde, a furanose is formed.

Pyranose

Those with six-membered rings resemble the organic molecule pyran and are termed pyranoses. The pyranose ring is formed by the reaction of the hydroxyl group on carbon 5 (C-5) of a sugar with the aldehyde at carbon 1. This forms an intramolecular hemiacetal.

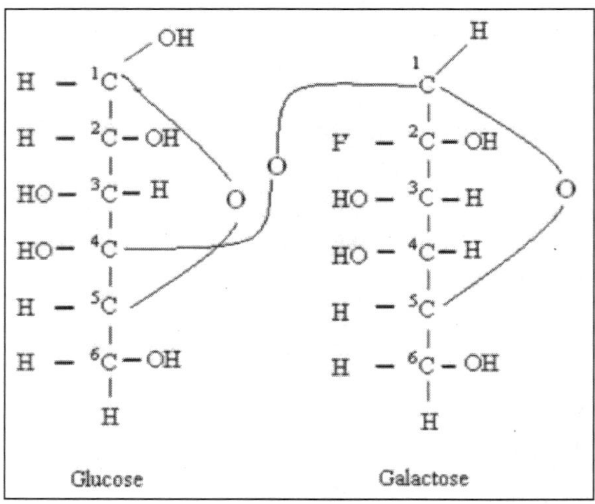

Figure 13.1: Fischer Structure of α Lactose [2].

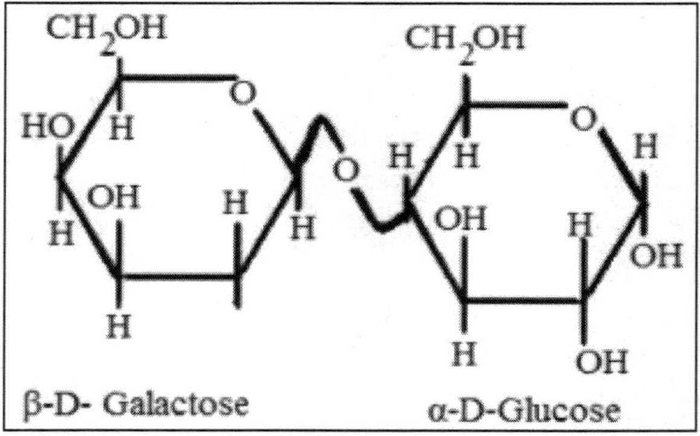

Figure 13.2: Haworth Structure of α Lactose [2].

The rings can open and re-close, allowing rotation to occur about the carbon bearing the reactive carbonyl yielding two distinct configurations (α and β) of the hemiacetals. The carbon about which this rotation occurs is the anomeric carbon and the two forms are termed anomers. In milk, lactose exists in two isomeric forms, called α- and β- lactose respectively [1,1b].

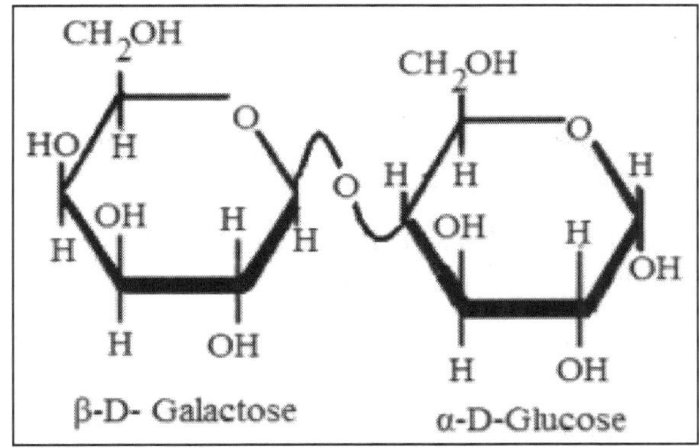

Figure 13.3: Structural Formula of α Lactose [3].

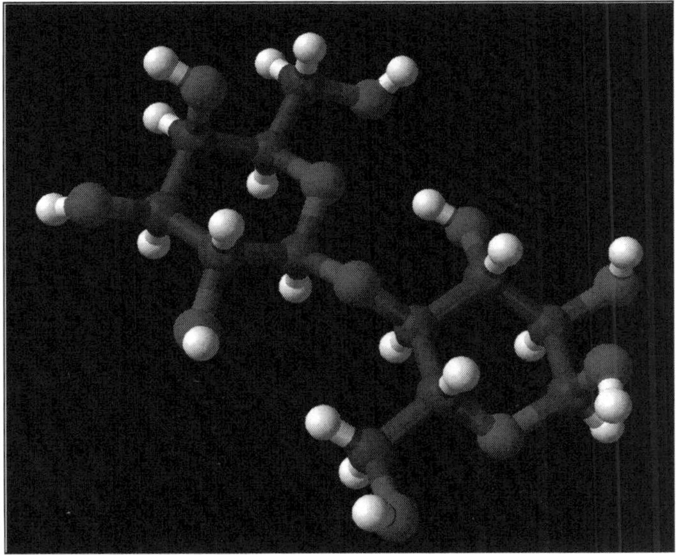

Figure 13.4: The Molecular Structure as Determined by X-ray Crystallography [4].

- ☆ The molecular structures of α- and β -lactose differ in the orientation of a hydrogen and a hydroxyl group on carbon atom no.1 in the glucose moiety [6].
- ☆ Carbohydrates (Lactose) can change spontaneously between the α and β configurations and this process known as mutarotation [1,1b,6]. The velocity of mutarotation is determined by factors like temperature, concentration and pH (acidity) of the solution [6].

Fischer Projection

When drawn in the Fischer projection, the α configuration places the hydroxyl attached to the anomeric carbon to the right, towards the ring.

Haworth Projection

When drawn in the Haworth projection, the α configuration places the hydroxyl downward.

Chair Form and the Boat form Conformations [1,1b,5]

The spatial relationships of the atoms of the furanose and pyranose ring structures are more correctly described by the two conformations identified as the chair form and the boat form.

a. The Chair Form [1,1b]

It is the more stable of the two. Constituents of the ring that project above or below the plane of the ring are axial and those that project parallel to the plane are equatorial. In the chair conformation, the orientation of the hydroxyl group about the anomeric carbon of α-D-glucose is axial and equatorial in β-D-glucose.

- ✰ Covalent bonds between the anomeric hydroxyl of a cyclic sugar and the hydroxyl of a second sugar (or another alcohol containing compound) are termed glycosidic bonds, and the resultant molecules are glycosides. The linkage of two monosaccharides to form disaccharides involves a glycosidic bond. Several physiologically important disaccharides are sucrose, lactose and maltose.

b. Boat Form [5]

An alternate conformation for a six-membered ring is called the 'boat'. In the boat conformation, two of the substituents – those on the 'bow' and the 'stern', are brought close enough to each other to cause steric strain [5].

Sources

1. *K.S. Sharma, Chemistry of milk, Agrimoon.com.*

1a. *M.K. Srivastava, Handbook on Analysis of Milk, CBS Publications, New Delhi, 2010. ISBN 13: 9788123926681.*

1b. *Lactose-Nomenclature and structure, Lesson-17, http: //ecoursesonline.iasri. res.in.*

2. *Robert Jenness and Stuart Patton, Principles of Dairy Chemistry, John Wiley and Sons, Inc., NY, 1959.*

2a. *J.M. Johnson and F.D. Conforti, Lactose-Physical and Chemical properties, Encyclopedia of Food Sciences and Nutrition (Second Edition), 2003.*

3. *Harrington, Journal of Dairy Science, 1934.*

4. *Wikipedia.*

5. *Conformations of cyclic organic molecules, https: //chem. libretexts.org, 2019*

6. *Lactose, Some basic properties and characteristics, DFE Pharma.*

A. Physical Properties of Lactose [1,1a,1c,2,3]

The major constituent present in the soluble portion of the milk is carbohydrates and more specifically the lactose. It influences several properties of milk. The various aspects of lactose will help in proper understanding of various phenomena occurring in milk during the milk processing. Lactose normally occurs naturally in either of two crystalline form α- monohydrate and anhydrous β or as an amorphous "glass" mixture of α- and β-lactose. Several other forms may be produced under special conditions. These designations refer to the configuration on the number one carbon of the glucose moiety. Two forms of Lactose are common, the alpha monohydrate and the beta anhydride form [1,1a,1c].

Commercial lactose often consists of about one part alpha and two parts beta. It is very hygroscopic especially what is called Lactose glass which is supercooled and supersaturated solution of lactose.

1. Physical Forms of Lactose [1,1a,1c,7]

Lactose normally occurs naturally in either of two crystalline form α-monohydrate and anhydrous β or as an amorphous "glass" mixture of α- and β-lactose. The difference is mainly in their solubility, optic rotation and melting point. Lactose in solid form can either be in a crystalline state or in an amorphous state. Crystalline lactose can exist in a number of distinct forms. Most well known are α-lactose monohydrate and β-lactose. Also, two crystalline anhydrous α-lactose types are known, a stable and an unstable (hygroscopic) form. Furthermore, a so-called mixed crystalline form is reported, containing both α- and β- lactose in a special crystal lattice [8].

- ☆ **Crystallinity** is the result of a highly ordered arrangement of the lactose molecules.
- ☆ **Amorphous** lactose lacks crystallinity and the arrangement of the lactose molecules is more or less random [8]

The physical properties of these two forms (One is α-lactose and the other is β-lactose) differ to some extent. The difference is mainly in their solubility, optic rotation and melting point. Lactose has two anomers (stereo-isomers), α and β, differing only in the configuration of the hydroxyl group at the anomeric hemiacetal carbon. α has an axial hydroxyl whereas β hydroxyl is equatorial [1,1c]. Lactose may appear in different crystal polymorphs, depending on the crystallization conditions and Each polymorph has its specific properties [2a].

i. α-Lactose [1,1c,2a,8]

α-lactose (α-Hydrate) (which crystallizes as a hydrate containing equimolar amounts of lactose and water) has very hard crystals and is not hygroscopic [2a].Ordinary commercial lactose is α-lactose monohydrate ($C_{12}H_{22}O_{11}$. H_2O). It is prepared by concentrating an aqueous lactose solution to super saturation and allow for crystallization to take place at a modern rate below 93.5 °C.

- ✰ The α-hydrate is the stable solid form at ordinary temperatures is indicated by the fact that the other solid forms change to the hydrate in the presence of a small amount of water below 93.5° C.
- ✰ It has a specific optional rotation of +89.40 (anhydrous weight basis) and a melting point of 201.6 °C.

The alpha form of lactose is responsible for crystallization. It forms a number of crystal shapes, depending on the conditions of crystallization; the most common type on fully developed is tomahawk-shaped.

α-*Anhydrous*

Anhydrous α-lactose may be prepared by dehydrating α-hydrate in vacuo at a temperature between 65 and 93.5 °C; it is stable only in the absence of moisture.

ii. β-lactose [1,1c,8]

This is the other isomeric form of lactose. It exhibits a specific rotation of +35° on anhydrous weight basis. Its melting point is 252.2°C. Above 93.5°C crystallization or drying of lactose solutions yields β anhydrate [1,1c].

- ✰ A hydrous β-lactose forms small crystals with a solubility up to 10 times higher than that of α-lactose. Because of this higher solubility, β-lactose is experienced sweeter than α-lactose. It is hygroscopic [2a].
- ✰ Crystals of pure β-lactose have a characteristic kite-like form. Particles with crystalline β-lactose are more brittle than α–lactose monohydrate crystals and they do not contain crystal water [8].
- ✰ β -Lactose is often referred to as anhydrous lactose [8]. The β-lactose crystals are usually smaller with a larger surface area.

2. Optic Property of Lactose [1,1c]

The specific rotation of a chemical compound [α] is defined as the observed angle of optical rotation 'α ' when plane-polarized light is passed through a sample with a path length of one decimeter and a sample concentration of one gram per millilitre. Therefore the specific rotation may be represented by the following formula -

$$[\alpha] = 100\ a/lc$$

where,

[α] = specific rotation at 20° C using D-line of sodium, a = degrees of angular rotation, l = length of tube in decimeter, c = concentration of substance in grams per 100 ml of solution.

- ☆ The specific rotation of a pure material is an intrinsic property of that material at a given wavelength and temperature [1,1c].
- ☆ Values should always be accompanied by the temperature at which the measurement was performed and the solvent in which the material was dissolved. Often the temperature is not specified; in these cases it is assumed to be room temperature [1,1c].
- ☆ The formal unit for specific rotation values is deg $dm^{-1}cm^{3}$per g but scientific literature uses just degrees. A negative value means levorotatory rotation and a positive value means dextrorotatory rotation [1b].

3. Solubility [1,1a,1c]

Lactose is about 17.8-18 per cent soluble in water at 25°C or relatively insoluble which is a problem in many dairy products, ice cream, sweetened condensed milk [1a]. Lactose is insoluble in alcohol and ether but is soluble in hot acetic acid. The low solubility of lactose, coupled with its capacity to form supersaturated solutions, is of considerable practical importance in the manufacture of concentrated milk products [2a].

- ☆ Lactose is freely soluble in water. However, the solubility of lactose is much lower than that of other common sugars.
- ☆ Solubility increases with increasing temperature. The solubility of α-lactose is more temperature-dependent than that of β-lactose. α- and β-lactose differ considerably in solubility and in the temperature dependence of solubility [2a]. α- Lactose is soluble only to the extent of 7 g per 100 g water at 20°C. β-lactose dissolves more readily (is about 10 times higher) than α-lactose at room temperature, as is apparent from their very different initial rates of solubility.
- ☆ Final solubility is the same for α- and β-lactose because of the mutarotation equilibrium that is eventually reached in solution. The particle size of the lactose influences its dissolving velocity. Coarse lactose crystals dissolve much slower than tiny lactose particles. Dissolving velocity, and hence particle size, does not alter final solubility [1,1c,8]. The solubility of lactose depends on the followings:
  - **i.** Final solubility of lactose depends on temperature.
  - **ii.** The initial solubility is the true solubility of the form. The increasing solubility with time is due to mutarotation.
- ☆ As some of the α form is converted to β form, the solution becomes unsaturated with respect to α, and more α -hydrate dissolves. This process continues until equilibrium is established between α and β in solution and no more α -hydrate can dissolve, thus establishing the final solubility.

- ☆ This solution is saturated with respect to α, but a great deal of β-lactose powder can be dissolved in it because of the greater initial solubility of the β form. The solution becomes saturated with α long before the saturation point of β is reached. However, additional β dissolving in such a solution upsets the equilibrium, and mutarotation takes place.
- ☆ Since the solution was already saturated with α, α formed by mutarotation will crystallize to re-establish equilibrium.
- ☆ Since β -lactose is much more soluble and mutarotation is slow, it is possible to form more highly concentrated solutions by dissolving β rather that α -lactose hydrate. In either case, the final solubility of the lactose in solution will be the same.
- ☆ Lactose solubility values at different temperatures are shown in following Figure [1,1c]:

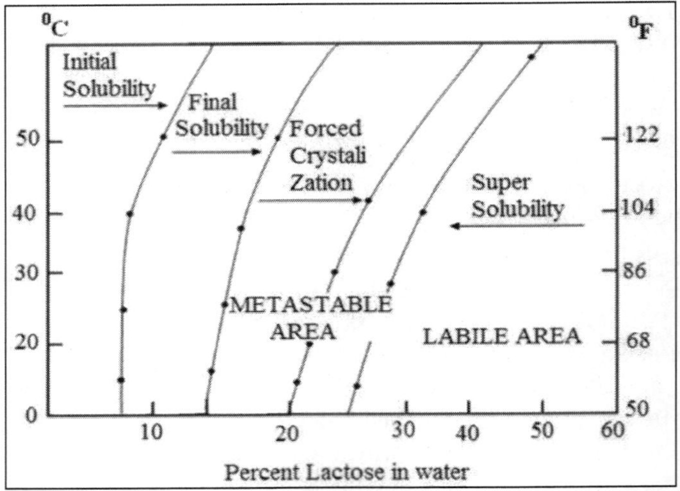

Figure 13.5: Lactose Solubility Curve.

- ☆ The solvent and the presence of salts or sucrose influence the solubility of lactose, as well as the rate of mutarotation.
- ☆ The solubility of lactose increases with increasing concentrations of several calcium salts-chlorides, bromide, or nitrate-and exceedingly stable, concentrated solutions are formed.
- ☆ Acetone also reduces the solubility of lactose based on which a procedure has been developed to recover the lactose from whey.

4. Equilibrium in Solution (Mutarotation) [1,1c,8]

Lactose is in the mutarotation equilibrium of two anomers in solution, the α and the β forms. Lactose exists in two forms *viz.*, α and β. By definition, α is the form with greater optical rotation in the dextro direction. The specific rotation of a substance is characteristic of that substance. Also important, besides the variables of the equation, are temperature of the solution, wavelength of the light source, and concentration of the solution.

- ☆ Regardless of the form used in the preparation of solution, the specific rotation will continue to change until +55.4 ° is reached at equilibrium. This is equivalent to the 37.3 per cent in α form and 62.7 per cent in β form. Since equilibrium rotation is the sum of the individual mix of α and β forms [1].

Factors affecting Rate of Mutarotation [1,8]

The rate of lactose mutarotation is influenced greatly by both temperature and pH [8].

- ☆ The rate is slow at low temperature but increases as the temperature rises, becoming almost instantaneous at about 75°C [1,1c].
- ☆ Lactose solutions strive after a state of equilibrium between the α and β form. At room temperature the equilibrium results in a ratio of about 40 per cent α-lactose and 60 per cent β-lactose. The fact that two forms of lactose exist which differ in molecular structure, has profound effects on various properties of lactose such as crystallization behaviour, crystal morphology, solid state properties and solubility [8].
- ☆ The presence of sugars and salts can also affect the rate of mutarotation. Although the effect is small in dilute solutions, a combination of salts equal to that found in solution in milk nearly doubles the rate of mutarotation. This catalytic effect is attributed primarily to the citrates and phosphates of milk. The presence of high levels of sucrose, on the other hand, has the opposite effect [1,1c].
- ☆ The effect of sucrose is only slight at concentration up to 40 per cent but as concentration increases above this level, mutarotation is rapidly decreased to about half the normal rate of the specific rotation [1,1c].

5. Crystallization of Lactose [1,1d,8]

The lactose crystallization plays a very important role in body and texture/ structure of some of the milk products. On crystallization lactose crystallizes with one molecule of water of crystallization ($C_{12}H_{22}O_{11}.H_2O$). The crystallization of lactose will influence the physical properties of the milk product. The knowledge of this physical property helps in obtaining quality products and helps to avoid several defects in the body and texture of these products [1].

These crystals are gritty in nature and sometime give gritty or sandy taste to some milk products like concentrated milk and ice-cream. Because lactose crystallizes slowly, it is often present in fresh milk products as a glass, crystallization takes place later on holding. **The crystallization results in the defect called sandiness. Milk sugar or Lactose present in milk as true solution in milk serum [1a].** The details related to the Lactose and crystallization of α Lactose and β Lactose are given in following table [2a]:

Table 13.1: Lactose and Crystallization of different Lactose

Sl.No.	Isomers of Lactose	Crystallization Conditions	Properties of Crystals
1	α-lactose	Usually, this crystallizes as a hydrate containing one molecule water of crystallization.	The crystals are very hard, slightly hygroscopic, often fairly large, and dissolve slowly. The water of crystallization is very strongly bound.
2	β Lactose	Above 93.5 °C, anhydrous β-lactose crystallizes from an aqueous solution.	It is not very hydroscopic, and it dissolves quickly; its solubility is good.

The principal factor governing the crystalline habit of lactose is the precipitation pressure, the ratio of actual concentration to solubility.

Effect of Precipitation Pressure [1,1d]

Low and high precipitation pressures affects the crystalline habit of lactose. Different relative growth rates on the crystal faces account for the various shapes observed. The rate of crystal growth increases rapidly as supersaturation (precipitation pressure) is increased.

- ☆ When the pressure is high and crystallization is forced rapidly, only prisms are formed.
- ☆ As precipitation pressure lessens, the dominant crystal form changes to diamond-shape plates, then to pyramids and tomahawks, and finally, in slow crystallization, to the fully developed crystal.

Crystallization of Lactose in Dairy Product [1,1d,8]

In dairy products, crystallization is more complex. The impurities(*e.g.* other milk components), as far as lactose is concerned, may interfere with the crystalline habit. As a result, the crystals tend to be irregularly shaped and clumped, instead of yielding the characteristic crystals obtained from simple lactose solutions.

- ☆ In some instances, the impurities may inhibit the formation of nuclei and thus retard or prevent lactose crystallization. The influence of a number of additives on growth rates has been studied; some additives resulted in marked retardation, whereas others accelerated growth on specific crystal faces [1,1d] The axes and the faces of the tomahawk crystal is diagrammatically shown in following figure:

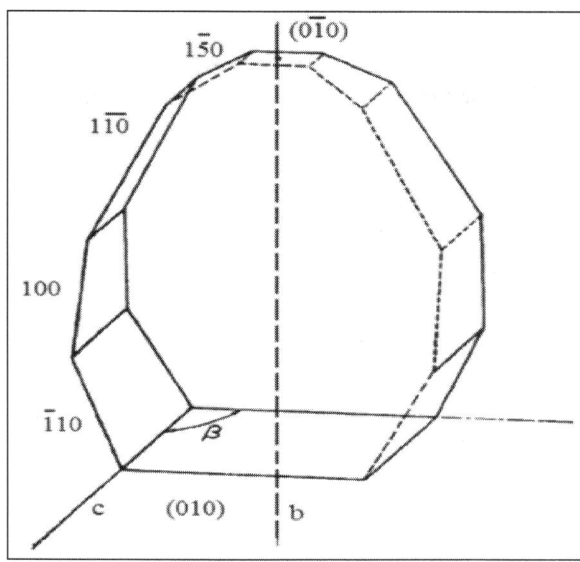

Figure 13.6: Tomahawk Crystal of α Lactose Monohydtrate [2].

- ☆ **Crystallization inhibitor** [1,1d]-Gelatin is an example of a crystallization inhibitor that reduces the growth rate to 1/3 to 3/4 of normal even at low gelatin concentrations. In highly supersaturated lactose solutions, however, gelatin cannot suppress nucleation, which explains its ineffectiveness in preventing sandiness in ice cream.
- ☆ Various marine and vegetable gums are currently in wide use in ice cream formulations to prevent the sandiness.
- ☆ Some carbohydrates actively inhibit the crystallization of lactose, where as others do not.
- ☆ **Crystallization accelerator [1,1d]**-Both methanol and ethanol accelerate crystallization by as much as 30 to 60 per cent even at low (1 per cent) concentrations, depending on which crystal face is being observed.
- ☆ The rate of lactose crystallization is also markedly increased at low pH (<1).
- ☆ Calcium chloride had the greatest growth-promoting effect; at the 10 per cent impurity level [1,1d].

6. Lactose Glass [1,1d,8]

It is also called as amorphous noncrystalline glass. When a lactose solution is dried rapidly(for instance by spray drying), its viscosity increases so quickly that crystallization cannot take place and a glassy lactose mass is obtained.

- The dry lactose is essentially in the same condition as it was in solution, except for removal of the water. This is spoken of as "concentrated syrup or an "amorphous"(noncrystalline) glass [1,1c,8]. This amorphous lactose contains both α- and β-lactose, the ratio being approximately the same as the equilibrium ratio of the original lactose solution at the drying temperature [8].
- Lactose glass is stable if protected from moisture, but since it is very hygroscopic, it rapidly takes up moisture from the air and becomes sticky. When the moisture content reaches about 8 per cent or a relative vapour pressure near 0.5, the lactose achieves a maximum weight; a discontinuity is observed in the sorption isotherm, and water is desorbed from the lactose.
- Amorphous lactose is formed when a solution (*e.g.* milk) is dried rapidly, as in a spray drier, or frozen. It is a very concentrated solution and it quickly dissolves or, rather, is diluted, on addition of water; but then,α-lactose hydrate may start to crystallize.
- It is present in the glassy state, which means that many properties, including hardness, density, and specific heat, are similar to those of crystalline sugar but that the packing of the molecules does not show perfect order [2a].
- If the water content of the amorphous lactose is low, say 3 per cent, crystallization may be postponed almost indefinitely; nucleation rate is negligible because of the extremely high viscosity of the "solution".
- The product is, however, very hygroscopic, and when moisture content rises to about 8 per cent, α-lactose hydrate starts to crystallize which helps to make very small crystals [1,1d,2a].
- But when crystallization of lactose caused by moisture uptake occurs in milk or whey powder, the result is caking; powder particles are cemented together by crystalline lactose, forming large and stony lumps [1,1d]. This caking is caused by the crystallization of lactose, as it causes the powder particles, largely consisting of lactose, to grow together (to sinter) [2a].
- At a higher temperature, crystallization can occur far more readily, a_w being higher; moreover, the viscosity of the highly concentrated lactose solution (essentially the continuous phase of the powder particles) is lower, leading to a more rapid nucleation, hence crystallization [2a].
- It is almost impossible to obtain pure crystals. For instance, α-hydrate usually contains a few per cent of β-lactose, and vice versa. The different forms mentioned are different crystal polymorphs (*i.e.* they have different crystal lattices).

7. Density [1,1d,8]

The specific weight of α-lactose monohydrate is 1.54 kg/l (at 20°C). Pure β-lactose is somewhat heavier (1.59 kg/l at 20°C). The density of lactose powder is much lower than the values cited above because of the presence of air between the individual lactose particles in the powder. Powder density is not as well defined a characteristic as is the true density (specific weight). It heavily depends on the granulometry (particle size distribution) of the powder [8].

☆ The densities of the various lactose crystals differ slightly from each other [1,1d].

- **i.** α-hydrate form is 1.540,
- **ii.** Anhydrous β is 1.589,
- **iii.** Anhydrous α is 1.544 and
- **iv.** Anhydrous αcrystallized from alcohol is 1.575.

Densities of lactose solutions are not linear functions of concentration [1,1d].

B. Biological Properties [1,1a,5]

- **i. Nutritive value [1a,4]-**When lactose is completely digested in the small intestine, its caloric value is 4 kcal/g, or the same as that of other carbohydrates. However, lactose is not always fully digested in the small intestine. Depending on ingested dose, combination with meals (either solid or liquid), and lactase activity in the intestines, the caloric value of lactose ranges from 2 to 4 kcal/g. Undigested lactose acts as dietary fiber. It also has positive effects on absorption of minerals, such as calcium and magnesium.
- **ii. The glycemic index [5,6]** - The GI of lactose is 46 to 65 [6]. For comparison, the glycemic index of glucose is 100 to 138, of sucrose is 68 to 92, of maltose is 105, and of fructose is 19 to 27 [6].
- **iii.** Lactose has relatively low cariogenicity among sugars [5,6]. This is because it is not a substrate for dental plaque formation and it is not rapidly fermented by oral bacteria [5,6]. The buffering capacity of milk also reduces the cariogenicity of lactose.
- **iv. Relative Sweetness [1,1a,1d,5]-** The sweetness of lactose is 0.2 to 0.4, relative to 1.0 for sucrose. [15] For comparison, the sweetness of glucose is 0.6 to 0.7, of fructose is 1.3, of galactose is 0.5 to 0.7, of maltose is 0.4 to 0.5, of sorbose is 0.4, and of xylose is 0.6 to 0.7 [5,6]. The relatives sweetness of sugars changes with the concentration. Therefore it is misleading to say that one sugar is so many times as sweet as another, because this will be true only at certain concentrations. The relative sweetness of some common sugars is presented in following table:

Table 13.2: Relative Sweetness of Sugars [3,4]

	Concentration (per cent w/v) to give Equivalent Sweetness			
Sl.No.	Lactose	Glucose	Fructose	Sucrose
1	1.9	0.9	0.4	0.5
2	3.5	1.8	0.8	1.0
3	6.5	3.6	1.7	2.0
4	6.5	3.8	-	2.0
5	6.0	3.2	-	2.0
6	15.7	8.3	4.2	5.0
7	14.9	8.3	4.6	5.0
8	13.1	7.2	4.5	5.0
9	25.9	13.9	8.6	10
10	0.00	14.6	-	10
11	20.7	12.7	8.7	10
12	27.8	17.2	12.8	15
13	34.6	20.0	13.0	15
14	33.3	21.8	16.7	20

- ☆ It should be noted that lactose is relatively sweeter at higher concentrations than at lower concentrations and is sweeter than is usually reported in reviews of food applications [1,1d].
- ☆ β-lactose appears to be somewhat sweeter than α-lactose, probably due to the fact that β-lactose dissolves somewhat quicker in the saliva of the mouth than α-lactose, hence reaching a higher concentration in the same period of time and thus giving rise to a higher sweetness sensation.
- ☆ Lactose has a clean and sweet taste without any after taste. The sweetness profile resembles that of sucrose. However, the relative sweetness of lactose is small (only 20 per cent) when compared to sucrose (100 per cent).
- ☆ Lactose is not as sweet as sucrose. Lactose is faintly sweet, about one-sixth the sweetness of cane sugar. When lactose is hydrolyzed by β-D-galactosidase (lactase), an enzyme that splits these monosaccharide, the result is increased sweetness, and depressed freezing point [1a].
- ☆ β- Lactose is sweeter than α-lactose but β form of lactose is not appreciably sweeter than the equilibrium mixture except when the concentration of lactose solution equals or is greater than 7 per cent.
- ☆ Since there is approximately 62.7 per cent of β form in the equilibrium mixture, a β-lactose solution differs less in sweetness from a solution in equilibrium than does α-lactose solution [1,1d].

Sources

1. *K.S. Sharma, Chemistry of milk, Agrimoon.com.*

1a. *M.K. Srivastava, Handbook on Analysis of Milk, CBS Publications, New Delhi, 2010. ISBN13: 9788123926681.*

1b. *Physical properties of Lactose, DC-4, Lesson 18, http: //ecoursesonline.iasri.res.in.*

1c. *Physical Properties of Lactose-Part-I, Lesson-18, http: //ecoursesonline.iasri.res.in.*

1d. *Physical Properties of Lactose-Part-II, Lesson-19, http: //ecoursesonline.iasri.res.in.*

2. *N.P.Wong, R. Jenness, E.H. Marth, and M. Keeney, Fundamentals of Dairy Chemistry, 1988, ISBN 978-0-8342-1360-9.*

3. *T.A.Nickerson, Fundamentals of Dairy chemistry, 1974.*

4. *B.H. Webb, A.H. Johnson, and J.A, Alford, J. A. 1965. Fundamentals of dairy chemistry. AVI Publ. Co., New York.*

5. *Wikipedia.*

6. *Schaafsma, Gertjan, 2008, "Lactose and lactose derivatives as bioactive ingredients in human nutrition". International Dairy Journal. 18 (5): 458–465.*

7. *A.M. Shendurse, C.D. Khedkar, Lactosein Encyclopedia of Food and Health, 2016.*

8. *Lactose, Some basic properties and characteristics, DFE Pharma.*

C. Chemical Reaction of Lactose [1,1a,1b2,4]

The carbohydrates undergoes extensive complex chemical changes during rigorous conditions of processing. It is well established fact and it is also observed that lactose undergoes similar changes even in mild conditions [1]. Lactose itself is very stable and inert from a chemical point of view. It has, apart from some special cases, no significant tendency to react with the drug or other components of a formulation. Some remarks on the chemical properties of lactose are useful however. The low hygroscopicity of crystalline lactose supports its virtual chemical inertness. Most chemical reactions of lactose occur noticeably only in aqueous environment [4]. The chemical reaction of lactose mainly involves following four sites [1,1b]-

i. The hemiacetal linkages between carbon 1 and 5 of the glucose moiety.

ii. β-1,4 glycosidic linkages between the two sugars.

iii. The hydroxyl groups of both the glucose and galactose units.

iv. The carbon to carbon bonds.

1. Oxidation [1,1a,1b]

The oxidation of lactose and extent to which lactose may be oxidized will vary. It depends up on the particular reagent, its concentration, and other reaction conditions. Hence by selection of conditions it is possible to derive oxidation products from lactose. These products range from relatively simple alterations of the reducing carbon in the glucose portion of the molecule to a carboxylic acid group to complete degradation with the end products being CO_2 and water. Lactose may undergoes vigorous oxidation, mild oxidation and biological oxidation depending upon the reaction conditions.

a. Vigorous Oxidation

More vigorous oxidation of lactose with dilute nitric acid ruptures the glycosidic linkage and produces dicarboxylic acid derivatives of the two sugars.

Lactose → D-glucaric acid + D-galactaric acid

(Saccharic acid) (Mucic acid)

The dicarboxyl derivative of galactose, galactaric or mucic acid, was formerly much used as a crystalline derivative for identification of galactose.

b. Mild Oxidation

Mild oxidation of lactose with such reagents as alkaline copper, iodine, or picrate forms lactobionic acid(4-0-β-D-galactopyranosyl-D-gluconic acid), which has a carboxyl group at C-1of the glucose unit.

- ✰ Lactose also can be oxidized to lactobionic acid by mild chemical dehydrogenation and by lactose dehydrogenase produced by certain species of bacteria of the genus Pseudomonas. The free aldehyde group of lactose can be oxidized to a carboxylic acid or lactobionic acid.
- ✰ Lactobionic acid readily forms lactones by esterification of the carboxyl with hydroxyls at C-4 or C-5.

$$\textbf{Lactose} \xrightarrow{-2H} \textbf{Lactobionic ə lactone} \xrightarrow{+H_2O} \textbf{Lactobionic acid}$$

Such compounds as lactobionic acid have a profound tendency to form lactones through inter esterification with hydroxyl groups of the number 4 or 5 carbon.

- ✰ Such oxidation reactions often are used to determine reducing sugars quantitatively. In methods for the quantitative measurement of lactose in which its reducing property serves as the basis of measurement where lactobionic acid is the resultant product.

c. Biological Oxidation

Biological oxidation of lactose to CO_2 and water can be brought about by mixed cultures of bacteria and protozoa obtained from sewage sludge.

Lactose → CO_2 and water

Such processes are useful in decomposing lactose-containing wastes from dairy factories.

2. Lactulose(4-O-β-D-Galactopyranosyl-D-Fructose) [1,1b]

It is isomerized lactose consisting of galactose and fructose. Lactulose is an epimer of lactose in which the glucose moiety is isomerized to fructose. Lactulose is a compound found in heated milk product in which the fructose moiety occurs predominantly in the pyranose and partly in the furanose form.

On strong Oxidation

Lactose → D-Glucaric acid + D- Galactaric acid

Saccharic acid   Mucic acid

- ☆ It is produced on heating milk to sterilizing conditions and is a commonly used index of the severity of the heat treatment to which milk has been subjected, *e.g.*, to differentiate in-container sterilized milk from UHT milk, it is not present in raw or HTST pasteurized milk. Its concentration in commercial evaporated milk will be up to 1 per cent.
- ☆ It is an isomer of lactose that is formed by molecular rearrangement, usually under alkaline conditions where by the terminal aldose residue of lactose is converted into a ketose.

Preparation

It can be produced under mild alkaline conditions via. the Lobry de Bruyn-Alberda van Ekenstein reaction and at a low yield as a by-product of β-galactosidase action on lactose.

i. Preparation of lactulose with calcium hydroxide has long been known but preparation of ketoses by this method is time consuming, yields are less than 20 per cent, and the keto sugar must be isolated from un-reacted starting materials, alkaline degradation products, and metal salts.

ii. A method used to prepare lactulose in nearly 90 per cent yield is by treatment of lactose with boric acid in an aqueous solution made basic by tertiary amines.

iii. Ultra filtration can also be used to obtain lactulose from sweet whey which can be made into a non hygroscopic lactulose by some treatments.

Properties

Lactulose is extremely soluble in water and polar solvents such as methanol.

- ☆ It is difficult to crystallize, especially when traces of other sugars are present.

- ☆ Lactulose is unstable in alkaline solution, degrading by alkaline peeling and β -elimination reactions to yield galactose, isosaccharinic acids, and other acid products.
- ☆ Lactulose is sweeter than lactose and about 60 per cent as sweet as sucrose
- ☆ Amines can bring about dehydration and degradation reactions. Lactulose is similar to sucrose in humectant's properties.

Important Uses

Lactulose has several important uses in the food and drug industries.

- ☆ There is much information on lactulose utilization in infant nutrition. The presence of lactulose in infant feeding encourages the development of Bifidobacterium bifidum in the intestinal flora, imitating flora in the guts of breast-fed infants.
- ☆ Some of the research workers suggested that lactulose could partially replace sucrose and corn sweeteners in intermediate-moisture foods. Only limited amounts could be tolerated in foods because of its laxative properties.

Concerns

There has been some concern about the possible laxative effects of lactulose, especially in infants; a low colonic pH might be a contributing factor to this effect.

This laxative is not metabolized by oral bacteria and not hydrolyzed by intestinal β-galactosidase and hence reaches the large intestine where it can be metabolized by lactic acid bacteria, including Bifidobacterium spp. and serves as a bifidus factor. So, lactulose has attracted considerable attention as a means of modifying the intestinal microflora, reducing intestinal pH and preventing the growth of undesirable putrefactive bacteria.

- ☆ It is currently believed that lactulose cannot be digested by human alimentary enzymes, so even lactose tolerated individuals cannot digest lactulose.

3. Hydrolysis of Lactose [1,1a,1b]

Lactose may be hydrolyzed to D-glucose and D-galactose with mineral acids, with cation exchange resins in the acid form, and with β-galactosidases (often called lactases). The β-1,4 linkage between the two sugar residues is much more resistant to hydrolysis than is the 1,2 linkage between glucose and fructose in sucrose. Treatments such as an hour at 90° C with 1.5 M HCI, or at 150° C with 0.1 M HCI, are required to hydrolyze lactose completely.

a. Acid Hydrolysis

Lactose is resistant to acid hydrolysis compared to other disaccharides such as sucrose.

- ☆ In fact, organic acids, such as citric acid, that easily hydrolyze sucrose are unable to hydrolyze lactose under similar conditions. This is useful in analyzing a mixture of these two sugars, because the quantity of sucrose can be measured by the extent of these changes in the optical rotation of reducing power as a result of mild acid hydrolysis [1,1b].
- ☆ Mineral acids (Hydrochloric acid *etc.*)also possesses the hydrolyzing capacity to split lactose into its two sugar units *viz.* glucose and galactose [1a].
- ☆ The speed of hydrolysis of lactose varies with time, temperature, and concentration of the reactant.
  - **i.** 5 to 40 per cent lactose solutions (w/w) can readily be hydrolyzed with 1 to 3 N hydrochloric acid or sulfuric acid.
  - **ii.** Ninety per cent of the lactose could be hydrolyzed to the constituent monosaccharides at relatively low temperatures (60°C) and long reaction times (up to 36 hr). Due to degradative side reactions producing high levels of off flavor and colour this process could not be used for whey concentrate.
  - **iii.** Lactose hydrolysis can also be brought about with 0.1 N hydrochloric acid in short reaction times at 121°C.
  - **iv.** Sulfonic acid-type ion exchange resins have been used to catalyze lactose hydrolysis. The resin was equally effective on lactose solutions and acid whey permeates [11b,]. The hydrolysis is carried out at temperatures ranging from 90° to 98°C. The advantages of this method are continuous operation, short reaction times, and no mineral acid to be removed from the hydrolyzed product. High temperature and low pH eliminate problems with microbial contamination.
- ☆ Best reaction rates were achieved with strong acid granular-type cation exchange resins with low degrees of cross linking.
- ☆ The formation of oligosaccharides during acid hydrolysis seem to be much less than during enzymatic hydrolysis [1,1b].

b. Enzymatic Hydrolysis [1,1b]

The hydrolysis of lactose using β-galactosidase (lactase) enzymes is most common method available. Lactose splits into glucose and galactose by the enzyme lactase found in small intestine as shown below [1a]-

Lactase

$$C_{12}H_{22}O_{11} + H_2O \rightarrow C_6H_{12}O_6 + C_6H_{12}O_6$$

There has been significant progress in this field, and several processes are almost commercially feasible. There are following three major approaches to enzymatic hydrolysis [1,1b]:

i. "Single use" or "throwaway" lactase systems.

ii. Lactase recovery systems based on membranes to retain the lactase for reuse.

iii. Immobilized systems in which the enzyme is physically or chemically bound to a solid matrix.

Lactases are suitable for industrial processing of whey or lactose. Several lactases are available.

i. The enzyme prepared from the yeast Kluveromyceslactis has a pH optimum between 6 and 7 and a temperature optimum of about 35°C.

ii. The lactase from K. fragilis has a pH optimum of 4.8 and a temperature optimum of about 50°C.

A batch processing operation is the simplest method of achieving enzymatic lactose hydrolysis but suffers from the disadvantage that a large amount of recoverable enzyme is needed. The following methods of enzymatic lactose hydrolysis are being used depending upon the irregular or regular uses-

a. For small users or manufacture on an irregular basis

For this type of uses, the single-use enzyme procedure is probably the method of choice. Membrane reactor systems in which the enzyme is recovered by ultra filtration of the reaction mixture after hydrolysis is complete have been developed.

b. For large and regular production

Lactose hydrolysis with immobilized systems is preferred when regular production of hydrolyzed syrups on a large scale is required.

☆ The best known of these is the Corning immobilized system. It uses lactase from Aspergillus niger covalently bound to a controlled-poresilica carrier. The particle size is 0.4 to 0.8 mm, the wet bulk density is 0.6, the activity is near 500 U/g at 50°C, and the optimal pH of operation is between 3.2 and 4.3.

i. The rate of hydrolysis is dependent on the mineral, lactose, and galactose concentrations, as well as on the temperature and pH.

ii. Inhibition of hydrolysis can be caused by galactose or sodium and calcium ions, so demineralization is often necessary.

- ☆ For immobilized systems (designed for long-term use) adequate techniques such as back flushing with water, acetic acid, milk alkali and detergents with bactericidal activity must be developed to ensure sanitary operations.
- ☆ Galactose is primarily involved in the formation of the oligosaccharides, which accounts for the lower concentration of free galactose than of free glucose during hydrolysis.
- ☆ The use of hydrolyzed lactose syrups has been proposed as an alternative sweetener to corn syrup solids.

4. Fermentation of Lactose [1,1a,1b,4]

Lactose is a carbohydrate and as such an easy substrate for micro-organisms, like bacteria, yeasts and moulds [4]. Lactose is metabolized by various microorganisms to compounds of lower molecular weight [1,1b].

a. By Lactic Acid Bacteria [1,1a,1b]

In dairy field the most important is lactic acid fermentation. Lactic acid bacteria are present everywhere. Lactic acid bacteria is important in the preparation of cultured products, butter and cheese.

- ☆ Lactose is a good substrates for the activity of microorganism hence the conversion of lactose to lactic acid is favoured by microorganisms and this causes sour taste to milk. At an acidity of 0.18 to 0.20 per cent milk curdles on heating.
- ☆ One of its most important functions is its utilization as a fermentation substrate. Lactic acid bacteria produce lactic acid from lactose, which is the beginning of many fermented dairy products.
- ☆ Because of their ability to metabolize lactose, they have a competitive advantage over many pathogenic and spoilage organisms. Their activity in milk intended for liquid consumption is not desirable while it is important in the preparation of cultured products, butter, and cheese.
- ☆ Some microorganism can produce as much as 1.5 per cent lactic acid in milk.

Homofermentative and Heterofermentative Bacteria

There are homo-fermentative and heterofermentative bacteria. The bacteria are classified as homofermentative (produce only lactic acid) and heterofermentative(produce lactic acid, acetic acid, alcohol and carbon dioxide).

i. Homofermentative

The bacteria are classified as homofermentative if they produce only lactic acid.

- ✰ Homofermentative bacteria have phosphotransferase system in the cell membrane that phosphorylates lactose at C-6 of the galactose moiety as it enters the cell.
- ✰ A β-D-phosphogalactosidase then hydrolyses the lactose-P to glucose and galactose-6-P.
- ✰ The glucose is phosphorylated at C-6 and metabolized to pyruvate via Embden-Meyerh of pathway of glycolysis. The galactose-P converted via a fructose isomer, tagatose-P to the triose-P stage of glycolysis and then to pyruvate.
- ✰ Other homofermentative bacteria such as S. thermophilus and Lactobacillus bulgaricus do not phosphorylate entering lactose but rather hydrolyse it with an intracellular β - galactosidase.

ii. Heterofermentative

If bacteria produce acetic acid, alcohol, and carbon dioxide along with lactic acid then they are called as Heterofermentative.

- ✰ These organisms lack aldolase enzyme to cleave the six carbon atom to two three carbon units in the Embden-Meyerhof pathway.
- ✰ They use a pathway known as hexose monophophates hunt pathway in which 6-phosphogluconic acid is formed from glucose and then it is decarboxylated and the resulting five carbon unit, ribose- 5-phosphate is split to acetyl phosphate and glyceraldehyde-3- phosphate. The former yields acetic acid and/or ethanol and the latter yields pyruvic acid and then lactic acid.

b. Fermentation by Yeast and Enzymes [1a]

Certain yeasts and enzymes have the ability to ferment lactose to alcohol and lactic acid which impart acid or alcoholic taste to fermented milk products.

$$\underset{\text{Lactose}}{C_{12}H_{22}O_{11}} + HOH \rightarrow \underset{\text{Glucose}}{C_6H_{12}O_6} + \underset{\text{Galactose}}{C_6H_{12}O_6}$$

$$2C_6H_{12}O_6 \rightarrow \underset{\text{Lactic acid}}{4\,CH_3CHOHCOOH}$$

5. Reducing Properties [1a]

Lactose is a reducing sugar due to the presence of a free aldehyde group and hence it reduces fehling solution.

6. Chemical Reactions Due to Heating of Lactose (Browning) [1,1c]

Milk products are sensitive to the effects of heat treatment given under conventional process and storage conditions because of an abundance of the following reactive functional groups:

i. Aldehyde group of lactose.

ii. ε -amino group of lysine.

iii. Other reactive N-containing groups *e.g.* indolyl group of tryptophan, imidazole group of histidine, guanidine group of arginine and theα-amino group of proteins and free amino acids.

The most important heat induced changes in dairy products that involve lactose are the changes associated with browning. Milk has naturally occurring protein with a high content of Lactose (reducing sugar). Lactose may isomerize or it may react with protein in milk. The reaction of lactose with the caseins and whey proteins of milk systems will be via the Maillard or non enzymatic browning reaction. It is also referred to as glycosylation of proteins or in case of lactose it is known as lactosylation of proteins.

A. Maillard Type Browning [1,1c,4]

The Maillard reaction is named after the French chemist Louis Camille Maillard. This reaction can be classified **as a non-enzymatic browning reaction**.

- ✰ In aqueous solution, lactose, as all reducing sugars (*e.g.* glucose, galactose, maltose, maltodextrines), tends to react with compounds possessing primary- and secondary amino-groups, such as proteins and peptides. This reaction is called the Maillard reaction [4].
- ✰ The Maillard reaction (nonenzymatic glycation) is a chemical reaction between amino group and carbonyl group. It is the extremely complex reaction that usually takes place during food processing or storage. In the case of milk, lactose reacts with the free amino acid side chains of milk proteins (mainly ε-amino group of lysine residue) to proceed to early, intermediate, and advanced stages of Maillard reaction and forms enormous kinds of Maillard reaction products [5].
- ✰ The Maillard-type browning, sugar-amino type is the most prevalent, since it requires relatively low energy of activation and is auto catalytic. Direct caramelization, on the other hand, has a rather high energy of activation and therefore is less important [1,1c].

The formation of caramel from sugar and milk and the change in colour during the preparation of condensed milk, are some common examples in which the Maillard reaction is responsible for the change in colours and flavours of dairy products.

Maillard type browning is some where desirable and some where undesirable as far quality of some dairy and food products, is concerned [1,1c,5].

☆ **Undesirable**- Some changes caused by Maillard reaction are unwelcome and reduces the quality of products. It is detrimental to the organoleptic, nutritional and functional qualities of the product and are therefore undesirable.

i. Over heating of milk during Curd (dahi)/sweet curd (Meetha Dahi), Kheer *etc.* causes the browning of milk which is demerit.

ii. The Maillard reaction shows various effects on milk proteins such as bioavailability, solubility, forming property, emulsifying property, and heating stability [5].

iii. Generally, in the Maillard reaction in milk, lactose mainly reacts with ε-amino group of lysine residue of milk proteins. Thus, the lysine loss by the Maillard reaction increases with a severity of heat treatment.

☆ **Desirable**- In some milk products, browning enhances the quality. Some examples are given below:

i. However this reaction is being utilized favourably in the preparation of products like khoa where milk is heated in the presence of sucrose to produce brown product with a pleasant flavour.

ii. Maillard reaction also plays an important role in the generation of flavour during the manufacture of ghee or clarified butter and milk chocolate. The Maillard reaction occurs in cooking of almost all kinds of foods, although the simple sugars and amino acids present produce distinctly different aromas.

Reaction Mechanism of Maillard Reaction [1,1c,3,6,7]

This reaction involves following major steps [7]-

i. The Maillard reaction mechanism begins with the formation of an N-substituted glycosamine (along with water) from the reaction between the amino group of the amino acid and the carbonyl group of the reducing sugar. Now, the glycosamine is transformed into ketosamines via Amadori rearrangement.

☆ **Formation of an N-substituted glycosamine**- The first step in the Maillard reaction involves the nitrogen atom of an amino compound and carbonyl group of an aldehyde, or ketone in food systems and the reactants are predominantly protein and reducing sugars. In some situations the carbonyl products of lipid peroxidation, vitamin C, free amino acids and ammonia are also important reactants. The reaction proceeds with the elimination of a molecule of water to form a Schiff s base which subsequently rearranges to for man N- substituted glycosylamine intermediate [1,1c].

- ☆ **Transformation of glycosamine into ketosamines via Amadori rearran-gement-** The amino acid carboxyl group plays an important role in the catalysis of the Amadori rearrangement [1,1c]. Thus the N-substituted glycosylamines derived from the amino acids are inherently unstable and they are either hydrolysed to the parent amino acid and reducing sugar or react via a spontaneous rearrangement to form the corresponding keto–(α-1-amino-1-deoxy- 2-ketose) or aldo (α-2-amino-2-deoxyaldose) derivative depending on whether the parent sugar is an aldose or a ketose respectively. The aldo to keto transformation is referred to as the Amadori rearrangement and the corresponding keto to aldo rearrangement as the Heynes rearrangement [1,1c].

ii. These ketosamines undergo further reaction via several pathways and can form reductones, butanedione, methylglyoxal, and several other products of short-chain hydrolytic fission.

- ☆ The protein-carbohydrate complex or its decomposition products result in the production of reducing substances, fluorescent substances, and disagreeable flavour materials. For example, 40 compounds were isolated and identified from a model system of casein and lactose that a product which had been stored at 80°C and 75 per cent relative humidity for 8 days to accelerate browning.
- ☆ Substances which are isolated are furans, lactones, pyrazines, pyridines, acetylpyrrole, amines, pyrrolidinone, succinamide, glutarimide, dicarboxylic acid, acetone, 2- galactose, D-tagatose, and lactulose [1,1c].

iii. The ketosamines can also go on to form melanoidins and other brown nitrogenous polymers. This imparts the characteristic brown colour to the food.

Hodge Maillard browning pathway given below helps in understanding the aforesaid reaction mechanism shown on next page Figure[1,1c].

Effect of Browning on Nutritive Value [1,1c]

Browning has detrimental effect on the nutritive value of food products through interaction of the free €-amino group of lysine in the proteins with carbohydrates and the resultant rearranged product.

- ☆ Destruction of essential amino acids, particularly lysine and probably histidine, has been shown to occur during the storage and browning of non fat dry milk with high moisture (7.6 per cent) content.
- ☆ Similar powders of low moisture (3.0 per cent) did not deteriorate in nutritive value during storage.

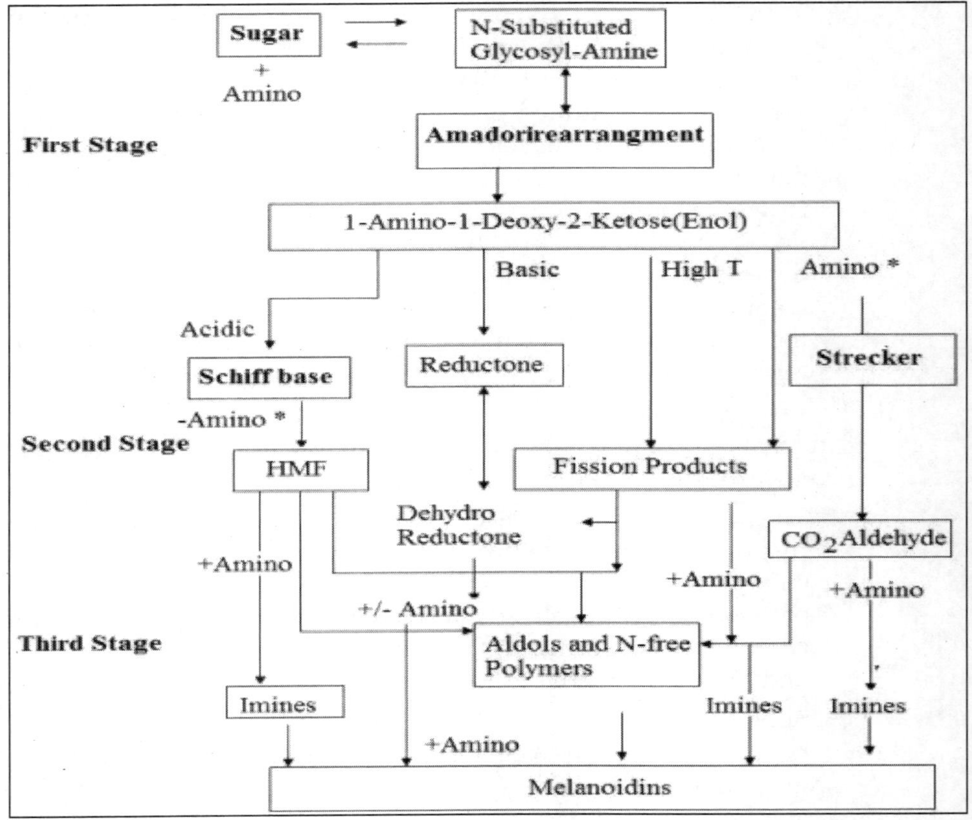

Figure 13.7: Maillard Browning Pathway [3].

- Reaction of β-lactoglobulin with lactose in the "dry state" (10 per cent moisture) resulted in various degrees of lysine destruction, depending upon temperature and heating times. Arginine, histidine, acidic and neutral amino acids were not damaged by the thermal treatments (0 to 90°C) in the presence of lactose.
- Lactose and casein are the two principal reactants in the browning of milk products, but dried whey products containing lactose also undergo browning. Roller-dried products showed significant losses in lysine content, due to the contact with the drums with high temperatures.

Inhibition of Browning Reaction in Milk/Milk Products [11c,]

In milk products, active sulfhydryl groups serve as natural inhibitors in retarding heat induced browning, but their mechanism is not understood.

- ☆ Sodium bisulfite, sulfur dioxide, and formaldehyde also inhibit browning in milk systems as well as in simpler mixture of amino acid and sugar solution.
- ☆ In actual practice, browning in dairy products is controlled by limiting heat treatments, moisture content, and time and temperature of storage.

Factors Influencing Maillard Reactions [1,1c,1a,4]

The overall rate and product profile of Maillard reaction in milk roducts/ foods are highly dependent on number of parameters. The following factors are most important factors that are affecting this reaction:

i. Reactants,

ii. Moisture content and Water activity,

iii. pH,

iv. Temperature.

i. Reactants

The nature and the molar concentration of the reactants have considerable impact on the rate and mechanism of the Maillard reaction.

- ☆ Low molecular weight reactants tend to react more readily than high molecular weight reactants.
- ☆ Glucose is more reactive than lactose and contribute to the increased rate of browning in lactose hydrolysed milks.
- ☆ The Maillard reaction can be controlled by changing the amount or the concentration of the reducing sugars present in the reaction environment.

ii. Moisture Content and Water Activity (aw)

Both moisture content and the *a*w of a food system exert a major influence on the Maillard reaction.

- ☆ **Moisture-**Water may influence the rate of reaction by controlling the viscosity of the liquid phase and by dissolution concentration or dilution of reactants.
- ☆ **Water activity** -At a very low aw value the proportion of the total reactants in solution is negligible and therefore the reaction rate is minimal. As the aw increases the concentration of remains constant provided excess solute is available to maintain saturated solution. However the total volume in which the reaction takes place increases. Since the Maillard reaction involves the production of water, the reaction is inhibited by environments that have high water activity.

iii. pH

The high pH speeds up the bottleneck of the Maillard reaction.

- ✰ The rate of Maillard reaction increases with increasing pH, maximum upto 9-10 depending upon the type of amino acids involved.
- ✰ Bases can catalyse the initial steps of the carbonyl amine reactions by removing the proton from the nucleophile increasing the nucleophilicity. The browning of pure Amadori product is also accelerated at alkaline pH values.
- ✰ pH exerts a considerable influence on the mechanism of the Maillard reaction by determining the type of enolization favoured (1,2 or 2,3-enolization) and hence the pattern of the Amadori compound degradation.

iv. Temperature

The rate of Maillard reaction is increased by high-temperature cooking because heat both increases the rate of chemical reactions and accelerates water evaporation [7].

v. Miscellaneous Factors

Phosphate, citrate, and phthalate buffers have been shown to accelerate the Maillard reaction.

- ✰ In the 5-7 pH range, phosphate has a dramatic effect on reaction rate (increasing up to 15 fold) as compared to that of phosphate free system values.

B. Caramelization and its Significance [1,7.8]

Caramelization is one of the most important types of browning processes in foods, together with Maillard reactions and enzymatic browning. As no enzymes are involved in the carmelization process, it is a non-enzymatic browning reaction. [8].

- ✰ Caramelization is a type of non-amino browning reaction. This process consists of heating sugar slowly to around 170°C [1].
- ✰ It is important to note that caramelization is not a Maillard reaction (it proceeds at higher temperatures and involves the pyrolysis of sugars). At higher temperatures, caramelization (the browning of sugars, a distinct process) and subsequently pyrolysis (final breakdown leading to burning) become more pronounced [7].
- ✰ Unlike the Maillard reaction, caramelization is pyrolytic, as opposed to being a reaction with amino acids.
- ✰ Caramelization is an entirely different process from Maillard browning, though the results of the two processes are sometimes similar to the naked eye (and taste buds).

- Caramelization may sometimes cause browning in the same foods in which the Maillard reaction occurs, but the two processes are distinct. They are both promoted by heating, but the Maillard reaction involves amino acids, whereas caramelization is the pyrolysis of certain sugars [7].
- As the carbohydrates are heated, the molecules break down and reform into compounds with a characteristic brown colour and flavour. It is a chemical decomposition of non-protein substances that occurs spontaneously at high temperatures and the reaction is known as pyrolysis.
- Caramelization is a complex, poorly understood process that produces hundreds of chemical products. Caramelization by heat during baking contributes to flavour and colour. However, it occurs to limited extent in milk and milk products and hence has limited significance [1,7].
- Caramelization reactions also result in the formation of flavours. Diacetyl is an important flavour compound, produced during the first stages of caramelization. Diacetyl is mainly responsible for a buttery or butterscotch flavour. Diacetyl is not only produced by caramelization, it can also be produced by bacteria in fermented products, such as beer and yoghurt [8].

Definition of Caramelisation [1,7]

Caramelization is the removal of water from a sugar, by heating. This process results in isomerization and polymerization of the sugars into various high molecular weight compounds. Compounds such as difructose anhydride may be created from the monosaccharides after water loss. The fragmentation reaction of sugars during heating would result in low molecular weight compounds that may be volatile and may contribute to flavour.

Polymerization reactions lead to large molecular weight compounds that contribute to the dark brown colour.

- Caramelization type of browning may be defined as the heat decomposition of sugars as a function of pH and buffers in the absence of amino compounds. It requires high order of activational energy.
- Caramelization is the oxidation of sugar by heating resulting in the production of nutty flavour and changing to brown colour. During the process, volatile chemicals are released, producing the characteristic caramel flavour.
- The types of reactions that occur during the process of caramelization are as follows:
  i. Sucrose inversion to fructose and glucose,
  ii. Condensation reactions,

iii. Intramolecular bonding,
iv. Isomerization of aldoses to ketoses,
v. Dehydration reactions,
vi. Fragmentation reactions and
vii. Unsaturated polymer formation.

Pyrolysis

Pyrolysis is the chemical decomposition of condensed substances that occurs spontaneously at high temperatures.

Effects on Caramelization [7]

The process is temperature-dependent. Specific sugars each have their own point at which the reactions begin to proceed readily. Impurities in the sugar, such as the molasses remaining in brown sugar, greatly speed the reactions.

- ☆ Caramelization reactions are also sensitive to the chemical environment, [4] and the reaction rate, or temperature at which reactions occur most readily, can be altered by controlling the level of acidity (pH).
- ☆ The rate of caramelization is generally lowest at near-neutral acidity (pH around 7), and accelerated under both acidic (especially pH below 3) and basic (especially pH above 9) conditions [7].

Caramel [1]

Caramel colouring is a dark, rather bitter-tasting liquid. It is the highly concentrated product of near total caramelization, which can be used as food colouring and apart from its addition in beverages such as cola.

- ☆ A good caramel should contain colloidal particles which do not precipitate during storage or to the products to which caramel is added.
- ☆ Though the control of viscosity of caramel is difficult manipulation of the temperature on the period of its contact with reagent, it is possible to obtain caramel of desired viscosity.
- ☆ The rate of dehydration is one of the most important parameter to influence the properties of caramel in general. The time of mutual contact of reagents and manipulation of the temperature may lead to caramel of desired viscosity.

Components of Caramel [1,7,8]

Caramelization or caramelisation is the browning of sugar, a process used extensively in cooking for the resulting sweet nutty flavour and brown colour. During caramelization several flavour components as well as polymeric caramels are produced. Caramels are complex mixture of various high molecular weight components. The brown colours are produced by the following three groups of polymers:

i. Caramelans ($C_{24}H_{36}O_{18}$).
ii. Caramelens ($C_{36}H_{50}O_{25}$), and
iii. Caramelins ($C_{125}H_{188}O_{80}$). [7].

The above three main components of caramel were isolated by using dialysis. The details of these components are given below [1]:

i. **Caramelan**: It was described as a brown brittle, deliquescent solid having a bitter taste. Caramelan melts between 136 °to 144 °C and it is readily water soluble.
ii. **Caramelen**: It is a brown substance much darker than caramelan and not deliquescent. It melts at 153.5°-154 °C.
iii. **Caramelin**: It exists in three modifications namely, soluble in cold water, soluble only in boiling water and insoluble in all ordinary solvents. They are infusible materials that are much darker than caramelan and caramelen.

Three high molecular weight components have been separated from caramel prepared from sucrose. They are melanoidin found in ammonia processed caramels seem to be the sole polymeric compound in caramel. The pH values of caramel constitute an important property of caramels. A high pH may indicate an incomplete burn or alkali present. Above pH 6.0 caramel is susceptible to the attack by molds and below pH 2.5 it quite readily resinifies [1].

Sources

1. *K.S. Sharma, Chemistry of milk, Agrimoon.com.*

1a. *M.K. Srivastava, Handbook on Analysis of Milk, CBS Publications, New Delhi, 2010. ISBN 13: 9788123926681.*

1b. *Chemical reaction of lactose-part -I, Lesson 20, http: //ecoursesonline.iasri.res.in.*

1c. *Chemical reaction of lactose-part -II, Lesson 21, http: //ecoursesonline.iasri.res.in.*

2. *B.H. Webb, A.H. Johnson, and J.A, Alford, J. A. 1965. Fundamentals of dairy chemistry. AVI Publ. Co., New York.*

3. *J.E. Hodge, "Dehydrated Foods, Chemistry of Browning Reactions in Model Systems". Journal of Agricultural and Food Chemistry. 1953, 1 (15): 928-43.*

4. *Lactose, Some basic properties and characteristics, DFE Pharma.*

5. *Tomoko Shimamura and Hiroyuki Ukeda, Maillard Reaction in Milk - Effect of Heat Treatment, 2012, DOI: 10.5772/50079.*

6. *Suzana Stojanovska1, Nevena Gruevska, Julijana Tomovska, Jasmina Tasevska, Aleksandar Krstanovski and Mirjana Menkovska, Maillard Reaction and Lactose Structural Changes during Milk Processing, Chemistry Research Journal, 2017, 2(6): 139-145.*

7. *Wikipedia.*

8. *Caramelization, Food Info, http: //www.food-info.net.*

Chapter 14

Milk Lipids

A lipid can be broadly defined as any molecule of intermediate molecular weight that contains a substantial portion of aliphatic or aromatic hydrocarbon; included are hydrocarbons, phospholipids, sterols and triacylglycerols. The fat content of milk is of economic importance because milk is sold on the basis of fat. The lipid (fat) fraction of milk is composed mainly of triacylglycerols, with minor contributions from diacylglycerols, monoacyl-glycerols, free (unesterified) fatty acids, phospholipids and sterols. Trace amounts of fat-soluble vitamins, β-carotene and fat-soluble flavouring compounds are also present [1]. The colour of fat depends upon its carotene content and varies with the species, breed and feed of the animal. The yellow colour of cow milk is because of carotene. Buffalo milk does not contain carotene [3].

- ☆ Lipids are a diverse and ubiquitous group of compounds which have many key biological functions, such as acting as structural components of cell membranes, serving as energy storage sources and participating in signaling pathways [1a].
- ☆ The common property of all lipids is their hydrophobic nature. Lipids (organic compounds) are nonpolar molecules. Lipids includes fats and oils, waxes, phospholipids, steroids (like cholesterol), and some other related compounds [2,3]. The butter, ghee, cheese, cholesterol and other steroids, phospholipids, waxes and fat-soluble vitamins, are examples of lipids.
- ☆ These lipids are being synthesized in adipose tissue, mammary gland and liver. Lipids are naturally occurring molecules from plants or animals. These are organic compounds which are soluble only in

nonpolar solvents (ether, benzene, hot alcohols, *etc.*) and insoluble in water because water is a polar molecule.

- ☆ The lipids in bovine milk are present in microscopic globules (average diameter 3-4 μm) surrounded by a membrane which maintains the integrity of the globule as **an oil-in-water emulsion**. If cool raw milk is kept for sometime without mixing there is tendency for the fat globules to cluster and rise at the surface forming cream layer. However, these globules are independent due to the presence of a protective fat globule membrane layer. The membrane prevents the merging of fat globules and formation of a continuous fat phase. The primary purpose of these lipids is to provide a source of energy to the newborn calf [2,3].
- ☆ The lipids in milk are unique for the species although they are termed as milk lipids or milk fat. This is because of the fact that the fatty acid composition and its biosynthesis is different in each species. Both the fat content of the milk and the fatty acid composition of the lipids can vary considerably as a result of changes in factors like species, breed, diet,stage of lactation, feed, season *etc.*
  - **i.** The fat content can vary from about 3.0 to 6.0 per cent, but typically is in the range 3.5 to 4.7 per cent in cow milk and in range of 4.8-6.0 per cent in buffalo milk.
  - **ii.** Changes in the composition of the fatty acids (*e.g.*: 16:0 and 18:1) can be quite marked and can lead to changes in physical properties of the fat.
  - **iii.** These changes make comparison difficult between different samples of milk fat and ideally comparisons should be made between cows in mid-lactation and fed on similar diets.

Importance of Milk Fat/Lipids [1,2,3]

i. Milk lipids play a positive role in the diet as a source of energy, fat-soluble vitamins and essential fatty acids. Milk lipids also contribute to the palatability of the diet.

- ☆ Lipids are better energy sources than sugars because they are less oxidized than carbohydrates. The role of milk fat in nutrition is its capability to yield approximately ~37 kJ per g (9 Kcal/g) apart from carrying the fat soluble vitamins *viz.*, A, D, E and K.
- ☆ The presence of significant amounts of essential fatty acids *viz.*, linoleic and arachidonic acids also play an important role in the nutrition of the new born [2,3].

ii. Milk lipids are very important as they confer distinctive nutritional, textural and organoleptic properties on dairy products, such as cream, butter, whole milk powder and cheese. Milk fat exhibits unique rheological properties and is largely responsible for the consistency and textural attributes of butter. The body and texture of the milk products is largely being influenced by this milk component [2,3].

iii. Milk fat melts over a wide temperature range, from –35 to + 38 °C with a substantial proportion of the fat melting between 10 and 20 °C. These melting characteristics (due to the complex mixture of triacylglycerols in milk fat), can markedly affect the functional properties of dairy products [1].

iv. The rich pleasing flavour contributed by the milk fat to the milk products render them acceptable by all the consumers and further no other fat can be used to duplicate it [2]. The desirable flavour of milk fat is one of its main attributes. The flavour profile is complex, with a large number of volatile compounds contributing to the overall aroma and taste [1]. In certain products and after certain processes, fatty acids serve as precursors of very flavourful compounds such as methyl ketones and lactones [3].

v. The conjugated linoleic acid present in milk fat may exert beneficial health effects [3].

vi. Two deteriorative reactions, namely oxidation and lipolysis, can occur in milk fat. These reactions produce off-flavours which can destroy the pleasant, delicate flavour of milk fat [1].

Nomenclature]1,4] -The fats and oils are made from two kinds of molecules-glycerol (a trihydroxy alcohol) and fatty acids attached to it by ester linkage. The ester of glycerol containing fatty acids are formerly known as triglycerides and now termed as triacylglycerol.

The practice of using common names for various chemical compounds was prevailing earlier but these names often caused confusion in correct identification of these compounds. An example is the use of caprylic to describe 1-octanol and 2-octanol and attempts to qualify the name with "primary" and secondary were not useful in identifying the compounds. The generally accepted guide-lines for lipid systematic names were initially defined by the International Union of Pure and Applied Chemists and the International Union of Biochemistry and Molecular Biology (IUPAC-IUBMB) Commission on Biochemical Nomenclature in 1976, and their recommendations are summarized on the IUPAC website [1a]. The IUPAC names have been given for the chemical compounds to overcome the difficulty in identification of the fatty acids.

Nomenclature of Fatty Acids [2,2a]

There are different types of nomenclature for the fatty acids.

A. The following three naming systems are also used for fatty acids:

i. Delta Nomenclature

Chemists use Delta Nomenclature. For this nomenclature, one needs to know 3 things- Example C18:1Δ9

Number of carbons in the fatty acid = 18

Number of double bonds = 01

Number of carbons from the carboxylic acid (alpha) end to the first carbon in the double bond(s) = 09

Fatty Acid with 18 carbons with a double bond at the ninth carbon atom. This is also an Omega 9 fatty acid.

ii. Omega Nomenclature

Nutritionists are more focused on the Omega system which focuses on the nutritional value of the fatty acid. The Omega Nomenclature is almost exactly the same as the Delta Nomenclature; the only difference is that carbons are counted from the methyl (omega) end instead of the carboxylic acid end, and the omega symbol is used instead of the delta symbol. For Omega Nomenclature one needs to know 3 things- Example C18:1

Number of carbons in the fatty acid = 18

Number of double bonds = 01

Number of carbons from the methyl end (aka Omega end) to the first carbon in the double bond closest to the methyl end = 09.

If it is a saturated fatty acid, then the omega nomenclature is not added to the end of the name. If it is an 18 carbon saturated fatty acid, then it would be named 18:0.

iii. Common Names

The common names of fatty acids are something that, for the most part, have to be learned/memorized. For example- Oleic acid

The Omega Nomenclature and common names are used more in the field of nutrition than the delta nomenclature when describing specific fatty acids. Some examples of common name and omega names of fatty acids are given in following table [2a]:

Table 14.1: Common and Omega Names of Fatty Acids

Saturated Fatty Acids		*Unsaturated Fatty Acids*	
Common Name	*Omega Name*	*Common Name*	*Omega Name*
Butyric acid	C4	Oleic Acid	C18:1 (n-9)
Lauric acid	C12	Linoleic Acid	C18:2 (n-6)
Myristic acid	C14	Alpha-linolenic Acid	C18:3 (n-3)
Palmitic	C16	Arachidonic Acid	C20:4 (n-6)
Stearic acid	C18	Docosahexanoic Acid	C22:6 (n-3)

B. Fatty acids are also classified depending upon their saturation and unsaturation with systematic names using IUPAC nomenclature as shown in following tables [1a,2]:

Table 14.2: Nomenclature of Fatty Acids on the Basis of Saturation

Sl.No.	*Fatty Acids*	*Parent Hydrocarbon*	*Name*
1	Saturated	With +oic	C 18:0 -Octadecanoic acid
2	Unsaturated	With one double bond +en	C 18:0 -Octadecenoic acid
		With two double bonds +dienoic	C 18:0 -Octadecadienoic acid
		With three double bonds +trienoic	C 18:0 -Octadecatrienoic acid

C. Another type of nomenclature is by using empirical formula and the name of the fatty acids as shown in following table:

Table 14.3: Systemic Nomenclature for Fatty Acids and their Empirical Formula

Sl. No.	*Fatty Acids*	*Carbon No.*	*Common Name*	*Systematic Name*	*Empirical Formula*
1	Saturated	C 16:0	Palmitic acid	n-Hexadecanoic acid	$CH_3(CH_2)_{14}COOH$
		C 18:0	Stearic acid	n-Octadecanoic acid	$CH_3(CH_2)_{16}COOH$
2	Unsaturated	C 16:1(9)	Pamitoleic acid	Cis-9-Hexadece-noic acid	$CH_3(CH_2)_5CH= CH (CH_2)_7COOH$
		C 18:1(9)	Oleic acid	Octadecenoic acid	$CH_3(CH_2)_7CH= CH (CH_2)_7COOH$
		C18:2 (9,12)	Linoleic acid	All cis-9,12-Octadecadienoic acid	$CH_3(CH_2)_4(CH= CHCH_2)_2(CH_2)_6COOH$
		C18:3 (9,12, 15)	Linolenic acid	All cis-9,12,15-Octadecatrienoic acid	$CH_3CH_2(CH= CHCH_2)_3(CH_2)_6COOH$
		C20:4 (5, 8,11,14)	Arachidonic acid	All cis-5,8,11,14-Eicosatetraenoic acid	$CH_3(CH_2)_4(CH= CHCH_2)_4(CH_2)_2COOH$

General Structure [2,3]

Lipids are the polymers of fatty acids that have a long, non-polar hydrocarbon chain with a small polar region containing oxygen. Triglycerides are the major type of lipid in milk fat. Triglycerides are composed of three fatty acids covalently bound to a glycerol molecule by ester bonds. General structure of Triglyceride (triacylglycerol, TAG or triacylglyceride) is an ester composed of a glycerol bound to three fatty acids. It is the main constituent of vegetable oil and animal fats [2,3].

The main milk lipids are a class called triglycerides which are comprised of a glycerol backbone binding up to three different fatty acids. Triglycerides are formed from a single molecule of glycerol, combined with three molecules of fatty acid. The glycerol molecule has three hydroxyl (OH-)groups. Each fatty acid has a carboxyl group (COOH-). In triglycerides, the hydroxyl groups of the glycerol join the carboxyl groups of the fatty acid to form ester bonds as shown below [3]-

CH_2OH		**RCH_2COOH**		**$CH_2O\text{-}OCCH_2R$**
|				|
CHOH	+	**RCH_2COOH**	⟶	**$CHO\text{- }OCCH_2R$**
|				|
CH_2OH		**RCH_2COOH**		**$CH_2O\text{-}OCCH_2R$**
Glycerol		**Fatty acids**		**Triglycerides of Fatty acids (Fat)**

CH_2OH			**$CH_2OOC_{17}H_{35}$**		
|			|		
CHOH	+	**$3C_{17}H_{35}COOH$**	**$CHOO\ C_{17}H_{35}$**	+	**$3H_2O$ Water**
|			|		
CH_2OH		**Stearic acid**	**$CH_2OOC_{17}H_{35}$**		
Glycerol		**(Fatty acid)**	**Stearin**		

$$
\begin{array}{l}
\qquad\qquad\qquad\quad H_2C-O-\overset{\overset{O}{\|}}{C}-R^1 \\
\qquad\qquad\qquad\qquad | \\
R^2-\overset{\overset{O}{\|}}{C}-O-C-H \\
\qquad\qquad\qquad\qquad | \\
\qquad\qquad\qquad\quad H_2C-O-\overset{\overset{O}{\|}}{C}-R^3
\end{array}
$$

Figure 14.1: Structure of Triacylglycerol [2,4].

R_1, R_2,and R_3 are alkyl chains of fatty acids. The three fatty acids R_1COOH, R_2COOH and R_3COOH may be different or same.

To understand the composition of various lipid classes and to locate the fatty acid in acylglycerol following terminologies are being used [2,3,4]-

i. To understand the composition of various lipid classes, it is essential to have an idea about the terminology used for explaining the position of the various fatty acids. An abbreviation of fatty acids is used for such purpose. According to this the first figure is the number of carbon atoms and the second number of double bonds. A convenient method of notation of fatty acids two numbers designating the carbon chain length and the unsaturation (Number of double bonds) [2,3.4]. For example-

a. Butyric acid is denoted as C4:0 with 4 carbons and saturated without any double bond.

b. Stearic acid is referred as C18:0 with 18 carbons and no double bond.

c. Oleic acid is denoted as C18:1. With 18 carbons and 1 double bond.

d. Linoleic acid is denoted as C18-2 with 18 carbons and 2 double bonds.

e. Linolenic acid is denoted as C18:3 with 18 carbons and 3 double bonds.

Normally the fatty acids percentage is not expressed on the weight basis but on molar percentage.

ii. **Stereospecific numbering [1a,2,]-** To locates the fatty acid in acylglycerols stereospecific numbering (Sn) is used. The use of the stereospecific numbering (sn) method is to describe glycerolipids and glycerophospholipids. The glycerol group is typically acylated or alkylated at the sn1 and/or sn2 position with the exception of some lipids which contain more than one glycerol group and archaebacterial lipids in which sn2 and/or sn3 modification occurs [1a]. If a glycerol molecule is drawn with the secondary hydroxyl to the left, the hydroxyl above it is Sn-1 and that below is Sn-3.

- ☆ The recent studies carried out on the whole milk fat composition revealed that the short chain acids are selectively associated with the sn-3 position. With increase in chain length of the acid there is greater tendency to be esterified at position Sn-1 [2,3].
- ☆ In general, the Sn-1 position binds mostly longer carbon length fatty acids, and the Sn-3 position binds mostly shorter carbon length and unsaturated fatty acids. For example - C4 - 97 per cent in Sn-3, C6 - 84 per cent in Sn-3 and C18 - 58 per cent in Sn-1 [2,3].

Types of Triglyceride (TG) [2]

There are following three types of Triglycerides (TG) -

i. **The first type of TG** -They are with 48- 54 carbon atoms composed of long chain 1,2 DGs containing 18:0,18:1 and 18:2.

ii. **2 type of TG** – They are containing 36-46 carbon atoms and the Sn-3 position acids composed of 4:0,6:0 and 8:0, these tri glycerides are enantiomers.

iii. **Type 3 TG** -The carbon 26- 34 the 1,2 DG containing medium chain fatty acids and the 3 position acids are short and medium chain acids in sn-3 position that are different from those in Sn-1 are also enantiomers.

- ☆ **Chain lengths** of the fatty acids in naturally occurring triglycerides can be of varying lengths, but 16, 18 and 20 carbons are the most common. Natural fatty acids found in plants and animals are typically composed only of even numbers of carbon atoms due to the way they are bio-synthesized from acetyl CoA. Bacteria, however, possess the ability to synthesize odd - and branched-chain fatty acids. As a result, ruminant animal fat contains odd-numbered fatty acids, such as 15, due to the action of bacteria in the rumen.

✰ Most natural fats contain a complex mixture of individual triglycerides. Because of this, they melt over a broad range of temperatures. Cocoa butter is unusual in that it is composed of only a few triglycerides, one of which contains palmitic, oleic, and stearic acids in order of concentration.

Sources

1. *M.W. Taylor, A.K.H. MacGibbon, Lipids- General Characteristics, in Encyclopedia of Dairy Sciences, 2002.*

1a. *Eoin Fahy, Dawn Cotter, Manish Sud, and Shankar Subramaniam, Lipid classification, structures and tools, Biochim Biophys Acta. 2011 Nov; 1811(11): 637–647.*

2. *K.S. Sharma, Chemistry of milk, Agrimoon.com.*

2a. *Fatty acid nomenclature, https: //courses.lumenlearning.com.*

3. *M.K. Srivastava, Handbook on Analysis of Milk, CBS Publications, New Delhi, 2010. ISBN 13: 9788123926681.*

4. *B.H. Webb, A.H. Johnson, and J.A, Alford, J. A. 1965. Fundamentals of dairy chemistry. AVI Publ. Co., New York,*

Composition of Milk Fat [1,2,3]

Milk fat is composed largely of triacylglycerols (about 98 per cent of the total fat), with minor contributions from phospholipids, cholesterol, free fatty acids, and mono- and diacylglycerols. There are also trace amounts of β-carotene, fat-soluble vitamins (A, D, E and K) and flavouring compounds [1]. The bulk of the milk fat exists in the form of small globules with a average size of approximately 2 to 5μm. This is water in oil type of emulsion; the surface of these globules is covered by a adsorbed layer of material commonly known as fat globule membrane. Small quantities of milk fat also occur in the milk serum in combination with proteins [2,3].

i. Composition of Lipids in Bovine Milk [1,2,4]

Fatty acids are pivotal components of all lipids found in milk. The bulk of milk lipids are triacylglycerols (TGs) which are 97-98 per cent of the total lipids found in pooled milk. Sterols mostly cholesterol and phospholipids are next in the quantity. The Diacylglycerols (DGs), monoacylglycerols (MGs) and free fatty acids are the products of lipolysis and are more in concentration as compared to the components which are present in traces. The cholesterol and phospholipids are cellular membrane material which accompanies the fat globule during extrusion from the secreting cell. The lipid composition of the bovine milk fat is presented in following table [14.4]:

Table 14.4: Lipid Composition of Bovine Milk Fat [1,4]

Sl.No.	*Lipids*	*Weight Per cent [4]*	*Amount (wt per cent) [1]*
1	Triacylglycerols	97-98	98.3
2	Diacylglycerols	0.28-0.59	0.3
3	Monoacylglycerols	0.016-0.038	0.03
4	Free sterols	0.22-0.41	0.3
5	Free Fatty acids	0.10-0.44	0.1
6	Phospholipid	0.2-1.0	0.8
7	Hydrocarbons	Traces	-
8	Sterol esters	Traces	-
9	Carotenoids	-	Trace
10	Fat Soluble Vitamins	-	Trace
11	Flavour compounds	-	Trace

Lipids in Milk [2,3]

In milk different classes of lipids are distributed in various phases and the following table [14.5] will give us an idea about their range of occurrence and their location in milk. It must be remembered here that the lipids mentioned in this represent a group of compounds and they should not be considered as a single constituent.

Table 14.5: Distribution of Lipids in Milk [6]

Constituents	*Range of Occurrence per cent*	*Location in Milk*
Triglycerides	98-99	Fat globule
Phospholipids	0.2-1.0	Fat globule membrane and milk serum.
Sterol	0.25-0.40	Fat globule, Fat globule membrane and milk serum.
Free Fatty acids	Traces	Fat globule and milk serum.
Waxes	Traces	Fat globule
Squalene	Traces	Fat globule
Fat soluble vitamins	Traces	Fat globule
Vitamin A	7.0-8.5 μg/g of fat	-
Carotenoids	8.0-10μg/g of fat	-
Vitamin E	2-50 μg/g of fat	-
Vitamin D	Traces	-
Vitamin K	Traces	-

It is evident from the above table that milk lipids are found in three distinctly different phases of milk namely fat globule, the membrane surrounding these globules and the milk serum.

Classification and Types of Lipids [1a, 1b,2]

1. Classification

Depending upon the nature, properties *etc.*, lipids can be classified in to the following categories-

a Saponifiable and Nonsaponifiable Lipids [1a]

The details are given below:

- ✫ A saponifiable lipid is part of the ester functional group. A saponifiable lipid contains one or more ester groups, which enables it to undergo hydrolysis in the presence of a base, acid, or enzymes. These include waxes, triglycerides, sphingolipids and phospholipids. Saponifiable lipids contain long chain carboxylic (of fatty) acids, that are linked to an alcoholic functional group through an ester linkage.
- ✫ A nonsaponifiable lipid cannot be disintegrated into smaller molecules through hydrolysis. Cholesterol, prostaglandins, *etc.* are the examples of this type of lipids. A non-saponifiable class is made up of "fat-soluble" A and E vitamins and cholesterol.

b. Polar and Non-polar lipids [1b,1c]

Lipid can also be categorised as non-polar and polar lipids.

- ✫ **Polar lipids** are utilized in membranes. It could form a barrier with an external water environment. *e.g.* sphingolipids and glycerophospholipids. The polar lipids contained higher levels of linoleic acid. The polar lipids are less saturated.
- ✫ **Nonpolar lipids**(namely triglycerides) are utilized as fuel and to store energy.

Palmitic, stearic, oleic and linoleic acids were the major fatty acids in both the fractions. Nonpolar lipids contained oleic acid as the predominant acid. Nonpolar lipids were more saturated than polar lipids.

Milk fat is encased in a polar lipid-containing tri-layer milk fat globule membrane (MFGM), composed of phospholipids (PLs) and sphingolipids (SLs). Milk PLs and SLs comprise about 1 per cent of total milk lipids [1c].

2. Types of Lipids [2,3,7]

In addition to the aforesaid major classes of lipids, these can also be broadly classified as simple lipids complex lipids, Neutral lipids and derived lipids.

A. *Simple Lipids*

The lipids which yields only fatty acids and glycerol upon hydrolysis. These are esters of fatty acids with various alcohols. Some examples are given below:

a. **fats and oils/Neutral Fat**- These are esters of three molecules of fatty acids with one molecule of glycerol and found in adipose tissue, butterfat, lard, fish oils, olive oil, corn oil, *etc.* This includes triacylglycerol, diacylglycerol and monoacyl-glycerol.

b. **Waxes**-These comprise of esters of fatty acids with long chain (high molecular weight) monohydric alcohol other than glycerol. It has industrial and medicinal importance.

B. *Compound/Complex Lipids*

These are esters of fatty acids containing groups (phosphate, nitrogenous base, carbohydrate and protein) in addition to alcohol and a fatty acid. The entire compound lipids are found mostly in the same places. The distribution between fat globule membrane and milk plasma (presumably in lipoprotein particles) is for all classes similar, but not exactly equal. Compound lipid are absent in the core of milk fat globule. Some examples of compound lipids are given below:

a. Phospholipids (Phosphatides)

Glycerophospholipids (Glycerol as alcohol), usually referred to as phospholipids. Found chiefly in animal tissues. These contain, in addition to fatty acids and alcohol, a phosphoric acid residue. Substituted fats, consisting of phosphatidic acid; composed of glycerol, fatty acids, and phosphoric acid bound in ester linkage to a nitrogenous base.

- ☆ The phosphoacylglycerols are lipids containing a phosphate group and hence these are called as phospholipids [2]. Milk contains phospholipids (0.8 per cent) which are mainly associated with the fat globule membrane. Several groups of phospholipids are normally present in milk.
- ☆ Phospholipids are closely associated with the milk proteins and milk fat. They are destroyed by heat and responsible for some of the oxidised flavour in milk. Phospholipids are excellent emulsifying agent and serve to stabilize milk fat emulsion [3].

    i. **Lecithin [2,3]**-Lecithin forms the important constituent of the fat globules membrane, consists of Nitrogen and phosphorous and contributes to the richness of flavour of milk and milk products. Lecithin is unstable in air especially in oxygen [3].

- ☆ Found in brain, egg yolk, and organ meats. Phosphatidyl choline or serine; phosphatide linked to choline; a lipotropic agent; important in fat metabolism and transport; used as emulsifying agent in the food industry [2].
- ☆ Lacithin is highly sensitive to oxidative changes, giving rise to oxidized/metallic flavours. **When milk is separated, a majority of the Lecithin goes to the cream where as during churning, most of the Lecithin goes to the butter milk** [3].

ii. **Cephalin [2]-**It occurs predominantly in nervous tissue. Phosphatidyl ethanolamine; Phosphatidyl serine where phosphatide linkage to ethanolamine or serine; plays a role in blood clotting.

iii. **Plasmalogen [1a,2,7]-** Plasmalogens were first described by Feulgen and Voit in 1924 based on studies of tissue sections. The plasmalogens are a class of glycerophospholipids which contain a vinyl-ether and an ester bond at the sn-1 and sn-2 positions, respectively, in the glycerol backbone. When a fatty acid is attached by an ether linkage at C_1 of glycerol in the glycerophospholipids, the compound is called plasmalogen [1a,2].

- ☆ Plasmalogens are a subclass of ether phospholipids that are commonly found in cell membranes in the nervous, immune and cardiovascular systems. Plasmalogen is plasmenyl-phospholipids, have an ether bond in position sn1 to an alkenyl group [1a,7].
- ☆ Plasmalogens play a crucial role as endogenous antioxidants, protecting other PL, lipid and lipoprotein particles from oxidative stress. This is due to the fact that the vinyl ether bond is preferably oxidized, while protecting the polyunsaturated fatty acids present in the sn-2 oxidation position [1a].
- ☆ It is a predominant component of the membrane phospholipids of many animal and microbial species. In animal tissues, the highest proportion of the plasmalogen form is usually in the phosphatidylethanolamine class with rather less in phosphatidylcholine, and commonly little or none is in other phospholipids such as phosphatidylinositol. In phosphatidylcholine of most tissues, a higher proportion is often of the O -alkyl rather than the O -alkenyl form. Found in brain, heart, and muscle [2].

The structure of Phospholipids, Lacithin, Cephalin and Plasmalogen are given below:

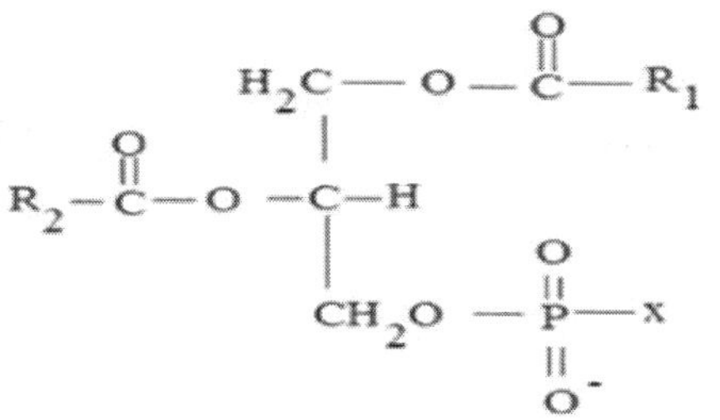

$$
\begin{array}{l}
H_2C - O - \overset{O}{\overset{\|}{C}} - R_1 \\
R_2 - \overset{O}{\overset{\|}{C}} - O - C - H \\
CH_2O - \overset{O}{\overset{\|}{P}} - x \\
\qquad\quad \| \\
\qquad\quad O^-
\end{array}
$$

Figure 14.2: General Structure of Phospholipids.

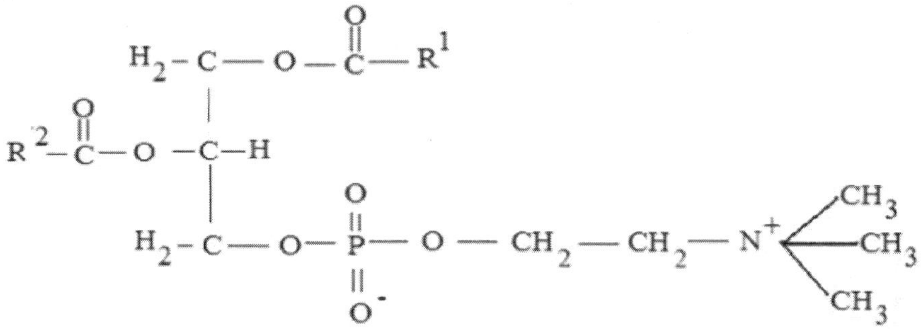

$$
\begin{array}{l}
H_2 - C - O - \overset{O}{\overset{\|}{C}} - R^1 \\
R^2 - \overset{O}{\overset{\|}{C}} - O - C - H \\
H_2 - C - O - \overset{O}{\overset{\|}{P}} - O - CH_2 - CH_2 - N^+(CH_3)_3 \\
\qquad\qquad\qquad \| \\
\qquad\qquad\qquad O^-
\end{array}
$$

Figure 14.3: Structure of Phosphatidylcholine (Lecithin).

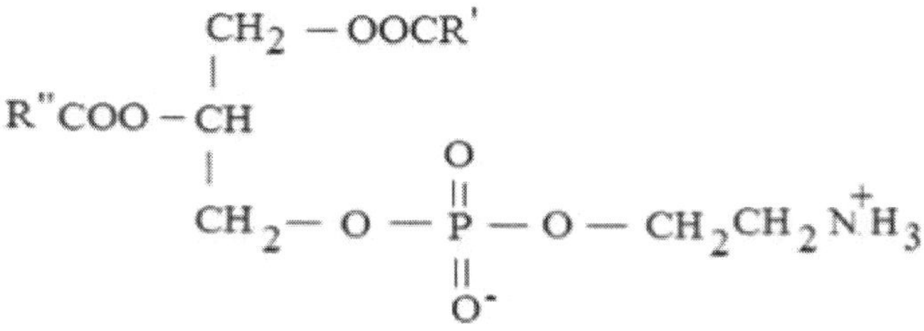

$$
\begin{array}{l}
CH_2 - OOCR' \\
R''COO - CH \\
CH_2 - O - \overset{O}{\overset{\|}{P}} - O - CH_2CH_2\overset{+}{N}H_3 \\
\qquad\qquad\quad \| \\
\qquad\qquad\quad O^-
\end{array}
$$

Figure 14.4: Structure of Ethanolamine (Cephalin).

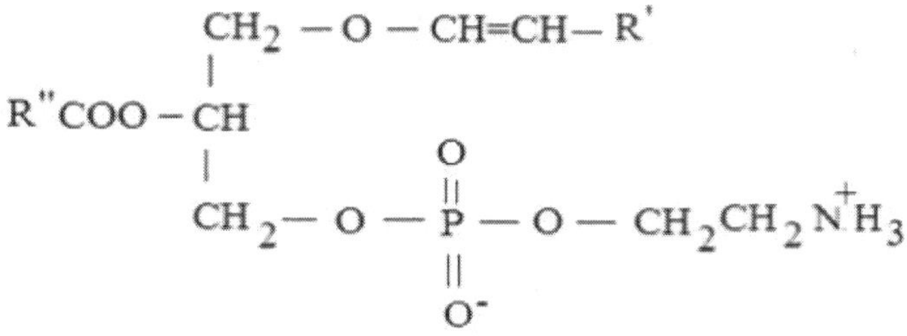

$$
\begin{array}{l}
CH_2 - O - CH{=}CH - R' \\
R''COO - CH \\
CH_2 - O - \overset{O}{\overset{\|}{P}} - O - CH_2CH_2\overset{+}{N}H_3 \\
\qquad\qquad\quad \| \\
\qquad\qquad\quad O^-
\end{array}
$$

Figure 14.5: Structure of Plasmalogen.

iv. **Lipositol [2]-** Found in brain, heart, kidneys and plant tissues together with phytic acid. Phosphatidyl inositol; phosphatide linked to inositol; rapid synthesis and degradation in brain; evidence for role in cell transport processes.

Inositol (Phosphatidylinositol) [1a,2,7] - Phosphatidylinositol is classified as a glycerophospholipid that contains a glycerol backbone, two non-polar fatty acid tails, a phosphate group substituted with an inositol polar head group [1a].

- ✰ Inositol (Phosphatidylinositol) is a negatively charged phosphor lipid (The phosphate group gives the molecules a negative charge at physiological pH.) and a minor component in the cytosolic side of eukaryotic cell membrane.
- ✰ Phosphatidyl-inositol consists of a family of lipids, a class of the phosphatidylglycerides. In such molecules the isomer of the inositol group is assumed to be the myo- conformer. Typically phosphatidylinositol form a minor component on the cytosolic side of eukaryotic cell membranes. It is important component of cell membranes.
- ✰ Inositol can be phosporylated to form phosphatidylinoistol phosphate, phosphati-dylinositolbisphosphate, phosphatidy-linositoltrisphosphate, all these are collectively called phosphoinositides(inositides) [1a,2].
- ✰ Phosphoinositides play important roles in lipid signalling, cell signalling and membrane trafficking [2].

v. **Sphingophospholipids (Sphingosine as alcohol) Sphingomyelin [1a,2,7]**: Sphingomyelin was first isolated by German chemist Johann L.W. Thudicum in the 1880. The major phosphosphingolipids of mammals are sphingomyelins (ceramide phosphocholines).

- ✰ Found in nervous tissue, brain, and red blood cells. Sphingomyelin consists of a phosphocholine head group, a sphingosine, and a fatty acid. Sphingosine is an amino alcohol present in sphingomyelins and do not contain glycerol. Sphingosine is attached by an amide linkage to a fatty acid to produce ceramide.
- ✰ Sphingosine containing phosphatide; upon hydrolysis it yields fatty acids, choline, sphingosine, phosphoric acid, and no glycerol. It is a source of phosphoric acid in body tissue [2].

b. Glycolipids [1a,2,7]

These Lipids contain a fatty acid, sphingosine and carbohydrate. The alcohol is sphingosine. Glycerol and phosphate are absent. Glycolipids are lipids with a carbohydrate attached by a glycosidic (covalent) bond. Glycolipids are distributed throughout animal, plant, and microbial cells.

- ☆ Their role is to maintain the stability of the cell membrane and to facilitate cellular recognition [1a,7]. They play important roles in providing energy and serve as markers for cellular recognition
- ☆ The neutral glycolipids of milk, a small fraction of the total lipids, are of potential biological importance.
- ☆ Glycolipids are found on the surface of all eukaryotic cell membranes.
- ☆ Glycolipids encompass a wide variety of compounds:
  - **a.** Glycosphingolipids (cerebrosides, gangliosides, sulfatides, and others),
  - **b.** Glycoglycerolipids.
  - **c.** Glycophosphospholipids.

Some of the glycolipids are discussed below:

i. Cerebroside [2,2a,7]-It is a common name for a group of glycosphingolipids called mono glycosylceramides which are an important component of muscle and nerve cell membranes.

- ☆ The cerebrosides are major glycosphingolipids of human milk. The cerebrosides appear to be primarily galactosylceramides (galactocerebrosides); glucosyl-ceramides (glucocerebrosides) are a minor component. Bovine milk glycol-sphingolipids differ from those of human milk, with bovine milk containing mainly NFA glucosylceramide (8 microM) and NFA lactosylceramide (17 microM); bovine milk contains little Gb3 or Gb4 [2a].
- ☆ They consists of a cerimide with a single sugar residue at the 1 – hydroxyl moiety. The sugar residue can be either glucose or galactose and accordingly these are named as glucocerebrosides and Galactocerebrosides. Galactocerebrosides are typically found in neural tissue while glucocerebrosides are found in other tissues Myelin sheaths of nerves, brain, and other tissues. Yields on hydrolysis of fatty acids, sphingosine, galactose (or glucose), but not fatty acids; includes kerasin and phrenosin or cerebronic acid [1a,2]

ii. Ganglioside [1a,2,7]- Milk gangliosides are part of the membrane fraction of the milk fat globule, which derives from the apical plasma membrane of the secretory cells in the lactating mammary gland.

- ☆ In human milk, the content and individual distribution of gangliosides changes during lactation, GD_3 being the most abundant ganglioside in colostrum, while in mature milk, GM_3 is the major individual species.
- ☆ It is found in Brain, nerve tissue, and other selected tissues, notably spleen. It is derivatives of cerebrosides and contains a

ceramide linked to hexose (glucose or galactose), neuraminic acid, sphingosine and fatty acids.

- ✰ A ganglioside is a molecule composed of a glycosphingolipid (ceramide and oligosaccharide) with one or more sialic acids (*e.g.* n-acetylneuraminic acid, NANA) linked on the sugar chain. The gangliosides are only a minor fraction of milk fat; they have several ketoses, including N-acetylneuraminic acid bound to the terminal oxygen.
- ✰ Recently, gangliosides have been found to be highly important molecules in immunology.

iii. **Sulfolipid [1a,2,7]**- Sulfolipids are a class of lipids which possess a sulfur-containing functional group. It is sulfur containing glycolipid and sulfate present in ester linkage to galactose. It is found in white matter of brain, liver, and testicle; also plant chloroplast. Mammalian brain and kidney are the richest sources of sulpholipids.

iv. **Proteolipids [1a,2,7]**- A proteolipid is a protein covalently linked to lipid molecules, which can be fatty acids or sterols. It is found in Brain and nerve tissue. Proteolipids are abundant in brain tissue, and are also present in many other animal and plant tissues.

- ✰ These are complexes of protein and lipids having solubility properties of lipids. They are not to be confused with lipoproteins.

C. *Neutral Lipids*

Uncharged lipids are neutral lipids.

a. *Sterols*

Sterols are steroids in which one of the hydrogen atoms is substituted with a hydroxyl group, at position 3 in the carbon chain. Sterols, such as cholesterol and its derivatives, are an important component of membrane lipids [4,5] along with the glycerophospholipids and sphingomyelins. Other examples of sterols are the bile acids and their conjugates.

Cholesterol [2,3]

It is found in egg yolk, dairy products, and animal tissues. It is a precursor of bile acids and Vitamin D3. It has one hydroxyl group at C3 and a double bond between C5 and C6. An 8 carbon aliphatic side chain is attached to C17. Cholesterol contains 5 methyl groups. More details are given under unsaponiofiable lipids.

The structures of Phosphatidylinositol, Sphingomyelin, Cerebroside and Cholesterol are given below:

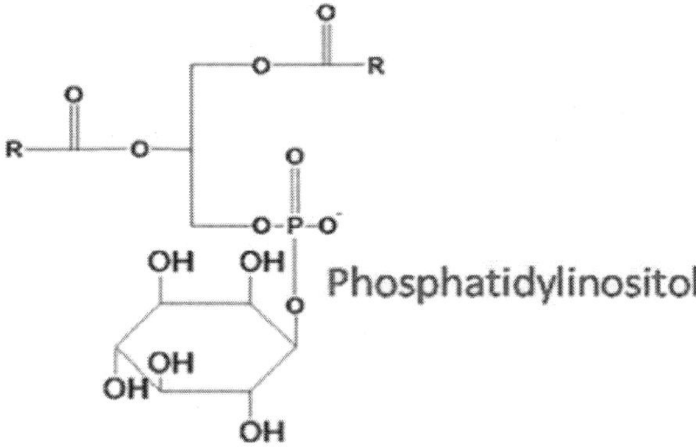

Figure 14.6: Structure of Phosphatidylinositol.

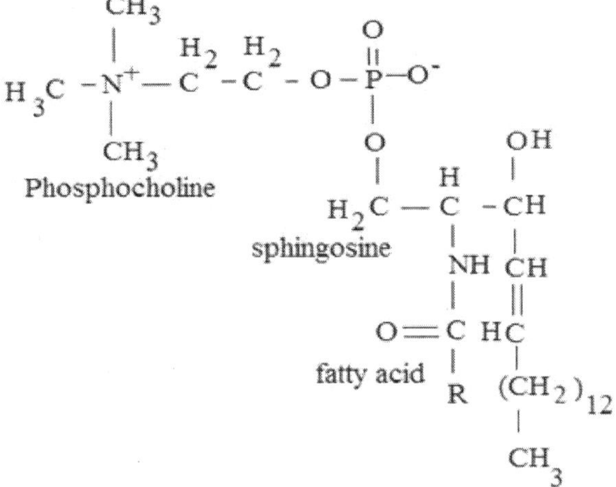

Figure 14.7: Structure of Sphingomyelin [5]

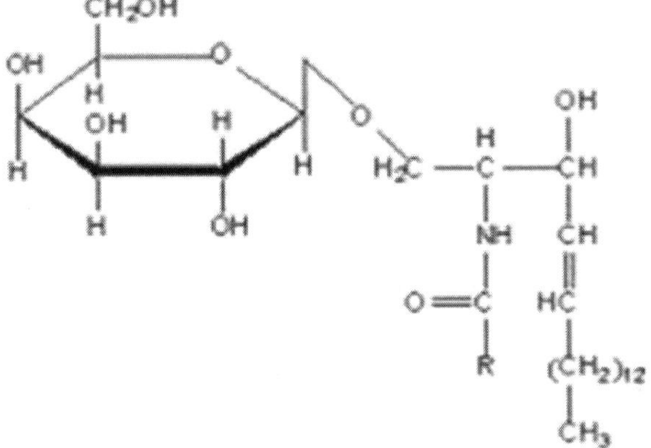

Figure 14.8: Structure of Cerebroside.

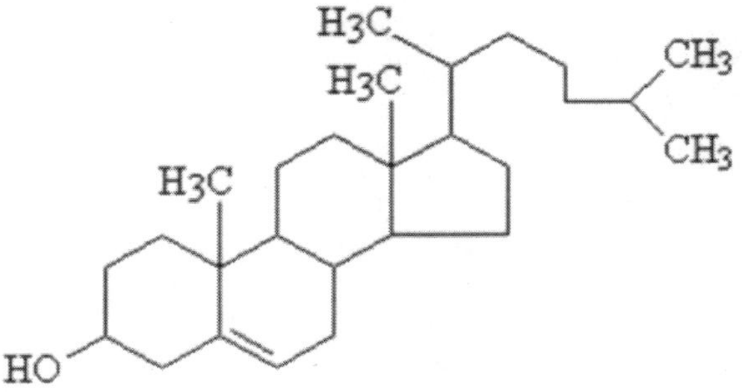

Figure 14.9: Structure of Cholesterol [5].

D. Derived Lipids [2,7]

These are the derivatives obtained on the hydrolysis of simple and compound lipids. *e.g.*, soluble vitamins, fatty acids, hydrocarbons, glycerol and other alcohols, ketone bodies. Derived lipids include the followings-

Fatty acids, glycerol, other alcohols,steroids, fatty aldehydes and ketone bodies, hydrocarbons, lipid-soluble vitamins, and hormones.

i. **Fatty acids-**They occur in plant and animal foods and also exhibit in complex forms with other substances. It is obtained from hydrolysis of fats and usually contains an even number of carbon atoms and are straight chain derivatives.

 ☆ A current designation is based on the position of the double bond, counting from the methyl (-CH_3) group, called the omega end. The most important omega fatty acids are:

 a. **Omega-6 fatty acid-** Linoleic acid (poultry, eggs, nuts, cereals, durum wheat, whole-grain breads, most vegetable oils, linseed oil, rapeseed oil).

 b. **Omega-3 fatty acid-** Linolenic acid (ALA) found in plant oils.

 c. **Eicosapentaenoic, and Docosahexaenoic acids-** both commonly found in marine oils.

Sources

1. *M.W. Taylor, A.K.H.MacGibbon, Lipids-General Characteristics, Encyclopaedia of Dairy Sciences, 2002.*

1a. *Wikipedia.*

1b. *H.S.Murali, M.S.Mohan, K. S. Manja, R. Sankaran, Polar and nonpolar lipids and their fatty acid composition of a few Fusarium species, October 1993, Journal of American oil Chemists' society.*

1c. *Liya Anto, Sarah Wen Warykas, Moises Torres-Gonzalez, Christopher N Blesso, Milk Polar Lipids- Underappreciated Lipids with Emerging Health Benefits, Nutrients, 2020 Apr 4; 12(4): 1001.*

2. *K.S. Sharma, Chemistry of milk, Agrimoon.com.*

2a. *D S Newburg and P Chaturvedi, Neutral glycolipids of human and bovine milk, Lipids, 1992, Nov; 27(11), 923-7.*

3. *M.K. Srivastava, Handbook on Analysis of Milk, CBS Publications, New Delhi, 2010. ISBN 13: 9788123926681.*

4. *S.Patton, S. and R.G.Jensen, Biomedical Aspects of Lactation, 1976, Pergamon Press, Oxford.*

5. *www. Lipid library. aocs.org.*

6. *Robert Jenness and Stuart Patton, Principles of Dairy Chemistry, John Wiley and Sons, Inc., NY, 1969.*

7. *Definition of Lipids. http: //courseware.cutm.ac.in.*

Fatty Acid Composition of Milk Lipids [1,2,3]

Milk and milk products have significant position in the human diet and this is primarily attributed to the body, texture and flavour of these foods. The fatty acids composition of the lipid plays a important role in the characteristics of the products. In bovine milk fat the triacylglycerols account for about 98 per cent of the milk fat. Of all edible fats, milk fat has the most complex fatty acid composition.

Fatty Acid Profile of Milk Fat

Milk fats (Both Buffalo and Cow milk fats) consist chiefly of the triglycerides of fatty acids, which make up 95-99 per cent of milk fat. The remaining portion of milk fat is composed of diglycerides (about 4.1 per cent in buffalo milk and 1.26-1.59 per cent in cow milk), monoglycerides (about 0.7 per cent in buffalo milk and 0.016-0.038 per cent in cow milk) [1,1c].

- ☆ The diacylglycerols, monoacylglycerol and free fatty acids (FFA) are mostly products of lipolysis, and the cholesterol and phospholipids are cellular membrane material which accompanies the fat globule during extrusion from the secreting cell.
- ☆ High, medium and low molecular weight triglycerides in buffalo milk occur in the proportion of 42.5, 17.1 and 40.5 per cent, respectively and corresponding values for cow milk fat are 52.9, 18.9 and 28.2 per cent, respectively. Free fatty acid content of buffalo milk fat is lower (0.22 per cent) as compared to that in cow milk fat (0.33 per cent). [1c].
- ☆ Milk also contains 7 per cent short-chain fatty acids (C4-C8), 15-20 per cent medium-chain fatty acids (C10-C14) and 73-78 per cent long-chain fatty acids (C16 or higher).

☆ The triacylglycerol are formed by the esterification of the hydroxyl groups of glycerol with the fatty acids. The fatty acids are composed of a hydrocarbon chain and a carboxyl group.

☆ The nature of these fatty acids varies vastly and has a significant influence on the chemical, physical and organoleptic properties of the fat [1,2].

☆ Common fatty acids in milk fat are [1]:

i. C16:0 palmitate or palmitic acid,

ii. C18:0 stearate or stearic acid,

iii. C18:1 oleate or oleic acid,C18:2 linoleate or linoleic acid, an essential fatty acid and.

iv. C18:3 linolenate or linolenic acid, an essential fatty acid.

Table 14.6: The Concentrations of Fatty Acids in 100gm of different Fresh Milk [2]

Sl.No.	*Fatty acids*	*Unit*	*Cow*	*Goat*	*Buffalo*	*Sheep*	*Human*	*Soy*
1	Mono-unsaturated	gm	0.965	1.109	1.787	1.724	1.658	-
2	Poly - unsaturated	gm	0.124	0.149	0.146	0.308	0.497	-
3	Cholesterol	mg	14	11	19	27	14	-
4	Total -saturated	gm	2.079	2.667	4.597	4.603	2.009	

Some facts regarding the fatty acids in milk fat are given below:

i. Several fatty acids have been identified in the milk fat but only few of them are present in significant quantity and are of nutritional, physical and chemical importance.

☆ The major fatty acids in milk fat are straight chain fatty acids that are saturated and have 4 to 18 carbons (C4:0, 6:0, 8:0, 10:0, 12:0, 14:0, 16:0, 18:0), monounsaturated fatty acids (C16:1, 18:1), and polyunsaturated fatty acids (C18:2, 18:3) [1c].

☆ Some of the fatty acids are found in very small amounts but contribute to the unique and desirable flavour of milk fat. For example, the C14:0 and C16:0 β-hydroxy fatty acids spontaneously form lactones upon heating which enhance the flavour of butter [1c]

ii. Further 80 per cent of the total fatty acids are distributed among the five fatty acids namely the oleic, palmitic, butyric, stearic and myristic acids.

iii. Nearly 50 of the total 80 fatty acids which are commonly encountered in the milk fat would make only 1 per cent of the total fatty acids.

iv. As many as 142 fatty acids have been earlier reported to be present in milk fat [2a]. Now recent studies show that milk fat contains more than 400 fatty acids most these fatty acids exist in traces while only 20 fatty acids are found in larger concentration and make up 90 per cent of the milk fat [1.1c].

v. The distribution of these fatty acids among the triglycerides present in such that milk fat is one of the most complex natural fat [2].

vi. The fatty acid composition of milk fat is not constant throughout the lactation cycle. The fatty acids that are 4 to 14 carbons in length are synthesized in the udder. Some of the C16 fatty acids are made by the animal and some come from the animal's diet. All of the C18 fatty acids come from the animal's diet [1c].

vii. **Arrangement of Fatty acids-**The fatty acids are arranged on the triglyceride molecule in a specific manner. Most of the short chain fatty acids are at the bottom carbon position of the triglyceride molecule, and the longer fatty acids tend to be in the middle and top positions. The distribution of the fatty acids on the triglyceride backbone affects the flavour, physical, and nutritional properties of milk fat [1c].

viii. **Melting of Milk fat**- The melting properties of milk fat are a result of the melting points of the individual fatty acids that make up milk fat and their arrangement on the triglyceride molecule.

Milk fat melts over a wide temperature range, from approximately - 40°C to 40°C. At refrigerator temperature butter is approximately 50 per cent solid, but is only about 20 per cent solid at room temperature, which is why it spreads more easily as the temperature increases.

Classification of Fatty Acids [1,2,3]

Fatty acids can be grouped in the followings [1-3]-

i. **Saturated and Unsaturated fatty acids**-These fatty acids are usually grouped on the basis of saturation (saturated, monounsaturated and polyunsaturated). Saturated fatty acids such as myristic, palmitic, and stearic make up two thirds of milk fatty acids. Oleic acid is the most abundant unsaturated fatty acid in milk with one double bond. Linoleic and Linolenic acids are also unsaturated acids. Nearly 70 per cent of the fatty acid of milk fat are saturated and 30 per cent are unsaturated [1,2]. Milk fat contains approximately 65 per cent saturated, 30 per cent monounsaturated, and 5 per cent polyunsaturated fatty acids [1c].

ii. **Straight chain and branched chain fatty acids**- Similarly fatty acids can also be grouped on the basis of geometric isomerism as straight chain, branched chain [1,2].

iii. **Short, medium and long chain fatty acids**-On the basis of chain length, fatty acids can be as called as short, medium and long chain. The chain length of fatty acids in milk fat varies from C_4 to C_{26} [1]. Short chain group of fatty acids have short chain *e.g.* C4 - butyric, C6 - caproic, C8- caprylic and C10-capric. The butyric fatty acid is specific for milk fat and is responsible for the rancid flavour when it is cleaved from glycerol by lipase action. Long chain group of fatty acids have long chain *e.g.* - C14 - myristic 11 per cent, C16 - palmitic 26 per cent, C18 - stearic 10 per cent and C18:1 - oleic 20 per cent. [2,3].

iv. **Cis and Trans fatty acids**-Polyunsaturated fatty acids are further classified as either cis or trans (geometric isomerism) or conjugated or non conjugated positional isomerism. The cis form of geometric isomer is the most common found in nature and approximately 5 per cent of all unsaturated bonds are in the trans position as a result of rumen hydrogenation [1,2].

—C— C—**C = C**—C—

Cis Carbon chain

—C—C—**C = C**—C—

Trans Carbon chain

v. **Even and odd Nos. fatty acids**-Depending upon the number of carbon atoms present in the fat can also be grouped as even or odd numbered fatty acids [1].

   The details related to fatty acids found in milk lipid and their nature, are given in Table 14.7.

vi. **Volatile and Non volatile fatty acids -** Volatile- Butyric and Caproic acid. Non Volatile - Myristic and Palmitic acids [2].

vii. **Water Soluble and insoluble fatty acids-** Soluble - Butyric and Caproic acid. Insoluble - Myristic and Palmitic acids [2].

Factors affecting Fatty Acid Composition [1,1a,2]

The composition of milk fat is not constant and is very variable. Milk fat composition is strongly influenced by stage of lactation, proportion of short chains is low initially and increases until at least 8 to 10 weeks into lactation. Milk fat composition is changed more by the amount and composition of dietary fat than any other dietary component. Seasonal and regional differences in milk fat composition are measurable, most likely because of local differences in feed supplies [1a].

- ☆ Buffalo milk fat contains appreciably higher butyric acid in its triglycerides in comparison to cow milk fat. However, other short-chain fatty acids (caproic, caprylic, capric and myristic) are lower in buffalo milk fat. The content of long-chain fatty acids (palmitic and stearic) is relatively higher in buffalo milk. The unsaturated fatty acid level of buffalo and cow milk is comparable [1c].

Table 14.7: Type of Fatty Acids in Milk Fat [2,2b]

S.N.	Formula of acid	Name of fatty acids	% by Wt. & M.P.	State at room temp.	Volatile/Non -volatile	Soluble /Insoluble
A-Saturated acid (Anoic acid) $C_nH_{2n+1}COOH$						
01	C_3H_7COOH Tetranoic acid	Butyric* (4:0)	3.5/-7.9 °C	Liquid	Volatile	Soluble
02	$C_3H_{11}COOH$ Hexanoic acid	Caproic (6:0)	2/-9.5°C	Liquid	Volatile	Soluble
03	$C_7H_{15}COOH$ Octanoic acid	Caprylic (8:0)	1/16.5°C	Liquid	Volatile	S.Soluble
04	$C_9H_{19}COOH$ Decanoic acid	Capric (10:0)	2/31.3 °C	Solid	S.Volatile	V.S.Soluble
05	$C_{11}H_{23}COOH$ Dodecanoic acid	Lauric (12:0)	2.5/48.5 °C	Solid	V.S.Volatile	Insoluble
06	$C_{13}H_{27}COOH$ Tetradecanoic	Mysteric (14:0)	10/58 °C	Solid	Non volatile	Insoluble
07	$C_{15}H_{31}COOH$ Hexadecanoic	Palmitic (16:0)	25 / 64 °C	Solid	Non volatile	Insoluble
08	$C_{17}H_{35}COOH$ Octadecanoic	Stearic (18:0)	10.3/69.3 °C	Solid	Non volatile	Insoluble
09	$C_{19}H_{39}COOH$ Bidecanoic	Arachidic (20:0)	0.5/77 °C	Solid	Non volatile	Insoluble
B-Unsaturated fatty acid with Enoic acid						
a-One Double bond C_nH_{2n-1} COOH -01	$C_{17}H_{33}COOH$ Octadecaenoic Mono	Oleic** (18:1)	33/14°C	Liquid	Non volatile	Insoluble
b-Two double bond C_nH_{2n-3} COOH -01	$C_{17}H_{31}COOH$ Octadecadienoic	Linoleic (18:2)	4/18°C	Liquid	Non volatile	Insoluble
c-Threedouble bond C_nH_{2n-5} COOH -01	$C_{17}H_{29}COOH$ Octadecatrienoic	Linolenic (18:3)	1/20 °C	Liquid	Non volatile	Insoluble

Note- S.-Slightly, V.S.-Very slightly, *- Hydrolytic Rancidity, ** Oxidative Rancid

The following factors affects the composition of milk fat:

a. Variations due to Feed [1,1a]

i. The important cause for variation is feed and both lipid and non lipid components affect milk fat composition [1]. At low rates of feeding, increased grain (fermentable starch) intake increases total milk production and milk protein content. At higher intakes increasing starch intake depresses milk fat percentage. When milk fat depression occurs, changes in milk Fatty acids composition also take place [1a].

ii. Consequently it will vary with the year region farming practice and so on.

iii. Since the fatty acids which are preformed from food fat is transferred to the mammary gland via the blood and lymph in the form of triglyceride and free fatty acids. Most of these fatty acids are having a chain length of 16 or more carbon atoms. Some of the fatty acids are being synthesized by the gland from the acetate and β hydroxyl butyrate produce by rumen bacteria. The fatty acids synthesized by utilizing the fatty acids producing rumen bacteria will have short or medium chain length C4 to C14 and part of the C16. Acetate contribute to the increments to all of the C4 to C14 acids and β hydroxyl butyrate is used primarily for the initial four carbon "primer" units of the most fatty acids synthesized.

iv. Low roughage diets diminish acetate and increase propionate production in the rumen. Milk produced by cow on low roughage diets may have only half of the fat content when compared to the milk from cows on high roughage diets and the proportion of the short and medium chain saturated acids which are synthesized from acetate is greatly diminished.

v. To a considerable extent the rumen microorganism hydrogenates the dietary fatty acids.

vi. Feeding highly unsaturated oil such as safflower oil increases the C18:2 content of the milk fat.

b. Variations Due to the Animal Factors [1a,1b,2]

Stage of lactation, season, and herds were all major sources of variation in fatty acid (FA)composition [1a]. The effects of animal factors, such as parity and stage of lactation, and the effects of Herd/Test Date(HTD) on the fatty acid composition of cow milk are well known. Lactation stage and HTD as important sources of variation in the fatty acid profile of buffalo milk also [1b].

i. Stage of Lactation [1b,2]

The percentages of short-chain and medium-chain fatty acids in buffalo milk increased in early lactation whereas long-chain fatty acids decreased. The only exception to this pattern was butyric acid, which linearly decreased from the beginning of lactation. Interestingly, the milk fatty acid profile in early lactation seems to be less affected by body energy reserve mobilization in buffaloes than in cows, probably due to a shorter period of negative energy balance [1b].

There are systematic changes in milk fat composition due to the stage of lactation and the energy needs of the animal. In early lactation, the animal's energy comes largely from body stores and there are limited fatty acids available for fat synthesis, so the fatty acids used for milk fat production are obtained from the diet and tend to be the long chain C16:0, 18:0, 16:1 and 18:2 fatty acids. In late lactation, more of the fatty acids in milk are formed in the mammary gland so that the concentration of the short chain fatty acids such as C4:0 and 6:0 are higher than in early lactation. These changes in fatty acid composition do not have a great impact on nutritional properties of milk, but may have some effect on processing characteristics for products such as butter [1c].

Effect of lactation period on the composition of fatty acids in milk fat can be easily understood by the following examples [1]-

i. Butyric acid (4:0) shows a maximum value during the first month of lactation, declining thereafter and becoming minimal at the end.

ii. Hexanoic acid (6:0) to Myristic (14:0), all had similar variations; the values increased during the first 4 to 8 weeks of lactation, remained relatively constant until the fifth or sixth month, and then decreased again until the end of lactation.

iii. There was little variation in 16:0 throughout lactation. Stearic (18:0) and oleic (18:1) acid contents were high in early lactation, then declining throughout the mid lactation and again increasing at the end of lactation.

ii. Parity- It is another well-known factor affecting milk production and MF composition in dairy cows. Parity, instead, had minor effects on the buffalo milk fatty acid profile [1b].

Table 14.8: Fatty Acid Profile of Bovine Milk Fat [4]

Sl.No.	*Fatty Acids and Notation*	*Mole per cent*	
		Buffalo	*Cow*
1	Butyric acid 4:0	11.4	8.8
2	Caproic acid 6:0	3.1	3.5
3	Caprylic acid 8:0	1.0	2.0
4	Capric acid 10:0	1.6	3.0
5	Lauric acid 12:0	2.6	3.8
6	Myristic acid 14:0	10.6	9.9
7	Palmitic acid 16:0	30.3	26.1
8	Stearic acid 18:0	10.5	9.1
9	Long chain saturated fatty acid 20:0-26:0	0.70	1.0
10	Monounsaturated acid 10:1-14:1	1.0	1.8
11	Oleic acid 16:1	3.6	3.6

Table 14.9: Fatty Acid Composition in Milk From Several Species [2]

Fatty Acid	Rabbit	Guinea pig	Mouse	Rat	Sow	Cow	Sheep	Goat
Short chain	Molar Per cent							
C4:0	<1	<1	<1	<1	*	11	8	8
C6:0	<1	<1	<1	<1	*	5	5	5
C8:0	37	<1	<1	6	*	1	4	4
Medium Chain								
C10:0	27	<1	7	17	*	3	6	13
C12:0	3	<1	12	13	*	3	5	7
C14:0	1	2	16	13	2	10	10	12
Long chain								
C16:0	8	34	33	28	28	23	22	24
C18:0	2	3	2	3	6	10	10	12
C18:1	9	31	14	13	35	29	22	17
C18:2	10	23	8	7	14	2	4	3
C18:3	1	4	<1	<1	<1	<1	<1	<1

*For the sow, the sum of C4:0 to C12:0 is 2 per cent total.

Structure of Fat Globule [1,2,3]

More than 95 per cent of the total milk lipid presents in the form of a globule. The globules are surrounded by a protein and phospholipid membrane that stabilizes the globules in the serum (water) phase of milk. Milk fat is predominantly present spherical droplets which range in diameter from less than 0.2 to 15 μm. The bulk of the fat is in globules 1-8 μm diameter.

- ☆ The uneven size distribution allows the larger globules to float in a process called creaming, thus resulting in a "cream line" at the top of the container. The size distribution found may depend greatly on the measurement method employed.
- ☆ Globule size can considerably altered by various treatments particularly homogenization.
- ☆ The fat globules of milk differ in composition. The size of the fat globule will also alter the composition. This is because the quantity of membrane lipids (predominantly phospholipids) per unit mass of fat is higher for smaller globules. The fat globules are enveloped by a thin layer (8 to 10 nm) called the fat globule membrane [1]. The average size of fat globules in buffalo milk is larger (4.15 to 4.60 μm) than in cow milk (3.36 to 4.15 μm). Milk fat globules fall in to the following three overlapping size distributions [1c]-

Table 14.10: Size Distribution of Milk Fat Globules [1c]

Sl.No.	Class	Diameter (μm)	Proportion of the Total Globule Population (per cent)	Fraction of Total Milk Lipid (per cent)
1	Small	Below 2	70-90	< 5
2	Intermediate	3-5	10-30	50
3	Large	8-10	0.01	1-4

☆ The properties of thin membrane are different from both milk fat and plasma. The native fat globule membrane (FGM) is comprised of apical plasma membrane of the secretary cell which continually envelopes the lipid droplets as they pass into the lumen and made up of protein and phospholipids (Lecithin) [2,3].

i. The membrane consists of a well organized sequence of different components arranged according to their polarity and hydrophobicity. This layer is predominantly proteinaceous in nature.

ii. The membrane constituents are very thinly spread over the surface of the fat globule and stabilize the fat in water emulsion. The breaking or disruption of the membrane also makes the lipids more susceptible to enzymatic action due to greater accessibility [1].

☆ The composition of the MFGM often varies with the method of isolation or preparation. The Average compositions of milk fat globules membranes is tabulated below:

Table 14.11: Average Compositions of Milk Fat Globules Membranes [1c]

Sl.No.	Component	Western Cow	Buffalo
1	Lipid content	35.7	38.5
2	Neutral lipids	65.7	74.4
3	Phospholipids	18.6	19.7
4	Triglycerides	45.5	52.7
5	Diglycerides	5.5	7.4
6	Sphingomyelin	3.2	4.3
7	Phosphatadyl serine	2.4	3.1
8	Protein (g/100 g fat globule)	1.8	1.36
9	Fatty acids		
	C 18:0 Stearic acid	24.1	34.4
	C16:0	23.8	22.4
	C18:1	14.5	14.5
	C14:0	14.4	12.6

i. A variety of substances are present in fat globule membrane (FGM). These include protein, phospholipids, phosphatase, complex of vitamin B_{12} binding proteins and a complex of riboflavin- phosphoric acid protein [2].

ii. Protein constitutes nearly 25 to 60 per cent of the total dry weight. 25 different enzymes have been reported and the isolated. MFGM is a rich source of enzymes including 5 Nucleotidase, alkaline phosphatase, acid phosphatase, aldolase, xanthine oxidase. Proteins of the MFGM could be in the form of polypeptides of varying molecular weight [1].

iii. There are also metals like copper, iron and zinc. In the presence of copper it causes oxidation of fat. The phospholipids are involved in the oxidation of milk. The membrane (FMG) helps the milk fat in keeping the milk fat from becoming a free oil when milk is heated and also protects against the action of Lipase enzyme [2].

- ☆ Lipids are the next major constituent of the MFGM and their content could vary from 0.5 to 1.2 mg per gram of protein. The lipid portion comprises of neutral lipids (50 to 80 per cent of total lipid) or phospholipids. Due to their amphiphilic nature, the phospholipids have an important role in maintaining the stability of fat emulsion. Glycerides constitute a major portion of the neutral lipids of the membrane. Other constituents like cerebrosides, gangliosides sialic acid, cytochrome and hexoses have been reported [1].
- ☆ The constituents of the MFGM are specifically oriented on the fat surface.

  i. The inner most layer consists of the high melting triglycerides in contact with the fat layer.

  ii. The next layer consists of the phospholipids with their hydrophobic portion oriented towards inwards while the hydrophilic portion oriented outwards. The phospholipids are interspaced with cholesterol, vitamin A *etc.*

  iii. The outermost layer is made up of protein enzymes *etc.* absorbed on to the surface. These proteins are in contact with plasma and are even capable of binding water to a some extent [1].

As per the **Stokes' Law**,the fat globules will cream due to the differences in densities between the fat and plasma phases of milk and hence if raw milk or cream is left to stand, it will separate. The creaming takes place faster in cold raw milk. The cream layer forms very rapidly, within 20 to 30 min., in cold milk. An immunoglobulin(IgM) forms a cryoglobulin complex with lipoproteins and this precipitates onto the fat globules and causes flocculation called cold agglutination [2].

Homogenization of milk prevents this creaming by decreasing the diameter and size of the fat globules as it breaks the fat globules in to smaller pieces. It also causes the formation of a **recombined membrane** which is much similar in density to the continuous phase.

- ☆ This Recombined membranes are very different than native FGM. Homogenization is the principal method for achieving stabilization of the fat emulsion in milk and it decreases the average diameter of fat globule and significantly increases the surface area.
- ☆ Some of the native FGM will remain adsorbed but it will not be able cover all of the newly created surface area. The surface tension raises to a high level of 15 mN/m just after disruption of the fat globule and amphiphilic molecules in the plasma quickly adsorb to the lipid droplet to lower this value. The adsorbed layers consist mainly of serum proteins and casein micelles [2]. The breaking or disruption of the membrane also makes the lipids more susceptible to enzymatic action due to greater accessibility [1].

Sources

1. *K.S. Sharma, Chemistry of milk, Agrimoon.com.*

1a. *D. L. Palmquist, A. Denise Beaulieu and D.M. Barbano, Feed and Animal Factors Influencing Milk Fat Composition, Journal of Dairy Science, Volume 76, Issue 6, 1993, Pages 1753-1771.*

1b. *S.Pegolo, G. Stocco, M. Mele, S. Schiavon, G. Bittante and A. Cecchinato, Factors affecting variations in the detailed fatty acid profile of Mediterranean buffalo milk determined by 2-dimensional gas chromatography, Journal of Dairy Science, Volume 100, Issue 4, April 2017, Pages 2564-2576.*

1c. *Y. Kotilinga Reddy, Yogesh Khetra and M.H. Sathish Kumar, Fat rich dairy products, www.AgriMoon.com.*

2. *M.K. Srivastava, Handbook on Analysis of Milk, CBS Publications, New Delhi, 2010. ISBN 13: 9788123926681.*

2a. *B.H. Webb, A.H. Johnson, and J.A, Alford, J. A. 1965. Fundamentals of dairy chemistry. AVI Publ. Co., New York.*

2b. *D. Ramaswamy, A.V. Shibu and H. Gopi, Dairy Technologist 's Handbook, 1999, ISBN-13: ý 978-8185860442.*

3. *Robert Jenness and Stuart Patton, Principles of Dairy Chemistry, John Wiley and Sons, Inc., NY, 1959.*

4. *M.P. Mathur, D.D. Roy and P. Dinakar, Text book of dairy chemistry, 2008, ISBN-13: þ: 978-8171640461, ICAR, Govt. of India.*

Properties of Milk Lipids [1,2,3]

Lipids perform three primary biological functions within the body: they serve as structural components of cell membranes, function as energy storehouses, and function as important signalling molecules. The individual lipid has several important properties. The composition of lipids in milk is already given in tables earlier under composition of lipids. Bioactive lipids in milk include triacylglycerides, diacylglycerides, saturated and polyunsaturated fatty acids, and phospholipids

Properties of Triglycerides [1,1b,2,3]

Triglycerides are the major lipid (97 to 98 per cent) as discussed earlier. The triglycerides can be hydrolysed under suitable conditions to yield component fatty acids and glycerol. It is observed that distribution of the fatty acids or groups of residues over the three positions is not random. Particularly the preference of C4: 0 and C6:0 for the 3 position is not random. Triglycerides may be a part of some lipoproteins. A kind of mixed micelles largely consisting of protein and compound lipids and thus be present in trace quantities in plasma [1].

- **i.** Position of the fatty acid residues in the triglyceride molecules affects physical properties particularly crystallization and it determines which fatty acids are predominantly liberated by enzymatic hydrolysis.
- **ii.** However the position of these fatty acids has least effect on the chemical reactivity. Triglycerides are not very reactive at room temperature. At higher temperature oxidative attack of the double bonds may occur. The Keto and hydroxyl groups may react.
- **iii.** The triglycerides are very a polar and not surface active. They act as a solvent for many other a polar substances. A small amount of water dissolves in liquid triglycerides (about 0.15 per cent at room temperature in liquid milk fat) but triglycerides do not dissolve in aqueous media at all.
- **iv.** The triglyceride composition in several milk products will be affected as in the case of butter milk which is obtained by churning of cream have a lower melting triglyceride than those of the whole milk fat which is reflected in a higher degree of unsaturated fatty acid residues.
- **v.** The fat can be fractioned by partial crystallization and separation resulting in obtaining fractions with different properties.

The some properties of triglycerides are discussed below:

A. Milk Fat Constants [1,2,3]

Physical and chemical constants of fats are helpful in characterizing the fat. The type of the fatty acids present in the fat can also be identified with the help of these constants. These constants would also enable to detect the adulteration qualitatively and some instances quantitatively. The average values of some physical and chemical constants of milk fat are given in Table 14.12:

Table 14.12: Physical and Chemical Constants of Milk Fat (Average) [2,3]

Sl.No.	*Constants*	*Values*
A	**Physical**	
1	R.I. at 40°C	1.4538
2	Density at 37°C	Ave. 0.91-0.93
3	Melting point	27-37°C
B	**Chemical**	
1	Saponification No.	220-240
2	R.M.Value	21-34
3	Polenski value	1.2-1.4
4	Iodine value	26-35
5	Kirschner value	19-26

The physico chemical properties of Cow and Buffalo milk fat are tabulated below:

Table 14.13: Physico Chemical Properties of Cow and Buffalo Milk Fat [1d]

Sl.No.	*Physico Chemical Properties*	*Cow Milk Fat*	*Buffalo Milk Fat*
1	**Physical Properties**		
	BR reading	41.05 - 42.40	41.00 - 43.50
	Refractive index	1.4498 -1.4530	1.4515 -1.4533
	Density g/ml	0.888 - 0.911	0.905 - 0.917
	Grain size mm	0.098 - 0.190	0.20 - 0.41
	Fat globule size μm	3.36 - 4.15	4.15 - 4.60
	Softening point range °C	33.5 - 35.9	34.3 - 36.3
	Softening point range °C	31.5 - 35.2	33.4 - 46.4
2	**Chemical Properties**		
	Acid value	0.26	0.17 - 0.352
	Unsaponifiable matter mg/100ml	414 - 450	392 - 398
	Saponification number	221.0 - 238.0	218.23 - 236.10
	Iodine value	27.70 - 37.32	27.00 - 33.90
	Reichert-Meissel value	24.60 - 29.70	27.83 - 35.50
	Polenske value	1.30 -1.80	0.70 - 1.60

i. Refractive Index

This property of milk fat is based on the principle that the degree to which the light waves pass through the liquid fat, which will be the characteristic feature of that milk fat.

- ☆ This physical constant for milk fat is determined by using the Abbey refractometer. The reading is usually obtained at 40°C. This instrument directly gives the refractive index. Butyro refractometer is also useful for determining the refractive index. The reading obtained with this instrument has to be converted in to refractive index or to be used as butyro refractometer reading (B.R. Reading).
- ☆ The reading for the milk fat ranges between 1.4527 and 1.4566.due to the large proportion of saturated glycerides and short chain acids in the bovine milk fat it is low in comparison to that of other fats and oils. The refractive index indeed of a fat is influenced by the both the molecular weight and the degree of unsaturation of the component fatty acids.
- ☆ The details related to values, instruments and method of determination of RI/BR of milk fat are discussed earlier under chapter -3 of this book.

ii. Saponification Number [1,2a,4,5]

The saponification number is equivalent to the number of milligrams of potassium hydroxide required to saponify one gram of fat. This value may range from 210 to 233 for milk fat.

- ☆ This value of a lipid indicates the average molecular weight of fatty acids present in it. The average molecular weight of a triacylglycerol is equals 168,300/SV.
- ☆ The formula useful for calculating the average molecular weight is-

$$SV = 56{,}100(45+14\,x)$$

where, x average number of C atoms per fatty acid residue.

- ☆ This value for milk fat is much higher for most other fats, oils with the exception of coconut and palm kernel oil.
- ☆ Saponification value is more useful in detecting the mineral values such as paraffin in ghee as these are not acted upon by alkali and such a sample does not form a homogenous solution upon saponification [1].
- ☆ Since the ghee or oil from a given source has a remarkably constant saponification value any deviation found in the test is an indication to adulteration *e.g.*-Milk Fat/Ghee-210-230, Soyabean oil-189-195, Coconut oil-253(245–260), Palm Kernel Oil -240–250, Soy and Corn Fats and Oils-200 and Rapeseed oil-168-181.
- ☆ Saponification value of Ghee or fat/oil is a very valuable test for the determination of adulteration. The average value of saponification value of butter fat is 230 which is distinctly higher than animal fat (*e.g.* Lard-197,Beef -195 and Mutton fat-193) and hence admixture of butter fat with animal fat causes decrease in saponification value [2a,4,5].

iii. Iodine Number [1,2,4,5]

Iodine number is the number grams of iodine absorbed by 100 g of fat under specified conditions. It gives the indication about the degree of unsaturation of the fatty triglycerides. This value is a measure for the unsaturated linkages present in a fat.

- ☆ Absorption of either iodine bromide (IBr) or iodine chlore (ICl) is used for measuring this value. One molecule of halogen compound absorbed by each unsaturated linkage and the absorption is expressed as equivalent number of iodine absorbed by 100g fat.
- ☆ The iodine value for milk fat ranges between 26 to 42 [1,2a].
- ☆ The test is a measure of unsaturation of a given fat or oil and done with Wij's solution. Since the degree of unsaturation is more or less characteristic to oil source, the test is routinely used for the determination of adulteration by other types of oils. Iodine values of ghee and some common edible oils and animal fats are given below [2a,4,5]-

  Butter oil: 26-38,Soybean oil: 120-143, Rapeseed oil: 94-120,Coconut oil: 125,Cotton seed oil:110, Mutton fat:40, Beef Fat:35-43 and Pig Fat:50-80.

iv. Reichert-Meissl Number [1,2,2a]

It is the number of milliliters of 0.1N sodium hydroxide or aqueous alkali solution required to neutralize the steam volatile water soluble fatty acids distilled from 5g of fat under the precise conditions specified in the method.

- ☆ This value is quite significant for milk fat, since it primarily measures the butyric acid and caproic acid content in the given fat. When compared to other fats this value for milk fat is high. This helps in the identification of milk fat from other fats.
- ☆ The determination of RM and P. Values give confirmatory information regarding the composition of fat. Under FSSAI,different values (From 21-28) of RM are prescribed for different states/union territories. Generally RM Value of milk fat varies from 25-34. Edible oils and fats (Soy and Corn Fats and Oils) have RM Value less than 1 where as Coconut oil and Palm Kernel Oil have RM Value 6 to 8 and 5 to 7 respectively.

v. Polenske Number [1,2a]

It is the number of milliliters of 0.1N sodium hydroxide or aqueous alkali solution required to neutralize the steam volatile water insoluble fatty acids distilled from 5g of fat under the precise conditions specified in the method.

- ☆ Caprylic, Capric and Lauric acids are steam volatile water insoluble acids responsible for Polenske value.Since most of the steam volatile fatty acids in milk fat are water soluble this value helps in identifying the presence of coconut oil content which contains higher proportion of these acids.

- ☆ Under FSSAI,different values (From 1.0-2.0) of PV are prescribed for different states/union territories The polenske value of milk fat varies from 1.2 to 3.3 where as animal fat has this value less than 1 and vegetable oil has value in between 10 to 15 *e.g.* Coconut Oil- 14-18, Palm Kernel Oil - 10-12 and Soy and Corn Fats and Oils- <1.

vi. Kirschner Value [1,2a]

It is the number of milliliters of 0.1 N aqueous alkali solution required to neutralize steam volatile water soluble fatty acids distilled from 5g of fat which forms water soluble silver salts distilled from 5g fat under specified conditions.

vii. Melting Point [1,2]

Triacyl glycerol molecule of a milk Fats are mixture of mixed fatty acids forming. There is no possibility to obtain a specific temperature to be regarded as melting point similar to pure chemical compounds. As such melting point of milk fat is the end of melting range.

- ☆ Milk fat is made up of several components which is a mixture where different constituents at different temperatures have their own melting point and other physical properties.
- ☆ The trans unsaturation increases melting points whereas odd-numbered and branched chains decrease melting points.
- ☆ There are several methods available for the determination of melting points. The following methods (details given under chapter-3 of this book) are used for the determination of melting point of milk fat-
  1. Mettler dropping point method,
  2. Falling ball method (softening point),
  3 Open capillary tube method and
  4. Slip point method.

The research work conducted by researchers revealed that the Mettler dropping method and softening point methods give reproducible results where as slip point method the results are poor. The average softening point for milk fat milk fats were 30.4°C for soft butter and 38.4°C for hard butter [1].

B. Crystallization Behaviour [1,1a,1b,2]

Crystallization is the physical transformation of a liquid, solution or gas to a solid crystal, which has an ordered internal arrangement of molecules, ions or atoms.

- ☆ Crystallization of milk fat is more intricate than that of most other fats due to its complexity both in terms of fatty acid composition and glyceride structure. The formation of crystals containing more than one triacylglycerol species, that is, mixed crystals, is common in milk fat [1b]. Milk fat contains a large number of triacylglycerols, making the process of crystallization complex [1].
- ☆ Crystallization behaviour of milk fat has been rationalized on the basis that it is capable of forming a series of mixed crystals with mutually overlapping melting ranges [1b].
  - **i.** Milk fat exhibits a melting range of - 40° to 40°C. Hence, both solid and liquid phases are present in milkfat at temperatures encountered during their storage and use [1b].
  - **ii.** Milk fat is liquid above 40°C and completely solid below -40°. Between these extremes milk fat is a mixture of crystals and oil, where oil is the continuous phase.
  - **iii.** The properties of milk fat are the average of the properties of its triacylglycerols.
- ☆ Crystals in fat globules generally cannot grow larger than the globule diameter. Most crystals are much smaller than globule size and they flocculate into a network, giving the globules rigidity.
- ☆ Crystallization starts with the formation of crystal nuclei in the molten fat as few molecules gather in molecular aggregates where the potential energy is reduced to a minimum [1].

The Crystallization of milk fat determines the physical stability of the fat globule and the consistency of high-fat dairy products. The milk fat crystals occur in alpha, β, β ′ 1, and β ′ 2, The alpha form is the least stable and is rarely observed in slowly cooled fat [2].

- ☆ Seasonal and regional variability in milk fat composition causes differences in crystallization behaviour.
- ☆ More detailed examination of the triacylglycerol profiles led to the conclusion that the relative contents of certain triacylglycerol combinations correlated well with crystallization behaviour [1a].
  - **i.** In particular, the ratio of higher-melting to lower-melting triacylglycerols could be used to predict crystallization behaviour.
  - **ii.** Higher ratios of higher-melting triacylglycerols led to higher crystallization rate [1a].

Milk Fat Crystallization and Texture of Dairy Products [1b,2]

Fat crystallization process plays a important and critical role in determining overall product quality. Butter, ghee, ice cream, *etc.*, are the dairy products where fat crystallization is important.

1. *Butter [1b]*

Butter is notoriously unspreadable at refrigeration temperature and is too soft or oily at room temperature. The narrow temperature range for spreadability is explained by the drastic change in the solid fat content of milk fat between 10° and 20°C and less viscous flow at high temperatures.

i. In making butter, crystallization of milk fat occurs in the fat globules as the cream is conditioned prior to churning. There are four different types of fat globules in cream, between the temperatures of 4° and 20°C, as differentiated primarily by the pattern of crystallization of milk fat. Type 4 fat globules (Globules have a thick crystalline shell at the periphery with irregular crystal aggregates and a few pools of liquid fat contained within) are the predominant form of globules left after churning in to butter. Thus, a higher content of Type 4 globules in the cream can significantly influence the consistency of butter.

ii. The composition of the lipid phase also impacts hardness of butter through an influence on crystal microstructure. However, through a control of crystallization of milk fat in cream and butter, the desired texture in the final product can be obtained.

- ☆ Crystallization conditions of butter influence the size of fat crystals, the solid to liquid fat ratio, formation of mixed crystals, the degree of fractionation between high and low melting components of milk fat and the relative proportion of primary and secondary bonds between crystals. These factors influence the phenomenon of setting in butter, where a substantial increase in firmness is found during the early stages of storage. This phenomenon has been related to continued crystallization and recrystallization (rearrangement of crystalline microstructure) of the milk fat [1b].

2. *Ghee*

Granulation in ghee is due to the crystallization phenomenon of the high melting triacylglycerols. The fatty acid profile of milk fat, temperature at which ghee is clarified, followed by the rate at which it is cooled, seeded and held under quiescent storage before packing, and temperature of ghee storage in the marketing network are various factors that contribute towards graininess in ghee [1b].

3. *Ice Cream*

Crystallization plays important role in quality and texture of ice cream. Therefore, it is important to control the crystallization of three ice cream components; ice, lactose and lipids [1b].

- ☆ Lipid crystallization also plays an important role in the development of structure during processing of ice cream [1b].

- ✰ The rate of crystallization and destabilization of the emulsion during freezing are particularly important for controlling the quality of the ice cream [1b].

C. Lipolysis and Rancidity of Milk Fat [1,2]

The development of any disagreeable odour and flavour in fats and oils causing spoilage is known as rancidity. Market milk and some milk products may have rancid flavour. Development of rancid flavour is due to the accumulation of proper concentration and type of free fatty acids which are released due to the hydrolysis of milk fat due to the action of lipases which are normally present in milk. Milk fat can be degraded by enzyme action, exposure to light, and oxidation and each of these processes proceeds through different mechanisms. Storing milk in opaque containers minimizes the degradation of milk fat caused by the exposure to light.

- ✰ Factors Leading to rancidity are- temperature, moisture content, air, light, metals, degree of unsaturation,absence of anti-oxidants.
- ✰ There are different types of rancidities in milk fat-Hydrolytic and Oxidative rancidities. The details related to these rancidities are given below:

1. Hydrolytic Rancidity

It is also known as **lipolytic rancidity**. The flavour defect commonly referred as **"rancidity"** or more specifically **"hydrolytic rancidity"** is caused primarily by the presence of a single enzyme "milk lipoprotein lipase" in milk. Lipases split fatty acids from the glycerol backbone of the triglyceride. Usually the action of lipase causes undesirable rancid flavours in milk. Pasteurization inactivates native lipase and increases the shelf life of milk. Presence of this enzyme in milk is ascribed to leakage from blood through mammary tissues rather than true secretion. Although, most of the milks contain sufficient amounts to cause rancidity, in practice it will not occur because the substrate (triacyl glycerol) and the enzyme are well partitioned. In addition to this there are several factors which affect the enzyme activity [1].

The hydrolysis of milk fat occurs naturally by the enzyme lipase. The hydrolysis can be brought about also by superheated steam or alcoholic NaOH. The reaction mechanism is given below [2]-

$$\begin{array}{l} CH_2\,OOC_{17}H_{35} \\ | \\ CHOOC_{17}H_{35} \\ | \\ CH_2OOC_{17}H_{35} \\ \textbf{Fat (stearin)} \end{array} \quad + 3NaOH \longrightarrow \begin{array}{l} CH_2OH \\ | \\ CHOH + 3C_{17}H_{35}COONa \\ | \qquad \textbf{Sodium Stearate} \\ CH_2OH \\ \textbf{Glycerol} \end{array}$$

$$2C_{17}H_{35}COONa + H_2SO_4 \longrightarrow 2C_{17}H_{35}COOH + Na_2SO_4$$

Sodium Stearate **Sulphuric Acid** **Stearic Acid** **Sodium Sulphate**

- ✰ All fats react in this manner but differ only in the number and relative proportion of fatty acids that are united to glycerol. If milk fat is treated in the above manner, about 16 fatty acids are obtained, some of them are soluble in water but most of them are insoluble and appear as white precipitate. Oleic, palmitic, butyric, stearic and myristic acids are the five major fatty acids. While many of these fatty acids are odourless and almost tasteless some of them such as butyric acid possess strong odour. The liberation of butyric acid leads to rancidity of fat caused by hydrolysis [2].
- ✰ Lipolysis takes place at an oil water interface. The factors which influence enzyme activity are given below [1]-
  - i. Surface area available,
  - ii. Permeability of the emulsion,
  - iii. Type of triglyceride employed,
  - iv. The physical state of the substrate (complete solid, complete liquid or liquid-solid),
  - v. Degree of agitation of the reaction medium.
  - vi. pH,
  - vii. Temperature,
  - viii. The presence of inhibitors and activators,
  - ix. concentration of the enzyme and substrate.
  - x. light.
- ✰ Lipolysis has been classified as spontaneous or induced [1].
  - a. **Spontaneous lipolysis-** It has been defined as lipolysis caused without apparent mechanical agitation.
  - b. **Induced lipolysis** - It is caused by some form of mechanical agitation of raw milk.

2. Oxidative Rancidity [1,1c,2]

Milk fat can also be degraded by a classical chemical oxidation mechanism *i.e.*, the attack on double bonds in the fatty acids by oxygen. It is also called **auto-oxidation**. Fats when in contact with oxygen produce products with undesirable characteristics. It is promoted by heat, light, ionizing radiations, catalyst and enzymes like lipoxygenase.

- ✰ Autoxidation refers to oxidations brought about by reactions with oxygen at normal temperatures, without the intervention of flame or electric spark [1c].

☆ Lipid oxidation in fluid milk and number of its products has been a concern of the dairy industry for the last several years. To prevent or to retard this defect low-temperature refrigeration of butter and butter oil, inert gas or vacuum packing of dry whole milk, are being used. The phospholipids, particularly the cephalins, have many unsaturated fatty acid residues compared to the neutral glycerides. They include polyenoic fatty acids [1]. This is important for lipid autoxidation.

☆ The rate of autoxidation is influenced by the followings [1]-

i. Complex composition of dairy products.
ii. The physical state of the product (liquid, solid, emulsion, *etc.*) and
iii. The presence of natural anti or pro oxidants, as well as processing, manufacturing and storage conditions tend to influence. These factors would also influence composition and percentage of products formed due to autoxidation.

Mechanism [1,1c]

Chemical reactions involved autoxidation of milk fat are grouped in to the following three categories-

i. **Initiation (formation of free radicals)-** The initiation step is often ill-defined and many agents have been proposed as radical initiators. The initial step in the autoxidation of unsaturated fatty acid and their ester is the formation of free radical. In the case of monounsaturated and non conjugated polyene fatty acids the reaction is initiated by the removal of hydrogen atom from the methylene group of adjacent to the double bond.

ii. **Propagation (free radical chain reaction)-** The resulting free radical stabilized by resonance adds oxygen to form peroxide containing free radicals these in turn react with another mole of unsaturated compound to produce two isomeric hydroperoxides in addition to free radicals capable of continuing the chain reaction.

iii. **Termination (terminal point-formation of non radical products)-** As the concentration of free radicals increases chain termination reactions become more important, these reduce the number of radicals by radical disproportionation or combination. Termination occurs when two free radicals interact to form a non- radical.

   ☆ As the degree of unsaturation (No. of double bonds) increases, more isomeric hydroperoxides are formed [1] *e.g.*:

   a. Oleic acid(Single double bond), having two methylene groups, gives rise to four isomeric hydroperoxides. The preferential points of attack in a polyenenon conjugated systems are the methylene groups located between the double bonds.

b. Autoxidation of linoleic (Two double bonds) and linolenic acids (Three double bonds) can lead to the formation of three and six isomeric hydroperoxides respectively by reacting with C_{11}methylene group linoleic acid and C_{11} and C_{14} methylene groups of linolenic acids.

c. In a polyunsaturated fatty acid methylene groups located other than those located between the double bonds can also involve in these reactions but to a lesser degree. In addition to formation of polyperoxides, carbon to carbon polymerization, and the formation of epoxides and cyclic peroxides are also possible in these reactions.

- Hydroperoxides formed due to autoxidation are unstable and they readily decompose forming the saturated and unsaturated aldehydes. The mechanism suggested for this is that it involves in cleavage of the isomeric hydroperoxide to the alkoxyl radical which undergoes carbon- to carbon fission to form aldehyde.
- Formation of other products such as unsaturated ketones saturated and unsaturated alcohols,saturated and unsaturated hydrocarbons and semialdehydes are also being observed.
- Saturated and unsaturated Aldehydes impart characteristic off-flavours in the products. The terms often used to characterize the flavour are "painty", nutty, melon-like, grassy, tallow, oily, card board, fishy, cucumber *etc.* [1].

D. Colour of Fat [1d]

The colour of fat depends upon its carotene content and varies with the species, breed and feed of the animal. The yellow colour of cow milk is because of carotene. Buffalo milk does not contain carotene.

Sources

1. K.S. Sharma, Chemistry of milk, Agrimoon.com.

1a. Y. Shi, C. M. Smith and R. W. Hartel, Compositional Effects on Milk Fat Crystallization, Journal of Dairy Science, Volume 84, Issue 11, November 2001, Pages 2392-2401.

1b. C.R. Rachana and B. Surendra Nath, Crystallization Texture of Milk Fat and its Importance in the texture of dairy Products - A Review, 2021, https: //www. researchgate.net.

1c. Wikipedia.

1d. Y.Kotilinga Reddy, Yogesh Khetra and M.H. Sathish, Fat rich dairy products, AgriMoon.com.

2. *M.K. Srivastava, Handbook on Analysis of Milk, CBS Publications, New Delhi, 2010. ISBN 13: 9788123926681.*

2a. *M.K. Srivastava, Handbook on Analysis of Milk and Milk Products, Write and Print Publications, New Delhi, 2016. ISBN 13: 9789384649258.*

3. *D. Ramaswamy, A.V. Shibu and H. Gopi, Dairy Technologist 's Handbook, 1999, ISBN-13: ý 978-8185860442.*

4. *Manual of Methods of Analysis of oils and fat (Lab manual -2), Directorate General of Health Services, GOI, New Delhi.*

5. *FSSAI Manual of methods of analysis of foods- Oils and Fats, Lab Manal-2, 2015.*

General Properties of Compound Lipids [1,2,3]

The compound lipids are less in their concentration but they play a very important role in the stabilization of the fat globule being an important constituent of the fat globule membrane. **The compound lipids ate defined as lipids which on hydrolysis yield phosphates, carbohydrates and nitrogenous compounds like sphingosine, ethanolamine, serine, carbohydrates, choline, *etc.* in addition to the glycerol and fatty acids.**

i. Phospholipids

These lipids contribute approximately 1 per cent to the total lipid found in bovine milk. The phospholipids in milk are synthesized by the mammary cells via pathways that are common to other mammalian cells. The major glycerophospholipids are phosphatidyl ethanolamine, phosphatidyl choline, phosphatidyl serine, phosphatidyl inositol.

- The total phospholipid content of buffalo milk fat averages 21.04 mg/100 ml of milk, whereas for cow milk the corresponding value is 33.71mg/100 ml. The following table shows the concentration of various fractions of phospholipids present in buffalo and cow milk [1a]-

Table 14.14: Concentration of Phospholipid Fractions in Buffalo and Cow Milk [1a]

Sl.No.	*Phospholipid Fraction*	*Concentration (mg/100ml)*	
		Buffalo Milk	*Zebu Cow Milk*
1	Lecithin	7.29	10.01
2	Cephalin	9.12	15.57
3	Sphingomyelin	4.63	8.13

- Phospholipids are also found as lipoprotein complexes in skim milk. The skim milk phase may containing 30-50 per cent of the phospholipids in milk.

Roles/Functions of the Phospholipids

Inspite of available in less quantity, phospholipids play following very important roles:

- ✰ In the formation and secretion of the milk fat.
- ✰ In the forming a stable colloidal suspension or emulsions in aqueous solution.
- ✰ Play an important role during storage and processing of milk due to their susceptibility to oxidation of the polyunsaturated fatty acids. Their bipolar molecules and the relatively high concentration of unsaturated fatty acids.

Composition

The composition of the phospholipids in milk is discussed earlier under composition of Lipids in bovine milk. The following table shows the composition of phospholipids in bovine milk:

Table 14.15: Phospholipids Composition of Bovine Milk [3]

Sl.No.	*Phospholipids*	*Moles per cent*
1	Phosphatidyl choline	34.5
2	Phosphatidyl ethanolamine	31.8
3	Phosphatidyl serine	3.1
4	Phosphatidyl inositol	4.7
5	Sphingomyeliun	25.2
6	Plasmlogen	3
7	Diphosphatidyl glycerol	Trace
8	Ceramides	Trace
9	Cerebrosides	Trace
10	Gangliosides	Trace

- ✰ In milk the glycerophospholipids are found predominantly in the diacyl form.
- ✰ However small amounts of plasmalogen which are the vinyl ether form of glycerophospholipid are also being reported.
- ✰ The studies carried out by some research workers revealed the followings:
  - **i.** 4.0 per cent of the phosphatidylethanolamine was in the ether form and 1.3 per cent of phosphatidylcholine was in this form.
  - **ii.** It was also observed that 1.3 to 2.5 per cent of bovine phospholipids are plasmalogens.

Some compounds that are being found as phospholipids in milk lipids which are suspected to be the degradation products or remnants of biosynthesis of lipid material in the mammary glands. Phosphotidic acid is a product formed by the action of phospholipase D on the phosphatidyl choline. The quantity of this product in milk is very low.

- ✰ The most abundant and best-known compound lipids in milk are the phospho-lipids or phosphatides, of which the phosphoglycerides makes up the major portion.
- ✰ Phospholipids or phosphatides are lipids which contain phosphorus esterified phosphoric acid. In other phosphoglycerids, the choline residue is replaced by ethanolamine (-CH_2-CH_2-NH_3^+),serine, or inositol (a cyclitol).

Amphiphilic Nature

The lipid becomes highly polar due to the two charged groups in the molecule where as the rest is very apolar, so that the molecule is amphiphilic and does not dissolve well in either water or fat.

- ✰ The phospholipids, however, form micelles in water and in fat, with the polar ends at the outside and inside, respectively.
- ✰ They are difficult to transform from the hydrated into the other form, and vice versa.
- ✰ Their amphiphilic nature makes the phospholipids very surface active. In milk, they are in contact with water and with proteins as "lipo proteins. "Little is known about the nature of the bonds in such complexes. The other details of Phospholipids – Lecithin, Cephalin *etc.* are also given under Compound Lipids of this chapter.

ii. Sphingolipids

In addition to phosphoglyceride, milk contains sphingolipids. Their common residue is sphingosine.

- ✰ In all of them, a fatty acid residue is bound to the.-NH group; on average these lipids contain very long-chain fatty acids.
- ✰ In sphingomyelin, a phosphocholine group (as in lecithin) is bound to the terminal oxygen; it is thus a phospholipid.
- ✰ The properties of the sphingolipids are comparable to those of the phosphoglycerides.
- ✰ The more details of Sphingolipids are also given under Compound Lipids of this chapter.

iii. Cerebrosides

In the cerebrosides or glycosyl ceramides a sugar group is bound. This is mostly glucose, and in milk lipids also lactose. Glucosyl and lactosyl (cerebrosides) compounds have been isolated from milk by some research workers. The quantities found by researchers are as follows:

a. 1.7 mg/100ml in the globule membrane. The membrane bound cerebrosides contained mainly acids of 20 to 25 carbons

b. 0.8 mg/10 ml in skim milk. Cerebrosides in skim milk contained 18 carbon or less.

☆ The more details of Cerebrosides are also given under Compound Lipids of this chapter.

Sources

1. *K.S. Sharma, Chemistry of milk, Agrimoon.com.*

1a. *Y. Kotilinga Reddy, Yogesh Khetra and M.H. Sathish Kumar, Rich Fat Dairy products, www.AgriMoon.com.*

2. *M.K. Srivastava, Handbook on Analysis of Milk, CBS Publications, New Delhi, 2010. ISBN 13: 9788123926681.*

3. *Wong et al., Fundamentals of dairy chemistry, 1988,*

Unsaponifiable Lipids [1,2,3]

The unsaponifiable matter of milk lipids is formed of sterols (mainly cholesterol, along with small amounts of 7-dehydrochole-sterol, lanosterol, dihydrolanosterol), hydrocarbons (alkanes, squalene),vitamins A, E, D), and alcohols [3]. The lipid compounds which are not saponifiable in the milk fat are considered under this fraction. The compounds in this fraction have unique properties hence the properties of these compounds needs a special mention [1,2]. The principal known compounds are cholesterol and related sterols, the fat soluble vitamins A,D,E and K, traces of squalene, unidentified waxes, and homologous series of n-alkylmethyl ketones containing odd number of carbons C3 to C15 [2].

Unsaponifiable Matter

Unsaponifiable matter is a group of lipids that don't break down to glycerols and free fatty acids in alkaline solutions. They consist of hydrocarbons, sterols, and aliphatic alcohols with high molecular mass, or can be imitation fat [4].

☆ When the milk fat is being saponified with an alcoholic base and ether extract obtained is known as unsaponifiable matter unsaponifiable lipids [2].

☆ The unsaponifiable matter consists of substances present in oils and fats which are not saponifiable by alkali hydroxides and are determined by extraction with an organic solvent of a solution of the saponified substance under examination [4].

The actual procedure for the saponification involves the following steps- [2,5]-

i. **Saponification**-When a weighed amount of fat in an excess of approximately 0.5N alcoholic base is heated for 1 hour would result in the decomposition of triglycerides and other ester materials decompose to fatty acid, salt, glycerol and small amount of other ester components.
ii. **Extraction-**The resulting soap solution is diluted with water and extracted under specified conditions with ethyl ether with the help of separating funnel.
iii. **Washing-**The ether extract is treated with some washing with successive quantity of water to remove impurities.
iv. **Evaporation**- The washed ether extract is evaporated and dried at 100-105°C for 30 minutes to constant weight. Cool in desiccator and weigh.

This dry residue represents the unsaponifiable matter which for milk fat generally falls within the range of 0.30 to 0.45 per cent by weight. These constituents are usually measured direct under mild conditions than those involved in the isolation of the unsaponifiable matter.

- ☆ As the above procedure involves heating for one hour in an alcoholic base, there is possibility of having some artifacts during the process along with the saponifiable matter.
- ☆ Shifting of double bonds of some of the unsaturated fatty acids is likely to occur during this process.

Sterols [2,3]

These compounds (sterols) are found in the unsaponifiable fraction of milk lipids and consists mostly of cholesterol with some lanosterol.

- ☆ Research workers who have come with dry saponification have reported that the content of sterols were 0.33 to 0.36 per cent of butter oil.
- ☆ Further investigations have revealed the presence of two new constituents named dihydrolanosterol and β sitosterol.
- ☆ Researchers were able to identify 17 sterols from bovine milk as trimethylsilyl ethers. Among these sterols five were identified and reported and their content as a per cent of the total sterol are given in Table 14.16.

Table 14.16: Main Sterols in Unsaponifiable Matter of Milk Lipids [2]

Sl.No.	Sterol	Per cent
1	Cholesterol	90.5
2	Campesterol	1.8
3	Stigmasterol	0.7
4	β sitosterol	0.2
5	Δ^5Avenasterol	0.03

- ☆ The aforesaid analyses were done by GLC mass spectrometry. The amount of cholesterol in whole milk fat was 13.49 ±1.01mg per 100 g of milk which contained 3.47± 0.74 g of fat.
- ☆ Detailed characterization of the unsaponifiable fraction of milk lipids by using cryogenically-modulated comprehensive 2D GC with dual MS/FID detection, was done by researchers and observed different sterols in different milch animals [3]. The results are given in Table 14.17:

Table 14.17: Different Sterols in Milk of different Milch Animals [3]

Sl.No.	Sterols	Buffalo (per cent)	Goat (per cent)	Butter (per cent)
1	Cholesterol	90.56	94.03	95.22
2	7-Dehydrocholesterol	0.97	0.59	0.49
3	Lathosterol	2.41	0.94	1.12
4	Desmosterol	0.90	0.39	0.23
5	Campesterol	0.62	0.23	0.40
6	Lanosterol	2.96	2.49	1.95

Cholesterol and its Properties [1,2]

Structure

Chemically, cholesterol is an organic compound belonging to the steroid family; its molecular formula is $C_{27}H_{46}O$ or $C_{27}H_{45}0H$. Cholesterol is a 27 carbon compound with a unique structure with a hydrocarbon tail, a central sterol nucleus made of four hydrocarbon rings, and a hydroxyl group. The center sterol nucleus or ring is a feature of all steroid hormones. It is a principal sterol of milk [1a,2].

Variation in Cholesterol in Milk [1,1b,2]

The amount of cholesterol in milk varies with the milk fat content in milk [1]. The cholesterol content of milk ranges from approximately 0.25 to 0.4 per cent by weight of the fat. Since the total unsaponifiable matter ranges from 0.3 to 0.45

per cent (by weight of the fat) giving an understanding that the unsaponifiable matter of milk fat is largely cholesterol [2].

The cholesterol content of milk is significantly affected by the species, breed, feed, stage of lactation and season of the year [1b].

i. Generally, the cholesterol content is higher in the western breeds of cattle (317-413 mg/100 g fat), followed by zebu (desi) cow (303-385 mg/100 g fat).

ii. It is lowest for buffalo (235-248 mg/100 g fat). Generally lowest at the beginning of the lactation period, the cholesterol content progressively rises to reach its highest level towards the end.

iii. It is quite high in colostrum, being 570 to 1950 mg per 100 g fat in the first milking after parturition, progressively declining to normal levels during subsequent milkings.

Biosynthesis

Cholesterol is biosynthesized by all animal cells and is an essential structural component of animal cell membranes.

- ☆ Cholesterol also serves as a precursor for the biosynthesis of steroid hormones, bile acid and vitamin D [1,1a]. It is also an important structural component of the animal cell membrane and can be biosynthesised by all animal cells. It is also known as cholesterin and cholesteryl alcohol. The formula for cholesterol is presented in following figure [1a,2]-

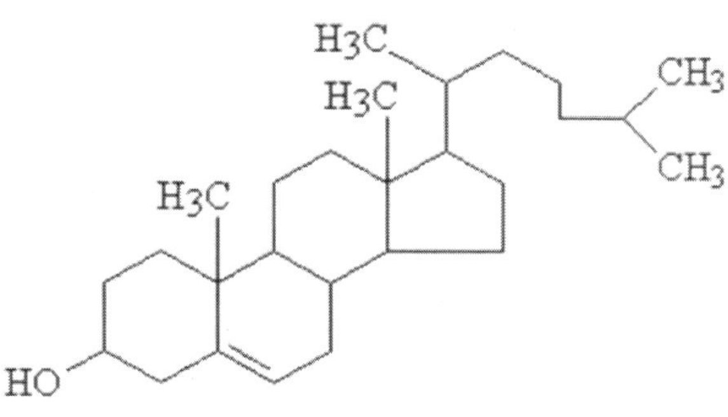

Figure14.10: Structure of Cholesterol.

- ☆ There are two types of cholesterol in the blood. **LDL(Low-density lipoproteins) or 'bad' cholesterol and HDL (High-density lipoproteins) or 'good' cholesterol**. Even though regular-fat milk naturally contains cholesterol, studies have shown that consuming dairy foods does not raise 'bad' cholesterol levels in the blood [1].

- ✰ Many of the methods used for the analysis of cholesterol, in milk are not specific for this sterol alone. There is a possibility that cholesterol values may include plant and animal hormone as well as other sterols all of which may be present in milk [2].
- ✰ The sterols are very a polar, but they easily associate with phospholipids, a small fraction of cholesterol is esterified, so this is, in fact, saponifiable. There are also conflicting results on the other, minor, sterols in milk [2].

Properties of Cholesterol [2,4]

Cholesterol is a yellow–white fatty chemical and appears to be present in true solution in the fat as part of the fat globules membrane complex and in complex formation with proteins in the non –fat portion of milk [2].Cholesterol has following important properties [2,4]-

- **i.** Cholesterol, a waxy substance that is present in blood plasma and in all animal tissues.
- **ii.** In its pure state it is a white (but yellow –white in milk),crystalline substance that is odourless and tasteless.
- **iii.** Cholesterol can be oxidized in various ways and cholesterol oxidized products (COP) are formed.
  - ✰ These oxidized products of cholesterol are found in animal foods including the dairy products.
  - ✰ A higher concentration of cholesterol oxidized products can be found only in processed dairy products exposed to harsh storage conditions where the impact of oxygen and light or oxygen and low water activity are associated [2].

Oxycholesterol or 5,6-epoxycholesterol is a form of oxidized cholesterol implicated in atherosclerosis. It is commonly formed from the reaction of fats and oxygen during high temperature cooking such as frying [1a,2].

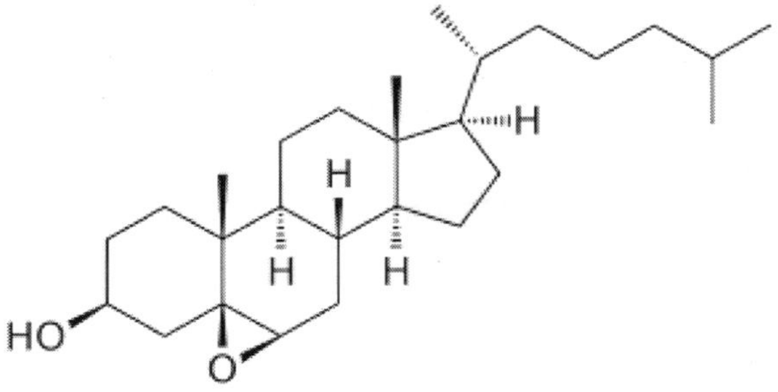

Figure 14.11: Oxycholesterol [1a].

Oxysterols are oxidized derivatives of cholesterol, which are important in many biological processes, including cholesterol homeostasis, sphingolipid metabolism, platelet aggregation and apoptosis.

Sources

1. *M.K. Srivastava, Handbook on Analysis of Milk, CBS Publications, New Delhi, 2010. ISBN 13: 9788123926681.*

1a. *Wikipedia.*

1b. *Y. Kotilinga Reddy, Yogesh Khetra and M.H. Sathish Kumar, Fat rich dairy products, www.AgriMoon.com.*

2. *K.S. Sharma, Chemistry of milk, Agrimoon.com.*

3. *Detailed characterization of the unsaponifiable fraction of milk lipids by using cryogenically-modulated comprehensive 2D GC with dual MS/FID detection, https: //www. shimadzu.com/an/industries/food-and-beverages.*

4. *Un-saponifiable matter, https: //www. webster. com/medical/unsaponifiable.*

5 *Determination of Unsaponifiable matter, https: //www.pharmaguideline.com.*

Vitamins [1,2,3]

Vitamins are organic substances essential for many biological processes. Vitamins are present in small amounts but these are vital for the health and growth [1].

- ☆ Vitamins may be defined as carbon compounds of diverse structures, which are not used for energy or fixed into the tissue frame work but which are required in minute amounts for the normal functioning of the living organisms.
- ☆ Different vitamins play different roles in the body, and a person requires a different amount of each vitamin to stay healthy. Vitamins are also recognized as nutrients for animals obtained from different foods *etc.*
- ☆ Most vitamins need to come from food because the body either does not produce them or produces very little. Vitamins are also synthesized by several other living organisms such as microorganisms like yeast and molds.
- ☆ Sun light also helps in the synthesis of vitamins like Vitamin D.
- ☆ Since milk being considered to be nearly complete food it is consumed by human beings of all ages.

Vitamins in Milk

Milk contains various fat soluble vitamins like vitamin A, D, E, K, along with water soluble vitamins like vitamin C and B complex vitamins and provitamins.

However, concentration of some of these vitamins is very less. The Tables 14.18 and 14.19 show the concentration of the vitamins in the milk,diseases caused by deficiency and vitamins in milk of different milch animals.

Table 14.18: Concentration of Vitamins in Milk and Diseases Caused by Deficiency [1,4]

Vitamin	*Contents per litre*	*Deficiency Diseases*
A (ug RE)	400	Night blindness
D (IU)	40	Ricket
E (ug)	1000	Sterility
K (ug)	50	Haemorrhagic Condition
B1 (ug)	450	Beriberi
B2 (ug)	1750	Chellosis, corneal opacity
Niacin (ug)	900	Disorder in stomach and nervous system
Pyridoxine B6 (ug)	500	-
Pantothenic acid (ug)	3500	-
Biotin (ug)Vitamin H	35	Dermatitis
Folic acid (ug)	55	Macrocyclic anaemia
B12 (ug)	4.5	Pernicious anaemia
C (mg)	20	Scurvy

Table 14.19: Concentration of Vitamins in Milk [3]

Fat Soluble Vitamins	*Concentration (Per Litre)*	*Water Soluble Vitamins*	*Concentration (Per Litre)*
Vitamin A	1590 IU*	Vitamin B_1	0.4 mg
Vitamin D	22.1 IU	Vitamin B_{12}	1.5 mg
Vitamin E	1.0 mg	Vitamin C	20 mg
Vitamin K	0.04 mg	Vitamin B_6 (Pyridoxin)	0.7 μg
		Pantothenic acid	3.0 mg
		Biotin	50 μg
		Niacin	0.2-1.2 mg
		Folic acid	1.0 μg
		Inositol	180 mg
		Choline	150 mg

* Aretinol equivalent (RE) is equal to 3.33 IU retinol or 10 IU β-carotene

The IU (abbreviation) stands for the international units which is the measurement for some vitamins. It is the same idea as milligrams, though they are not equivalent. These units exist in order to account for the fact that certain preparation of the same vitamin is not bioequivalent. Regardless of the preparation 1 IU will provide the same biological activity. *e.g.*, 1 IU of vitamin E equals the biological equivalent of about 0.667 mg d-alpha-tocaopherol or 1 mg of dialphatocopherol acetate. Though the mass of each preparation is different, the biological activity is the same.

Table 14.20: The Concentration of Vitamins in 100 gm of different Type of Fresh Milk [1]

Sl.No.	Vitamins	Units	Cow	Goat	Buffalo	Sheep	Human	Soy
1	Vitamin C	mg	0.9	1.3	2.3	4.2	5.0	0.0
2	Thiamine	mg	0.038	0.048	0.052	0.065	0.014	0.161
3	Riboflavin	mg	0.162	0.138	0.135	0.355	0.036	0.070
4	Niacin	mg	0.084	0.277	0.091	0.417	0.177	0.147
5	Pantothenic acid	mg	0.314	0.31	0.192	0.407	0.223	0.048
6	Vitamin B-6	mg	0.042	0.046	0.023	0.06	0.011	0.041
7	Folate	mcg	5	1	6	7	5	1.5
8	Vitamin B-12	mcg	0.36	0.07	0.36	0.71	0.05	0.0
9	Vitamin A	IU	126	185	178	147	241	32
10	Vit A (Retinol)	mcg	28	56	53	44	60	-
11	Vitamin D	IU	40000	12000	-	-	4000	-
12	Vitamin E	mg_ATE	0.1	0.09	-	-	0.9	0.01

Classes of Vitamins [1,2]

Milk includes both fat soluble vitamins and water soluble vitamins.On the basis of their solubility vitamins are distinguished into following two classes-

i. Class A-Fat Soluble Vitamins

Vitamins A, D, E and K are fat soluble vitamins. Vitamin A is derived from retinol and β -carotene. Because milk is an important source of dietary vitamin A, fat reduced products which have lost vitamin A with the fat, are required to supplement the product with vitamin A [1].

ii. Class B-Water Soluble Vitamins

The B-complex and vitamin C (Ascorbic acid) are water soluble vitamins. Milk is also an important source of dietary water soluble vitamins. The B-complex vitamins includes thiamine(B1), riboflavin (B2), niacin, pantothenic acid, pyridoxine (B6), biotin, choline, inositol, folic acid, p-aminobenzoic acid and Cyanocobalamin (Man made BI2). There is also a small amount of vitamin C (ascorbic acid) present in raw milk but is very heat-labile and easily destroyed by pasteurization. All of the above vitamins are present in milk, however, vitamin A and B_2 group vitamins are present in appreciable amounts.

1. Vitamin-A (Retinol) [1,1a,2]

Vitamin A is an organic (carbon-containing) compound that is required to keep the body's metabolism running smoothly. Vitamin A is a fat-soluble vitamin and that means it is absorbed into the body with dietary fat, and then stored in the body's own adipose tissue. First discovered in 1909, vitamin A is also known as retinol and retinoic acid. Two forms of vitamin A are available in the human diet-

i. **Preformed vitamin A** (retinol and its esterified form, retinyl ester).

ii. **Provitamin A carotenoids**.

Preformed vitamin A is found in foods from animal sources, including dairy products, fish, and meat (especially liver). By far the most important provitamin A carotenoid is beta-carotene; other provitamin A carotenoids are alpha-carotene and beta-cryptoxanthin [1a].

- ✰ The chemical names of vitamin A are retinol, retinal, and "the four carotenoids,
- ✰ Vitamin A is a primary alcohol derivative of a polymer consisting of four isoprene units. It is formed by the cleavage of provitamin A carotenoids, polymers of eight isoprenes, of these β carotene has the highest vitamin A potency because it consists of identical halves and thus yields two retinol molecules, while other carotenoids yield only one [2].
- ✰ Plants synthesize carotenoids but not retinol, animals cannot synthesize carotenoids but can split them to retinol (vitamin A).
- ✰ Vitamin A levels are frequently expressed in International Units (I.U.), although this unit is officially no longer accepted. Vitamin A is now measured in mcg RAE (Retinol activity Equivalent). One I.U. equals 0.344μg of crystalline vitamin A acetate, or 0.300 μg vitamin A alcohol; or 0.600 μg β-carotene. Current usage is to express vitamin A potency as retinol equivalents (RE) [2].

1 RE = 1 μg Retinol
= 6 μg β carotene
= 12 μg other provitamin A carotenoids
= 3.33 IU from retinol
= 10.00 IU from β carotene

To convert IU to mcg RAE, use the following [1a]-

1 IU retinol = 0.3 mcg RAE

1 IU supplemental beta-carotene = 0.3 mcg RAE

1 IU dietary beta-carotene = 0.05 mcg RAE

1 IU dietary alpha-carotene or beta-cryptoxanthin = 0.025 mcg RAE

☆ Vitamin A is present in milk as retinol, retinol ester and carotene. The content depends strongly on the amount of carotenoids in the feed.

i. The ratio of retinol to carotene in milk varies with breed and species of animals.

ii. There are several forms of provitamin A, belonging to the carotenoid pigments. The most important one is β-carotene and some of the pigments that can be derived from it are of practical importance. These are β-apo-8′ carotenal and β-apo-8′-carotenoic acid ethyl ester. For example, Apo carotenal (R = CHO) and apo-carotenoic acid ester (R=COOC2H5). Other provitamins are α-and γ-carotene s and cryptoxanthin.

Table 14.21: Vitamin A and Carotene Content in some Foods [3]

Sl.No.	*Foods*	*Vitamin A (IU/100g)*	*Carotene (mg/100g)*
A	**In Milk and Milk Products**		
1	Milk	110-307	0.01-0.06
2	Butter	2363-3452	0.43-0.77
3	Cheddar cheese	553-1078	0.07-0.71
B	**In Non Veg Foods-Meat, Fish, Egg *etc.***		
4	Egg boiled	165-488	0.01-0.15
5	Beef(Grilled Sirloin)	37	0.04
6	Herring (canned)	178	0.07
C	**In Vegetables**		
7	Tomato (canned)	0	0.5
8	Peach	0	0.34
9	Cabbage	0	0.3
10	Broccoli (boiled)	0	2.5
11	Spinach(boiled)	0	6.0

Factors affecting Stability of Vitamin A [1a,2]

i. Vitamin A is relatively stable to heat in the absence of oxygen, because of the highly unsaturated character of the molecule, It is quite susceptible to oxidation, especially in the presence of sunlight or artificial light [2].

ii. Vitamin A is unstable in the presence of mineral acids but stable in alkali. Vitamin A and the carotenoids have good stability during various food processing operations [2].

iii. Losses may occur at high temperatures in the presence of oxygen. These compounds are also susceptible to oxidation by lipid peroxides, and conditions favouring lipid oxidation will also result in vitamin A breakdown.

iv. Pasteurization of milk does not result in vitamin A loss but exposure to light will result in loss. It is essential, therefore, that sterilized milk be packed in light impervious containers [2]. Pasteurized milk showed non-significant decrease, however, boiling and sterilization (showed a significant difference in vitamin A content in comparison to control [1a].

vi. Loss of vitamin A (per cent) was significantly greater in iron-fortified milk suggesting that iron might have played a role in accelerating the degradation of vitamin A [1a].

v. Possible losses during storage of foods are more affected by duration of storage than by storage temperature.

vi. Vitamin A was light sensitive and its degradation significantly increased with an increase in the intensity of light exposure [1a]. **It has been found that vitamin A added to milk, is more easily destroyed by light than the native vitamin A**. This is not because natural and synthetic vitamin A is different, but because these two types of vitamin A are dispersed differently in the milk. The form in which vitamin A is added to food products may influence its stability. Vitamin A in beadlets form is more stable than that added as a solution in oil. The beadlets are stabilized by a protective coating. If this coating is damaged by wetting with water, the stability of the vitamin is greatly reduced [2].

Importance of Vitamin-A

Vitamin A is very important vitamin which plays significant role in different functions. The details are discussed below:

i. Vitamin A is involved in immune function, vision, reproduction, and cellular communication [1a]. It is essential for night vision and for healthy skin, hair, nails, and bones. Vitamin A is an antioxidant.

ii. Vitamin A is important in the diet as it is necessary for growth, health and reproduction. It keeps the epithelial tissues healthy and thus aids in preventing infections [2].

iii. Vitamin A is critical for vision as an essential component of rhodopsin, a protein that absorbs light in the retinal receptors, and because it supports the normal differentiation and functioning of the conjunctival membranes and cornea [1,1a]. It has an important role in the vision of mammalian organisms. It acts to initiate vision from light energy [2]. The deficiency of Vitamin-A may cause night blindness and keratomalacia, which causes the clear front layer of the eye to grow dry and cloudy.

iv. Vitamin A aids in maintaining normal glandular functions such as maintenance of adrenal cortex and steroid hormones synthesis.

Vitamin A is present as carotene in plant and must be converted to vitamin A for its function.

v. Vitamin A is a growth factor and it also supports cell growth and differentiation, playing a critical role in the normal formation and maintenance of the heart, lungs, kidneys, and other organs [1a].

2. Vitamin D [1b,1c,2]

The chemical names of Vitamin -D are ergocalciferol, cholecalciferol. Vitamin D is a group of fat-soluble secosteroids responsible for increasing intestinal absorption of calcium, magnesium, and phosphate, and many other biological effects.

- ✰ Several forms (vitamers) of vitamin D exist. The two major forms are vitamin D2 or ergocalciferol, and vitamin D3 or cholecalciferol. In humans, the most important compounds in this group are vitamin D3 (also known as cholecalciferol) and vitamin D2 (ergocalciferol) [1c].
- ✰ Vitamin D without a subscript refers to either D2 or D3, or both, and is known collectively as calciferol [1c]. It includes a number of compounds with antirachitic activity.
- ✰ Ergocalciferol (D_2) occurs in plant material and is formed by UV irradiation of ergosterol. Cholecalciferol (D_3) is formed by UV irradiation of 7-dehydro-cholesterol in animals particularly in the skin.
- ✰ Milk contains both forms of this vitamin, D_2 coming from the feed and D_3 from their radiated skin of the cow.
- ✰ The recent research shows, that the active form of vitamin D is identified as 1,25-Di-hydroxycholicalciferol which triggers the synthesis of calcium binding proteins for the calcification of bones. Thus this vitamin prevents disease called rickets in children and osteomalacia in pregnant women and osteoporosis in old.
- ✰ The potency is sometimes expressed as international units where, 1 IU is 0.25ng of cholecalciferol. Average value of this vitamin in bovine milk is 13.0 and 31.2, I.U. for winter and summer milk respectively.

Factors affecting the Potency of Vitamin D [2]

The potency of the vitamin D in milk can be increased by the followings-

i. Exposure of cow to sunlight
ii. Feeding higher levels of irradiated plant material
iii. Irradiating milk
iv. Fortification in milk with vitamin D.

Stability of Vitamin D in Milk

It is stable in milk, not affected by pasteurization, boiling or sterilization. Storage of fluid milk for 30 months at -17.8°C, irradiated evaporated milk for 2 to 3 years or of frozen butter for over 2 years resulted in little or not loss of the vitamin.

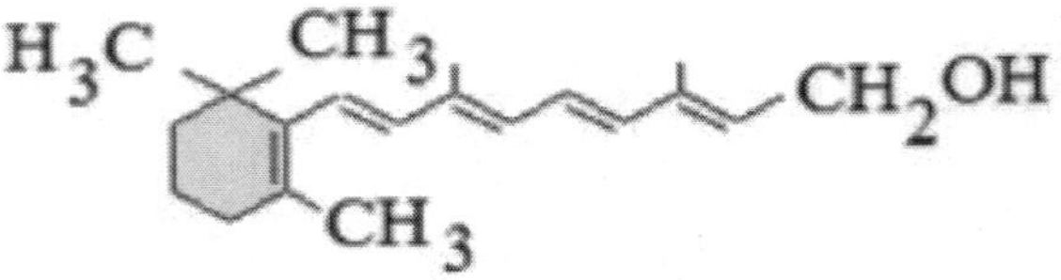

Figure 14.11: Structure of Retinol

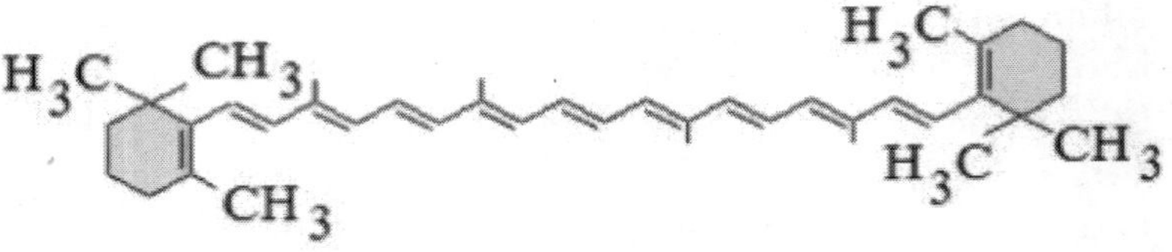

Figure 14.12: Structure of β- Carotene

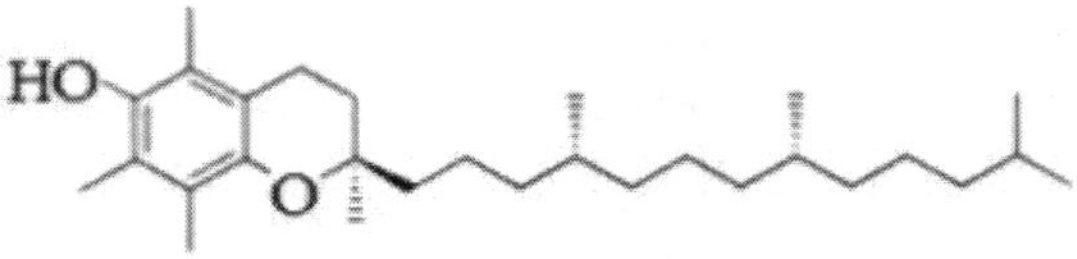

Figure 14.13: Structure of Vitamin E

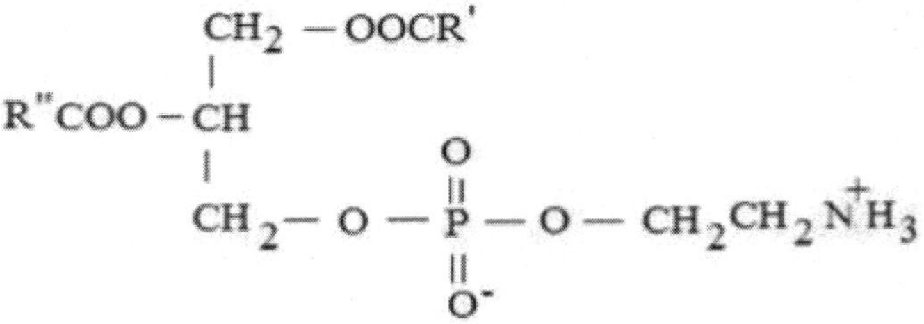

Figure 14.14: Structure of Ethanolamine

Importance [1b,2]

This Vitamin also plays very important role in making strong bones and preventing rickets. It is necessary for the healthy mineralization of bone. The deficiency of Vitamin-D may cause rickets and osteomalacia, or softening of the bones.

i. Vitamin D is necessary to aid in calcium and phosphorus retention so as to make strong bones and prevent rickets. Getting enough vitamin D helps to protect children from rickets and older adults from osteoporosis. Vitamin D works in concert with calcium to help build and maintain strong bones. It increases tubular preabsorption, and citrate blood levels [1b,2].

ii. It maintains and activates alkaline phosphatase in bones and serum calcium and phosphorus level.

iii. Vitamin D is sometimes called the sunshine vitamin because the ultraviolet radiation of the sun can convert an active precursor of the skin (7-dehydrocholesterol) to vitamin D [2].

iv. Children need vitamin D for their growth and to a lesser extent mothers require this vitamin during pregnancy and lactation.

v. Vitamin D is important to the body in many other ways as well. Muscles need it to move, for example, and nerves need it to carry messages between the brain and every body part. In addition, a growing body of evidence supports the potential benefits of vitamin D. For example, it may help to reduce the risks of high blood pressure, diabetes, cardiovascular disease and cancer [1b]

3. Vitamin E [1,2]

The chemical names of Vitamin-E, are tocopherol, tocotrienol. The lipophilic Vitamin E (tocopherol) is an important fat-soluble vitamin that plays several roles in the body. Vitamin E was discovered in 1922, isolated in 1935 and first synthesized in 1938.

- ☆ Vitamin E is a group of eight fat soluble compounds that include four tocopherols and four tocotrienols. [1c].
- ☆ Vitamin E naturally exists in eight different forms (alpha-, beta-, gamma-, and delta-tocopherols and tocotrienols) each of which has slightly different activity in the body. The most potent of these and the principal one in milk is α-tocopherol $C_{29}H_{50}O_2$ this is strong reductant and serves as an antioxidant protecting lipids from oxidation [2].
- ☆ Alpha-tocopherol is considered the most active natural form because it is the preferred form of vitamin E transported and used by the liver. The most biologically active form of the vitamin- E is called alpha-tocopherol [2a].
- ☆ Alpha-Tocopherol was the major form of vitamin E (>85 per cent) where as gamma-tocopherol and alpha-tocotrienol were present to a lesser extent.
- ☆ As the fat contents of milk products decreased from 11 to 0.3 per cent, the vitamin E contents decreased. Raw milk as compared to non fat milk had both higher and higher total lipids. Vitamin E, cholesterol, and total lipids increased as cream was added back to non fat milk during production.

☆ The vitamin E content is rather low. [1,2b]. The vitamin E content of human milk is about ten times that of cow milk [2].

Factors affecting Vitamin E in Milk [2.2c]

Vitamin E content in milk depends upon the followings-

i. It depends to some extent on feed and breeds of milch animals [2,2c].
ii. Summer milk has higher concentration than winter milk.
iii. It is stable to heat but may be partly destroyed by intensive illumination in the presence of $O_{2.}$
iv. The vitamin E content of milk is not influenced by pasteurisation and evaporation but a loss of 9 per cent may occur after drying and reconstitution [2c].

Importance of Vitamin E [2,2a]

The deficiency of Vitamin -E, is rare, but it may cause hemolytic anemia in newborns. This condition destroys blood cells.

i. It is a strong antioxidant, prevents oxidation of unsaturated fatty acids and vitamin A in intestinal tract and body tissues. Its most important role is as an antioxidant that works closely with its partners Selenium and Vitamin C to protect the body from the oxidative stress of exercise, illness and certain medical conditions [2,2a]. Its antioxidant activity helps prevent oxidative stress, an issue that increases the risk of widespread inflammation and various diseases.
ii. It acts in oxidation reduction reactions, plays role in human nutrition and is associated with reproductive factor, necessary for nutrition of muscles and better utilization of vitamin A [2].
iii. Vitamin E also helps keep the immune system strong. It aids the body in properly using vitamin K, helping maintain normal blood viscosity [2a].
iv. It maintains integrity of vascular system and central nervous system.
v. Vitamin E is a detoxifying agent and maintains kidney, tubules, lungs, genital structure, liver and red blood cell membranes [2].
vi. Vitamin E may be beneficial at reducing UV damage to skin.
vii. Vitamin E is used for treating vitamin E deficiency, which is rare, but can occur in people with certain genetic disorders and in very low-weight premature infants.
viii. Too much vitamin E (over 10,000 IU per day) can be dangerous and may increase the risk of, as well as interfere with, the uptake of Vitamin A [2a].

4. Vitamin K [1c,2]

Vitamin K refers to structurally similar, fat-soluble vitamers found in foods and marketed as dietary supplements. The human body requires vitamin K for post-synthesis modification of certain proteins that are required for blood coagulation or for controlling binding of calcium in bones and other tissues [1c].

- ✰ It is present only in traces in milk if at all. Human needs for this vitamin are supplied from consumption of plant materials containing it and by microbial synthesis in the digestive track [2].

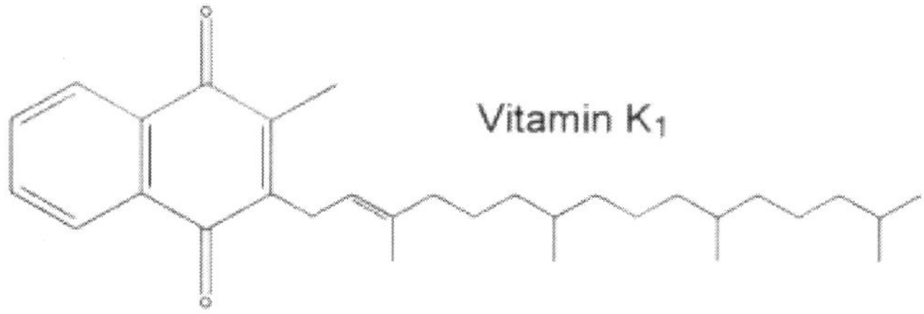

Vitamin K_1

Figure 14.15: Vitamin K_1 (phylloquinone) – both forms of the vitamin contain a functional naphthoquinone ring and an aliphatic side chain. Phylloquinone has a phytyl side chain.

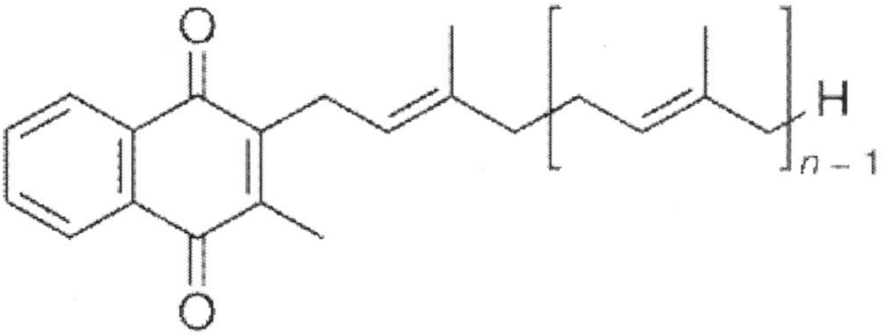

O O H $n-1$

Figure 14.16: Vitamin K_2 (menaquinone). In menaquinone, the side chain is composed of a varying number of isoprenoid residues. The most common number of these residues is four, since animal enzymes normally produce menaquinone-4 from plant phylloquinone.

- ✰ Chemically, the vitamin K family comprises 2-methyl-1,4-naphthoquinone (3-) derivatives. Vitamin K includes two natural vitamers: vitamin K_1 (phylloquinone) and vitamin K_2 (menaquinone). Vitamin K_2, in turn, consists of a number of related chemical subtypes, with differing lengths of carbon side chains made of isoprenoid groups of atoms [1c].
    - i. Plant-sourced forms are primarily vitamin K_1. It is primarily from plant especially leafy green vegetables. Small amounts are provided by animal-sourced foods.
    - ii. Animal-sourced foods are primarily vitamin K_2.

Table 14.22: Vitamin K (Animal-sourced K_1) in Milk and Milk Products [1c]

Milk and Milk Products	*K_1 (μg/measure)*	*K_2-MK-4 to MK-7 (μg/100g)*
Cheese diced, 1/2 cup	1.4-1.7	—
Cheese Hard	—	8-10
Cheese Soft	—	3.6
Yogurt whole milk, 1.0 cup	0.4	—
Milk whole or low fat, 1.0 cup	0.2	—
Human milk, per liter	0.85-9.2 (median 2.5)	—
Milk, whole	—	0.9
Milk, skim	—	—
Yogurt, whole milk	—	0.9
Butter	—	15

Deficiency [1c,5]

i. Low levels may cause an unusual susceptibility to bleeding, or bleeding diathesis.

ii. **Causes of Deficiency**-The most common causes of vitamin K deficiency are insufficient dietary intake, inadequate absorption, and decreased storage of the vitamin due to liver disease, but it may also be caused by decreased production in the intestines [5].

Normal diets are usually not lacking in vitamin K and hence, primary deficiency is uncommon in healthy children and adults. One exception is infants, who are at an increased risk of deficiency regardless of the vitamin status of the mother during pregnancy and breast feeding because of poor transfer of the vitamin to the placenta and low amounts of the vitamin in breast milk [1c].

Factors affecting Vitamin K [6]

Food production practices, bacterial strain selection in particular, are likely to provide a relevant means of altering the Vitamin K (MK) content of fermented foods.

i. **Fermentation** can also be used to improve the nutritional qualities of food products, which can include increasing Vitamin K (MK) content.

   ☆ Current knowledge indicates that dairy products are poised to be primary contributors to dietary MK intake, and industrial application of MK-producing bacteria in fermented dairy products may provide an effective means of increasing MK in the food supply.

- ☆ Vitamin K_2 levels in dairy products are coming from the bacteria used during fermentation.
- ☆ Fermented products like buttermilk, cheese and yoghurt will be the main dairy sources for vit K_2. The amount of vit K_2 in the product depends for example on the type of bacteria used, the fermentation conditions, the type of dairy products and the fat levels in the product.

ii. **Pasteurization** has no direct effect on vit K_2, but it does affect the activity of the bacteria present in the product. Since some bacteria produce vitamin K_2 and others use vitamin K_2 the net effect of pasteurization will again depend on many factors.

Effect of Processing on Milk Vitamins [1]

Milk processing *e.g.* Pasteurisation and Sterilization have effects on vitamin contents. Pasteurisation carried out with reasonable care has no effect on vitamin A. carotene, riboflavin and a number of remaining vitamin B and vitamin D. Of the remainder, a 10 per cent loss of thiamine and 20 per cent loss of ascorbic acid may be expected. Milk sterilization increases the loss of thiamine and ascorbic acid to 30-50 per cent and 50 per cent respectively, though the remaining vitamins are little effected.

Sources

1. *M.K. Srivastava, Handbook on Analysis of Milk, CBS Publications, New Delhi, 2010. ISBN 13: 9788123926681.*

1a. *Vitamin A-Fact Sheet for Health Professionals, National Institute of Health, U.S.*

1b. *Why Vitamin D in milk is good choice, https: //milklife.com/articles/nutrition/how-to-get-benefits-of-vitamin-d-in-milk.*

1c. *Wikipedia.*

2. *K.S. Sharma, Chemistry of milk, Agrimoon.com.*

2a. *Vitamin E: Synthetic or Natural?, https: //www.Poulingrain.com.*

2b. *S Kaushik, R Wander, S Leonard, B German and M G Traber, Removal of fat from cow's milk decreases the vitamin E contents of the resulting dairy products, Lipids, 2001 Jan; 36(1): 73-8.*

2c. *Assefa Bezie. The Effect of Different Heat Treatment on the Nutritional Value of Milk and Milk Products and Shelf-Life of Milk Products. A Review. Dairy and Vet Sci J. 2019; 11(5): 555822.*

3. *M.P.Mathur, D.D. Roy and P. Dinakar, Text book of Dairy Chemistry.*

4. *Manual of Dairy chemistry, University of Guelph.*

5. *Vitamin K Deficiency, https: //labtestsonline.org.*

6. *Barbara Walther, J. Philip Karl, Sarah L. Booth, and Patrick Boyaval, Menaquinones, Bacteria, and the Food Supply: The Relevance of Dairy and Fermented Food Products to Vitamin K Requirements, Adv Nutr. 2013 Jul; 4(4): 463-473.*

Chapter 15

Salts in Milk

The incineration of milk leaves the ash which equal to 0.70-0.80 per cent of the weight of milk. The ash includes all the minerals but not all of its salt components. Organic salts, particularly citrates disintegrate during incineration. Calcium citrate is the most abundant in fresh milk followed by Potassium dihydrogen orthophosphate, Sodium chloride *etc.* [1].

- The salt composition in milk cannot be expressed either as an equivalent to inorganic or organic substance because of the fact that some salts exist as organic substances also. It is also not possible to express them as ionic substances as a considerable part of these minerals are present in non ionized form.
- The salts of the milk include those constituents,that are present as ions or in equilibrium with ions, except hydrogen and hydroxyl ions. The ash content of the milk also does not represent the salt composition since during ashing of milk, some of the salts are lost due to volatilization. It is also possible that some non salt substances get included in ash [2].
- Minerals and trace elements in milk occur as inorganic ions and form complexes with proteins, peptides, carbohydrates, fats and small molecules. The main mineral binder or chelators of calcium are the caseins, alphas 1-casein, alphas 2-casein, beta casein and kappa casein, but also whey proteins and lactoferin bind specific minerals like calcium, magnesium, zinc, iron, sodium and potassium [1c].
- Colostrum is high in Ca, Cl,Mg,P, and Na but low in K. Mastitis milk contains as much as three times the normal chlorine content. This milk also contains high Ca and low K. The salt balance of milk refers to the

chemical balance between the calcium and magnesium on the one hand and the phosphates, citrates and carbonates on the other hand. Any disturbance of one of these upsets the balance and reduces the heat stability of milk [1,3].

Importance of Salts

Salts are important from nutritional point of view. They also largely determine the physicochemical state of the casein micelles, which in turn influences heat stability of the milk.

- ✰ Some metals acts catalysts in oxidation of milk fat leading to oxidative rancidity. Certain metals like Ca, P, Mg, Mo, Co, Fe, *etc.* are components of metallo proteins such as lactoferrin, lactoperoxidase, xanthine oxidase and parts of phospholipids and certain vitamins.
- ✰ The minerals in human, cow and buffalo milk plays an important role in the availability of these minerals for the young ones. They also play an important role in the digestibility of milk proteins [2].
- ✰ All minerals, considered to be essential to the human diet, are present in milk, so milk as major source of nutrients in child and adolescent nutrition is a good source of essential minerals [1c].
- ✰ All 22 minerals considered to be essential and important to the human diet are present in milk. These minerals make following three families of salts [1]:
  - i. **Sodium (Na), Potassium (K) and Chloride (Cl)-**These free ions are negatively correlated to lactose to maintain osmotic equilibrium of milk with blood. Salt contents of milk is about 0.90 per cent by weight.
  - ii. **Calcium (Ca), Magnesium (Mg), Inorganic Phosphorous (Pi) and Citrate**- consists of 2/3 of the Ca, 1/3 of the Mg, 1/2 of the P(i), and less than 1/10 of the citrate in colloidal form and present in the casein micelle.
  - iii. **Diffusible salts of Ca, Mg, citrate, and phosphate salts**-pH dependent and contribute to the overall acid-base equilibrium of milk.

Factors affecting Minerals/Salt in Milk [1b,1c,2]

The mineral concentration/content in raw milk vary according to the following different factors:

- i. **Factors related to Milch animals**-Such as lactation period/stage of lactation, animal species, breed and health status, Udder infection.
- ii. **Factors related to Processing**-The processing conditions can have a significant effect on content and ratio of mineral compounds in milk.

iii. **Factors related to environment-**Season, climate, environmental conditions during pasture, climate to post-milking handling, transportation.

iv. **Factors related to feed-** Feeding, dietary composition of animal feed.

Composition of Minerals [1,2]

Minerals occur in all foodstuffs as well as in milk and dairy products. Sometimes minerals are classified in to following two groups [1c]-

i. **Macro-minerals-**Sodium, Potassium, Chloride, Calcium, Magnesium and Phosphorus.

ii. **Trace elements -**Iron, copper, zinc, manganese, selenium, iodine, chromium, cobalt, molybdenum, fluoride, arsenic, nickel, silicon and boron.

All of them are present in milk at some concentration [1c]. Not all of the salt constituents are found in the dissolved state in milk. Minerals or milk salts, are traditional names for a variety of cations and anions within bovine milk. Calcium, phosphate, magnesium, sodium, potassium, citrate, and chloride are all included as minerals and they typically occur at concentration of 5-40 mM [1a]. The most important salts of bovine milk are given in a Table 15.1

Table 15.1: The Mineral Content of Fresh Milk [1,4]

Mineral	*Content per litre*	*Mineral*	*Content per litre*
Sodium (mg)	350-900	Fluoride (ug)	30-220
Potassium (mg)	1100-1700	Selenium (ug)	5-67
Chloride (mg)	900-1100	Cobalt (ug)	0.5-1.3
Calcium (mg)	1100-1300	Chromium (ug)	8-13
Magnesium (mg)	90-140	Molybdenum (ug)	18-120
Phosphorus (mg)	900-1000	Nickel (ug)	0-50
Iron (ug)	300-600	Silicon (ug)	750-7000
Zinc (ug)	2000-6000	Vanadium (ug)	tr-310
Copper (ug)	100-600	Tin (ug)	40-500
Manganese (ug)	20-50	Arsenic (ug)	20-60
Iodine (ug)	260		

☆ Calcium, magnesium, phosphate, and citrate are partitioned between the solution phase and the colloidal casein micelles. For analytical purposes, partition of the salt constituents can be achieved by equilibrium dialysis or by pressure ultra filtration.

☆ The salt constituents in the dissolved state (ultra filterable or diffusible) interact with each other to form various complexes. The concentrations of each of these constituents can be calculated (with suitable computer programs) from a knowledge of their several interaction or association constants [2]. The following tables show the proportion of the several constituents found in the dissolved and diffusible state:

Table 15.2: Overall Composition of Minerals in Bovine Milk [2]

Sl.No.	*Constituent*		*Concentration*	
	Ions	*Mineral*	*Mean (mg/100g)*	*Range (mg/100g)*
1	Cationic	Sodium	58	47-77
		Potassium	140	113-171
		Calcium	118	111-120
		Magnesium	12	11-13
2	Anionic	Total Phosphorus	74	61-79
		Inorganic Phosphorus	63	52-70
		Ester Phosphorus	11	8-13
		Chloride	104	90-127

Table 15.3: Concentration of Minerals (in Totality and in Dissolved state) in Cow and Buffalo Milk [1,6]

Constituents	*Phase (Unit)*	*Concentration*	
		Cow Milk Mean	*Buffalo Milk Mean ± Std Error*
Calcium	Total (mg/100gm)	123.0	175.5±1.15
	Dissolved (per cent)	39.2	20.93±0.939
Magnesium	Total (mg/100gm)	12.0	19.25±0.381
	Dissolved (per cent)	70.9	50.64±0.843
Citrate (As Citric acid)	Total (mg/100gm)	160.0	163.6±2.60
	Dissolved (per cent)	91.56	85.72±1.52
Phosphate (as P)	Total (mg/100gm)	95.0	97.34±1.28
	Dissolved (per cent)	44.10	32.98±0.65
Chloride	Total (mg/100gm)	119.0	62.84±0.933
	Dissolved (per cent)	100.0	99.51±0.454
Phosphorous	Total (mg/100gm)	141.0	103.0±0.913
	Dissolved (per cent)	96.22	95.15±0.322
Sodium	Total (mg/100gm)	58.0	44.60±0.402
	Dissolved (per cent)	94.0	95.10±0.439
Copper	Total (ppm)	0.130	0.219
Iron	Total (ppm)	0.47	1.24

Table 15.4: The Concentrations of Minerals in 100gms of different Types of Milk [7]

Sl.No.	Minerals	Units	Cow	Goat	Buffalo	Sheep	Human	Soy
1	Calcium Ca	mg	119	134	169	193	32	4
2	Iron Fe	mg	0.05	0.05	0.12	0.10	0.03	0.58
3	Magnesium Mg	mg	13	14	31	18	3	19
4	Phosphorous P	mg	93	111	117	158	14	49
5	Potassium k	mg	152	204	178	137	51	141
6	Sodium Na	mg	49	50	52	44	17	12
7	Zinc Zn	mg	0.38	0.30	0.22	0.54	0.17	0.23
8	Copper Cu	mg	0.011	0.046	0.46	0.046	0.052	0.12
9	Manganese Mn	mg	0.004	0.018	0.018	0.018	0.026	0.17
10	Selenium	mg	2	1.4	-	1.7	1.8	1.13

- Actually, phosphate is present in five classes of compounds:
  Inorganic dissolved, Inorganic colloidal, Water soluble, Esters, Ester bound in caseins, and Lipid.

These can be determined by making the following analyses:

- Total P in the dry or wet ashed sample-I,
- Lipid Pin digested Rose-Gottlieb extract-II,
- Dissolved P in digested ultrafiltrate-III,
- In organic dissolved P in undigested ultrafiltrate-IV and
- Acid soluble P in undigested 12.5 per cent trichloroacetic acid filtrate-V.

Then-Inorganic dissolved P = IV, Inorganic colloidal P = V – IV, Water soluble ester P = III – IV, Casein P = I - (II + V) and Lipid P = II. The total inorganic phosphate (Pi) is, of course, V.

- Na, K, and C1 are primarily present as free ions but Ca, Mg, phosphate, and citrate are distributed throughout many complexes; those in the highest concentration are CaCit-, Mg Cit-, H_2PO_4, HPO^{4-} and $CaHPO_4$. The calculation yields Ca^{2+} and Mg^{2+} concentrations of 2.0 and 0.8 mm respectively.

Trace Elements [1,2]

A large number of trace elements are found in milk. Although these elements are very less quantitatively they have significant influence in the various properties of milk and also play a significant role in human nutrition. These elements have been detected in normal bovine milk by spectroscopic and chemical analyses. They include a large number of metals, the metalloids such as, B, and Si, and the halogens F, Br, and I.

☆ Breast milk usually has adequate mineral and trace element contents for feeding full-term infants, with the exceptions of fluoride, for which supplementation of infants is recommended, and of selenium in some countries, such as Finland and New Zealand, where maternal intakes are low. However, breast milk selenium contents have increased in these countries in recent years due to increased maternal selenium intakes.

☆ The reported concentrations of the trace elements exhibit large variation. (*e.g.*, I, Mo, Zn), The concentration of elements like I, Mo, and Zn in the milk depends markedly on the diet consumed by the cow. The concentrations of some of them are increased by contamination With utensils and equipment to which milk is exposed while handling and processing.

☆ Some high values for the concentrations of certain trace elements may have resulted from contamination during laboratory analysis.

☆ In addition to the major salt constituents as discussed above, the elements listed in the following table are also present in milk as trace elements:

Table 15.5: Concentration of Trace Minerals in Bovine Milk [2]

Sl.No.	*Element*		*Concentration (µg/litre)*	
	Name	*Symbol*	*Range*	*Typical Value*
1	Aluminium	(Al)	150-1000	500
2	Arsenic	(As)	30-60	-
3	Barium	(Ba)	-	Trace
4	Boron	(B)	100-1000	300
5	Bromine	(Br)	500-20000	-
6	Cadmium	(Cd)	1-30	-
7	Cesium	(Cs)	-	Trace
8	Chromium	(Cr)	5-80	15
9	Cobalt	(CO)	0.4-1.0	0.5
10	Copper	(Cu)	10-200	75
11	Flourine	(F)	70-220	-
12	Iodine	(I)	10-1000	-
13	Iron	(Fe)	100-1500	300
14	Lead	(Pb)	20-80	40
15	Lithium	(Li)	-	Trace
16	Manganese	(Mn)	20-100	50
17	Mercury	(Hg)	-	Trace
18	Molybdenum	(Mo)	20-120	-

Sl.No.	Element		Concentration (µg/litre)	
	Name	Symbol	Range	Typical Value
19	Nickel	(Ni)	0.30	-
20	Rubidium	(Rb)	100-3400	-
21	Selenium	(Se)	4-1200	12
22	Silicon	(Si)	-	~1400
23	Silver	(Ag)	15-50	45
24	Strontium	(Sr)	40-500	170
25	Tin	(Sn)	-	Trace
26	Titanium	(Ti)	-	Trace
27	Vanadium	(V)	-	Trace
28	Zinc	(Zn)	2000-5000	3300

Molybdenum

It appears to be found exclusively in xanthine oxidase and Co in vitamin B_{12}.

Iron and Copper

It is an essential component of xanthine oxidase, lacto-peroxidase, and catalase. About half of the total Fe and 10 per cent of the Cu are in the fat globule membrane. Copper has been studied extensively in relation to oxidation of milk lipids.

Zinc

The trace metal present in highest concentration in milk is Zn; its concentration of 3.5 mg/liter is about 3 per cent of that of Mg, the major salt constituent present in lowest concentration. About 85 per cent of the Zn is associated with casein micelles. Alkaline phosphatase, a Zn-containing enzyme, is located primarily in the fat globule membrane but accounts for only a small fraction of the total Zn.

Manganese

It is required for fermentation of citrate by certain lactic acid bacteria, and with some milks the bacterial formation of diacetyl in cultures is inhibited by lack of sufficient Mn. Apparently, iodine is present in milk solely in the form of iodide ion; its concentration depends markedly on the amount consumed by the cow.

Distribution of some Salts between Casein Micelles and Serum

The casein micelles contain undissolved calcium phosphate and a little citrate, it is often called "colloidal calcium phosphate". Some cations, notably $Ca^{2+}$and Mg^{2+}, are associated with the negatively charged proteins. Small quantities of other ions (*e.g.* Cl^{-}) also may be bound to the proteins.

☆ Almost all of the salt is in the serum or in the casein micelles, and very little is bound to the fat globules. The distribution of salt between various casein micelles in serum is presented in the following table:

Table 15.6: Distribution of Salts between Casein Micelles and Serum [5]

Sl.No.	Compound	Milk (mg/100g)	Serum (mg/100g)	Dry Casein (mg/g)
1	Sodium (Na)	48	49	0.9
2	Potassium (K)	143	145	3.3
3	Calcium (Ca)	117	40	31.0
4	Magnesium (Mg)	11	8	1.5
5	Inorganic phosphate	203	116	37.0
6	Citrate	175	173	5.6

☆ The salt in the casein micelles can be dissolved by lowering the pH to about 4.6 or lower, and also, though slowly and incompletely, by dialysis against water or a calcium free solution.

Phosphorus

i. Milk contains phosphorus in many forms. It is present as orthophosphate, but part of it is bound to organic components like serine and threonine residues of casein, molecules of hexoses and glycerol, phospholipids, *etc.*

ii. Phosphorus is also found in the hydroxyapatite structure of the colloidal calcium phosphate in the micellar structure of casein.

Sulphur

The sulphur content of milk is about 0.36 g per kg, but most of it is in the amino acid residues methionine and cysteine of the proteins. About 10 per cent is present as inorganic sulphate. There is, of course, a significant variation between lots of milk.

Physical Properties of Milk and Salts [1,2]

The salt present in milk effects the physical properties of milk in following ways-

i. The electrical conductivity is largely determined by the salt solution.

ii. The pH of milk dialysate is a function of its salt composition and temperature. The presence of the proteins in milk does affect this pH, of course, as it affects the salt composition.

iii. The titratable acidity of milk depends both on its salt composition (largely phosphate content) and its protein content.

iv. The freezing point of milk follows from the total concentration of dissolved substances especially chloride and lactose.

Sources

1. *M.K. Srivastava, Handbook on Analysis of Milk, CBS Publications, New Delhi, 2010. ISBN 13: 9788123926681.*

1a. *Wikipedia.*

1b. *Physical Equilibrium among Milk Salts, http: //ecoursesonline.iasri.res.in.*

1c. *Nada Vahcic, Mirjana Hruskar, Ksenija Markovic, Mara Banovic and Irena Colic Baric, Essential minerals in milk and their daily intake through milk consumption, Mljekarstvo, 60(2), 77-85, 2010.*

2. *K.S. Sharma, Chemistry of milk, Agrimoon.com.*

3. *M.P. Mathur, D.D. Roy and P. Dinakar, Text book of Dairy Chemistry, 2005.*

4. *Manual of milk of University of Guelph.*

5. *Davis and White, J. Dairy. Res, 1960.*

6. *N. K. Roy and D.C. Sen, Text Book of Practical Dairy chemistry.*

7. *Fred Peaquod and Department of U.S. Agriculture.*

Physical Equilibrium among Milk Salts [1,2,3]

Milk contains several elements but all of them are not entirely in a soluble state. Some minerals exist in colloidal and ionic state at the normal pH of milk.

- ✰ The compounds which are in soluble condition play an important role in keeping various milk constituents in stable condition.
- ✰ A balance exists between the components which are in soluble state and those which are in colloidal state. This makes physical equilibrium among milk salts.

Salt Solution in Milk [1,1b,3]

The dissolved salts of milk are phosphate, citrate, chloride, sulphate, bicarbonate, sodium, potassium, magnesium calcium.

- ✰ Sodium and potassium in milk are present entirely in soluble state. It is necessary to, have physical equilibria among the milk salts for stability of milk, especially during heat processing.
- ✰ Colloidal particles of casein contain predominantly calcium, magnesium, phosphate and citrate.
- ✰ The chloride and sulphate are entirely present as the free ions Cl^- and SO_4^{2-} at the pH of milk.
- ✰ The salts of the various weak acids (phosphates, citrates, and carbonates) are distributed among various ionic forms.

☆ Calcium and magnesium form soluble ions with citrate, phosphate and bicarbonates.

- $H©^{-2} \rightarrow ©^{-2} + H^{+}$
- $©^{-2} + Ca^{+2} \rightarrow Ca©^{-}$
- $Ca\,©^{-} + H^{+} \rightarrow CaH©$
- $2\,Ca\,©^{-} + Ca^{+2} \rightarrow Ca_3©_2$

where, © =Citrate.

☆ A considerable portion of the calcium is held in the form of complex soluble ions such as $Ca©^{-}$. Similar complex ions are also formed with phosphate $CaPO_4^{\ -}$ and with bicarbonate $CaHCO_3^{\ +}$.

☆ The formation of such complex ions has the effect of reducing the concentration of calcium and magnesium ions in the solution. Only calcium salts are included but a similar scheme would apply to the magnesium salts.

Salt Balance [1,1b,3]

Salts in milk exist in colloidal and soluble form. Not all of the salt constituents are found in the dissolved state in milk. A large amount of salt is present in casein micelles structure and is in equilibrium with salts in the soluble phase of milk.

☆ A balance exists between the components which are in soluble state and those which are in colloidal state [1b].

☆ During acidification/fermentation of milk some salts gets solubilized from the colloidal state and come in to the soluble state. This destabilizes the casein micelle structure and tends to precipitate out as happens during curd setting of milk and preparation of acid casein.

☆ It also happens that the casein micelle stability is lost during heating of milk at high temperature such as in UHT processing, sterilization and boiling of milk. This is due to the in solubilization of colloidal calcium phosphate and final loss in the stability of milk, this in turn affects the quality of dairy products.

☆ Such changes in the salt balance also take place during manufacture of condensed milk, evaporated milk and milk powder.

The salt balance in milk is defined by the following equation:

$$\text{Salt Balance} = \frac{Ca^{+2} + Mg^{+2}}{\text{Citrate}^{-3} + PO_4^{-3}}$$

☆ It has been observed that the balance between certain salts such as Ca^{+2} and Mg^{+2} with citrate and phosphate in the colloidal and soluble phase plays a pivotal role in the stability of casein micelles structure.

☆ It appears a delicate balance cations and anions in the colloidal and soluble state is important in determining the stability of milk.

i. The monovalent cations such as Na^+ have a dispersing effect, while divalent and trivalent cations such as Ca^{+2} and Mg^{+2} have an aggregating effect.

ii. During heat processing of cow milk and buffalo milk the heat stability would be governed by the concentration of above ions in addition to other factors.

iii. Therefore, addition of sodium citrates and phosphate improves the heat stability of milk.

Colloidal salts remain in equilibrium with the dissolved salt (diffusate form). Various treatments of milk may cause transfer of salts from one phase to the other. Various methods have been used to study the partition of salts between dissolved and colloidal phases.

a. Two thirds of the calcium and one third of the magnesium and on half of the phosphorus are colloidal in normal milk.

b. It is presumed that all the colloidal phosphorus is present in the micellar calcium phosphate and it appears that all the colloidal citrate is similarly located where as colloidal calcium and magnesium are partly in corporate in milk colloidal phosphate and partly bound in a more direct manner to casein.

c. A small portion of colloidal calcium is bound to α lactalbumin. β lactoglobulin can also bind calcium and magnesium and there is other minor calcium binding protein.

d. Nearly all the sodium, potassium and chlorine are diffusible.

e. The principle dissolved salt constituents or milk consists of phosphate, citrate, chloride, sulphate, bicarbonate, sodium, potassium, magnesium and calcium. At pH 6.6 of milk sodium and potassium are not found in any combination with other constituents and are present as the cations Na^+ and K^+ while the chloride and sulphate being anions of strong acids are present as free ion Cl^- and SO_4^{-}

f. However the salts of the weak acids phosphate citrates and carbonates are distributed among various ionic forms.

g. Calcium and magnesium form soluble complex ions with citrate, phosphate, and bicarbonates.

Changes in Salt Balance

The following figure shows the changes in salt balance due to various treatments of milk:

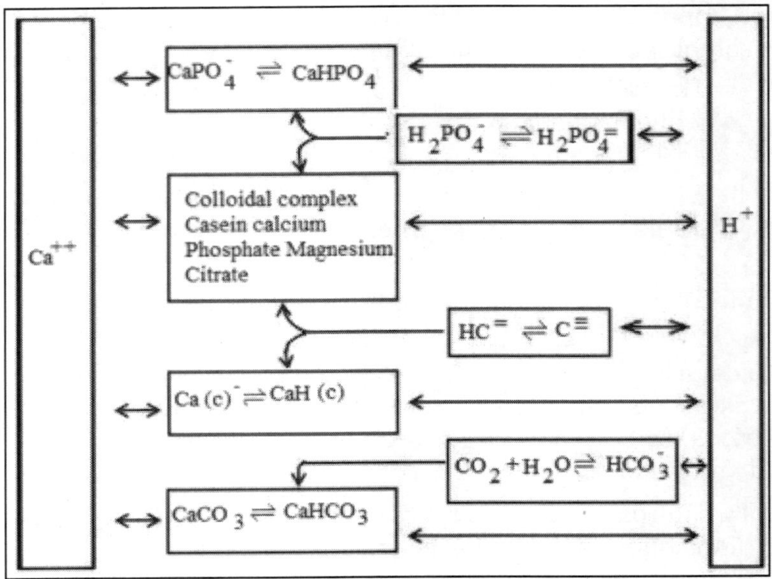

Figure 15.1: Equilibrium among Milk Salt [3].

Factors Influencing the Salt Equilibria [1,3]

The following various factors effect salt equilibria:

i. Temperature.
ii. Variation in acidity.
iii. Variation in carbon dioxide content.
iv. Concentration and
v. Sequestering agents and ion exchangers.

Acidity [1,3,5]

There will be pronounced shift in the salt equilibria with the addition of acid to milk whether directly or indirectly by bacterial action. Decrease in the pH withdraws the calcium and phosphate from colloidal particles until at about pH 5.2 all the calcium and phosphate is in the dissolved state. The acid production at elevated temperatures is the principal factor leading to the heat coagulation of milk.

- A variety of other heat-induced changes probably contribute to heat coagulation under normal circumstances but these can be overridden if the pH is maintained close to normal values by occasional neutralization of heat-induced acidity [5].

Carbon Dioxide Content [1,3]

Milk as secreted by the cow contains about 20 mg of CO_2 from milk is accelerated by heating and agitation. Removal of CO_2 would affect the balance in the rest of the system.

- ☆ It would be expected that the removal of CO_2 and the consequent rise in pH would be reflected in a shift in calcium phosphate from the dissolved to the colloidal state and probably also in a shift of calcium ions activity per 100ml or about 10 per cent by volume.
- ☆ This gas is rapidly lost from milk owing to the low content in the air. The loss is irreversible under ordinary conditions of handling.

Temperature [1,3,4]

The temperature will shifts the balance among the various forms as milk is subjected to various cooling and heating treatments after it is drawn from the cow at 37°C.

- ☆ With the rise in the temperature the dissolved will be transferred to colloidal phase. Similarly lowering the temperature below that at which the milk is drawn would cause a transfer of calcium and phosphate from the colloidal particles to the dissolved state [1,3].
- ☆ Cold storage increased soluble calcium and phosphate concentrations, and warm-up of the milk restored the initial ionic equilibria.
- ☆ More drastic heat treatments (80-90 degrees C) caused precipitation of tricalcium phosphate and calcium citrate [4].

Concentration [1,3,6]

As the milk is concentrated there is a tendency for calcium phosphate and calcium citrate to accumulate in the colloidal particles because the solubility is exceeded. As these materials are insolubilized, hydrogen ions are liberated, lowering the pH. The net result is an increase in concentration of citrate and phosphate in both the dissolved and colloidal state [1,3].

- ☆ The influence of milk concentration on the heat stability (determined as heat coagulation time (HCT) at 130°C) as well as the salt balance of buffalo milk was studied.
  - i. It was observed that buffalo milk was more stable than bovine milk when heated in its fluid state with HCT values of 32.3 min as compared to 31.3 min for cows' milk.
  - ii. However, concentration caused a greater destabilization of buffalo milk (HCT = 2.9 min) as compared to bovine milk (HCT = 6.6 min). The greater decrease in the HCT of buffalo milk may be attributed to a greater disruption of the salt equilibrium and a larger decrease in pH caused during concentration [6].

Sequestering Agents and Ion Exchangers [1,2.3]

In order to stabilize the milk or even to improve the utility of milk for a particular purpose it is desirable to treat milk so as to alter its ionic balance.

- ☆ The best known example for this is the addition of phosphate or citrate to stabilize milk against subsequent heat coagulation.
- ☆ This practice is common during the manufacture of evaporated milk, Addition of phosphate or citrate result in the binding of more of the calcium in the form of soluble complexes and decreasing activity of the calcium ions.

Sequestering Agents [1,2,3,7]

A sequestering agent is an organic compound capable of linking metal ions or molecules together to form complex ring-like structures known as chelates. Sequestering agents are used to link undesirable metal ions together to form a stable structure that does not readily decompose [2].

- ☆ Soluble reagents which thus tie up a particular ions are called sequestering agents and are said to sequester that ion. Ethylenediamine tetra acetic acid is on such sequestering reagent. It is very good reagent for divalent and polyvalent cations [1,3].
- ☆ The most important chelating agent in the food industry is EDTA (Ethylenediaminetetraacetic acid). EDTA is used as a stabilizing agent in the food industry. Chelating, or sequestering, agents protect food products from many enzymatic reactions that promote deterioration during processing and storage. These agents bind to many of the minerals that are present in food (*e.g.*, calcium and magnesium) and are required as cofactors for the activity of certain enzymes [7].

Ion Exchangers [1,1a,3,]

Ion exchange is a reversible interchange of one kind of ion present on an insoluble solid with another of like charge present in a solution surrounding the solid with the reaction being used especially for softening or making water demineralised, the purification of chemicals and separation of substances. Ion exchange is widely used in the food and beverage industry, *etc.* [1a].

Ion Exchange in Milk/Milk Products

Treatment of milk with ion exchange column will remove both anions and cations from milk by a process called "mono bed resin" treatments. It is mixture of cation exchange resin in the hydrogen form and an anion exchange resin in the hydroxyl form.

- ☆ When a salt containing solution is passed through such a mixture of resins both anions and cations are absorbed by the resin and hydrogen and hydroxyl ions are released which immediately combine to form water.

- ✰ By such treatments milk can be deionized to any extent desirable in a single pass through a resin bed or by batch wise treatment with a mixture of the two resins [1].
- ✰ The some examples of application of ion exchange in dairy industry, are given below [1b,8]:
    - i. **Milk [1b]**-Researchers have reported a method for the determination of Ca^{2+} and Mg^{2+} activity in milk using an ion-exchange resin. From this early work, it was established that with suitable resins exchange of salts was possible in milk, permeate and whey. This included the exchange of Ca and Mg from micelles by using Na^+ or K^+.
    - ii. **Ca-adjusted milk [1b]**-There is evidence that reduction of Ca^{2+} would also reduce the deposit formation that may occur during storage of UHT milk. Thus there is a need to establish commercial techniques to produce Ca-adjusted milks that could then be used as raw materials for industrial applications or product development work.
    - iii. **Ca-reduced single cream [1b]**- Cationic resins of the Zerolit 236 type have been used successfully in the production of Ca-reduced single cream. Such treatment prior to UHT processing was shown to increase physical stability when it was used as a coffee whitener.
    - iv. **Demineralised Whey [8]**-Ion exchange is used in the dairy industry in obtaining lactic acid for demineralization of whey, remove radioactive elements from raw milk, high heat stability of milk intended for sterilization to reduce its acidity. Whey is passed first through a cation exchange resin and then through anion exchanger. Cation binds cations present in the whey minerals, highlighting the appropriate acid anions are bound anion. Whey has high ash (Two time higher than milk) and bitter-salty taste which causes barrier to the widespread use of whey for food. After d demineralization,whey becomes more useful.

Sources

1. *K.S. Sharma, Chemistry of milk, Agrimoon.com.*

1a. *Wikipedia.*

1b. *Heva Ranjith, David Maw and Mike J. Lewis, Production of calcium-reduced milks using ion exchange, 1999, J Dairy Res 66(1): 139-44.*

2. *M.K. Srivastava, Handbook on Analysis of Milk, CBS Publications, New Delhi, 2010. ISBN 13: 9788123926681.*

3. *Jenness and Patton, Principles of dairy chemistry, 1959.*

4. *M Schmitt, F Saulnier, L Malhautier, G Linden, Effect of temperature on the salt balance of milk studied by capillary ion electrophoresis, J Chromatogr, 1993 Jun 25; 640(1-2): 419-.*

5. *P.F. Fox, Heat stability of Milk, Ir, J.Fd.Sci, Technol, 5, 1-11, 1981.*

6. *Mukul Tayal and Jagveer S. Sindhu, Heat stability and salt balance of buffalo milk as affected by concentration and addition of casein, Journal of Food Processing Preservation, 1983.*

7. *P. Michael Davidson, Co-editor of Food Additives, https: //www.britannica.com.*

8. *Prosekov A. Ju, Ul'rih E.V. and Budrik V.G., On ways demineralization of whey, http: //www.science-sd.com.*

Milk and Metals

The corrosive properties of milk are due not only to various mineral salts in solution but also due to its protein content. Acid/Acidic milks are more corrosive. High acidity of milk together with sugar,the effect is likely to be greater. The corrosive action of milk on metals is by no means a new observation. It has long been known that milk in contact with iron and copper will not only acquire a metallic taste, but corrode these metals readily. The corrosion of metals by milk is one of the most serious corrosion problems facing modern civilization and industry [1a].

- ✰ The solubility of metals in milk and other dairy products is having a great interest. After milking the milk is stored in a container which are usually made up of metals. As soon as milk comes in the contact of the metal, it starts reacting with the metal. If the container is of good quality, the milk will remain good and if it is not then, it is difficult to maintain the quality of milk and because of this, the metallurgy of the dairy plant is very different from the other industries.
- ✰ A slight change in the composition of the metal greatly affects the working quality of the equipment and wholesomeness of milk and milk products. Aeration favours corrosion and corrosion is greatest at the milk-air junction [1].

Milk [1]

i. The passage of any particular metal into milk may have two undesirable consequences in addition to corrosion. It may impart a bitter, metallic flavour and initiate the cycle of changes leading to an oxidized taints.

ii. Small traces of heavy metals, nickel, manganese, chromium in general and copper in particular exercise a powerful catalytic effect to develop an oxidative deteriorations. These metals find their way into milk through the worn coolers in which the tinned surface is no longer intact.

iii. Milk is thus in contact with tin and copper at the same time and copper dissolves under the influence of protein and form copper-protein complex, which become absorbed by the fat globules and in this manner, the metal is brought into contact with the fat.

Milk Products [1]

i. Cream, butter and whole milk powder are similarly sensitive to the catalytic effect of copper
ii. In case of whole milk powder, concentration of copper as low as 4 ppm has a strong catalytic effect on the oxidation reactions and again this is assisted by presence of moisture.

Milk Contact Surface and Metallic Contamination [1,2,3]

Food contact surfaces comprise all surfaces that may come into contact with food products during production, processing, and packaging.

☆ An oxidized flavour frequently occurs in milk and in practically all of its products. It is often designated as 'metallic' because its origin can usually be traced to the exposure of the milk to metal. This flavour varies in intensity in the different stages of its development, but usually the sequence of flavours is as follows: astringent, papery, metallic, metallic and oily, and, finally, tallowy [1b]. The following observations were made during study conducted [1b] on interaction of sweet milk with different metals:

  i. Copper, and copper alloys showed weight losses when exposed to sweet milk and produced oxidized flavours in such milk.
  ii. Tin-plated copper or copper alloys were unsatisfactory on account of mechanical wearing away of the plating,
  iii. Chromium-plated copper was less satisfactory in this regard than tin-plated copper,
  iv. Pure nickel showed high weight losses and became badly clouded when exposed to sweet milk. This metal often induced oxidized flavours in milk, and.
  v. The chromium alloys showed little or no weight losses and remained clear in appearance, but in certain milks induced slight oxidized flavours.

☆ Action of Milk on certain metals would dissolve small amount of the metal and form metallic salts which would give rise to a metallic taste to the milk. Some of the salts in milk have a catalytic action on the oxidation of milk fat resulting in the development of an oxidized flavour. These metals are said to taint milk [2].

The following observations were made on the basis of the several studies conducted on effect of milk on metals [1a]:

i. Copper, naturally present in small amounts in milk and other foods, is a physiologically necessary and important constituent.
ii. The natural copper content of milk varies between 0.3 and 0.7parts per million.
iii. Copper, nickel and copper-nickel alloys, when exposed to milk during pasteurization, may bring about an increase above 0.7 p.p.m, or a decrease below 0.3 p.p.m, in the copper content of the milk.
iv. This change in the copper content of the milk is more pronounced at elevated temperatures.
v. When milk in contact with nickel apparatus loses copper, this metal is precipitated out and nickel goes into solution.
vi. Nickel corrodes in milk more readily in the absence of oxygen.
vii. Copper corrodes in milk more readily in the presence of oxygen.
viii. High chromium nickel (18-8- 18 parts of chromium and 8 parts of nickel) iron alloys, "Hyblum," chromium plated copper and chromium plated nickel are very resistant to corrosion by milk and are satisfactory for dairy equipment on this account.

Action of Milk on Metals and Influencing Factors [1a,2,3]

The factors which influence the degree of action by milk on the metal are given below:

i. **Temperature of milk [1a,2,5]**- Various studies on the effect of temperature on metal solubility in milk have been reported. The temperature-solubility relationship has been studied most extensively for copper. Increasing the temperature of milk increases the acidity, consequently this factor may be one of the explanations why the corrosion rate of metals in milk increases at the higher temperatures.
ii. **Period of contact [2,3]**- As contact period of milk with metal increases, generally interaction increases.
iii. **Cleanliness and polish of metal [2,3]**- In spite of several precautions wide differences were found in the corrosion of nickel which were undoubtedly due to differences in the condition of the surface. Cleanliness and polish of metal play important role in corrosion caused by milk.
iv. **Amount of free air in milk [2,3,5]**-The viscosity decreases and the pH decreases with a higher temperature. Above all, the amount of oxygen dissolved in the milk is decreased considerably. In the presence of oxygen, nickel has a tendency to form a protective film of oxide which resists corrosive action. When the oxygen content of the milk drops

below a certain value, the corrosion of the nickel increases. At the same time as the oxygen or air in the milk is eliminated, any copper that is present in the milk has a tendency to pass from the cupric (Cu^{2+}) to the cuprous (Cu^{+}) state.

v. **Acidity of milk and Presence of sugar [2,3,5]-** The acidity has been generally assumed to be the most important factor in metal corrosion. The corrosion was much higher in ripened milk than in sweet milk but less than in the lactic acid solution. Acid/Acidic milks are more corrosive. High acidity of milk together with sugar the effect is likely to be greater [2,5].

Selection of metals for Dairy Equipment [2,3]

Milk will come in contact with metallic surface during the processing, storage and transportation of milk and hence the interaction between the two cannot be avoided. It is therefore essential to take in to consideration aforesaid aspect while choosing the metals for designing various equipment with which milk will be in direct contact.

The following important aspects need to be considered to meet the challenges arises in such situations-

i. **Non-tainting**- The metal should be non tainting when it is being used to make the equipment for processing, storage and transportation of milk. Aluminium will not taint the milk and being a Soft metal it is very easy to use. This metal is non tainting and is resistant to corrosive action by acid. Chromium and its alloys metal is non-tainting but expensive.

ii. **Non-toxic** - Arsenic, Cadmium, Hexavalent Chromium, Lead and Mercury are toxic metals. In dairy industry the metals which should not imparts any toxicity to milk, are being used for handling of milk.

iii. **Insoluble (in milk or its products)-** It is recommended that the metals used form making dairy equipment for milk handling, should insoluble in milk.

iv. **Highly resistant to corrosion**-The metal should be highly resistant to corrosion by milk and cleaning and cleaning and sanitizing agents. Chromium and its alloys is resistant to corrosive action by acid and alkaline cleaners.

v. **Meet the quality standards of regulatory authorities-**Metal to be used for milk processing should meet the quality standards prescribed by National (BIS, FSSAI *etc.*) and International Agencies (Codex *etc.*).

In addition to the above captioned important aspects, following aspects need to be considered:

i. Easy to clean and keep bright.

ii. Light yet strong,
iii. Good agents of heat transfer,
iv. Good in appearance throughout its use,
v. Low in cost.
vi. Non-absorbent and
vii. Durable.

It is not possible to get any single metal which can meet the entire above requirement, hence alloys of different metals like stainless steel is recommended for use in the dairy industry. The most satisfactory combination of metal that could be used is 18:8 stainless steel and aluminium alloy presently.

Characteristics of different Metals and their Alloys (for Dairy Equipment) [1,2,3,4]

The different metals and their alloys are recommended for application in making of different dairy equipment for milk handling. On its journey from the farm to the consumer, milk comes into contact with the walls of the equipment in which it is being processed or transported. The final product must reach the consumer meticulously clean and free from all germs. It is for this reason that the dairy industry has employed stainless steel for more than sixty years. The compact and smooth surface of stainless steel prevents bacterial contamination and facilitates cleaning and disinfection.

1. Aluminium and its Alloys [1,2,3]

One of the common single metal, second best to SS, is used in dairy. This metal and its alloys are often use in the preparation of milk cans, milk pails, linings for tanks and tankers, *etc.*

- ☆ Aluminium will not taint the milk and being a Soft metal it is very easy to work with this metal. If the metal is impure its durability is reduced. However if aluminium alloyed or anodized it is much more durable. It is non toxic and not imparts any flavour to milk. Aluminium is moderately soluble, but does not affect flavour materially. It can be easily repaired.

Advantages and Disadvantages

The greatest advantage of this metal is its light weight. The disadvantage of this metal is difficulty in cleaning due to its porosity, corrosion when alkaline dairy detergents and sanitizers are used. It is difficult to solder this metal.

2. Copper and its Alloys [1,2]

Milk utensils in villages are commonly made of earthenware, zinc plated or galvanized iron, brass or copper. In the normal milk, copper is bound with the fat as such this metal is not active chemically. Milk when comes in contact with copper, it will dissolve in milk and would exist in active form as copper ions.

- ☆ This metal is mostly used for the preparation of Milk pails, coolers, vats, strainers, pipe fittings, milk pumps, pasteurizer coils, *etc.* (used for tinning only).
- ☆ Due to the action of milk it will cause taints in the milk or the dairy products. For preventing such effect the surface is coated with tin. Due to the action of milk on this metal a green corrosion product is formed which is toxic for human consumption [2].

Advantages

i. The greatest advantage of this metal is that it is a Soft metal and it is easy to work with this metal. In order to increase the usefulness of this metal it could be used by having tin coating on the surface and this makes the metal more durable.

ii. Retinning of the metal is cheap and it will increase the makes it reusable for a considerable period.

iii. It high conductivity promotes rapid heat transfer and its cost is also reasonable.

3. Low-carbon Steel [2]

This low carbon steel is widely used for bodies of tanks, vats, bottle-washers, conveyors, *etc.*

4. Stainless Steel (18:8) [1,2,4a,4]

Today dairy equipment is made of stainless steels because it is easy to clean, disinfect and does not react with the lactic acids formed by fermenting milk.

- ☆ This alloy of iron is most widely used for all milk/dairy product contact surfaces. The pure iron taints milk/dairy product while stainless steel does not taint. The Rust caused to this metal in its pure state is slightly toxic.

Advantages

As this is a tough metal it will presents problems in fabrication but in spite of that SS has also following advantages:

i. The advantages in the use of stainless steel are highly resistant to corrosion by common acids and alkalis.

ii. Stainless steel is highly durable in the preparation of various dairy equipment.

iii. Takes high polish and therefore easy to.

iv. Corroded by brine and chlorine solution.

v. Welding has to be done to repair cracks *etc.*

- ✰ Stainless steel is widely present in the food industry as the main component of work surfaces, cookware, pasteurizers, homogenizers, separators, decanters, mixing, process and storage tanks, fittings, valves, pumps, and pipework. Based on the high resistance to corrosion by acidic or sulfur dioxide-containing foods and cleanability,stainless steel should be preferred to **AISI 302 or AISI 304**.
- ✰ Although naturally passivated by air or other oxidizers, stainless steel needs to undergo additional surface treatments (pickling, electropolishing, and mechanical cleaning) to improve its strength [4].
- ✰ Both the equipment and facilities used in the dairy industry today are made from austenitic stainless steels. The composition of 304 and 316 SS are given in following table [a]:

Table 15.7: Composition of 304 and 316 SS

Sl.No.	*ASTM*	*EN*	*Type*	*Composition (per cent)*
1	304	1.4301	Austenitic	18Cr, 9Ni
2	316	1.4401	Austenitic	17Cr, 10Ni, 2Mo

The table below gives a rough idea of the grades encountered for the different equipment used in the dairy industry [4a].

Table 15.8: Grades of SS and Dairy Equipment

Sl.No.	*Equipment*	*End use*	*Grades*
1	Refrigerated storage tank	All dairy products	304
2	Centrifuge, Pasteurizer	Milk, Yoghurt, Cream, Butter	304, 316
3	Plate and tubular heat exchangers	Milk, Cheese, Cream, Butter, Yoghurt	316
4	Packaging machine	Milk, Cream, Yoghurt	316
5	Ultra filtration equipment	Cheese	316
6	Maturation tank	Cheese, Ice Cream, Cream, Butter	304, 316
7	Cheese racks	Cheese	304
8	Other equipment	All dairy products	304, 316

5. Nickel and its alloys [2,3a,6].

This metal is used as a coating for milk/dairy product contact surfaces of pasteurizing vats coolers *etc.* Ni-alloy used in freezing chamber of ice-cream freezers, cylinders and plungers of homogenizers, *etc.*

- ✰ A nickel alloy called Inconel (80 Ni, 14 Cr and 6 Fe) is preferred to S.S. in that it can be more readily soldered where soldering is preferred to welding. However, this is not used with brine (salt solution).

Advantages

i. It has very slight effect on milk flavour although the most soluble in milk among dairy metals. It is mildly toxic.

ii. An alloy with iron makes it very tough and is quite difficult to handle during fabrication. However this metal is much more durable than tin coating, but it is more expensive.

iii. Lactic acid causes corrosion of this metal however it is not corroded by the alkaline washing powders.

iv. It has good corrosion resistance.

v. High Toughness and ductility.

Disadvantages

i. The highest disadvantage in the use of this metal is that it is more costly when compared with the chromium and tin.

ii. Not mix with cheap alloying elements.

6. Tin [2,3a,6]

This metal is used mainly as a coating for milk/dairy product contact surfaces of cans, vats, *etc.*

- ☆ It will not taint milk although quite soluble in it. This metal is too soft to be used for any kind of equipment.
- ☆ Tin coating is not durable as it wears off readily by corrosion, abrasion, *etc.* However re-tinning process not at all difficult.
- ☆ Tin is quite soluble in milk, but luckily it does not cause off flavours. It can be removed by refining [3a].

7. Chromium and its Alloys [1,2]

This metal is more useful for the coating on various types of equipment. It is mostly used for the milk/dairy product contact surfaces.

- ☆ This metal is non tainting and is resistant to corrosive action by acid and alkaline cleaners. The most important characteristic of this metal is that it is very expensive. This metal is useful for the coating on various types of equipment.

Corrosion Control [1,2,3,7]

Corrosion prevention and control (CPC) entails the characteristics of a system design to preclude or reduce corrosion, materials selection, non-destructive inspections for corrosion detection, coatings, finishes, cleaning materials and washings, repairs, and other maintenance activities [7].

☆ From the practical point of view corrosion cannot be entirely prevented in dairy equipment; however its rate can be controlled to a large extent. The following measures will help in preventing corrosion of stainless steel surface [2]-

i. Keep the surface clean,
ii. Permits surface to air dry, whenever possible,
iii. Use cleaners and sanitizers in the lowest concentration and for the shortest duration that will do the desired cleaning job.
    a. It is the common observation in dairy industry that the stainless steel which is used extensively in the dairy industry also will form an invisible film of chromium oxide on its surface when the stainless steel is dry and exposed to the atmosphere. This film protects the surface forms corrosion.
    b. When the film breaks or wears away, and the active metal gets exposed and corrode more easily.
    c. As the chlorine and its compounds are very corrosive it is advisable that equipment should be sanitized with chlorine solutions, preferably just before it is to be used, so as to avoid pro-longed contacts, and thus corrosion (pitting).

Corrosion Prevention

For better corrosion prevention in SS, dairymen should follow the good practices given below [8]-

i. **Selection of appropriate metal**-The most effective way to prevent corrosion is to "engineer it out" by selecting the appropriate material for the food application.
ii. **Training-**Coordination with the supplier of equipment or cleaning systems and materials is essential in planning corrosion-related training.
iii. **Preventative maintenance programs-**It should address the potential implications of corrosion on food contact surfaces through routine equipment inspection, with follows:up procedures for handling corrosion before it gets severe.
iv. **Application proper concentration of Cleaner/Sanitizer**- Food-handlers, by mistake, using sanitizers at inappropriate concentrations, following the erroneous assumption that "if a little is good, more is better," or allowing stainless-steel components to "soak" in a sanitizer solution for prolonged periods of time. T he supplier's recommendations be followed when using sanitizers, as well as detergents.

Milk Contact Surface and Metallic Contamination [1,2,3]

It is established that milk normally contains metallic elements like aluminium, copper, iron, manganese, silicon, zinc and others.

- ☆ The concentration of these metals is very low in normal milk but they may be added to milk during processing due to the action of milk with the contact surfaces on these metals.
- ☆ Milk utensils in our villages are commonly made of earthenware, zinc plated or galvanized iron, brass or copper. The product contact surfaces and other portions of modern processing equipment are increasingly being made of stainless steel.

  i. **Copper-**In the normal milk copper is bound with the fat as such this metal is not active chemically. Milk when comes in contact with copper it will dissolve in milk and would exist in active form as copper ions.

  - ☆ In the presence of direct sunlight and/or air the copper ions catalyze the hydrolysis of milk fat which an irreversible reaction resulting in off flavour development.
  - ☆ This flavour defect is more in high fat products. Copper salts formed due to the reaction of its ions with milk slats or lactic acid are greenish in colour, quite bitter to taste and is toxic.

  ii. **Iron**

  a. In normal milk iron is present in a chemically bound form. It is largely related to the fat globule membrane and the enzyme proteins. Similar to the copper, milk also dissolves iron when it comes in contact with it and will be forming free ions resulting in the development of off flavours and toxic compounds. Iron will also stain the product contact surface.

  b. A common practice prevailing when utensils, cans or buckets made of galvanized iron are to be used for handling milk and milk products, to coat the contact surface with zinc (Zinc plating) but upon long usage these vessels become defective due to wearing off the plating exposing the iron to be in contact with the milk directly.

  c. Special consideration is to be given to the reaction of these metals with milk when they are used for preparing the equipment and containers for handling milk and milk products.

  d. Development of off flavours, discoloration of the products, corrosion of the vessels, formation of toxic compounds, are some of the consequences when milk contact surface is made with these metals.

iii. **Stainless steel-** It is an alloys containing chromium, manganese silicon and nickel. The most satisfactory materials for the product surfaces is the stainless steel containing 18 per cent chromium, 8 per cent nickel in a low carbon steel. The processing equipment made with surfaces with this alloy are easily cleaned.

Sources

1. *Soma Maji, Milk contact surfaces, www.coursware, cutm.ac.in.*

1a. *Colin G. Fink and Frederick A. Rohrman, The corrosion of metals by milk, Division of Electrochemistry, Columbia University, New York.*

1b. *E. S. Guthrie, C. L. Roadhouse and G. A. Richardson, Corrosion of metals by milk and its relation to the oxidized flavours of milk, Hilgardia, 1931, 5(14): 425-453.*

2. *K.S. Sharma, Chemistry of milk, Agrimoon.com.*

3. *Utkarsh P. Deshmukh, Milk contact surface and metallic contamination, https: // www.slideshare.net.*

3a. *Properties of milk, chapter-2, Dairy and Food Engineering, College of Agricultural Engineering and Technology, http: //ouat.nic.in.*

4. *Dewangan AK, Patel AD and Bhadania AG, Stainless Steel for Dairy and Food Industry: A Review, 2015, Journal of Material Science and Engineering 04(05).*

4a. *ISSF Stainless Steel in the Dairy Industry, International Stainless Steel Forum, ISBN 978-2-930069-64-7, https: //www.worldstainless.org.*

5. *H. T. Gebhardt and H. H. Sommer, The solubility of metals in milk-. Submerged corrosion tests of various metals in milk, https: //www.journalofdairyscience.org.*

6. *Non Ferrous Metal – Nickel and Ni alloys, Department of Mechanical Engineering, NITTTR, Chandigarh, https: //www.slideshare.net.*

7. *M.W. Schleider, Corrosion Control in Industry, 2009, Business strategies for corrosion control in fleet maintenance, https: //www.sciencedirect.com.*

8. *Timothy Rugh, 2019, Address corrosion in 'stainless' steel, https: //www. dairyfoods.com.*

Part 3

Chemistry of Dairy Products

Chapter 16

Chemical Composition and Legal Standards of Milk Products

A. Composition of Milk Products

1. Malai

The indigenous product similar to cream is malai. Malai refers to cream or to cream like preparations such as unsweetened rabari or sar. Malai and Sar resemble cream in some respects.

- ✰ It is normally obtained by hand skimming milk that was heated and held undisturbed while slowing cooling. Malai is not normally made in dairy plants. It is usually made in houses, in shops and by the halwai *etc.*

2. Cream [1,1a,2,2a,4]

Cream is a dairy product composed of the higher-fat layer skimmed from the top of milk before homogenization. Cream skimmed from milk may be called **"sweet cream"** to distinguish it from cream skimmed from whey, a by-product of cheese-making. Whey cream has a lower fat content and tastes more salty, tangy and "cheesy" [2a].

- ✰ Cream is a fat rich component and known from immemorial as the fatty layer that rises up when left undisturbed. Cream is produced by the centrifugal separation of milk. During separation, two streams are created, a highly concentrated milk fat stream termed cream and a non-fat stream of skim milk that can be further processed into evaporated or dried milk products. Cream and skim milk are the most common products obtained by skimming or by separating milk.

- ☆ Cream is produced by separation of unhomogenised whole milk, Standard, Toned milk or Double Toned milk. The fat content of the cream resulting directly from the separator typically ranges from 35-45 per cent. Creams of different fat levels can be produced by standardizing (addition of a low fat stream to a higher fat stream to achieve a desired fat level) with skim milk. Higher fat creams can also be produced by further separating the high-fat stream (these creams are usually industrial products) [1,2,4].
- ☆ Creams are **oil-in-water emulsions** and therefore when it is standardized to various fat levels with skim milk, it must be homogenized to reduce the fat globule size to increase the stability (prevent fat separation). Proteins from the milk migrate to the fat globule membranes to act as emulsifiers. During homogenization, various grades of viscosity in cream products can occur. Usually, lower pressures are used than those used for milk products.
  - i. High fat whipping cream is not normally homogenized (unless UHT pasteurized), due to the excessive viscosity created and degraded whipping functionality.
  - **ii.** On the other hand, homogenization of high fat cream is utilized in the production of spoonable whipped cream products used as dessert toppings.
  - **iii.** Homogenization is conducted for UHT pasteurized whipping cream in order to prevent fat separation which increases with UHT pasteurization [2,4].

Composition [1,1a,2,3]

Cream can be defined in following three ways:

- **i.** That portion of milk, which is rich in milk fat.
- **ii.** That portion of milk into which has been gathered and which contains large portion of milk fat.
- **iii.** When milk fat is concentrated into a fraction of the original milk that portion is called cream.

- ☆ The cream mainly contains fat and Non Fat components. The milk fat content of the cream as it comes from the modern industrial cream separator may very from 30-80 per cent.
- ☆ Cream is rich in fat soluble vitamins A,D,E and K,the content of which depends on the fat level in cream. Cream containing 25 per cent fat contributes heat energy to the human diet at a rate of 225 kcal/100gm.

☆ Cream contains Non fat milk components in proportions to the amount of non fat fraction which is present and which came from the original milk. The non fat fraction of whey cream is similar to the whey.

Cream generally contains all the constituents of milk but non-fat constituents are inversely proportional to the fat content. The details related to the composition of cream is given below [1a,2]:

Table 16.1: Composition of Cream

Sl.No.	*Parameters*	*Range Value per cent*	*Values in Cream (30 per cent fat)*	*Values in Cream (with 50 per cent fat)*
1	Water per cent	45.45 - 68.2	64.0	44.43
2	Fat per cent	25- 60	30.0	50.0
3	Protein per cent	1.69 - 2.54	2.4	1.69
4	Lactose per cent	2.47 -3.71	3.5	2.47
5	Ash per cent	0.37- 0.56	0.4	0.37
6	Total solids per cent	31.8 - 54.55	36.0	55.57
7	Solid not fat (SNF) per cent	4.55-6.80	6.3	4.53

☆ SNF of cream can be calculated by taking the ratio of water to SNF of milk. If ratio of water to SNF of milk is 10:1, then the approximate SNF in cream = (100-F/11) where F is fat content of cream [1a,2].

Different quality/regulatory agencies have prescribed quality standards for different types of creams. Details are given under legal quality standards.

Classification of Cream [1a,2]

The fat in cream varies from 18-85 per cent. As the fat percentage in cream increase the components of milk in cream gradually decreases. The SNF content constitutes lower proportions than present in milk. Broadly, cream may be classified into two groups-

i. **Market Cream**-The cream used for direct consumption and.

ii. **Manufacturing or Industrial Cream**- It is used in production of various milk products.

The details of different types of cream following under these groups are given below:

a. *Fat per cent in different Types of Creams*

The details related to the fat per cent in different types of creams are given in Table 16.2.

Table 16.2: Fat per cent in different Types of Creams

Sl.No.	Types of Cream	Per cent Fat	Sl.No.	Types of Cream	Per cent Fat
1	Table cream	20-25	2	Light cream	20-25
3	Coffee cream	30-45	4	Whipping cream	30-45
5	Heavy cream	65-85	6	Plastic cream	65-85

b. *Description of different Types of Creams [2,5]*

The following types of cream are relevant to market and industrial purposes:

Table 16.3: Types of Creams

Sl.No.	Types of Creams	Description of creams
1	Table cream	Fat percentage- 18 to 22 per cent fat and used for eating directly. Various food preparations are made from this cream as fruit cream *etc.*
2	Light Cream or Thin Cream	Fat percentage -About 20 to 25 and used for table purpose and for manufacturing of cream cottage cheese *etc.*
3	Coffee Cream	Fat percentage- 18 to 25 per cent, light or thin type fresh cream and used mostly in coffee making. If this cream has any lactic acid, it becomes coagulated after adding to coffee. These cream coagulates are known as Cream feathers. This feathering may be prevented by addition of sodium citrate to this slight acidic cream. If the water which used in coffee making has more calcium, accelerate the cream feathering. The addition of 0.05 per cent sodium citrate also reduces the hardness and thickness of cream plug in cream bottles.
4	Heavy Cream	Fat percentage-30 to 40 per cent and used in production of butter and ice cream. Sometimes it can be used in paneer production.
5	Plastic Cream	Fat percentage-30-40 per cent fat re-separated to obtain the cream of 80 per cent (Fat rich cream or plastic cream). Mostly this type of cream is used for ghee production. This cream can be produced directly from milk by an especially designed plastic cream separator.
6	Whipped Cream	It is prepared by incorporating air in the normal cream (30-40 per cent fat). The air bubbles are stabilized by protein adsorption. The fat globules are assemble around these air bubbles. As whipping continues, these air cells subdivided into smaller one each with adsorbed layer and in numerable fat globules. At the maximum stiffness, these smaller air cells (nuclei) are surrounded by films of adsorbed protein which is so thin that drainage is at a minimum and the structure of dry foam is set up. i. The favourable temperature for whipping is 10 ºC. ii. The cream pasteurization and sugar addition declines the whipping rate in cream. iii. Some of the dairy plants used Nitrous oxide gas for cream whipping and. iv. Whipped cream is used in cakes and ice cream *etc.* for decorative purpose.

Sl.No.	Types of Creams	Description of creams
7	Sour Cream	This type of cream are produced by inoculating sweet, pasteurized and homogenized cream with a starter culture containing lactic acid producing and aroma producing microorganism and incubated to proceed fermentation. This cream has 0.6 per cent lactic acid, clean flavour and smooth texture. Sometimes rennet can be added @ 0.03 ml/gal cream to get more firm product. Sour cream should be stored at 40 ºF or less temperature to check the further increase in acidity.
8	Clotted cream	Milk is scalded for fifteen minutes at about 85-90 ºC in shallow pan and allowed to form a cream layer. Considerable evaporation occurs from the cream layer and on cooling over a period of twenty four hours, the clotted cream may be removed. Difference between ordinary cream and clotted cream -The decreased ratio of SNF/ water and relatively higher proportion of protein in the clotted cream.
9	Frozen Cream	Pasteurize the ordinary cream (40- 50 per cent fat) at 77 ºC for 15 minutes and packing in paper, plastic or tin containers after cooling at 4 ºC or below. Freeze quickly this packaged cream and store at –12 ºC or below. This stored cream is used during shortage in preparation of ice-cream *etc.* During storage the fat globule membrane may ruptured by ice crystals. Therefore, this cream tends to 'Oil-off ' on thawing. This 'Oiling-off' impairs the whipping characteristics of the ice cream.
10	Reconstituted Cream	By vigorous emulsification of unsalted plain butter in milk or separated milk or reconstituted milk or condensed milk, the fat globules are dispersed and coated with a layer of adsorbed protein, the resulted product is closely resembles with cream and known are reconstituted cream.
11	Sterilized or Canned Cream	Homogenize the 20 per cent cream at 80 ºC using pressure 175 kg/cm2 in the first stage and 35 kg/cm2 in the second stage and cooled to 16°C immediately. Tri-sodium phosphate is added @ 0.2 per cent as a stabilizer to the cream. This cream is packaged in tin cans or bottles and sterilized in following manner- i. Coming up time 15 Minutes to 114± 1 ºC. ii. Holding time 15 Minutes to 114± and iii. Cooling time 15 Minutes to room temperature.
12	Whey Cream	It is slightly differs in composition from regular fresh cream obtained from milk. Separation is sometimes used to remove fat from cheese whey and resulted cream is known as whey cream.
13	Synthetic/Artificial Cream	In this cream, butter fat is substituted with margarine fat and resulted product is known as synthetic cream. In production of synthetic cream, the refined oils are carefully blended to have some physical cream, the refined oils are carefully blended to have some physical properties resemble to butter fat. Synthetic cream may have some emulsifying agents such as egg yolk, vegetable lecithin and glycerol-Mono or di-stearated *etc.*

Uses of Cream [1a,2,4]

Cream is a very versatile product and can be used straight from the package or as an ingredient in many food products [2,4]. Cream is added as an ingredient to a large number of commercial food products including canned soup, dried bakery mixes, *etc.* In this context the cream serves as a source of dairy fat and its structural properties are of little importance. There are, however, two major product groups *viz.*, alcoholic cream liqueurs and cream cakes, desserts *etc.*, where cream is both a major ingredient and a major determinant of the properties of the complete product [1a].

a. Industrial Uses [2,4]

- **i.** The most common use of cream are for butter, ghee and Ice cream making. Cream is a key component in ice cream, providing the source of milk fat and functionality.
- **ii.** Other dairy products such as cheeses and yogurts- added to standardize fat content.
- **iii.** Cream fillings - whipping cream is typically used in decadent cream fillings for desserts,Cheesecakes

b. Uses at Retailer and Consumer Levels

- **i.** The major consumer cream products include coffee/table cream, whipped and whipping cream and sterilized cream. One of the most common uses of cream is adding it to coffee, tea, hot chocolate and other beverages where a richer flavour and texture and a characteristic colour are desired.
- **ii.** Table cream and whipping cream are often used as toppings to fresh fruit, desserts and cereals.
- **iii.** Cream can be found as an ingredient in a wide variety of products. Some common uses of creams are -

- ☆ Confectionary products – such as fondant centers and truffles.
- ☆ Bakery products – fresh and dried bakery mixes.
- ☆ Alcoholic cream liqueurs and beverages.

Characteristics of Cream [2,4]

Cream has following characteristics:

- **i. Appearance and Flavour-** Cream should appear white to off-white with smooth consistency and varying degrees of viscosity depending on the fat content (typically increased viscosity as the fat content increases). Cream should be homogeneous without separation. The flavour and odour of cream should be rich, clean and sweet with no off-flavours.

ii. **Storage-**Cream should be stored at refrigeration temperatures (4-7°C). The typical shelf life of pasteurized cream is one to two weeks and a few months for UHT pasteurized cream. However, once the container of UHT pasteurized cream is opened it should be treated as pasteurized cream. As cream ages, its lactic acid content increases and can result in curdling. Freezing destabilizes the oil-in-water emulsion of creams and therefore the fat will separate upon thawing.

☆ Separated cream cannot be used for whipping, but can be used in products such as "cream soups" where flavour is the critical factor.

iii. **Varieties-** There are a whole range of creams based usually on the milk fat content. For industrial purposes, creams can be produced at specific fat levels, however availability is dependent on the volume usage. The details related to the different varieties of creams are given under legal standards.

Physico-Chemical Properties [1a,2]

Some important physico-chemical properties relevant to cream processing and consumption are given below:

A. **Specific Gravity-** Specific gravity of cream is inversely proportional to the fat percentage.

B. **Titratable Acidity (TA)-**There is an inverse relationship between the per cent fat and per cent titratable acidity. Fresh cream has lower acidity percentage than milk.

C. **Viscosity**- It is the resistance offered by the liquid to flow. Consumer judges the richness of cream by its viscosity. The following factors effect the viscosity of cream:

   i. **Fat percentage**-Higher the fat per cent greater is the viscosity.

   ii. **Temperature**- Higher the temperature lower is the viscosity.

   iii. **Homogenization-** It increases the viscosity of cream. Single stage homogenization increases viscosity more than the double stage homogenization.

   iv. **Separation conditions**-Higher temperature of separation lowers the viscosity.

   v. **Ageing-**The viscosity of cream increases with storage period.

   vi. **Cooling-**Cooling of cream increases the viscosity.

   vii. **Clumping**-It refers to the tendency of fat globules to adhere loosely to one another and form clumps. Clumping depends on fat globule size, temperature (maximum at 7 °C), agitation and method of separation. The greater is the clumping, greater will be the viscosity.

D. **Whipping Quality**-Whipping means emulsion of gas or foam production by beating of cream. Whipped cream has remarkable stability and used in cakes, ice creams and for decorative purposes. The optimum aging period is 24 hours at 4°C. Homogenization, acidity and stabilizers in cream reduces the whipability of cream.

Functional Properties [2,4]

Cream and other milk fat products (such as butter) are very different than fat from other animal and vegetable sources due to the presence of an aqueous phase, proteins and phospholipids. Milk fat exists in the form of globules dispersed in an aqueous phase of milk. Surrounding the fat globules is a membrane made up of varying proportions of protein, phospholipids and glycerides. The phospholipids in the membrane are of primary importance and account for much of the emulsifying properties of the fat globule membrane.

i. The absorption of casein, proteins which are within the aqueous phase of cream, onto the milk fat globule surface when the globule membrane is disrupted (*e.g.* as a result of shear or homogenization), allows for production of variations in viscosity which can impart different functionality in food products.
ii. Caseins promote interaction with other molecules such as calcium, creating a new structure within the cream and therefore changing the textural properties.

The principle functional attributes of cream are:

a. **Texture-** Primarily due to the milk fat present in cream and the fat-protein network created during homogenization, cream adds rich, smooth viscosity and mouthfeel to many food products such as soups and sauces.
b. **Flavour-** Creams unique natural flavour adds richness to many foods and enhances other flavours in food products. Due the narrow melting range of the milk fat in cream, cream provides quick release of flavours and taste sensations.
c. **Fat Soluble Ingredient Carrier-** The milk fat in cream acts as a carrier of fat soluble vitamins and aids in the even distribution of other fat soluble ingredients.
d. **Emulsification-** Creams natural proteins act as emulsifiers and aid in emulsification, aeration, foaming and overrun to give products such as whipped cream and ice cream a smooth and stable texture.
e. **Whitening Colour-**Cream adds a whitening effect in products such as coffee. The colouring power of cream depends on the suspension of particles within the cream. Large particles in cream such as fat globules and casein reflect and scatter light creating a whitening effect.

f. **Browning-** Cream contributes to the browning of cooked foods through the Maillard browning reaction of the proteins and lactose found in cream. The Maillard reaction will also contribute to brown flavours in cooked food products containing cream.

Sources

1. *J.N.Warner, Principles of Dairy Processing.*

1a. *Y. Kotilinga Reddy, Yogesh Khetra and M.H. Sathish Kumar, Fat Rich Dairy Products, http: //www. agrimoon.com.*

2. *M.K. Srivastava, Handbook on Analysis of Milk Products. Write and Print Publications, New Delhi, 2016. ISBN: 9789384649258.*

2a. *Wikipedia.*

3. *www.dairyforall.ccom.*

4. *Milk ingredient, Canadian Dairy Commission, http: //www.milkingredients.ca.*

5. *Aneja, R.P., Mathur, B.N., Chandan, R.C. and Banerjee, A.K. 2002. Technology of Indian Milk Products. A Dairy India Publication, Delhi, India.*

3. Butter [White/Cooking (Deshi) and Table (Creamery)]

Butter is one of the most highly concentrated forms of fluid milk. Butter may be defined as fat concentrate and it is produced by churning cream until the fats separate from the liquid (buttermilk) and the butter is in a semi-solid state. Butter is essentially the fat of the milk.

- Butter is a product containing predominately fat (~80 per cent) and a little quantity of SNF besides added ingredients such as salt. However, compositional variations can occur depending on the type of butter, which may be covered under various classification schemes.
- The two most common types of butter are available. It is usually made from sweet cream and is salted and can also be made from acidulated or bacteriologically soured cream and salt less (sweet). An indigenous type of butter is commonly sold in India which is called as Desi Butter. This product deteriorates very soon and is usually made for ghee making.

The details related to the Physical and Chemical composition of butter are given below [1,2]:

a. Physical Composition

The butter making changes the nature of emulsion (inversion of the original cream emulsion) *i.e.* in milk,fat is in water whereas in butter, water is in fat. Fat globules vary from 0.1-10 micron in diameter. The fat globule membrane is comprised of surface active materials: phospholipids and lipoproteins. Fat globules typically aggregate in three ways:

i. Flocculation.
ii. Coalescence
iii. Partial coalescence.

- ☆ The final product *i.e.* Commercial butter has 80-82 per cent milk fat, 16-17 per cent water, and 1-2 per cent milk solids other than fat (curd). It may contain salt, added directly to the butter in concentrations of 1 to 2 per cent. Unsalted butter is often referred to as "sweet" butter. This should not be confused with "sweet cream" butter, which may or may not be salted. Reduced-fat, or "light," butter usually contains about 40 per cent milk fat.
- ☆ The complicated physical structure of butter has continuous, amorphous phase of solid fat surrounds not only the water droplets, but also air bubbles, intact fat globules, and highly ordered crystals of milk fat that have grown during the cooling process. The proportion of continuous or "free" fat can vary from 50 per cent of the total to nearly 100 per cent, and it has a direct influence on the behavior of butter. The more fat there is in discrete globules or crystals, the harder and more crumbly the butter, even to the point of brittleness. A preponderance of free fat, on the other hand, makes for a malleable butter that softens readily and may even weep some liquid fat in the process. The difference is a matter of both large-scale and molecular arrangements. In a mass where the free fat merely fills the small interstices between globules and crystals, the texture will be largely that of the separate particles and it takes more energy to separate the molecules ordered in a crystal than it does to disrupt an already disordered phase of the same molecules

    i. Mostly crystalline butter will be relatively stiff and not as smooth as mostly amorphous butter.
    ii. The ideal butter, lies somewhere between the two extremes and it is obtained by manipulating the cooling process.
    iii. Due to its high water content(the size of the water droplets in the water-in-oil emulsion being less than 6 µm), butter is subject to rapid microbiological spoilage at normal temperatures.

b. Chemical Composition

Chemically butter fat consists essentially of a mixture of triglycerides (with small amounts of mono- and diglycerides, phospholipids, glycolipids, and lipoproteins), particularly those derived from fatty acids, such as palmitic, oleic, myristic, and stearic acids.

- ☆ The fatty acid composition of butter fat varies according to the producing animal's diet. The triglycerides (98 per cent of milk fat) are of diverse composition with respect to their component fatty acids, approximately 40 per cent of which are unsaturated fat.Firmness varies with chain length, degree of unsaturation, and position of the fatty acids on the glycerol. The typical breakdown of fatty acids in butter is given below [2]:

Table 16.4: Fatty Acids in Butter

Fatty Acids	*Percentage*	*Fatty Acids*	*Percentage*
Oleic acid	31.9	Caproic acid	1.9
Mysteric acid	19.8	Capric acid	1.6
Palmitic acid	15.2	Caprylic acid	0.8
Stearic acid	14.9	Linoleic acid	0.2
Lauric acid	5.8	Linolenic acid	0.1
Butyric acid	2.9	-	-

Table 16.5: Fatty Acids in Unsaturated and Saturated Fat (Butter)

Unsaturated		*Saturated, 51.368g*	
Fatty Acids	*USDA Average/100g*	*Fatty Acids*	*USDA Average/100g*
Monounsaturated, g	21.021	4:0 Butyric, g	3.226
16:1 Palmitoleic, g	0.961	6:0 Caproic, g	2.007
18:1 Oleic, g	19.961	8:0 Caprylic, g	1.19
Polyunsaturated, g	3.043	10:0 Capric, g	2.529
18:2 Linoleic, g	2.728	12:0 Lauric, g	2.587
18:3 Linolenic, g	0.315	14:0 Myristic, g	7.436
-	-	16:0 Palmitic, g	21.697
-	-	18:0 Stearic, g	9.999

☆ Butter also contains protein, calcium and phosphorous (about 1.2 per cent) and fat-soluble vitamins A,D and E. Although there are over 120 different compounds that contribute to butter's unique flavor, the five primary factors responsible for butter's flavor include-fatty acids, lactones, methyl ketones, diacetyl and dimethyl sulfide.

Classification of Butter [1,3,4]

The classification of Butter may be done on the basis of method of manufacturing,end use, on treatment given to cream and salt content.

1. *Classification Based on the Manufacturing Practice (as followed by FSSAI)*

a. **Pasteurized cream butter/Pasteurized Table butter**: This is made usually from pasteurized sweet cream. Such butter usually has a milder flavour than that made from similar cream not pasteurized.

b. **Desi butter**: The butter obtained by traditional process of churning dahi or malai as practiced at domestic levels.

The details related to the quality standards given under FSSAI, are discussed under Legal/Quality standards.

2. *Classification Based on End Use (as followed by BIS)*

a. **Table Butter**: the product made from pasteurized cream obtained from cow or buffalo milk or a combination thereof with or without ripening with the use of standard lactic culture, addition of common salt, annatto or carotene as colouring matter and diacetyl as flavouring agent.

b. **White Butter**: the product made from pasteurized cream obtained from cow or buffalo milk or a combination thereof without ripening and without addition of any preservative including common salt, any added colouring matter or any added flavouring agent.

The details related to the quality standards given under BIS, are discussed under Legal/Quality standards.

3. *Classification Based on Acidity of Cream Used for Butter making*

On the basis of the acidity of cream, butter can be classified into the following three groups:

a. **Sweet cream butter** (made from non-acidified cream): This includes butter in which no bacterial culture have been added to enhance diacetyl content) having pH of ≥6.4 (acidity of the churned cream does not exceed 0.2 per cent).

b. **Mildly acidified butter** (made from partially acidified sweet cream): Having pH in the range of 5.2 to 6.3.

c. **Sour cream butter** (made from ripened cream which has more than 0.2 per cent acidity): Having pH ≤ 5.1from cream.

4. *Classification Based on Salt Content*

a. **Salted butter**: Butter to which salt has been added. It is added to improve flavour and keeping quality of butter.

b. **Unsalted butter**: This type of butter contains no salt. It is usually prepared for manufacturing other products such as ghee and butter oil.

Sources

1. *M.K. Srivastava, Handbook on Analysis of Milk Products. Write and Print Publications, New Delhi, 2016. ISBN: 9789384649258.*

2. *USDA National Nutrient Database for Standard Reference, Release 20 (2007), Wisco sin Milk Marketing Board, Medison.*

3. *Y. Kotilinga Reddy, Yogesh Khetra and M.H. Sathish Kumar, Fat Rich Dairy Products, http: //www. agrimoon.com.*

4. *Aneja, R.P., Mathur, B.N., Chandan, R.C. and Banerjee, A.K. 2002. Technology of Indian Milk Products. A Dairy India Publication, Delhi, India.*

Nutritional Value of Butter [1,2,3,4]

Butter fat contains 97-98 per cent triglycerides and approximately 0.2 per cent cholesterol. A typical serving of butter (1 tablespoon/14g) contains only 30mg of cholesterol and only 0.56-1.26g of trans fat (4-9 per cent). Current recommendations suggest not exceeding an intake of 300mg/day for dietary cholesterol and keeping trans fat intake as low as possible.

- ☆ Humans have been eating butter and animal fat a lot longer than they have been abstaining from them. Butter plays an essential role in maintaining our health. It is stated that all animal fat, especially butter, is bad for us because it is saturated. Today, in India there is also a growing awareness about the negative values of butter. In India 70-75 per cent of the milk is buffalo milk, which is rich in saturated fats and hence most of the butter is made from buffalo milk.
- ☆ Butter is particularly vilified because it has more cholesterol than other fats, yet recent scientific studies show that its bad reputation is undeserved. Every fat is a combination of both saturated and unsaturated fatty acids.
  - i. Butter is about 50 per cent saturated, but all other animal fats are more unsaturated than they are saturated.
  - ii. Scientists now know that, except in rare individuals, dietary cholesterol does not influence blood cholesterol, and fat from ruminants (animals that chew their cud) contains valuable nutrients that maintain health and prevent disease.
- ☆ Every cell in our body needs fat, and that dietary fat is a corner stone of good health.
  - i. Our brain and hormones rely on fat to function, and fat supports our immune system, fights disease, and protects our liver.
  - ii. Fat promotes clear skin and healthy hair, regulates our digestive system, and leaves us feeling sated after a meal.
  - iii. Fat is the body's preferred fuel, providing us with more than twice the amount of energy as the same quantity of carbohydrates and protein.
  - iv. It helps the body absorb nutrients, calcium, and the fat-soluble vitamins A, D, E, and K.
  - v. Fat and protein are found together in nature because it's the fat that helps us digest the protein.

☆ Trans fats (man-made)fats, produced when liquid or polyunsaturated fats are made solid by adding hydrogen, are difficult for our body to process, so it stores them.

i. They adversely affect our cholesterol levels by increasing **bad cholesterol (LDL)** and lowering **good cholesterol (HDL),** and they interfere with insulin production, promoting diabetes and obesity.

ii. Conjugated linoleic acid (CLA), a natural trans fat found in butter, is beneficial to our health, credited with fighting cancer, and preventing weight gain and heart disease.

☆ Butter is considered as high calorie food as 100 g of butter provides almost 700 Kcal. Butter is extremely rich in minerals like Calcium, Phosphorus and Potassium. It has good amount of sodium and small amounts of fluoride, selenium, zinc and magnesium. It is also rich in Vitamin A, Vitamin E, Riboflavin, Niacin and Pantothenic acid. It also has Vitamin K, Folate and Vitamin B12 in small amounts.

☆ The fatty acid composition of Butter is important for several reasons. The significant amount of short-chain fatty acids contributes to butter's quality as a softer fat with a lower melting point, which ensures a quick flavor release when melting. The details related to the nutritional values are given below:

Table 16.6: Nutritional Values of Butter

Sl.No.	*Nutrients/100g*	*Salted*	*Unsalted*	*Salted, Whipped*
1	Energy, kcal	717	717	717
2	Water, g	15.87	17.94	15.87
3	Protein, g	0.85	0.85	0.85
4	Total Lipids (Fat), g	81.11	81.11	81.11
	Saturated Fat, g	51.37	51.37	50.49
	Monounsaturated Fat, g	21.02	21.02	23.43
	Polyunsaturated Fat, g	3.04	3.04	3.01
	Trans Fat, g	2.98	2.98	2.9
	Cholesterol, mg	215	215	219
5	Total Carbohydrates, g	0.06	0.06	0.06
6	Ash, g	2.11	0.04	2.11
7	Vitamins			
	Vitamin A, IU	2499	2499	2499
	Vitamin E(alpha-tocopherol), mg	2.32	2.32	2.32
	Folate, ug	3	3	3
	Niacin, mg	0.04	0.04	0.04
	Riboflavin, mg	0.034	0.03	0.03
	Thiamin, mg	0.005	0.005	0.005

Sl.No.	Nutrients/100g	Salted	Unsalted	Salted, Whipped
	Vitamin B6, mg	0.003	0.003	0.003
	Vitamin D, IU	56	56	56
	Pantothenic acid, mg	0.11	0.11	0.11
	Vitamin B12, ug	0.17	0.17	0.13
	Vitamin K, ug	7	7	7
8	Minerals			
	Calcium, mg	24	24	24
	Iron, mg	0.02	0.02	0.16
	Magnesium, mg	2	2	2
	Phosphorous, mg	24	24	23
	Potassium, mg	24	24	26
	Sodium, mg	576	11	827
	Zinc, mg	0.09	0.09	0.05
	Selenium, ug	1	1	1

Sources

1. *M.K. Srivastava, Handbook on Analysis of Milk Products. Write and Print Publications, New Delhi, 2016. ISBN: 9789384649258.*
2. *"Fat: An Appreciation of a Misunderstood Ingredient, With Recipes" by Jennifer McLagan (Ten Speed Press, 2008) and WebExhibits.*
3. *http: //www.nutralite.com/spread_facts.htm.*
4. *USDA National Nutrient Database for Standard Reference, Release 20 (2007) Wiscosin Milk Marketing Board, Medison.*

4. Desi Ghee [1-4]

Ghee is called Ghrita or Ghrittham in sanskrit, ghee in Hindi and 'Neyi'in Tamil. **Ghee is a fat rich dairy product of Indian origin and its western equivalent is butteroil.**

- ☆ Desi Ghee is clarified butter and it is manufactured by prestratification method from butter or cream of the milk by raising the temperature to 110-120° C, when the product with mild flavour is obtained. When butter is melted at approximately 100 degree C, the water evaporates and ghee is produced. The main constituent of Ghee is fat.
- ☆ The fats in ghee are mainly Triglycerides (97 per cent). Triglyceride is a lipid or neutral fat and is the form in which fat is stored in the body.
  a. The other constituents in ghee are Di-glycerides, monoglycerides, ketoacid glycerides, free fatty acids, steroles, vitamin A, traces of vitamin E K *etc.*

b. The glycerides in the ghee are Butyric acid, caproic, coprylic and capric acids, linoleic acid, stearic acid palmitic acid *etc.* The following facts about the fat enable us to understand health benefits of ghee:

   i. Fats can be categorized into **two major categories**-Saturated and unsaturated.

   ii. **Saturated fats** can further be divided into long-chain fatty acids and short-chain fatty acids.

   iii. Short-chain fatty acids are assimilated and metabolized so that they release energy.

   iv. Long-chain fatty acids (most **animal fats**) cannot be completely metabolised by the body and can lead to cancer and **blood clots**

   v. **Unsaturated fats** can be further categorized into two kinds-**Monounsaturated** and P**olyunsaturated**.

   vi. **Monounsaturated fats,** such as olive oil, are healthy, while **polyunsaturated fats** are not. The latter become oxidized and create free radicals, which damage the cells of the body.

☆ Ghee contains both Monounsaturated Fats and Saturated Fat. Most of the saturated fats of Ghee are short-chain fatty acids and that renders it easily digestible. Ghee also contains up to 27 per cent monounsaturated fatty acids and only 4 per cent to 5 per cent polyunsaturated fatty acids. Our body needs both saturated and unsaturated fats and hence the aforesaid fatty acids profile of ghee is ideal for Human Health.

☆ One hundred grams of ghee contains 100 g of fat and 600 µg of carotene. Carotene is an yellow or orange pigment occurring in four forms – alpha, Beta, Gamma and delta. of these four, the most important form is Beta - Carotene, which is converted in the body to Retinol (vitamin A). Ghee has beta-carotene and vitamin E, both known anti-oxidants that counter the effects of free radicals.

☆ Ghee starts to solidify below 63°F, starts to melt above 99°F and its smoking point is 375°F.

☆ **Except Ghee and cod liver oil and other fish oils, no other plant oil or fat contains vitamin A.**

   i. This vitamin helps bubrication of the eyes and protects them. It keeps the epithelial cells in order.

   ii. Ghee repairs the mucus lining of the stomach and evens out the acid balance in there.

   iii. Ghee has been held in high esteem for its nutritional and medicinal values and characteristic flavor and pleasant aroma. Ghee with intense "Burnt Flavor"(obtained by clarification at 120-140°C) is very much liked in eastern and southern states of the India.

☆ Ghee could be in liquid, semisolid and some time in solid state based on the storage temperature. Ghee made from buffalo milk is whitish with greenish tinge and that of cow milk is golden yellow colour. It is usually prepared form cow milk, buffalo milk or mixed milk.

Composition [1,5]

Ghee is a complex lipid of glycerides (majorly triglycerides), free fatty acids, phospholipids, sterols, sterol esters, fat soluble vitamins, carbonyls, hydrocarbons, carotenoids, (only in ghee derived from cow milk). Its detailed chemical composition is given below:

Table 16.7: Chemical Composition of Ghee [6]

Sl.No.	Constituents	Buffalo Milk Ghee	Cow Milk Ghee
1	Fat per cent	99 – 99.5	99 – 99.5
2	Moisture (per cent)	<0.5	<0.5
3	Carotene(mg/g)	-	3.2-7.4
4	Vitamin A(IU/g)	17-38	19-34
5	Cholesterol (mg/100g)	209 – 312	302 – 362
6	Tocopherol(mg/g)	18 – 31	26-48
7	Free fatty acid (per cent)	2.8	2.8

☆ Milk fat contains at least 500 fatty acids and fatty acid derivatives with 4 -20 or more carbon atoms in their chain.

i. The fatty acid may saturated or unsaturated and usually contains an even number of carbon atoms. Composition of fatty acid also varies between buffalo milk fat and cow milk fat.

ii. Following table shows The detailed composition of buffalo milk and cow milk fatty acid are given below:

Table 16.8: Fatty Acid Composition of Ghee [6]

Sl.No.	Fatty Acids Per cent	Cow Milk Ghee	Buffalo Milk Ghee
1	Butyric	3.2	4.4
2	Caproic	2.1	1.5
3	Capric	2.6	1.3
4	Lauric	2.8	1.8
5	Myristic	11.9	10.8
6	Palmitic	30.6	33.1
7	Stearic	10.1	12.0
8	Oleic	27.4	27.2
9	Linoleic	1.5	1.5
10	Linolenic	0.6	0.5

Sources

1. *M.K. Srivastava, Handbook on Analysis of Milk Products. Write and Print Publications, New Delhi, 2016. ISBN: 9789384649258.*
2. *S.K. Ayurveda, Chennai, India, Ayurveda.com.*

2a. *Shubhra Krishan,Journalist from India, now based in Colorado Springs, USA.*

3. *Ayurveda Training Program, Rajasthan visit.com.*
4. *V.K.Bansal, Indian Dairy man, Nov.2010.*
5. *Y. Kotilinga Reddy, Yogesh Khetra and M.H. Sathish Kumar, Fat Rich Dairy Products, http://www. agrimoon.com.*
6. *R.P. Aneja et al., Technology of Indian milk products, Dairy India publication. Section 3.4: Fat rich dairy products, page 186-187.*

Heat-acid Coagulated Milk Products [1,1a,2]

i. Channa

Channa is used as an intermediate base for variety of milk based Bengali sweets especially in Rosogollas and it is also called as India's traditional soft, cottage cheese.

- ✰ Chhana is like paneer, except some whey is left and the mixture is beaten thoroughly until it becomes soft, of smooth consistency, and malleable but firm.
- ✰ Chhana contains fairly high level of fat and protein as well as some minerals(Ca and P).
  - i. It is good source of fat soluble vitamin A. It's nutritive value is attributed to the presence of whey proteins which are good source of essential amino acids.
  - ii. The chemical composition of buffalo milk chhana and cow milk chhana is tabulated below [1a]:

Table 16.9: Chemical Composition of *Chhana*

Sl.No.	Constituents Per cent	Buffalo Milk Chhana	Cow Milk Chhana
1	Moisture	51.7	53.4
2	Fat	29.7	24.8
3	Lactose	2.3	2.2
4	Proteins	14.4	17.4
5	Ash	1.9	2.1
6	pH	5.4	5.7

ii. Paneer

It is an unaged, acid-set, non-melting farmer cheese prepared by curdling heated milk with citric acid (lemon juice) or other non-rennet food acid, and then removing the whey and pressing the curdled material into a dry unit.

☆ The details of composition of cow milk paneer, Buffalo milk paneer and average composition of paneer are tabulated below [1,1a]:

Table 16.10: Composition of Paneer

Sl.No.	*Constituents/ 100 gm*	*Cow Milk Paneer [1]*	*Buffalo Milk Paneer [1]*	*Proximal Composition of Paneer [1a] per cent*
1	Energy (kcal)	265	292	
2	Fat g	20.8	23.0	23.41 (50.84 on dry matter basis)
3	Proteins g	18.3	13.4	18.23
4	Carbohydrates g	1.2	7.9	2.40 (Lactose)
5	Minerals g	2.6	-	1.90 (Ash)
6	Calcium mg	208	-	-
7	Phosphorus mg	138	-	-
8	Total Solids	-	-	46.04

Paneer contains reasonably good amounts of fat and cholesterol. It would be better to avoid it for those with hypertension and diabetes due to its high fat content. It can however be used in small quantities for such patients one or twice a week. It is suitable for all age groups. Paneer is nutritious food made mostly from the milk of cows but also of other mammals, including, buffalo, goats, sheep, *etc.*

Nutritional Values of Channa and Paneer [1,2]

Both Channa and Paneer have following nutritional values:

Channa

The details of nutritive value of channa are given below [2]

Table 16.11: Nutritive Value of *Channa*

Biological value-88	Calorific value (Kcal) -1300-1700,	Digestibility coefficient- 92
PER value- 3.1	NPR- 5.25	NPU -71.5

Paneer

Paneer has following nutritional values [3,4]:

Table 16.12: Nutritive Value of Paneer

Serving Size- 85g

Ingredient	*Quantity*	*Per cent Daily Value**	*Ingredient*	*Quantity*	*Per cent Daily Value**
Total Carbohydrates	3.0 g	1	Protein	20 g	40
Sugars	1 g	-	Sodium	20 mg	1
Total Fat	23 g	35	Dietary fibre	0g	-
Saturated	15 g	75	Cholesterol	90mg	30
Trans fat	0.0 g	0.0	Energy total From fat	300 kcal 210 kcal	

Note: Established per cent of calories from Fat- 69 per cent, carbohydrates- 4 per cent and Protein -26.7 per cent.

Sources

1. *M.K. Srivastava, Handbook on Analysis of Milk Products. Write and Print Publications, New Delhi, 2016. ISBN: 9789384649258.*

1a. *M. Ranganadham, Sathish Kumar M. H., Devraja H. C. and F. C. Garg. Traditional Dairy Products, www.AgriMoon.com.*

2. *R.P. Aneja, B.N. Mathur, R.C. Chandan and A.K. Banerjee, Technology of Indian Dairy Products.*

3. *Neesha Bukht Choksy, Consultant Nutritionist, Doctor, NDTV and 4-Diet Nut.in.*

Fermented Milk Products [1,1a]

Fermentation is one of the simplest ways of preserving milk constituents for human consumption and it gives an acid taste to milk which is particularly refreshing in worm climate and also imparts certain therapeutic benefits originally absent in milk.

- ✰ Fermented dairy products have assumed prominent position in human diet in many regions of the world. Fermentation leads to partial breakdown of milk constituents and increases the digestibility of cultured milk products.

Nutritional Values of Fermented Milk Products [1,1a,5]

India is one of the largest milk producer in world from the total milk production, only 7 per cent milk is utilized for the production of fermented milk products [5].

i. In 7 per cent mainly include curd (Dahi), sweetened concentrated curd (Shrikhand) and stirred curd (Lassi). Fermented milk and Probiotic yogurt showed stimulate cellular immunity which contains an agent such as L. casei.

ii. Fermentation is an enhancement of nutritional value and increase in sensory properties, an affordable and economical technology which introduce a new variety of food product. Fermented foods have great nutrition and health-enhancing properties.

- ☆ It also helps in elimination and detoxification of objectionable compounds present in the unprocessed form of food.
- ☆ Fermented milk products can contribute to boosting the immune system, helps in restoring the balance of intestinal microflora, helps in digestion and improve health disorders and prevention from gastrointestinal infections.
- ☆ Health benefits of fermented food product depend upon the microorganism used during fermentation.
- ☆ Help to decrease in serum cholesterol levels and antimutagenic activity in the human body.
- ☆ Fermentation of milk is beneficial for ingestion by lactose intolerant human being and the patients experiencing atherosclerosis.

Curd (or Dahi) [1,1a,2]

Dahi is the common term used in India for any type of sour milk. It is typical product of lactic acid fermentation of milk and it is also called as yoghurt of the Indian subcontinent, known for its characteristic taste and consistency.

- ☆ The word dahi seems to be derived from the Sanskrit word **dadhi**, one of the five elixirs, or panchamrita, often used in Hindu ritual. Dahi also holds cultural symbolism in many homes in the Mithilanchal region of Bihar.

  i. It is found in different flavours, sour dahi **(tauk doi)** and sweet dahi **(mee-sti or podi doi)** are two famous flavours.

  ii. In India, it is often used in cosmetics mixed with turmeric and honey. Sour dahi is also used as a hair conditioner by women in India.

  iii. Dahi is also known as **Thayiru** (Malayalam), **doi** (Assamese, Bengali), **dohi** (Oriya), **perugu** (Telugu), **Mosaru** (Kannada), **Thayir** (Tamil), **or Qəzana a pəəner** (Pashto).

  iv. Quality and consistency vary very much in India. In Northern India dahi is of thicker consistency than in Southern India where dahi is generally in diluted form [1].

- ☆ Dahi is produced from heat treated milks after inoculation with certain species of lactic acid bacteria added to milk in the form of starter culture.
  - i. Lactic acid bacteria added multiply, grow and produce lactic acid, acetic acid and carbon dioxide by utilizing lactose present in milk.
  - ii. Some bacteria uses citric acid of milk to produce certain volatile organic compounds mainly diacetyl, which is mainly responsible for flavor of dahi.
  - iii. Judicious combination of acid producing and flavour producing microorganisms in the starter helps in the production of Dahi with a firm body and good flavour [1a].
- ☆ The preparation of dahi from pasteurized or boiled milk is advisable. If dahi is prepared from raw milk then no starter is added and it is allowed to sour naturally. If dahi is prepared from pasteurized or boiled milk then a little portion of dahi of previous day is added as starter and then milk is allowed to curdle at room temperature. In India as dahi is prepared with less care hence its quality varies greatly from place to place.
- ☆ The chemical composition of dahi made from skim milk and whole milk is tabulated below [1,1a]:

Table 16.13: Composition of Dahi

Sl.No.	Constituents Per cent	Skim Milk Dahi	Whole Milk Dahi
1	Water	90- 91	85-88
2	Fat	0.05-0.1	5-8
3	Proteins	3.3-3.5	3.2-3.4
4	Lactose	4.7-5.3	4.6-5.2
5	Lactic acid	0.5-1.1	0.5-1.1
6	Minerals (Ash)	0.7-0.75	0.7-0.75

Nutritional Values of Dahi [1]

If one has food, supplemented with 250 g of dahi a day then the status of **thiamine** improves.

- i. Dahi can also be consumed by people who suffer from **lactose intolerance**.
- ii. **Dahi** also increases the pyruvic and lactic acids among children on a typical poor rice diet.
- iii. As dahi is made from fermentation of milk, so a greater amount of **phosphorus and calcium is made available to the digestive system**

by their precipitation in the lower intestines due to the acid condition induced by Lactobacillus sp. Thus, dahi in its different forms as lassi, kadhi, shrikhand, *etc.* also contributes significantly to improve the nutritive contents of an average Indian diet.

Misti Doi [1,2]

Mishti Doi (Bengali) or Mitha Dahi (Oriya) is a sweet dahi or sweet yogurt. Sweet dahi or Misti doi (dahi with sugar) is a fermented milk with creamish to brown colour, smooth texture, pleasant flavour and firm consistency.

i. This product is preferred dessert in eastern part of India *e.g.*Orissa **and** West Bengal and in Bangladesh.
ii. It is also known as payodhi and lal dahi. It is made with milk and sugar. Corn sugar/syrup, maltose or beet sugar can also be used as sweetener in place of cane sugar.
iii. It is prepared by boiling milk until it is slightly thickened, sweetening it with sugar, either **gura** (brown sugar) or **khajuri gura** (date molasses), and allowing the milk to ferment overnight.
iv. Earthenware is always used as the container for making mitha dahi because the gradual evaporation of water through its porous walls not only further thickens the yoghurt, but also produces the right temperature for the growth of the culture.
v. Very often the yoghurt is delicately seasoned with a hint of ***elaach*** (cardamoms) for fragrance.
vi. Misti doi is coloured and flavoured commonly with caramel. Synthetic flavours like cardamom, caramel,rose and vanilla may also be used with and without nuts and fruits.
vii. Misti doi stored below 10°C retains good quality without any significant change in flavour and microbiological quality for about 3 days and storage of misti doi at 4°C increases it's shelf life to 5 days.

Chakka and Shrikhand [1,2,3]

Chakka

It is an Indian strained yoghurt, In south Asia, regular unstrained yogurt (dahi), made from cow or buffalo milk, is often sold in disposable clay pots. Kept for a couple of hours in its clay pot, some of the water evaporates through the clay's pores. But true strained yogurt (*chakka*) is made by draining dahi in a cloth.

- ✰ Average composition of Chakka made from standardized milk is as below [1a]:

Table 16.14: Composition of Chakka

Sl.No.	Constituents	Per cent	Sl.No.	Constituents	Per cent
1	Water	71-72	2	Fat	11.5
3	Proteins	12.5	4	Lactose (on DM basis)	2.5-3.5
5	Total Ash	1.0	6	Acidity (per cent LA max.)	2.1

Shrikhand [1,2,3,5]

Shrikhand is indigenous ethnic fermented milk food assumes special importance due to pleasant taste. It is an Indian sweet dish made of Chakka (strained yogurt).

i. It is semi-solid, soft, sweetish/sour fermented milk product, prepared from cow, buffalo or mixed milk dahi. Lactic fermented curd is obtained by the associative action of microorganisms on the milk constituents.

ii. It is important desserts in Gujarati and Maharashtrian cuisine and is also served with Gujarati thali sometimes as a sweet dish. Preparation of this dish is very simple but it takes some time to process yogurt properly. The Dahi (yogurt) is tied in a cotton cloth and left under pressure to drain off most of the water. In the past, it used to be hung from a wall to achieve the desired thick and creamy product. The Chakka (strained yogurt) and sugar are mixed thoroughly. The cardamom and saffron are then added and mixed. A popular variation of shrikhand in Maharashtra is **Amrakhand,** which is shrikhand mixed with mango pulp and made homogeneous with a blender.

iii. In a few parts of Gujarat, another variant of shrikhand, Matho is also very popular. The preparation method is almost the same, but some fresh fruits are always added in Matho.

Chemical composition of shrikhand is given in following table [1a]:

Table 16.15: Composition of Shrikhand

Sl.No.	Constituents	Per cent	Sl.No.	Constituents	Per cent
1	Water	35-40	2	Fat (On DM basis)	8.5-9.0
3	Proteins(On DM basis)	9.5-12.5	4	Lactose(On DM basis)	2.5-3.5
5	Sucrose (On DM basis)	70-74	6	Total Ash	0.6-1.0
7	Acidity(per cent LA max.)	1.0-1.3	8	pH	4.3-4.4

☆ Average values of product made by the mechanized process in the organized sector ,are given below [1b]:

Table 16.16: Average Values of Shrikhand

Sl.No.	*Constituents Per cent*	*Traditional Shrikhand*	*Industrial Shrikhand*
1	Water	40.52	35.11
2	Fat (On DM basis)	8.27	9.25
3	Proteins(On DM basis)	9.53	12.53
4	Lactose(On DM basis)	3.74	2.59
5	Sucrose (On DM basis)	76.25	70.00
6	Total Ash	0.53	0.683
7	Acidity(per cent LA max.)	1.075	1.310
8	FFA (per cent Oleic acid)	0.395	0.265
9	Soluble Nitrogen per cent	0.29	0.286
10	pH	4.1	4.33

☆ There are several reports about the determination of effect of addition of fruits, vegetables and condiments incorporation in Shrikhand. Shrikhand contain the properties of fermented foods along with the goodness of added fruits and condiments. Addition of different additives leads to the enhancement of colour, textural properties, nutritional value, shelf life and along with consumer acceptance.

Yoghurt [1,2,4]

Yogurt or yogourt (Turkish-Yoğurt) or Yoghurt, is a dairy product manufactured by bacterial fermentation of milk. Fermentation of lactose by these bacteria ("yoghurt cultures") produces lactic acid, which acts on milk protein to give yoghurt its texture and its characteristic tang.

☆ Cow milk is most commonly used to make yoghurt, but milk from buffalo, goats, sheep, camels and yaks is also used. Soya yoghurt, a non-dairy yoghurt alternative, is made from milk, this is not an animal product, being made from soy beans.

Nutritive Value of Yoghurt

Yoghurt is nutritionally rich in protein, calcium, riboflavin, vitamin B6 and vitamin B_{12}. It has nutritional benefits beyond those of milk.

i. People who are moderately lactose-intolerant can consume yoghurt without ill effects, because much of the lactose in the milk precursor is converted to lactic acid by the bacterial culture.

ii. The full fat Yoghurt has following nutritive values:

Table 16.17: Nutritive Values of Yoghurt

Serving Size-100 g [2,4]

Ingredient	*Quantity*	*Ingredient*	*Quantity*
Carbohydrates	4.7 g	Protein	3.5 g
Sugars	4.7 g	Vitamin A equiv.	27 μg (3 per cent)
Fat	3.3 g	Riboflavin (Vit.B_2)	0.14mg (9 per cent)
Saturated	2.1 g	Calcium	121 mg (12 per cent)
Monounsaturated	0.9 g	Energy	61 kcal

Sources

1. *M.K. Srivastava, Handbook on Analysis of Milk Products. Write and Print Publications, New Delhi, 2016. ISBN: 9789384649258.*

1a. *M. Ranganadham, Sathish Kumar M. H., Devraja H. C. and F. C. Garg. Traditional Dairy Products, www.AgriMoon.com.*

1b. *Shrikhand, dairy-technolog.com, 2014.*

2. *Wikipedia.*

3. *http: //amchirecipes.com/content/shrikhand.*

4. *USDA Nutrient database.*

5. *Aneja, R.P., Vyas, M.N., Nanda, K. and Thareja, V.K. 1977. Development of an Industrial process for the manufacture of shrikhand. J. Food Sci. Technology.14: 159-163.*

Lassi [1,4]

In India, buttermilk is known with a local name as Lassi, also known as stirred Dahi. It is a ready-to-serve fermented milk beverage popular in India particularly in summer months.

- ☆ Good quality lassi should have creamy consistency, smooth texture, slight acidic sweet taste, viscous liquid and white colour with yellowish tinge. Mild acidic flavour and sweetish taste of lassi make it a refreshing soft drink.
  - i. It can be flavoured either with salt or sugar and other condiments or spices like ginger, coriander, and mint, depending on regional consumer's preferences.
  - ii. Lassi is obtained from pasteurized whole milk or partly skim milk, cultured with lactic and aroma/flavour producing organisms.
- ☆ Originally Lassi is a product of rural part of India and it is a by-product of Dahi obtained while the separation of makkhan by churning. It was manufactured at cottage level earlier but now salted Lassi and

sweetened lassi are commercially available in urban market packaged aseptically in UHT boxes/milk cartons/sachets [1a].

☆ Lassi is a white to creamy-white viscous liquid with a sweetish, rich aroma and pleasant mild acidic taste. The chemical composition of lassi depends on the type of milk, initial composition of milk, level of concentration of milk solids and the sugar level. The details of chemical composition of lassi is given under chapter-18.

Condensed and Evaporated (Sweetened, Unsweetened and Skimmed) Milk [1,2,3,4]

Concentrated milk is milk in which solids have been concentrated by the removal of water.

i. Several processes are being used to remove water from milk *e.g.* Evaporartion, Sublimation, Partial freezing followed by removal of Ice crystals.
ii. The evaporation method is common method for removal of water from milk.

☆ Condensed milk is a milk product obtained by evaporating part of water of whole milk, or fully or partly skimmed milk, with or without the addition of sugar.

   i. The term 'condensed milk' is commonly used when referring to full cream sweetened condensed milk whereas the term evaporated milk is generally used while referring to full cream unsweetened condensed skim milk.
   ii. Skimmed milk products are known as sweetened condensed skim and unsweetened condensed skim milk respectively. The two terms, condensed milk and sweetened condensed milk, have become synonymous. The Standardisation of milk for the manufacturing of condensed milk is highly required and it is being done by using the following Fat/SNF or Fat/TS ratio-

Unsweetened Condensed Milk

Fat: SNF ratio = [8:(26-8)] or 8: 18 or 1kg fat for 2.25 kg of SNF or 0.44444 kg of Fat for 1 kg of SNF.

Unsweetened Condensed Skimmed milk

Fat: SNF ratio = [0.5:(20-0.5)] or 0.5: 19.5 or 1kg of Fat for 39 kg of SNF or 0.0256kg of Fat for 1 kg of SNF.

History [2]

As per Marco Polo, the Tartars were able to condense milk. They carried milk paste which was used by mixing water. However, this probably refers to the soft Tartar curd which can be made into a drink ("ayran") by diluting it, and therefore refers to fermented, not fresh, milk concentrate. Nicolas Appert introduced condensed milk in France in 1820 and Gail Borden, Jr. in the United States in 1853. In 1864, Gail Borden's New York Condensed Milk Company constructed the New York Milk Condensery in Brewster, New York. The first Canadian condensery was built at Truro, Nova Scotia, in 1871. In 1899, E. B. Stuart opened the first Pacific Coast Condensed Milk Company. In 1911, Nestlé constructed the world's largest condensed milk plant in Dennington, Victoria, Australia.

Advantages of Condensed/Evaporated Milk [3]

This type of milk has several advantages:

i. The higher concentration of milk solids resulted from concentration (Evaporation/Condensation) provides some protection against spoilage by certain chemical actions or microbiological agents. The greater the proportion of the water removed, the more the protection.
ii. The reduction of weight and volume of milk by concentration saves handling, transportation, storage and packing costs.
iii. Condensed/Evaporated milk makes the sweet making easier and cost effective.

Uses of Evaporated Milks

Evaporated milk is a multipurpose, convenient dairy product ready for every milk use. Pouring directly from the can, evaporated milk can be used in following countless applications-

i. creaming coffee or tea.
ii. poured on cereals and fruits.
iii. providing consistency in meat patties and loaves,
iv. coatings for baked or fried meats, and.
v. in place of milk in the manufacture of candies, frostings and pies.

Food and Nutritive Value of Condensed Milk [1,2]

The nutritive value of the sweetened and unsweetened condensed milk is very high.

i. Both of them are rich in fat and fat soluble vitamins *viz.* A, D, E and K and body building proteins, bone forming minerals and energy giving lactose.
ii. While sweetened condensed milk is especially high in energy giving sucrose because of the added sugar, evaporated milk is suitable for infant feeding since it makes a soft curd and is easily digested.

iii. The Details related to presence of nutrients in condensed milk are given in following table:

Table 16.18: Nutrients in Condensed Milk

Sl. No.	Type of Condensed Milk	Water	Total solids	Fat	Milk Solids Not Fat	Protein	Lactose	Ash	Total Milk Solids	Sucrose
1	Condensed milk	26.0	74.0	9.0	22.0	8.3	12.2	1.5	31.0	43.0
2	Evaporated milk	31.0	69.0	9.0	22.0	8.3	12.2	1.5	31.0	—
3	Skim sweetened	29.0	71.0	0.5	25.5	9.3	14.0	2.2	26.0	45.0

Sources

1. *M.K. Srivastava, Handbook on Analysis of Milk Products. Write and Print Publications, New Delhi, 2016. ISBN: 9789384649258.*

1a. *Ramana, B.L.V. Standardization of a method of manufacture of lassi with enhanced shelf stability. M.Sc. Thesis. NDRI Deemed University, Karnal, India 1994.*

2. *Wikipedia.*

3. *J.N.Warner, Principles of Dairy Processing.*

4. *Aneja, R.P., Mathur, B.N., Chandan, R.C. and Banerjee, A.K. 2002. Technology of Indian Milk Products. A Dairy India Publication, Delhi, India.*

Desiccated Milk Based Products/Non Curdled Milk Products

Khoa/Mawa [1,4]

Khoa is reputed as versatile intermediate base for several sweets *i.e.* Burfi, Peda, Kalakand, Milk cake, Pantua. It is also called as Khoya, Khava, Kava, Khawa, mawa and Palghoa. Khoa or Mawa is made by reducing milk in an open pan over heat.

☆ It contains fairly large amount of Fat (energy source),Proteins (muscle building), Lactose and Minerals (bone forming) and fair amount of Vitamins A and D and water soluble Vitamin B.

i. Khoa is a rich source of energy(458 kcal/100gm).Dhap, Pindi and Danedar khoa are available in the market.

ii. The shelf life of khoa is 2-4 days at ambient temperature and 3 weeks under refrigerated condition. There is a potential of achieving a 120 days refrigerated shelf life by using modern technologies.

Chemical Quality of Khoa [2,3]

Wide variations exist in chemical quality of cow and buffalo milk khoa. The market samples of khoa show wide variations in chemical composition. The details of chemical quality of Buffalo milk khoa and Cow milk khoa are given in following table [2,3]:

Table 16.19: Chemical Quality of Buffalo and Cow Milk Khoa

Sl.No.	Constituents Per cent	Buffalo Milk Khoa	Cow Milk Khoa
1	Moisture	22.34	30.93
2	Total Solids	77.66	69.07
3	Fat	32.20	22.0
4	Proteins	17.70	19.10
5	Lactose	23.70	24.10
6	Ash	3.71	3.72

Chemical Composition of Three Varieties of Khoa

The average chemical composition of three varieties of khoa is tabulated below [3]:

Table 16.20: Chemical Composition of different Types of Khoa

Sl.No.	Constituents per cent	Types of Lhoa		
		Pindi	Dhap	Danedar
1	Total Solids	65.0	56.0	61.5
2	Fat	22.5	19.4	20.9
3	Proteins	19.5	17.1	18.4
4	Lactose	19.4	16.7	18.9
5	Ash	3.5	2.9	3.4

Nutritional Value of Khoa [1]

Khoa contains fairly large quantities of muscle building proteins, bone forming minerals and energy giving fat and lactose.

i. In concentrated form, it has most of the fat soluble vitamins and many water soluble vitamins contained in the original milk.

ii. Khoa made from whole cow milk and skimmed milk has following nutritional values:

Table 16.21: Nutritional Values of Khoa

Serving Size-100g

Sl.No.	Parameters/100g	Khoa made from Cow Milk(i)	Khoa Made from Skimmed Milk(ii)
1	Energy(Kcals)	413- 421	206
2	Moisture(g)	25	-
3	Protein(g)	20	22.3
4	Fat(g)	26	1.6
5	Mineral(g)	4	-
6	Carbohydrates(g)	20.5- 25.0	25.7
7	Fibre(g)	-	-
8	Calcium(mg)	956	990
9	Phosphorous(mg)	613	-
10	Iron(mg)	-	2.7

Sources

1. *M.K. Srivastava, Handbook on Analysis of Milk Products. Write and Print Publications, New Delhi, 2016. ISBN: 9789384649258.*
2. *M. Ranganadham, Sathish Kumar M. H., Devraja H. C. and F. C. Garg. Traditional Dairy Products, www.AgriMoon.com.*
3. *Physico- chemical changes in milk constituents during manufacture and storage of Indigenous products, courseware.cutm.ac.in.*
4. *Aneja, R.P., Mathur, B.N., Chandan, R.C. and Banerjee, A.K. 2002. Technology of Indian Milk Products. A Dairy India Publication, Delhi, India.*

Ice Cream and Kulfi

Ice Cream (Originally iced cream) [1,2,4]

It is a frozen dessert made from dairy products, such as milk and cream, combined with flavorings and sweeteners, such as sugar, and possible other ingredients.

- ☆ A smooth, sweet, cold food prepared from a frozen mixture of milk products and flavorings, containing a minimum of 10 per cent milk fat and eaten as a snack or Dessert.
  - i. The composition of ice cream may vary depending on whether it is an economy brand satisfying minimal requirements, a trade brand of average composition, or a premium brand of superior composition.

ii. The components by weight of an average-composition ice cream are 12 per cent fat, 11 per cent nonfat milk solids, 15 per cent sugar, and 0.3 per cent vegetable gum stabilizer.

The Nutritional Value of Ice cream [1,2]

The nutritional value of ice cream varies according to the amount of milk or cream it contains. Other common ingredients of ice cream include added sugar, stabilizers, emulsifiers and flavorings, none of which contain nutrients. Ice cream may have the following composition:

i. Greater than 10 per cent milkfat and usually between 10 per cent and as high as 16 per cent fat in some premium ice creams.
ii. 9 to 12 per cent milk solids-not-fat: this component, also known as the serum solids, contains the proteins (caseins and whey proteins) and carbohydrates (lactose) found in milk.
iii. 12 to 16 per cent sweeteners: usually a combination of sucrose and glucose-based corn syrup sweeteners.
iv. 0.2 to 0.5 per cent stabilizers and emulsifiers and
v. 55 per cent to 64 per cent water which comes from the milk or other ingredients. These compositions are percentage by weight.

- Since ice cream can contain as much as half air by volume, these numbers may be reduced by as much as half if cited by volume. In terms of dietary considerations, however, the percentages by weight are more relevant. Even the low fat products have high caloric content. Higher fat ice creams are easily consumed to excess and may cause weight gain. Also, this type of easily-digestible food is a popular choice for binge-eaters.
- Ice Cream Contains Vitamins A, B_2 and B_{12}. The more gourmet varieties that includes butterscotch, fudge, chocolate pieces, or cookie dough bits provide approximately 300 or more calories per half cup where as straight vanilla or chocolate variety though that is not as rich in flavor provides 125 calories per half a cup mark.
- Ice cream does not provide a whole lot of nutrients. The tiny amount of calcium found in it, but more importantly it is high in sugar and also saturated fat depending on the variety. Ice cream provides following nutrition:

Table 16.22: Nutritive Values of Ice Cream

Nutrition Fact - Serving -1/2 cup Calories- 240(1003kj)

Sl.No.	Nutrients	Quantity in per Serve	Per cent Daily Value
1	Fat — Total Fat/Sat. Fat	16 g/11 g	25/55
2	Cholesterol	75 mg	25
3	Total Carbs - Dietary Fibre/Sugars	21 g 0g/19g	7 0
4	Protein	2.5g	-
5	Sodium	60 mg	3
6	Calcium	4 mg	-
7	Potassium	150mg	-

Kulfi [1,3]

Kulfi is Indian version of ice cream that contains delicate spicy flavours of spices such as cardamom, saffron, cinnamon. Traditionally, Kulfi is not whipped during freezing resulting in somewhat of solid dessert similar to frozen custards.

The details related to the nutritional value of kulfi are given below:

Table 16.23: Nutritional Value of Kulfi

Nutritional Fact- Number of Servings- 5 (~60 g/serv.) [3]

Sl.No.	Nutrients	Quantity in per Serve	Per cent Daily Value
1	Total Calories (kilo calories)	266.7	13
2	Total Calories from Fat (kilo calories)	175.9	-
3	Total Fat (g)	19.9	31
4	Total Saturated Fat (g)	10.7	53
5	Total Transfat (g)	3.9	-
6	Cholesterol (mg)	53.7	18
7	Sodium (mg)	59.8	2
8	Carbohydrate (g)	18.1	6
9	Dietary Fiber (g)	0.9	4
10	Sugar (g)	10.4	-
11	Protein (g)	5.7	-
12	Vitamin A (IU)	525.2	11
13	Vitamin C (mg)	1.5	3
14	Calcium (mg)	168.6	17
15	Iron (mg)	0.4	2

Table 16.24: Nutritive Value of Ice Creams [5]

One Serving (12cup, 4.4 fl oz, 120 mL, or Approximately 66 g) of Ice Cream.

Sl.No.	*Parameters*	*Types of Ice crams*							
		Vanilla Ice cream		*Vanilla Ice cream*		*Vanilla Ice Cream*		*Orange Sherbet*	
		Fat 4 per cent	*Per cent Daily Value**	*Fat 10 per cent*	*Per cent Daily Value**	*Fat 16 per cent*	*Per cent Daily Value**	*Fat 2 per cent*	*Per cent Daily Value**
1	Total calories, Kcal	123.4	-	133.6	-	174.7	-	135.8	-
2	Calories from fat Kcal.	34.1	-	64.0	-	106.6	-	17.3	-
3	Total fat g	3.8	6	7.1	11	11.8	18	1.9	3
4	Saturated fat g	2.4	12	4.4	22	7.4	37	1.2	6
5	Cholesterol mg	12.2	4	29.5	10	43.8	15	7.1	2
6	Total Carbohydrate g	19.5	6	15.7	5	16	5	29.5	10
7	Sugars g	18.4	-	14.8	-	15.1	-	27.8	-
8	Protein g	3.5	7	2.4	5	2.1	4	1.1	2
9	Dietary fiber g	0	0	0	0	0	0	0	0
10	Vitamin A (IU)	143.4	2	269.3	6	448.4	8	93.1	0
11	Vitamin C mg	0.5	0	0.4	0	0.3	0	1.9	4
12	Sodium mg	70.2	3	57.6	2	54.1	2	44.4	2
13	Calcium mg	118.3	10	87.2	8	75.6	8	52	6
14	Iron, mg	0.1	0	0.1	0	0.1	0	0.2	0

Per cent Daily value is based on 2000 calorie diet.

Sources

1. *M.K. Srivastava, Handbook on Analysis of Milk Products. Write and Print Publications, New Delhi, 2016. ISBN: 9789384649258.*
2. *Wikipedia.*
3. *Food-india.com..*
4. *R.P. Aneja, B.N. Mathur, R.C. Chandan, and A.K. Banerjee, 2002. Technology of Indian Milk Products. A Dairy India Publication, Delhi, India.*
5. *Nutritive Value of Foods, Home and Garden Bulletin Number 72, October 2002. Available at www.nal.usda.gov/fnci/foodcomp.*

Milk Powders [1,2,6]

Milk Powder is a dairy product manufactured by evaporating milk to dryness. Pasteurized milk is first concentrated in an evaporator to about 50 per cent milk solids. The resulting concentrated milk is sprayed into a heated chamber where the water almost instantly evaporates, leaving fine particles of powdered milk solids. The drying of milk serves the following purposes-

- **i.** To preserve the milk - milk powder has a far longer shelf life than liquid milk and does not need to be refrigerated, due to its low moisture content.
- **ii.** To reduce the bulk of milk for economy of transportation.

Skimmed Milk Powder (Nonfat dry milk) is a product that is created when water is removed from pasteurized skim (nonfat milk). By depriving micro-organisms of the water which they require in order to develop, drying allows milk powder to have a shelf life of up to 3 months to 1 year if stored properly.

- ✰ **Milk powder** generally contains fat (Partially Skimmed and Whole milk powder), lactose, milk proteins and minerals in the same relative proportions as the fresh milk from which it was made.
- ✰ Dry milk powder must be reasonably uniform in composition, white- or cream-coloured, free from the brown or yellow colour (typical of overheated product) or any other unnatural colour and and must also be substantially free from brown specks. Its flavour and odour must be sweet, clean and free from rancid, thallowy, fishy, cheesy, soapy or other objectionable flavours and odours.

Nutritional Values of Milk Powders [3-8]

Milk powders contain all twenty standard amino acids (the building blocks of proteins) and a good source of soluble vitamins and minerals(US Dairy Export Council).

i. As per USAID, the typical average amounts of major nutrients in the non- reconstituted milk are (by weight) 36 per cent protein, 52 per cent carbohydrates (predominantly Lactose), calcium 1.3 per cent, potassium 1.8 per cent.

ii. Indian Skim milk powder has average Milk Fat 1 per cent, Milk Protein 35 per cent,Carbohydrates 51 per cent, Minerals 7 per cent and Calorific Value(approx.) 350 kcal/100 gm.

iii. Inappropriate storage conditions (high relative humidity and high ambient temperature) can significantly degrade the nutritive value of milk powder [5].

iv. Commercial milk powders are reported to contain oxysterols- oxidized cholesterol [6] in higher amounts than in fresh milk (up to 30µg/g, versus trace amounts in fresh milk [7]. The oxysterol free radicals have been suspected of being initiators of atherosclerotic plaques.

Skim Milk Powder contains following nutritional values and functional Properties:

Table 16.25: Nutritional Values of Skim Milk Powder

Sl.No.	*Ingredients - Major (per cent)*			
1	Moisture	3.0 - 4.0	Lactose	49.5 - 52.0
	Fat	0.6 - 1.25	Ash	8.2 - 8.6
	Protein	34.0 - 37.0		
2	*Essential Amino Acids (g/100g Protein)*			
	Isoleucine	2.19	Methionine	0.91
	Leucine	3.54	Phenylalanine	1.75
	Lysine	2.87		
3	*Minerals (mg/100g)*			
	Calcium	1248.00	Iron	0.40
	Sodium	494.00	Magnesium	110.00
	Potassium	1674.00	Zinc	4.08
	Phosphorus	993.00		
Sl.No.	*Minerals (mg/100g)*			
4	Whey Protein –N_2 Index (WPNI)			
	High heat and High stability (HHHS)/ High heat (HH)		Not more than 1.5 mg/g	
	Medium-Heat (MH)		Between 4.00 - 5.99 mg/gm	
	Low-Heat (LH)		Not less than 6.0 mg/gm	

Functional Properties [1,2,5,6]

Dairy ingredients likes SMP perform a wide range of key functions in prepared foods, including emulsification, thickening, gelling and foaming. In addition to the aforesaid key functions, they also provide nutrition and desirable sensory characteristics to the manufactured foods. Extensive worldwide food research has focused on understanding milk protein interaction in these products in order to optimize the manufacturing process and to control product functionality in food applications.

- Addition of SMP to a food formulation provides a variety of desirable benefits since it contains approximately 35 per cent of milk proteins and 50 per cent lactose. The some of the notable functional benefits of SMP are given below:

Table 16.26: Functional Properties of Skim Milk Powder

Functional Property	*Mode of Action*	*Product Use*
Emulsification	Formation and stabilization of fat emulsions (proteins acting as oil/ water interfaces)	Salad dressings, Sausages (meat emulsions),Soups, sauces, Dairy beverages,Baked goods
Foaming/Whipping*	Formation of a stable film of surrounding air bubbles (proteins acting as air-water interfaces)	Ice creams (Soft-serve and hard frozen), Cakes, mousses, Choose a powder that is free of antifoaming agents
Browning/Color/ Flavour	Lactose and proteins undergo Maillard caramelization	Baked goods (doughs, cakes, muffins)
Thickening/Water-binding/Gelling	Addition of milk solids to increase viscosity and gel strength by trapping water lactose and proteins show high water holding capacities)	Poudings,Yogurts (set types), Sauces, Baked goods, Dairy beverages
Solubility	i. High-Heat (HH) = Low Solubility	Prevents loaf volume depression in bakery products, High moisture absorbing quality desirable in meat,confectionary and bakery products.
	ii. Medium-Heat (MH) = Medium solubility	Used in ice cream, dry mixes and others products like confections and frozen desserts in which water absorption and flavour are important.
	iii. Low-Heat (LH) = High solubility	Possesses optimum sensory qualities,Ideal for use in dairy products and beverages, Most common source of instant skim milk powder.

Application/Uses of Milk Powders [1,2,6]

Milk powders or Dried milk is used widely as a food ingredient. High nutritional quality combined with its versatile and multi-functional properties of SMP make it useful for following:

A. Application in Dairy Industry

i. Reconstitution (reconstituted milk is the liquid milk obtained by adding water to skim milk powder).

ii. Recombination (recombined milk is the liquid milk obtained by adding water to skim milk powder and adding milk fat separately in such a quantity that the desired fat content is achieved),

iii. As a substitute for mother's milk in baby foods (infant formulas),

iv. Producing ice creams, low-fat set-type yogurts, frozen desserts, fresh cheeses, and cultured skim milk and

v. Producing healthy dairy beverages.

B. Application in Bakery

i. Mixing into dough to increase the volume of the bread and improve its water-binding capacity, allowing the bread to remain fresh for a longer period of time.

ii. Mixing into pastry dough to make it crisper and

iii. It is useful as a substitute for eggs in bread and pastries.

C. Confectionery/Chocolate Industry

Useful in producing different types of yeasty and delicious milk chocolates in the chocolate industry.

D. Application in Food Industry

Useful in producing sausages and various types of ready-cooked meals in the food industry and catering trade.

E. Application in Animal Feed Industry

Useful in producing animal feed.

F.

Milk Powder is also widely used in various sweets such as the famous Indian milk balls known as Rasagolla and popular Indian sweet delicacy (sprinkled with desiccated coconut) known as Chumchum (made with skim milk powder).

G.

Milk Powder is also a common item in UN food aid supplies, fallout shelters, warehouses, and wherever fresh milk is not a viable option.

Sources

1. *M.K. Srivastava, Handbook on Analysis of Milk Products. Write and Print Publications, New Delhi, 2016. ISBN: 9789384649258.*
2. *Wikipedia.*
3. *Milk Powder Nutritional Information, US Dairy Export Council.*
4. *USAID Commodities Reference Guide, Non Fat Dry Milk (Jan 2006).*
5. *Okamoto and R. Hayashi (1985) "Chemical and Nutritional Changes of Milk Powder Proteins under Various Water Activities" Agric. Biol. Chem., Vol.49 (6), pp1683-1687.*
6. *Advanced Dairy Chemistry- Volume 2 - Lipids by P.F. Fox and P. McSweeney, Birkhäuser, 2006, p655.*
7. *Toxins in Food by W.M. Dabrowski and Z.E. Sikorski, CRC Press, 2004, p 296.*
8. *R.W. Hubbard, Y.Ono and A. Sanchez (1989) Prog Food Nutr Sci. Vol.13(1) pp 17-44.*

Cheese and Processed Cheese [1,1a,2]

Cheese is the most important dairy product because of concentrated form of milk constituents, mainly fat, casein and insoluble salts together with water in which small amounts of soluble salts, lactose and lactalbumin are also found and it works as balancing wheel of dairy industry because of the followings:

i. Provides best method for conserving and preserving milk solids.
ii. Provides an outlet for the utilisation of seasonal/regional surplus milk solids.
iii. It is a nutritionally balanced and convenience food.
iv. Manufacturing is amenable to modern technology.
v. Availability of various varieties to meet the consumer's demand,
vi. Cheese is more compact and has a longer shelf life than the milk from which it is made. The long storage life of cheese allows selling it when markets are more favorable.

Cheese is a food consisting of proteins and fat from milk, usually the milk of cows, buffalo, goats, or sheep and other mammal milk.

i. Cheese can be defined as a fresh or motured solid or semi solid product produced by coagulation of the milk protein casein and hence it is also called as fermented and coagulated dairy product.
ii. Coagulation is effected either by means of lactic acid produced by microorganism or by the addition of the enzyme rennet or by both or by direct acidification.

iii. The solids are then separated and pressed into final form.

iv. Some cheeses also contain molds, either on the outer rind or throughout. For a few cheeses, the milk is curdled by adding acids such as vinegar or lemon juice. Most cheeses are acidified to a lesser degree by bacteria, which turn milk sugars into lactic acid, then the addition of rennet completes the curdling. Some vegetarians avoid any cheese made from animal-based rennet.

v. Most widely available vegetarian cheeses are made using rennet produced by fermentation of the fungus Mucor miehei. Vegetarian alternatives to rennet are available; most are produced by fermentation of the fungus Mucor miehei, but others have been extracted from various species of the Cynara thistle family.

vi. Paneer and Channa could also be classified as cheese. Cheese is one of the most used ingredients for breakfast around the world. It is a milk based solid food and

vii. Cheeses are eaten both on their own and cooked in various dishes; most cheeses melt when heated. Cheese has served as a hedge against famine and is a good travel food. It is valuable for its portability, long life, and high content of fat, protein, calcium, and phosphorus.

Types of Cheese [1,1a]

Hundreds of types of cheese exist all over the world. The types of cheese exist due to the usage of the milk from different mammals, specific species of molds and bacteria and also varying the aging length.

i. Other processes are also used to prepare different types of cheese.

ii. Other factors which determine the type of cheese is the diet of the animal the milk is taken from. The diet can include herbs, spices and wood smoke.

Cheese cannot be categorized on a single categorization concept. Therefore, there are various systems used for the categorization of cheese. Some factors taken into consideration while classifying cheese are the length of aging, the methods of making it, the curd and the various processes relating to the curd, the kind of milk, the fat content and the whether the texture of the cheese is hard or soft. The yellow to red color of many cheeses is a result of adding annatto.

Factors in Categorization of Cheeses [1,2]

No one categorization scheme can capture all the diversity of the world's cheeses. In practice, no single system is employed and different factors are emphasised in describing different classes of cheeses. The Length of aging,Texture,Methods of making, Fat content, Kind of milk and Country/ Region of Origin are the important factors relevant to the categorization of cheeses. The details of categorisation based on the aforesaid factors are given below:

i. Categorisation on Aging

a. **Fresh cheeses [1,1a]**-The most common type of cheese is Fresh Cheese(curdling of milk and draining). Some examples are -Cottage Cheese, Cream Cheese,Mascarpone, Mozzarella, Italian Ricotta, Romanian Ca°, **Romanian Urda**,Neufchâtel (the model for American-style cream cheese), Chevre,Whey,Provencal Brousse, Corsican Brocciu,Romanian Urda,Greek Mizithra, Norwegian Geitost,Corsican cuisine,Paneer,Queso Fresco are the examples of fresh cheese. Whey cheeses are fresh cheeses made from the **whey** discarded while producing other cheeses., Brocciu is mostly eaten fresh, and is as such a major ingredient in Corsican cuisine, but it can be aged too. Traditional **pasta filata** cheeses such as **Mozzarella** are also fresh cheese.

b. **aged cheeses**.

ii. Categorisation on Texture

a. Usually aged cheeses or ripened cheeses are further classified according to texture *i.e.* Hard cheeses, Semi firm cheeses, Semi soft cheeses. Categorizing cheeses by firmness is a common but in exact practice. The lines between "soft", "semi-soft", "semi-hard", and "hard" are arbitrary, and many types of cheese are made in softer or firmer variations. The factor controlling the hardness of a cheese is its moisture content which is dependent on the pressure with which it is packed into molds and the length of time it is aged. Cheeses that range in texture from semi-soft to firm include Swiss-style cheeses like Emmental and Gruyère(Bacteria give such cheeses their holes also contribute to their aromatic and sharp flavors) and **Gouda, Edam, Jarlsberg Cantal** (this type of cheeses are ideal for melting and are used on toast for quick snacks).Harder cheeses have a lower moisture content than softer cheeses. They are generally packed into molds under more pressure and aged for a longer time. Cheeses that are semi-hard to hard include the familiar Cheddar, Cheshire and Gloucester. **Hard cheeses** - Grating cheeses such as Parmesan and Pecorino Romano, are quite firmly packed into large forms and aged for months or years.

b. **Soft-ripened and Blue-vein cheese**-Certain types of soft cheeses have mold on the rind that encourages the cheese to ripen closest to the rind first, then the ripening progresses towards the middle of the cheese. Such cheeses often have a runny texture closest to the rind and are firmer in the middle. There are three main categories of cheese in which the presence of mold is a significant feature-soft ripened cheeses, washed rind cheeses and blue cheeses. **Soft-ripened cheeses** are those which begin firm and rather chalky in texture but are aged from the exterior inwards by exposing them to mold(Penicillium candida **or** P. camemberti. Brie and Camembert are made by allowing white mold to grow on the outside of a soft cheese. **Washed-rind cheeses** are soft in character and ripen inwards like those with white molds; however, they are treated differently. Washed rind cheeses are periodically cured in a solution of saltwater brine and other mold-bearing agents. Blue cheese is created by inoculating a cheese with Penicillium roqueforti or Penicillium glaucum. This is done while the cheese is still in the form of loosely pressed curds, and may be further enhanced by piercing a ripening block of cheese with skewers in an atmosphere in which the mold is prevalent. The mold grows within the cheese as it ages. These cheeses have distinct blue veins which gives them their name *e.g.* Roquefort, Gorgonzola, and Stilton,

iii. Categorization on Process and Methods of making Cheese

Processed cheese is another type of cheese and it is prepared by adding emulsifying agents, milk, preservatives, more salt and food coloring to traditional cheese. Velveeta and yellow American cheese are the most popular types of processed cheese.

iv. Categorisation on Milk content/Fat Content

Some cheeses are categorized by the source of the milk used to produce them or by the added fat content of the milk from which they are produced. Cheese is generally made from cows' milk, many parts of the world also produce cheese from goats and sheep. Feta style cheeses, for example, being made from goats' milk in Greece and of sheep and cows milk elsewhere. Roquefort, produced in France and **Pecorino Romano**, produced in Italy, are made from ewes's milk. Double cream cheeses are soft cheeses of cows' milk which are enriched with cream so that their fat content is 60 per cent or, in the case of triple creams, 75 per cent. The reduced-fat, fat-free cheeses, Imitation cheese are also examples of such type of cheeses.

v. Categorization on Region and Country

Africa, Asia, North America, **cream cheese**,Central America,South America, Europe (**Italian cheese, cheese Spanish**), **USA cheese**, Australia and South Pacific cheese are also available.

Composition of Cheese [1c]

The approximate composition of some cheeses are tabulated below:

Table 16.27: Composition of some Cheeses

Sl. No.	Types of Cheese	Moisture Per cent	Fat Per cent	Protein Per cent	Ash Per cent Salt Free	Salt	Minerals		Energy (Calories/ 100g)
							Ca	P	
1	Cottage Uncreamed	79.5	0.3	15.0	0.8	1.0	0.1	0.15	200
2	Cheddar	37.5	32.0	25.0	2.0	1.5	0.86	0.60	398
3	Gouda	38.5	28.5	25.5	2.5	1.5	-	-	377
4	Swiss	39.0	28.0	27.0	2.0	1.5	0.90	0.75	370
5	Mozzarella	54.0	18	22.1	2.3	0.7	-	-	290
6	Roquefort	39.5	32.0	22.0	2.0	4.0	0.65	0.45	385
7	Brick	41.0	31.0	22.0	1.2	1.8	-	-	370
8	Limburger	45.5	28.0	22.0	2.0	2.1	0.50	0.40	345

Nutritional Value for Cheese [1,3-7]

Cheese is a delicious, nutritious and versatile dairy food. It contains a high concentration of essential nutrients relative to its energy level. The precise nutritional composition of cheese, is determined by the following multifactorial parameters:

a. The type of milk used (species, breed, stage of lactation, and fat content)
b. The manufacturing and ripening procedures.

Generally cheese is rich in the fat and casein constituents of milk, which are retained in the curd during manufacture. It contains relatively small amounts of the water soluble constituents (whey proteins, lactose, and water soluble vitamins), which partition mainly into the whey.

Advantages

i. In general, cheese supplies a great deal of calcium, protein, and phosphorus. A 30-gram serving of Cheddar cheese contains about 7 grams of protein and 200 milligrams of calcium. Nutritionally, cheese is essentially concentrated milk: it takes about 200 grams of milk to provide that much protein, and 150 grams to equal the calcium [1,3].

ii. Some studies claim to show that Cheddar, Mozzarella, Swiss and American cheeses can help to prevent tooth decay because of the followings- a-calcium, protein, and phosphorus in cheese may act to protect tooth enamel., b-Cheese increases saliva flow, washing away acids and sugars,

iii. Cheese may have an antibacterial effect in the mouth [1,4,5].

Cheese has following nutritional values [1,3]:

Table 16.28: Nutritional Values of Cheese

Sl.No.	Parameters	Values	Sl.No.	Parameters	Values
1	Calories	380.00	18	Thiamin (mg)	0.06
2	Protein (g)	26.93	19	Riboflavin (mg)	0.30
3	Fat Total (g)	27.80	20	Niacin (mg)	0.09
4	Carbohydrate (g)	5.38	21	Vitamin B6 (mg)	0.08
5	Fiber - Total (g)	0.00	22	Folate - Total (mcg)	6.00
6	Sugar - Total (g)	1.32	23	Food - Folate (mcg)	6.00
7	Calcium (mg)	791.00	24	Folate-DFE (mcg_DEF)	6.00
8	Iron (mg)	0.20	25	Vitamin B12 (mcg)	3.34
9	Magnesium (mg)	38.00	26	Vitamin A (IU)	830.00
10	Phosphorus (mg)	567.00	27	Retinol (mcg)	214.00
11	Potassium (mg)	77.00	28	Vitamin E (mg)	0.38
12	Sodium (mg)	192.00	29	Vitamin K (mcg)	2.50
13	Zinc (mg)	4.36	30	Fat - Saturated (g)	17.78
14	Copper (mg)	0.04	31	Fat-Monounsaturated(g)	7.27
15	Manganese (mg)	0.01	32	Fat- Polyunsaturated (g)	0.97
16	Selenium (mg)	18.20	33	Cholesterol (mg)	92.00
17	Vitamin C (mg)	0.00			

Disadvantages [1,6,7]

Cheese potentially shares milk's nutritional disadvantages as well. The Center for Science in the Public Interest describes cheese as America's number one source of saturated fat, adding that the average American ate14 kg (30 lb) of cheese in the year 2000, up from 5 kg (11 lb) in 1970. Their recommendation is to limit full-fat cheese consumption to 57 g (2 oz) a week. Whether cheese's highly saturated fat actually leads to an increased risk of heart disease is called into question when considering France and Greece, which lead the world in cheese eating (more than 400 g/14 oz a week per person, or over 20 kg/45 lb a year) yet have relatively low rates of heart disease. This seeming discrepancy is called the French Paradox, the higher rates of consumption of red wine in these countries is often invoked as at least a partial explanation.

Sources

1. *M.K. Srivastava, Handbook on Analysis of Milk Products. Write and Print Publications, New Delhi, 2016. ISBN: 9789384649258.*

1a. *Aneja, R.P., Mathur, B.N., Chandan, R.C. and Banerjee, A.K. 2002. Technology of Indian Milk Products. A Dairy India Publication, Delhi, India.*

1c. *S. K. Kanawjia and Yogesh Khetra, Cheese Technology, AgriMoon.com.*

2. *Wikipedia.*

3. *Nutritional data from CNN Interactive, Retrieved October 20, 2004.*

4. *National Dairy Council. "Specific Health Benefits of Cheese". Retrieved October 15, 2005.*

5. *The Pharmaceutical Journal, Vol. 264 No 7078 p48 January 8, 2000 Clinical.*

6. *Center for Science in the Public Interest (2001). "Don't Say Cheese". Retrieved October 15, 2005.*

7. *McGee, p 67. More food poisonings in Europe are caused by pasteurized cheeses than raw-milk.*

B. Legal Quality Standards of Milk Products

For maintaining the quality and to prevent adulteration of different food products including milk and milk products certain standard specifications have been laid down.

a. Regulatory/Quality Agencies

Different Regulatory/Quality control agencies have prescribed quality specifications for milk and milk products in India. Some such agencies are:

1. Food Safety and Standards Authority of India [1,2]

It is a statutory body established under the Ministry of Health and Family Welfare, Government of India. The Food Safety and Standards Authority of India (FSSAI) is a statutory body under Food Safety and Standards Act, 2006. The Food Safety and Standards Act (FSS)-2006 is the primary law for the regulation of food products. Standards framed by FSSAI are prescribed under Food Safety and Standards (Food Product Standards and Food Additives) Regulation, 2011, Food Safety and Standards (Packaging and Labelling) Regulation, 2011 and Food Safety and Standards (Contaminants, Toxins, and Residues) Regulations, 2011.

The FSS Act is a bucket for all the older laws, rules and regulations for food safety. The FSS Act took following older acts into one umbrella-

Prevention of Food Adulteration Act, 1954,Fruit Products Order, 1955,Meat Food Products Order, 1973,Vegetable Oil Products (Control) Order, 1947,Edible Oils Packaging (Regulation) Order 1988,Solvent Extracted Oil, De- Oiled Meal and Edible Flour (Control) Order, 1967 and Milk and Milk Products Order, 1992.

The FSSAI has prescribed standards for the following:

Dairy products and analogues,Fats, oils and fat emulsions,Fruits and vegetable products,Cereal and cereal products, Meat and meat products, Fish and fish products,Sweets and confectionery, Sweetening agents including honey, Salt, spices, condiments and related products, Beverages, (other than dairy and fruits and vegetables based),Other food product and ingredients, Proprietary food, Irradiation of food,Fortification of staple foods *i.e.* vegetable oil, milk, salt, rice and wheat flour/maida.

Earlier Directorate General of Health Services, Ministry of Health and Family Welfare, Government of India has prescribed quality standards, which are known **as Prevention of Food Adulteration Act (PFA Standards).**

2. Bureau of Indian Standards (BIS)- [1,3]

The Bureau of Indian Standards (BIS) is the National Standards Body of India under Department of Consumer affairs, Ministry of Consumer Affairs, Food and Public Distribution, Government of India. It is established by the Bureau of Indian Standards Act, 2016 for the harmonious development of the activities of standardization, marking and quality certification of goods and for matters connected therewith or incidental thereto. BIS has been providing traceability and tangibility benefits to the national economy in a number of ways – providing safe reliable quality goods; minimizing health hazards to consumers; promoting exports and imports substitute; control over proliferation of varieties *etc.* through standardization, certification and testing. The new name *i.e.* BIS came into existence from 1st April, 1987 under the BIS Act 1986.

The organisation was formerly the **Indian Standards Institution (ISI),** set up under the Resolution of the Department of Industries and Supplies No. 1 Std.(4)/45, dated 3 September 1946. The ISI was registered under the Societies Registration Act, 1860.

3. Agriculture Produce Grading and Marketing, Directorate Ministry of Agriculture Government of India (AGMARK Standards) [1,4]

In order to have a systematic marketing of Agricultural Produce on the basis of well defined quality, Indian Legislature in 1937 passed an act known as Agriculture Produce (Grading and Marking) Act, 1937. Rules under this Act are called-General Grading and Marking Rules, 1937. Rules have been revised in 1988 and are called General Grading and Marking Rules, 1988.Grading of Agricultural items under these rules is called AGMARK GRADING or Agmark certification. Among dairy products, only ghee and creamery butter (Table butter or salted butter) are graded under Agmark.

Out of these standard specifications, the specification prescribed by FSSAI under Food Safety and Standard Act-2006 and Food Safety and Standard (Food Product Specification and Food Additives) Regulation -2011 are mandatory or legal standards while the standards prescribed by AGMARK or BIS are obligatory/optional/quality standards.

Sources

1. *Wikipedia.*
2. *FSSAI website.*

3. *BIS website.*
4. *Agmark website.*

b. Quality Standards (FSSAI, BIS and Agmark)

FSSAI has prescribed quality standards for milk and milk products. Quality standard for raw (Cow, Buffalo,Sheep, Goat,Camel *etc.*) and processed milk (Skim milk, Double Toned, Toned, Standardised milk and Full cream milk) are given in part -2 under Chemistry of milk.

Quality Standards for Milk Products (FSSAI, BIS and Agmark)

Food Safety and Standards (Food Products Standards and Food Additives) Regulation-2011

Milk Product" means a product obtained by processing of milk, which may contain food additives and other ingredients functionally necessary for the milk product as permitted in these regulations and shall include the following, namely:

i. Cheese and Processed cheese.
ii. Chhana, skimmed-milk chhana and Paneer,
iii. Condensed milk-sweetened and unsweetened and condensed skimmed milk-sweetened and unsweetened,
iv. Cream.
v. Curd, skimmed milk curd, Dahi,
vi. Ghee, butter oil,
vii. Ice-cream,
viii. Infant milk food,
ix. Khoa,
x. Malai,
xi. Milk derivatives such as whey proteins, casein, lactose *etc.*
xii. Milk ices, milk lollies, kulfi,
xiii. Milk powder, skimmed milk powder, partly skimmed milk powder,
xiv. Table butter and white butter,
xv. yoghurt,
xvi. any other product as may be specified in these regulations.

Earlier FSSAI has prescribed Quality standards for milk and milk products under Food Safety and Standards (Food Products Standards and Food Additives) Regulation -2011 issued on 1.8.2011 but later on this organisation has also prescribed standards for essential composition (raw materials),Food Additives, Contaminants, Toxins and Residues, Hygiene, Labelling and method of sampling along with quality standards of products (Codex has already given standards of milk and milk products on aforesaid pattern) in compendium of Food Safety and Standards(Food Products Standards and Food Additives) Regulation -2011 (Version-XIV -25.03.2021).

1. Malai

As per FSS [(Food Products Standards and Food Additives) Regulation-2011)]

Malai means the product rich in butter fat prepared by and cooling cow or buffalo milk or a combination thereof. It shall contain not less than 25.0 per cent milk fat.

As per Version-XIV of Above Regulation

"Malai" means the product rich in milk fat prepared by boiling and cooling of cow milk, buffalo milk or milk of any other species as defined under this regulation or a mixture thereof. It is characterized by presence of insoluble mass, principally fat and denatured protein, formed on heating and cooling of milk.

The products shall conform to the microbiological requirements specified in Appendix 'B' of these regulations. For products covered under this standard, specific food additives specified in Appendix 'A' of these regulations may be used and only within the limits specified.

2. Cream

As per FSS [(Food Products Standards and Food Additives) Regulation-2011)]

Cream including sterilised cream means the product of cow or buffalo milk or a combination thereof. It shall be free from starch and other ingredients foreign to milk. It may be of following three categories, namely-

- i. **Low fat cream**-containing milk fat not less than 25.0 per cent by weight.
- ii. **Medium fat cream**-containing milk fat not less than 40.0 per cent by weight.
- iii. **High fat cream**-containing milk fat not less than 60.0 per cent by weight.

Note-Cream sold without any indication about milk fat content shall be treated as high fat cream.

As per Compendium-Version-XIV of Above Regulation

"Cream" means the fluid product comparatively rich in fat, in the form of an emulsion of fat-in-skimmed milk, obtained by physical separation from cow milk, buffalo milk or milk of any other species as defined under this regulation or a mixture thereof.

The product shall contain minimum 10.0 per cent. (m/m) milk fat. Acidity of the finished products, other than fermented and acidified creams, should not be more than 0.15 per cent (as lactic acid).

Types of Cream

a. **"Reconstituted cream"** means cream obtained by reconstituting milk products with or without the addition of potable water and with the same end product characteristics as the product described under cream.
b. **"Recombined cream"** means cream obtained by recombining milk products with or without the addition of potable water and with the same end product characteristics as the product described under cream.
c. **"Prepared creams"** means the milk products obtained by subjecting cream, reconstituted cream or recombined cream or any combination of these, to suitable treatments and processes to obtain the characteristic properties as specified below:
   i. **"Pre-packaged liquid cream"** means the fluid milk product obtained by preparing and packaging cream, reconstituted cream or recombined cream, or any combination of these for direct consumption or for direct use as such.
   ii. **"Whipping cream"** means the fluid cream, reconstituted cream and recombined cream or any combination of these, that is intended for whipping. When cream is intended for use by the final consumer the cream should have been prepared in a way that facilitates the whipping process.
   iii. **"Cream packed under pressure"** means the fluid cream, reconstituted cream and recombined cream or any combination of these that is packed with a propellant gas in a pressure-propulsion container and which becomes Whipped Cream when removed from that container.
   iv. **"Whipped cream"** means the fluid cream, reconstituted cream or recombined cream in to which air or inert gas has been incorporated without reversing the fat-in-skimmed milk emulsion.
   v. **"Fermented/cultured/sour cream"** means the milk product obtained by fermentation of cream, reconstituted cream or recombined cream, by the action of suitable microorganisms that results in reduction of pH with or without coagulation. Where the content of (a) specific microorganism(s) is(are) indicated, directly or indirectly, in the labelling or otherwise indicated by content claims in connection with sale, these shall be present, viable, active and abundant in the product to the date of minimum durability.

If the product is heat treated after fermentation the requirement for viable micro-organisms shall not apply.

vi. **"Acidified cream"** means the milk product obtained by acidifying cream, reconstituted cream or recombined cream, or any combination of these, by the action of acids or acidity regulators, or both to achieve a reduction of pH with or without coagulation.

The products shall conform to the microbiological requirements specified in Appendix 'B' of these regulations. For products covered under this standard, specific food additives specified in Appendix 'A' of these regulations may be used and only within the limits specified.

3. Standard for Milk Fat Products

a. Milk Fat, Ghee, Butter Oil, Anhydrous Milk Fat and Anhydrous Butter Oil

A. FSSAI

As per FSS (Food Products Standards and Food Additives) Regulation -2011)

Ghee- Ghee means the pure clarified fat derived solely from milk or curd or from desi (cooking) butter or from cream to which no colouring matter or preservative has been added. Baudouin test shall be negative. The standards of quality of ghee produced in a State or Union Territory specified in the following tables shall comply. FFA and Moisture limits are same (FFA- Max 3.0 per cent and Moisture Max.-0.5 per cent) as given in Table 16.30

Milkfat/Butter oil and Anhydrous Milk fat/Anhydrous Butter oil means the fatty products derived exclusively from milk and/or products obtained from milk by means of process which result in almost total removal of water and milk solids not fat. It shall have pleasant taste and flavour free from off odour and rancidity. It shall be free from vegetable oil/fat, animal body fat, mineral oil, added flavour and any other substance foreign to milk. It may contain food additives permitted in these regulations including Appendix A. It shall conform to the microbiological requirements prescribed in Appendix B. It shall conform to the requirements given in Table 16.29.

Table 16.29: Quality Standards for Milk Fat, Butter Oil, Ghee *etc.*

Sl. No.	*Parameter*	*Milk Fat, Butter Oil*	*Anhydrous Milk Fat, Anhydrous Butter Oil*	*Ghee*
1	Moisture, maximum, per cent, (m/m)	0.4	0.1	0.5
2	Milk fat, minimum, per cent, (m/m)	99.6	99.8	99.5
3	Butyro-refractometer Reading at 40 °C	40.0 to 44.0	40.0 to 44.0	As per table for Ghee below
4	Reichert Meissl Value, minimum	28.0*	28.0*	

Sl. No.	Parameter	Milk Fat, Butter Oil	Anhydrous Milk Fat, Anhydrous Butter Oil	Ghee
5	Polenske Value	1.0-2.0	1.0-2.0	-
6	FFA as Oleic Acid, maximum, per cent	0.4	0.3	3.0
7	Peroxide Value (Milliequivalent of Oxygen/Kg fat), maximum	0.6	0.3	-
8	Baudouin Test	Negative	Negative	Negative

Note: *As per FSS(Food Products Standards and Food Additives) Regulation-2011.
Reichert Meissl Value, minimum for milk fat/Butter oil and Anhydrous Milk Fat.
Anhydrous Butter Oil shall be 24.

As per Compendium Version-XIV of Above Regulation

These are fatty products derived exclusively from milk or products obtained from milk, or both, by means of processes which result in almost total removal of water and milk solids-not-fat. Ghee has especially developed flavour and physical structure as a result of its method of manufacturing. The standards of quality of milk fat, butter oil, anhydrous milk fat, anhydrous butter oil and ghee shall conform to the following requirements:

Table 16.30: Quality Standards of Ghee

States and U. T	BR at *40 °C	R.M. Value (min)	Polenske Value
Bihar, Chandigarh, Delhi, Punjab, Haryana (Areas other than cotton tract areas), West bengal (Areas other than Bishnupur sub div), Sikkim, Jharkhand.	40-43	28	-
Manipur, Meghalaya, Mizoram, Arunachal Pradesh, Orissa, Nagaland, Tripura, Assam, Goa, Kerala, Himanchal Pradesh, U.P., J and K, Maharashtra (Areas other than cotton tract areas), Rajasthan (Areas other than jodhpur areas), Haryana(cotton tract areas), Lakshadeep. Uttarakhand.	40-43	26	-
Karnataka (Belagaum distt), Madhya Pradesh (Areas other than cotton tract areas), Pondicherry, Chhatishgarh.	40-44	26	-
Andhra Pradesh/Telangana, Dadar and Nagar Haveli, Karnataka (Areas other than Belagaum distt.)	40-43	24	-
Andaman and Nicobar Island, Tamil Nadu	41-44	24	-
Gujarat (Areas other than cotton tract areas), Daman and Diu,	40-43.5	24	-
Gujarat (cotton tract areas), Madhya Pradesh (cotton tract areas), Maharashtra (cotton tract areas), Rajasthan (Jodhpur div.), West Bengal (Bishnupur div.)	**41.5-45**	**21**	**-**

For products covered under this standard, specific food additives specified in Appendix 'A' of these regulations may be used and only within the limits specified.

B. As per AGMARK and IDF Standards

Agmark and IDF have also prescribed quality standards for ghee. As per these standards, ghee shall meet the quality standards given in Table 16.31

Table 16.31: Agmark and IDF Standards for Ghee

Sl.No.	Parameters	AGMARK	IDF
1	Colour	White with or without yellowish or greenish Taint and uniform throughout	-
2	Body and Texture/Physical Structure	The solid phase should be of well defined, Granular Structure	Must consists of a mixture of higher softening point fats in crystal form,dispersed through lower softening point fats in liquid form
3	Taste and Flavour	Should be pure, clarified milk fat only, shall have natural sweet, and pleasant odour. Shall have an agreeable taste and be free from rancid or objectionable flavour. On melting, it should be clear, transparent and should be free from sediments or foreign colouring matter.	It should not have an objectionable taste and odour which does not meet edible requirement.
4	Moisture per cent Max.	0.3	0.3
5	Free Fatty Acid per cent (oleic acid) Max	Special grade (Red Label) Max – 1.4 General grade (Green Label) Max – 2.5 Standard grade-Max.-3.0	0.3
6	Fat per cent Min.	-	99.6
7	Baudouin Test	Negative	-
8	B.R. Reading at 40 °C	40-43	-
9	Peroxide Value,Max Meq. of oxygen/Kg of fat	-	0.5
10	Reichert Meissl Value	28	-
11	Polenske value	1.0-2.0	-
12	Phytosterol acetate test	Negative	-

Sl.No.	Parameters	AGMARK	IDF
13	Copper and Iron Max.	-	0.05ppm/0.2ppm
14	Neutralising Substances	-	Traces
15	Food Additives	-	Permitted-NaOH, Na_2CO_3, $NaHCO_3$
16	Antioxidants	-	a. Gallated-100mg/Kg b. Any combination of propyl octyl and dedecyl gallate with BHA or BHT-200mg/Kg

Source: Agmark and IDF Standard booklets.

For cotton tracts areas such as part of Saurshtra and Madya Pradesh following standards are applicable.

Table 16.32: AGMARK Standards for Ghee Produced in Cotton Tract Areas

Sl.No.	Parameters	Special Grade	
		Winter	Summer
1	Baudouin Test	Negative	Negative
2	Butyro-refractomer reading at 40°C	41.5-43.0	42.5-45
3	Reichert Meissl value	Not less than 23.0	Not less than 21.0
4	Polenske value	0.5-1.2	0.5-1.0
5	Moisture content	Not more than 0.3 per cent	Not more than 0.3 per cent
6	Percentage of Free Fatty Acid (as oleic acid)	Not more than 1.4	Not more than 2.5

General and Standard grade have Percentage of Free Fatty Acids (as Oleic acid) shall not exceed 2.5 and 3.0 respectively. According to the law it is not compulsory for every trader and manufacturer, to get his produce certified under AGMARK symbol. Presently it is only a voluntary scheme of the Government.

C. *As Per CODEX STAN 280-1973 (Formerly CODEX STAN A-2-1973. Adopted in 1973-Revision 1999-Amendment 2006)*

Ghee is a product exclusively obtained from milk, cream or butter, by means of processes which result in almost total removal of water and non-fat solids, with an especially developed flavour and physical structure. It has made from Raw materials. Milk and/or products obtained from milk and have Permitted ingredients. Starter cultures of harmless lactic acid producing bacteria. It contains following:

Table 16.33: CODEX Standards for Ghee

Parameters	*Ghee*
A. Composition	
Minimum milkfat (per cent m/m)	99.6
Maximum water (per cent m/m)	-
B. Other Quality Factors	
Maximum free fatty acids (per cent m/m as oleic acid)	0.4
Maximum peroxide value (milli-equivalents of oxygen/kg fat)	0.6
Taste and odour	Acceptable for market requirements after heating a sample to 40–45°C
Texture	Smooth and fine granules to liquid, depending on temperature

The details related to the Food Additives,Contaminants, Hygiene, Labelling and Methods of sampling and Testing under codex are as per codex booklet.

b. Table Butter and White Butter

A. FSSAI

As per FSS (Food Products Standards and Food Additives) Regulation -2011)

Butter means the fatty product derived exclusively from milk of Cow and/ or Buffalo or its products principally in the form of an emulsion of the type water-in-oil. The product may be with or without added common salt and starter cultures of harmless lactic acid and/or flavour producing bacteria.

Table butter shall be obtained from pasteurised milk and/or other milk products which have undergone adequate heat treatment to ensure microbial safety. It shall be free from animal, body fat, vegetable oil and fat, mineral oil and added flavour. It shall have pleasant taste and flavour free from off flavour and rancidity. It may contain food additives permitted in these Regulations including Appendix A. It shall conform to the microbiological requirements prescribed in Appendix B.

As per Compendium Version-XIV of Above Regulation

"Butter" means the fatty product principally in the form of an emulsion of the type water in-oil derived exclusively from milk or milk products, or both. Butter may be of following types:

i. Table butter and.
ii. White butter/Cooking butter.

Table butter shall be made from pasteurised cream. The product shall conform to the compositional specifications provided in the Table 16.34:

Table 16.34: Quality Standards for Butter

Sl.No.	*Parameter*	*Table Butter*	*White Butter/ Cooking Butter*
1	Moisture, maximum, per cent, (m/m)	16.0	-
2	Milk fat, minimum, per cent, (m/m)	80.0	76.0
3	Milk solids-not-fat, maximum, per cent, (m/m)	2.0*	-
4	Common salt, maximum, per cent, (m/m)	3.0	-

Notes: i. Where butter is sold or offered for sale without any indication as to whether it is table butter or white butter, the Standards of table butter shall apply. ii. The extracted fat from butter shall meet the standards for Reichert Meissl value and Butyro-refractometer reading as prescribed for ghee. iii. * As per FSS (Food Products Standards and Food Additives) Regulation -2011, Milk solids-not-fat, maximum, per cent, (m/m) shall be not more than 1.5 per cent.

The products shall conform to the microbiological requirements specified in Appendix 'B' of these regulations. For products covered under this standard, specific food additives specified in Appendix 'A' of these regulations may be used and only within the limits specified.

B. *As per BIS (IS:13690-1992)*

The following definitions are given for Table butter and White butter:

1. **Table butter -** It means the product made from pasteurized cream obtained from pasteurized cow or buffalo milk or combination thereof with or without ripening with the use of standard lactic culture, addition of common salt(as per IS:253-1985) and annatto or carotene (as per IS:2557-1963 and IS:3841-1966 respectively) as colouring matter and diacetyl (Not more than 4 ppm and as per IS:3507-1966) as flavouring agent.
2. **White butter-** It means the product made from pasteurized cow or buffalo milk or combination thereof or from pasteurized cream obtained from cow or buffalo milk or combination thereof without ripening and without addition of any preservatives including common salt, any added colouring matter or any added flavouring agent.

Both types of butter shall be clean, smooth, firm at 15.5°C and uniform colour and shall not show sign of curd. The product shall be free from other animal fats, wax, mineral oils and vegetable oils and fats and added extraneous matter. The flavour and aroma of butter shall be pleasant.

C. As per Agmark

Agmark has prescribed standards for Organoleptic as well as chemical quality of Table butter.

Organoleptic Standards-Flavour and Aroma

Clean, Pleasant and Free from objectionable taint or rancidity. Body and Texture-Firm at 60 ° C. Should not be greasy and oily. Should be compact and show a uniform fine granular surface on breaking.

Colour-Uniform, should not show streakiness, mottling, stains or signs of curd. Only annatto is permitted.

Chemical Standards

Given in Table 16.35.

D. As per CODEX STAN 279-1971 (Formerly CODEX STAN A-1-1971. Adopted in 1971. Revision 1999. Amendment 2003, 2006)

Butter is a fatty product derived exclusively from milk and/or products obtained from milk, principally in the form of an emulsion of the type water-in-oil. It has made from Raw materials-Milk and/or products obtained from milk and have Permitted ingredients-Sodium chloride and food grade salt,Starter cultures of harmless lactic acid and/or flavour producing bacteria and Potable water. The details related to the Food Additives,Contaminants, Hygiene, Labelling and Methods of sampling and Testing under codex are as per codex booklet.

The detailed quality norms for butter under Codex, FSSR-2011, Agmark and BIS are given in Table 16.35.

I. Quality Norms Related to Chemical Composition

Given in Table 16.35.

II. Quality Norms Related to Microbiological Quality

FSSR-2011 and BIS have formulated the standards for microbiological quality of Pasteurized Table (Creamery) butter and White butter are shown in Table 16.36.

4. Standard for Fermented Milk Products

a. Dahi/Fermented Milk Products

A. FSSAI

As per FSS (Food Products Standards and Food Additives) Regulation -2011

Dahi or curd means the product obtained from pasteurised or boiled milk by souring, natural or otherwise, by a harmless lactic acid culture or other harmless bacterial culture may also be used in conjunction with lactic

Table 16.35: Quality Norms of Table Butter (Chemical Quality)

Sl. No.	*Butter*	*CODEX STAN 279-1971)*	*FSSR-2011*	*Agmark*	*BIS IS:13690-1992*	
					Table	*White*
1	Milk Fat Content per cent m/m (Min)	80	Table-80 Deshi-76	80	80	82
2	*Water/Moisture Content (Max) per cent m/m	16	Table-16	16 uniformly distributed	16	16
3	Milk Solids–not-Fat Content or curd (Max) per cent m/m	2.0	1.5 (FSSR-2011) 2.0 (As per version XIV of FSSR-2011	1.0	1.0	1.5
4	Acidity (Lactic acid) per cent max.	—	—	0.15	0.15	0.06
5	Permitted Ingredients per cent, Max.	Sodium Chloride and Food grade salt, Starter culture of harmless lactic acid and or flavour producing bacteria, Potable water.	Salt-3.0 (No other preservative) Desi- Nil Starter culture of harmless lactic acid and or flavour producing bacteria	3.0	Salt-2.5	—

Sl. No.	Butter	CODEX STAN 279-1971)	FSSR-2011	Agmark	BIS IS:13690-1992	
					Table	White
6	Food Additives- Max. Level					
	i. Colours:					
	β-Carotene (synthetic) Carotenes (natural extracts)	25 mg/kg, 600 mg/kg	Annatto and Carotene colours are allowed.	Only annatto is permitted	Only annatto is permitted	—
	Annatto extracts	20mg/kg-biaxin/norbixin basis				
	β-apo-carotenal	35mg/kg				
	β-apo-8'-carotenoic acid, methyl or ethyl ester	35mg/kg				
	ii. Acidity Regulators:		Table Butter	—	—	—
	Sodium phosphates	2g/kg	0.2 per cent by wt.			
	Sodium carbonate	Limited by GMP	0.2 per cent by wt.			
	Sodium hydrogen carbonate	Limited by GMP	0.2 per cent by wt.			
	Sodium Hydroxide	Limited by GMP	0.2 per cent by wt.			
	Calcium Hydroxide	Limited by GMP	0.2 per cent by wt.			
	iii. Flavouring agent:	—	4.0 ppm	—	—	—
	Diacetyl (Total). Max	—	Deshi-Nil			
	iv. Foreign Fat	—	Free from animal fat, vegetable oil, wax and mineral oil *etc.*			
07	Heavy Metals-Lead	As per CAC0.5mg/kg	—	—	—	—
08	Pesticide residues	As per CAC	—	—	—	—

acid bacteria cultures for souring. Dahi may contain added cane sugar. Dahi shall have the same minimum percentage of milk fat and milk solids-not-fat as the milk from which it is prepared. Where dahi or curd is sold or offered for sale without any indication of class of milk, the standards prescribed for dahi prepared from buffalo milk shall apply. Milk solids may also be used in preparation of this product.

Table 16.36: Microbiological Quality for Butter

Microbiological Standards	*FSSR-2011*		*BIS (IS:13690-1992) (both for Table and White)*
	Sampling Plan	*Count*	
Total count/gm Max	m	10,000/g	-
	M	50,000/g	
Coliform count/gm	m	10/g	05 Max/ml.
	M	50/g	
E. coli	M	Absent/g	-
Salmonella	M	Absent/25g	-
Shigella	M	Absent/25g	-
Staph. aureus	m	10/g	-
	M	50/g	
Y and M Count	m	20/g	Below 20/ml
	M	50/g	
Anaerobic spore Count	M	Absent/g	-
Listeria monocytogens	M	Absent/g	-

As per compendium Version-XIV of Above Regulation

Standard for Fermented Milk Products -This Standard applies to fermented milks, including, heat-treated fermented milks, concentrated fermented milks and composite milk products based on these products in conformity with the definitions given below:

Fermented Milk is a milk product obtained by fermentation of milk, which may have been manufactured using other permitted raw material, by the action of suitable microorganisms and resulting in lowering of pH with or without coagulation (iso-electric precipitation). Fermented milk may be heat treated after fermentation. The raw material used shall be subjected to a heat treatment as defined in the General Standard for Milk and Milk Products. Certain fermented milks are characterised by specific starter culture(s) used for fermentation as shown in Table 16.37.

Table 16.37: Specific Starter Culture(s) for Fermentation

Sl.No.	*Fermented Products*	*Specific Starter Culture*
1	Dahi (Curd)	Lactic acid bacteria
2	Yoghurt	Symbiotic cultures of Streptococcus thermophilus and Lactobacillus delbrueckii sub sp. bulgaricus.
3	Alternate Culture Yoghurt	Cultures of Streptococcus thermophilus and Lactobacillus species
4	Acidophilus milk	Lactobacillus acidophilus.

Other harmless microorganisms than those constituting the specific starter cultures specified above may also be added.

Fermented milks shall have a minimum milk protein content of 2.9 per cent (m/m) and minimum titrable acidity of 0.45 per cent (m/m as lactic acid) unless otherwise specified. In case of Flavoured Fermented Milks and Drinks based on Fermented Milks, these specifications apply to the Fermented Milk Part unless otherwise specified;

Plain Dahi shall have the same minimum percentage of milk fat and milk solids-not-fat as that of the milk, as provided for in the Standard for Milk, from which it is prepared. Where plain Dahi is sold or offered for sale without any indication of class of milk, the Standards prescribed for Dahi prepared from mixed milk shall apply;

Flavoured fermented milk are composite milk products, as defined in sub-regulation 2.1.1, obtained from fermented milks and which contain a maximum of 50 per cent (m/m) of permitted non-dairy ingredients including flavourings. The non-dairy ingredients can be mixed prior to or after fermentation: Provided that flavoured dahi shall only be sold in pre-packaged form.

Drinks based on fermented milk are composite milk products, as defined under "General Standards for Milk and Milk Products" obtained by mixing fermented milks as described with potable water with or without the addition of whey, other milk and milk products, other permitted non-dairy ingredients and flavours. Drinks based on fermented milk contain a minimum of 40 per cent (m/m) fermented milk. Other microorganisms than those constituting the specific starter cultures may be added. Drinks based on fermented milk include products such as **lassi, chhaach, buttermilk,** ***etc.***

The products shall conform to the microbiological requirements specified in Appendix 'B' of these regulations. For products covered under this standard, specific food additives specified of these regulations may be used and only within the limits specified.

B. *As per BIS (IS 9617:1980)*

It is product obtained by lactic acid fermentation through the action of single or mixed strains of lactic acid bacteria or by lactic fermentation accompanied by alcoholic fermentation by yeast from milk with or without addition (Not more than 25 per cent by mass of final product) of sugar, Flavouring essence (permitted under PFA/FSSR-2011), Flavouring extract and various artificial flavour and stabilisers(not more than 0.2 per cent and permitted under PFA/FSSR-2011).Dahi shall confirm to the requirements of milk fat and MSNF as given in PFA/FSSR-2011. It shall have a pleasing bouquet flavour resulting from the blend of a clean delicate some what aromatic odour and a pronounced though clean acid taste. It shall be free from undesirable flavour *e.g.* Bitter, Coarse due to over ripening, Flat(Lack of flavour), Metallic and yeasty. It should be firm, solid and uniform with negligible whey separation before shaking. Dahi shall be plain or flavoured.

Chemical Standards (Under BIS) and microbiological standards (under FSSR-2011 and BIS) are given below:

Table 16.38: Chemical and Microbiological Standards for Dahi

<table>
<tr><th>Sl.No.</th><th>Quality Standards</th><th>FSSR-2011</th><th>BIS IS:9671-1980</th></tr>
<tr><td rowspan="3">1</td><td colspan="3">Chemical Standards</td></tr>
<tr><td>Acidity, as Lactic acid per cent by mass</td><td>-</td><td>0.6 to 0.8</td></tr>
<tr><td>Phosphatase Test</td><td>-</td><td>Negative</td></tr>
<tr><td rowspan="10">2</td><td colspan="3">Microbiological Standards</td></tr>
<tr><td>Total count/gm</td><td rowspan="4">As per Appendix- B of regulation</td><td>-</td></tr>
<tr><td>Coliform count/gm</td><td>10(Max)</td></tr>
<tr><td>E. coli</td><td>-</td></tr>
<tr><td>Salmonella</td><td>-</td></tr>
<tr><td>Shigella</td><td rowspan="5">As per Appendix- B of regulation</td><td>-</td></tr>
<tr><td>Staph. Aureus</td><td>-</td></tr>
<tr><td>Y and M Count</td><td>Not more than 100/g</td></tr>
<tr><td>Anaerobic spore Count</td><td>-</td></tr>
<tr><td>Listeria monocytogens</td><td>-</td></tr>
</table>

b. *Chakka and Shrikhand*

A. *FSSAI and BIS*

i. *Chakka*

As per FSS (Food Products Standards and Food Additives) Regulation-2011

Chakka means a white to pale yellow semi-solid product of good texture and uniform consistency obtained by draining off the whey from the Yoghurt obtained by the lactic fermentation of cow's milk, buffalo's milk, skimmed milk and recombined or standardised milk which has been subjected to minimum heat treatment equivalent to that of pasteurisation. It shall have pleasant Yoghurt/Dahi like flavour. It shall not contain any ingredient foreign to milk. It shall be free from mouldness and free from signs of fat or water seepage or both. It shall be smooth and it shall not appear dry. It shall not contain extraneous colour and flavours. Chakka when sold without any indication shall conform to the standards of Chakka.

As per BIS (IS 9532:1980)

Chakka is semi-solid product obtained by draining off the whey from the yoghurt obtained by the lactic acid fermentation of cow's milk, buffalo's milk, skimmed milk,reconstituted skimmed milk and recombined or standardised milk which has been subjected to minimum heat treatment equivalent to that of pasteurisation. Chakka shall not contain any ingredient foreign to milk. It shall be free from signs of fat or water seepage or both and mouldiness and objectionable flavour/odour and coarseness. It shall be white or pale yellow in colour, of good texture and uniform consistency and have pleasant yoghurt (Dahi) flavour. No extraneous colour shall be added.

Chakka shall also meet the following Chemical and Microbiological standards under FSSR-2011 and BIS:

Table 16.39: Chemical Standards

Sl. No.	*Parameters*	*FSSR-2011*			*BIS(IS9532: 1980)*	
		Chakka	*Skimmed Milkchakka*	*Full Cream Chakka**	*Chakka*	*Skimmed Milkchakka*
1	Total Solids per cent by weight, Min	30	20	28	30	20
2	Milk Fat (On Dry basis) per cent by weight	Min 33	Max 05	Min 38	Min 33	Max 05
3	Milk Protein (On Dry basis) per cent by weight, Min	30	60	30	37	60
4	Titrable acidity (as LA) per cent by weight	Max 2.5	Max 2.5	Max 2.5	Max 2.5	Max 2.5
5	Total Ash (On Dry basis) per cent by weight	Max 3.5	Max 5.0	Max 3.5	Max 3.5	Max.5.0

Table 16.40: Microbiological Standards

Microbiological Standards	*FSSR-2011*	*BIS* (IS 9532:1980)*
Total count/gm Max	As per Appendix B of the regulation	-
Coliform count/gm		10 Max
E.Coli		-
Salmonella		-
Shigella		-
Staph. Aureus		-
Y and M Count		20 Max
Anaerobic spore Count		-
Listeria monocytogens		-

* For both types of Chakka

ii. Shrikhand

As per FSS (Food Products Standards and Food Additives) Regulation-2011

Shrikhand means the product obtained from chakka or Skimmed Milk Chakka to which milk fat is added. It may contain fruits, nuts, sugar, cardamom, saffron and other spices. It shall not contain any added colouring and artificial flavouring substances.

As per BIS (IS 9532:1980)

Shrikhand is an indigenous sweet, made from mixing Chakka(conforming to the requirement mentioned in BIS Chakka)with other ingredients (as mentioned incase of BIS Chakka) like sugar and condiments (such as cardamon and saffron). When milk fat is added to make this product from skimmed milk chakka, it should be free from microorganism which are harmful to human health. It shall not contain any materials foreign to milk except those permitted. No artificial colouring material should be added to Shrikhand. It should not be too thick or too thin and its consistency should be very smooth. There should not be any syrup pockets or separation in the body of the Shrikhand. It shall be free from signs of free fat or syrup separation, objectionable flavour/odour. It shall have smooth texture and uniform consistency. Shrikhand shall also meet the following Chemical and Microbiological standards given under FSSR-2011 and BIS:

Table 16.41: Chemical Standards

Sl.No.	*Parameters*	*Shrikhand*		
		FSSR-2011		*BIS (IS9532: 1980)*
		Shrikhand	*Full Cream Shrikhand*	
1	Total Solids per cent by weight	Not less than 58	Not less than 58	Not less than 58
2	Milk Fat (On Dry basis) per cent by weight	Not less than 8.5 (for fruit shrikhand 7).	Not less than 10	Not less than 8.5 (7.0 incase of fruit shrikhand)
3	Milk Protein (On Dry basis) per cent by weight	Not less than 9 (for fruit shrikhand 06).	Not less than 7	Not less than 10.5 (for fruit shrikhand also)
4	Titrable acidity(as LA) per cent by weight	Not more than 1.4	Not more than 1.4	Not more than 1.4
5	Total Ash (On Dry basis) per cent by weight	Not more than 0.9	Not more than 0.9	Not more than 0.9
6	Sugar as sucrose (On Dry basis) per cent by weight	Not more than 72.5	Not more than 72.5	Not more than 72.5

Table 16.42: Microbiological Standards

Microbiological Standards	*FSSR-2011*	*BIS (IS 9532:1980)*
Total count/gm Max	As per Appendix B of the regulation	-
Coliform count/gm		10(Max)
E.Coli		-
Salmonella		-
Shigella		-
Staph. Aureus		-
Y and M Count		50/g (Max.)
Anaerobic spore Count		-
Listeria monocytogens		-

c. *Yoghurt*

As per FSS (Food Products Standards and Food Additives) Regulation-2011

Yoghurt means a coagulated product obtained from pasteurised or boiled milk or concentrated milk, pasteurised skimmed milk and/or pasteurised cream or a mixture of two or more of these products by lactic acid fermentation through the action of Lactobacillus bulgaricus and Steptococcus thermophilus. It may also contain cultures of Bifidobacterium bifidus and Lactobacillus acidophilus and other cultures of suitable lactic acid producing harmless bacteria and if added a declaration to this effect shall be made on the label. The microorganisms in the final product must be viable and abundant. It may contain milk powder, skimmed milk powder, unfermented buttermilk, concentrated whey, whey powder, whey protein, whey protein concentrate, water soluble milk proteins, edible casein, and caseinates manufactured from pasteurised products. It may also contain sugar, corn syrup or glucose syrup in sweetened, flavoured and fruit yoghurt or fruits in fruits yoghurt. It shall have smooth surface and thick consistency without separation of whey. It shall be free from vegetable oil/fat, animal body fat, mineral oil and any other substance foreign to milk. The product may contain food additives permitted in these regulation. It shall conform to the microbiological requirements prescribed in Appendix B of regulation. It shall conform to the following requirements:

Table 16.43: Quality Standards for Yoghurt

Sl. No.	*Parameters*	*Yoghurt (FSSR-2011)*				
		Yoghurt	*Partly Skimmed*	*Skimmed*	*Sweetened/ Flavoured*	*Fruit*
1	Milk Solids Not Fat per cent m/m,Not less than	8.5	8.5	8.5	8.5	8.5
2	Milk Fat per cent m/m	Min. 3.0	Min. 0.5 to Max 3.0	Max.0.5	Min. 3.0	Min. 1.5
3	Sugar per cent m/m Not less than	-	-	-	6.0	6.0
4	Protein per cent m/m Not less than	3.2	3.2	3.2	3.2	2.6
5	Titrable Acidity (as LA) per cent by weight	0.85-1.2	0.85-1.2	0.85-1.2	0.85-1.2	0.85-1.2
6	Bacterial (Specific Lactic acid) Count/gm Min.	1000000	1000000	1000000	1000000	1000000.

Provided further that the type of Yoghurt shall be clearly indicated on the label otherwise standards of plain Yoghurt shall apply. The Yoghurt subjected to heat treatment after fermentation at temperature not less than 65 degree C shall be labelled as Thermised or Heat Treated Yoghurt and shall conform to the above parameters except the minimum requirement of specific lactic acid producing count per gram.

As per BIS (IS:12898-1989)

Yoghurt is fermented milk product, where cultures of Lactobacillus bulgaricus and Streptococcus thermophilus are mainly employed for the fermentation process. In addition the yoghurt may also contain of Bifidobacterium bifidus and Lactobacillus acidophilus and if added, the declaration to this effect shall be made on the label. It shall be made from essential raw materials *e.g.* milk and cultures with optional ingredients *e.g.* Milk powder, Skimmed milk powder, Fermented butter milk, concentrated whey, whey powder *etc.*,Sugars, corn syrup or glucose syrup(for Sweetened Flavoured and Fruit Yoghurt), fruits, permitted colour (for Flavoured and Fruit Yoghurt) and permitted stabilisers. BIS has prescribed quality standards for different types of Yoghurts. The product shall have a smooth surface with no whey separation and a custard like consistency. It shall have a typical nut like flavour and a pleasant distinctive sourish taste. The flavour and taste of flavoured yoghurt shall be typical of the flavour used with a mild to light sourish tinge. The different types of Yoghurts shall meet the following specifications:

Table 16.44: BIS Standards for Yoghurt

Sl. No.	*Parameters*	*Yoghurt (under IS:12898-1989)*					
		Plane	*Low Fat*	*Skim-med*	*Swee-tened*	*Flavou-red*	*Fruit*
1	Total Milk Solids per cent by mass Min.	13.5	12.0	11.0	13.5	13.5	10.0
2	Milk Fat per cent by mass.	Min. 3.0	Max. 1.5	Max. 0.5	Min. 3.0	Min. 5.0	Min 1.5
3	Sugar per cent by mass Min.	-	-	-	6.0	6.0	6.0
4	Protein per cent by mass. Min.	3.2	3.2	3.2	3.2	3.2	2.6
5	Titrable Acidity (as LA) per cent by mass.	0.8-1.2	0.8-1.2	0.8-1.2	0.8-1.2	0.8-1.2	0.8-1.2
6	Special Lactic acid Bacterial count/ gm.Min	1000000	1000000	1000000	1000000	1000000	1000000
7	E.coli count/gm	Absent	Absent	Absent	Absent	Absent	Absent
8	Staph. aureus/gm	Absent	Absent	Absent	Absent	Absent	Absent
9	Salmonella and Shigella	Absent	Absent	Absent	Absent	Absent	Absent

As per Compendium Version-XIV of Above Regulation (FSSR-2011)

Concentrated Fermented Milk is fermented milk, the protein of which has been increased prior to or after fermentation.

i. **Chakka** means the fermented and concentrated milk product obtained by (partial) removal of the whey from plain dahi or plain yoghurt or by any other process which leads to a product of same composition and characteristics. It shall have white to pale yellow colour and uniform semi-solid consistency. It shall not be moldy and shall be free from signs of free fat and water. It shall be smooth and not appear dry. The milk from which dahi or yoghurt is prepared for manufacturing chakka shall be subjected to a heat treatment as defined under General Standard for Milk and Milk Products (2.1.1). Chakka shall meet the specifications given in FSSR-2011 as given above.

ii. **Shrikhand** means the semi-soft concentrated composite milk product obtained from chakka, or skimmed milk chakka to which milk fat and sugar is added or by any other process which leads to a product of same composition and characteristics. It may also contain permitted non-dairy ingredients. Shrikhand shall meet the specifications given in FSSR-2011 as given above.

iii. **Yoghurt-** Yoghurt (including Flavoured Yoghurt) and Flavoured Dahi shall conform to the following compositional specifications:

Table 16.45: Quality Standards for Yoghurt

Parameter	*Yoghurt and Flavoured Dahi*	*Partly skimmed Yoghurt and Flavoured Partly Skimmed Dahi*	*Skimmed Yoghurt and Flavoured Skimmed Dahi*
Milk Fat, per cent, (m/m)	Not less than 3.0 and not more than 15	More than 0.5 and Less than 3.0	0.5 (maximum)
Milk solids-not-fat, minimum, per cent, (m/m)	8.5	8.5	8.5
Milk protein*, minimum, per cent, (m/m)	2.9	2.9	2.9
Titratable acidity, minimum, per cent (as lactic acid)	0.6	0.6	0.6

* Protein content is 6.38 multiplied by the total nitrogen determined

Note: i. When sold without any indication, the product shall conform to the Standards of 'Yoghurt' or 'Flavoured Dahi', as appropriate. The term 'flavoured' covers sweetened, flavoured and fruit variants, labelled in accordance with following- type of dahi, yoghurt, chakka or shrikhand shall be always declared on the label or otherwise if the product is not pre-packaged such declaration to be given on the container from which product will be offered to the consumer; ii. For the use of probiotics in dairy products; the 'Indian Council Medical Research Guidelines for Evaluation of Probiotics in Food shall be followed.

The products (Chakka, Shrikhand and Yoghurt) shall conform to the microbiological requirements specified in Appendix 'B' of these regulations. For products covered under this standard, specific food additives specified in Appendix 'A' of these regulations may be used and only within the limits specified.

5. Heat-acid Coagulated Milk Products

a. *Chhana or Paneer*

A. *FSSAI*

As per FSS (Food Products Standards and Food Additives) Regulation -2011

Chhana or paneer means the product obtained from the cow or buffalo milk or a combination thereof by precipitation with sour milk, lactic acid or citric acid. It shall not contain more than 70.0 per cent moisture and the milk fat content shall not be less than 50.0 per cent of the dry matter. Milk solids may also be used in preparation of this product.

Low Fat Paneer or Chhana

Provided that paneer or chhana when sold as low fat paneer or chhana, it shall conform to the following requirements-

i. Moisture Not more than 70.0 per cent.

ii. Milk fat Not more than 15.0 per cent of dry matter.

Provided further that such low fat paneer/chhana shall be sold in sealed package only and shall bear proper label declaration as provided in regulation 2.4.5 (39) of Food Safety and Standards (Packaging and Labelling) Regulations, 2011. Paneer and Chhana shall meet the microbiological requirements given in Appendix- B of the regulation.

As per Compendium Version-XIV of Above Regulation

Chhana or Paneer means the product obtained from any variant of milk**, with or without added milk solids, by precipitation with permitted acidulants and heating. The product shall conform to the compositional specifications provided in the Table 16.46.

Table 16.46: Quality Standards for Chhana and Paneer

Parameter	*Chhana or Paneer*	*Medium fat Chhana or Paneer*	*Low fat Chhana or Paneer*
Moisture, maximum, per cent, (m/m)	65.0 (for Chhana) 60.0 (for Panner)	65.0 (for Chhana) 60.0 (for Panner)	70.0 (for Chhana) 70.0 (for Panner)
Milk fat, per cent, (m/m), dry matter basis	50.0 (minimum)	More than 20.0 and less than 50.0	20.0(maximum)

The products shall conform to the microbiological requirements specified in Appendix 'B' of these regulations. For products covered under this standard, specific food additives specified in Appendix 'A' of these regulations may be used and only within the limits specified.

B. As per BIS

a. Paneer (IS:10484-1983/1998)

Paneer is an important indigenous product prepared by the combined action of acid coagulation and heat treatment of buffalo milk or cow milk or combination thereof. Paneer shall be clear and free from dirt, surface discolouration, insects and rodents contamination and adulterants. It shall not have any free moisture. It shall have a pleasant odour and characteristic mild acidic flavour, closely knit smooth texture, firm cohesive and spongy body.

b. Channa (IS 5162:1980 -first revision)

The solid product obtained by the acid coagulation of milk which has been subjected to minimum heat treatment equivalent to that of pasteurisation followed by subsequent removal of whey. The coagulant used shall be sour chhana whey, Lactic acid or Citric acid. Chhana shall not contain any ingredient foreign to milk and shall be of good texture and uniform consistency. Chhana shall be free from signs of fat or water seepage, or both, mouldiness and objectionable flavour/odour and coarseness. It shall be white or pale yellow in colour and have pleasant curdy flavour. No extraneous colour shall be added.

Chhana/Paneer shall have the following standards under FSSR-2011 and BIS:

Table 16.47: BIS and FSSAI Quality Standards for Chhana and Paneer

Sl. No.	Parameters	FSSR-2011			BIS	
		Chhanna or Paneer	Skim milk Chhanna or Paneer	Paneer (IS: 10484-1983/1998)	Chhana IS:5162-1980	
					Skim	Chhana
1	Moisture per cent by mass (not more than or Max.)	70	70	60	60	65
2	Milk Fat per cent by mass. Min (on DM basis)	50	15 (As per fourth amendment-2005)	50	5 Max.	50
3	Titratable acidity (as Lactic acid) per cent Min	-	-	0.50	-	-
4	Protein per cent by mass. Min (on DM basis)	-	-	-	30	25
5	Ash per cent by mass. Max. (on DM basis)	-	-	-	5.0	5.0
6	Microbiological Total count/gm Max	50000/g	50000/g	$5x10^5$	-	-

Sl. No.	Parameters	FSSR-2011			BIS	
		Chhanna or Paneer	Skim milk Chhanna or Paneer	Paneer (IS: 10484-1983/1998)	Chhana IS:5162-1980	
					Skim	Chhana
	Coliform count/ gm(Max.)	90 (Max.)	90 (Max.)	90	-	-
	E.Coli	Absent in 1 g	Absent in 1 g	-	-	-
	Salmonella	Absent in 25 g	Absent in 25 g	-	-	-
	Shigella	Absent in 25 g	Absent in 25 g	-	-	-
	Staph. Aureus	100/g (Max.)	100/g (Max.)	-	-	-
	Y and M Count (Max)	250/g (Max.)	250/g (Max.)	250/g	-	-

6. Evaporated/Condensed Milk and Milk Products

A. *FSSAI*

As per FSS (Food Products Standards and Food Additives) Regulation-2011

Evaporated Milk means the product obtained by partial removal of water from milk of cow and/or buffalo by heat or any other process which leads to a product of the same composition and characteristics. The fat and protein content of the milk may be adjusted by addition and/or withdrawal of milk constituents in such a way as not to alter the whey protein to casein ratio of the milk being adjusted. It shall have pleasant taste and flavour free from off flavour and rancidity. It shall be free from any substance foreign to milk. It may contain food additives permitted in these regulation. It shall conform to the microbiological requirements prescribed in Appendix B of regulation. It shall conform to the following requirements:

Table 16.48: FSSAI Quality Standards for Evaporated Milk

Sl. No.	Parameters	Evaporated milk			
		Evaporated	Partly Skimmed	Skimmed Milk	High Fat Milk
1	Milk Fat per cent m/m	Not less than 8	Not less than 1and Not more than 8	Not more than 1	Not less than 15
2	Milk Solids per cent m/m	Not less than 26	Not less than 20	Not less than 20	Not less than 27
3	Milk Protein in Milk Solids not Fat per cent m/m	Not less than 34	Not less than 34	Not less than 34	Not less than 34

Sweetened Condensed Milk means the product obtained by partial removal of water from milk of Cow and/or Buffalo with the addition of sugar or a combination of sucrose with other sugars or by any other process which leads to a product of the same composition and characteristics. The fat and/or protein content of the milk may be adjusted by addition and/or withdrawal of milk constituents in such a way as not to alter the whey protein to casein ratio of the milk being adjusted. It shall have pleasant taste and flavour free from off flavour and rancidity. It shall be free from any substance foreign to milk. It may contain food additives permitted in these regulations. It shall conform to the microbiological requirements prescribed in Appendix B of this regulation. It shall conform to the following requirements:

Table 16.49: FSSAI Quality Standards for Sweetened Condensed Milk

Sl. No.	*Parameters*	*Sweetened Condensed Milk*			
		Sweetened Condensed	*Partly Skimmed*	*Skimmed Milk*	*High Fat Milk*
1	Milk Fat per cent m/m	Not less than 9	Not less than 3 and Not more than 9	Not more than 1	Not less than 16
2	Milk Solids per cent m/m	Not less than 31	Not less than 28	Not less than 26	Not less than 30.
3	Milk Protein in Milk Solids not Fat per cent m/m	Not less than 34	Not less than 34	Not less than 34	Not less than 34

As per Compendium Version-XIV of Above Regulation

Evaporated Milk means the product obtained by partial removal of water from milk by heat or any other process which leads to a product of the same composition and characteristics. The fat and protein content of the milk may be adjusted, only to comply with the compositional requirements as given in following table, by addition or withdrawal of milk constituents in such a way as not to alter the whey protein to casein ratio of the milk being adjusted. The product shall conform to the compositional specifications provided in the Table 16.50.

The products shall conform to the microbiological requirements specified in Appendix 'B' of these regulations. For products covered under this standard, specific food additives specified in Appendix 'A' of these regulations may be used and only within the limits specified.

Table 16.50: Quality Standards for Evaporated Milk

Parameter	*Evaporated Milk*	*Evaporated Partly Skimmed Milk*	*Evaporated Skimmed Milk*	*Evaporated High Fat Milk*
Milk fat, per cent, (m/m)	7.5 (minimum)	More than 1 and Less than 7.5	1.0 (maximum)	15.0 (minimum)
Milk solids, minimum, per cent, (m/m)	25.0	20.0	20.0	26.5
Milk protein* in milk solids not fat, minimum, per cent, (m/m)	34.0	34.0	34.0	34.0

Sweetened Condensed Milk is the product obtained by partial removal of water from milk with the addition of sugar or a combination of sucrose with other sugars, or by any other process which leads to a product of the same composition and characteristics. The fat or protein content or both of the milk may be adjusted, only to comply with the compositional requirements as given in following table of this Standard, by addition or withdrawal of milk constituents in such a way as not to alter the whey protein to casein ratio of the milk being adjusted. The product shall conform to the compositional specifications provided in the table below:

Table 16.51: Quality Standards for Sweetened Condensed Milk

Parameter	*Sweetened Condensed Milk*	*Sweetened Condensed Partly Skimmed Milk*	*Sweetened Condensed Skimmed Milk*	*Sweetened Condensed High Fat Milk*
Milk fat, per cent, (m/m)	8.0 (minimum)	More than 1.0 and less than 8.0	1.0 (maximum)	16.0 (minimum)
Milk solids, minimum, per cent, (m/m)	28	24	24	-
Milk solid not fat, minimum, per cent, (m/m)	-	20.0	-	14.0
Milk protein* in milk solids-not-fat, minimum, per cent, (m/m)	34.0	34.0	34.0	34.0

The products shall conform to the microbiological requirements specified in Appendix 'B' of these regulations. For products covered under this standard, specific food additives specified in Appendix 'A' of these regulations may be used and only within the limits specified.

B. As per BIS

Condensed milk- BIS has specified quality norms for Condensed milk, partly skimmed and skimmed condensed milk under IS 1166:1986 (second revision). The quality norms for Condensed milk and Sweetened condensed skim milk are given below:

Table 16.52: BIS Quality Standards for Condensed and Sweetened Condensed Milk

Sl.No.	Characteristics	Standards	
		Condensed Milk	Skim Sweetened
1	Total milk solids (per cent wt.) Min	31.0	26.0
2	Fat (per cent wt.)	Not less than 9.0	Not more than 0.5
3	Sucrose (per cent wt.) Min	40	40
4	Acidity (per cent lactic) Max	0.35	0.35
5	Bacterial count (per g.)	500	500
6	Coliform count (per g.)	-ve	-ve
7	Yeast and Mould count (per g.) Max	10	10

C. As per Codex Standards

Codex standards for Evaporated Milk and Sweetened Condensed milk are given below:

Evaporated Milk-Under CODEX STAN 281-1971 (Formerly CODEX STAN A-3-1971. Adopted in 1971. Revision 1999)

Evaporated milks are milk products which can be obtained by the partial removal of water from milk by heat, or by any other process which leads to a product of the same composition and characteristics. The fat and/or protein content of the milk may have been adjusted, only to comply with the following compositional requirements by the addition and/or withdrawal of milk constituents in such a way as not to alter the whey protein to casein ratio of the milk being adjusted. It is made from **Raw Materials**-Milk and milk powders, cream and cream powders, milkfat products and may have **Permitted ingredients-** Potable water and Sodium chloride. The following milk products are allowed for protein adjustment purposes-

i. Milk retentate- Milk retentate is the product obtained by concentrating milk protein by ultrafiltration of milk, partly skimmed milk, or skimmed milk,

ii. Milk permeate- Milk permeate is the product obtained by removing milk proteins and milk fat from milk, partly skimmed milk, or skimmed milk by ultrafiltration; and.

iii. Lactose.

Evaporated milk shall also contain the following:

Table 16.53: Codex Standards for Evaporated Milk

Sl. No.	Type of Milk	Evaporated Milk			
		Evaporated	Skimmed	Partly Skimmed	High Fat
1	Milk Fat per cent m/m	Min. 7.5	Max.1	More than 1 and less than 7.5	Min.15
2	Milk Solids ** Min. per cent m/m	25	20	20	11.5
3	Milk Protein in Milk Solids ** per cent m/m	34	34	34	34

Note: ** The milk solids and milk solids-not-fat contents include water of crystallization of the lactose.

Sweetened Condensed milk-Under CODEX STAN 282-1971 (Formerly CODEX STAN A-4-1971. Adopted in 1971. Revision 1999)

Sweetened condensed milks are milk products which can be obtained by the partial removal of water from milk with the addition of sugar, or by any other process which leads to a product of the same composition and characteristics. The fat and/or protein content of the milk may have been adjusted, only to comply with the following compositional requirements by the addition and/or withdrawal of milk constituents in such a way as not to alter the whey protein to casein ratio of the milk being adjusted. It is made from **Raw materials-**Milk and milk powder, cream and cream powders, milkfat products and may have **Permitted ingredients-** Potable water, Sugar, Sodium chloride. In this product, sugar is generally considered to be sucrose, but a combination of sucrose with other sugars, consistent with Good Manufacturing Practice, may be used. The following milk products are allowed for protein adjustment purposes:

i. Milk retentate- Milk retentate is the product obtained by concentrating milk protein by ultrafiltration of milk, partly skimmed milk, or skimmed milk,
ii. Milk permeate- Milk permeate is the product obtained by removing milk proteins and milk fat from milk, partly skimmed milk, or skimmed milk by ultrafiltration and
iii. Lactose* (Also for seeding purposes)

For all sweetened condensed milks the amount of sugar is restricted by Good Manufacturing Practice to a minimum value which safeguards the keeping quality of the product and a maximum value above which crystallization of sugar, may occur.

Sweetened Condensed milk shall also contain the following:

Table 16.54: Codex Standards for Sweetened Condensed Milk

Sl.No.	Type of Milk	Sweetened Condensed Milk			
		Sweetened Condensed	Skimmed	Partly Skimmed	High Fat
1	Milk Fat per cent m/m	Min. 8.0	Max.1	More than 1 and less than 8.0	Min.16
2	Milk Solid Not Fat** Min. per cent m/m	-	-	20	14
3	Milk Solids ** Min. per cent m/m	28	24	24	-
4	Milk Protein in Milk Solids** per cent m/m	34	34	34	34

** The milk solids and milk solids-not-fat content include water of crystallization of the lactose.

Both Evaporated milk and Sweetened Condensed milk may contain the following food additives:

Table 16.55: Food Additives in Evaporated and Sweetened Condensed Milk

Type of Additives	INS No	Name of Additive	Maximum Level
Acidity regulators	170 339 340 341	Calcium carbonates Sodium phosphates Potassium phosphates Calcium phosphates	2 g/kg singly or 3 g/kg in combination, expressed as anhydrous substances
	450 451 452	Diphosphates Triphosphates Polyphosphates	2 g/kg singly or 3 g/kg in combination, expressed as anhydrous substances
	500 501	Sodium carbonates Potassium carbonates	
Firming agents	508 509	Potassium chloride Calcium chloride	2 g/kg singly or 3 g/kg in combination, expressed as anhydrous substances
Stabilizers	331 332 333	Sodium citrates Potassium citrates Calcium citrates	2 g/kg singly or 3 g/kg in combination, expressed as anhydrous substances
Thickener	407	Carrageenan	150 mg/kg
Emulsifier	**322**	**Lecithins**	**Limited by GMP**

7. Desiccated Milk Based Products/Non Curdled Milk Products

Khoa

A. FSSAI

As per FSS (Food Products Standards and Food Additives) Regulation-2011

Khoa by whatever variety of names it is sold such as Pindi, Danedar, Dhap, Mawa or Kava means the product obtained from cow or buffalo or goat or sheep milk or milk solids or a combination thereof by rapid drying. The milk fat content shall not be less than 30 per cent on dry weight basis of finished product. It may contain citric acid not more than 0.1 per cent by weight. It shall be free from added starch, added sugar and added colouring matter. Khoa shall meet the microbiological requirements given in Appendix- B of the regulation.

As per Compendium Version-XIV of Above Regulation

Khoa by whatever name it is sold such as Khoa or Mawa or any other region specific popular name means the product obtained by partial removal of water from any variant of milk with or without added milk solids by heating under controlled conditions. The product shall conform to the compositional specifications provided in the Table 16.56

Table 16.56: Composition of Khoa

Parameter	*Khoa*
Total solids, minimum, per cent, (m/m)	55.0
Milk fat, minimum, per cent, (m/m), dry matter basis	30.0
Total ash, maximum, per cent, (m/m)	6.0
Titratable acidity (as per cent lactic acid), maximum, per cent	0.9

It shall be free from added starch and added sugar.

The extracted fat from Khoa shall meet the standards for Reichert Meissl value, Polenske value and Butyro-refractometer reading as prescribed for ghee.

The products shall conform to the microbiological requirements specified in Appendix 'B' of these regulations. For products covered under this standard, specific food additives specified in Appendix 'A' of these regulations may be used and only within the limits specified.

B. As per BIS (IS:4883-1980)-Specification of Khoa

All three types of khoa shall be made from fresh, sweet,clean milk which is in every way fit for human consumption and milk shall also be free from adulterants, preservatives, colostrum and foreign matter. Khoa shall be white to yellow with a tinge of brown in colour, have uniform texture and consistency, characteristic flavour, free from objectionable flavour, sign of free fat, mouldiness and water seepage. The following quality standards for three types of khoa are specified:

Table 16.57: BIS Quality Standards for Khoa

Sl. No.	*Types of Khoa*	*Total Solids per cent by Mass (min)*	*Fat (Dry matter) per cent by mass (min.)*	*Total Ash Per cent by Mass (Dry Matter) (min)*	*Titrable Acidity (as L A) Per cent by Mass (max)*
1	Dhap	55	37	6	0.6
2	Pindi	65	37	6	0.8
3	Danedar	60	37	6	0.9

The following microbiological standards are prescribed under **IS:4883-1980:**

Table 16.58: Microbiological Standards for Khoa

Sl.No.	*Types of khoa*	*Yeast and Mould Count/g (max.)*	*Coliform Count/g (Max,)*
1	Dhap	50	90
2	Pindi	50	90
3	Danedar	50	90

8. Dairy based Desserts/Confections

A. *FSSAI*

As per FSS (Food Products Standards and Food Additives) Regulation-2011

a. Ice Cream, Kulfi, Chocolate Ice Cream or Softy Ice Cream (hereafter referred to as the said product) means the product obtained by freezing a pasteurized mix prepared from milk and/or other products derived from milk with or without the addition of nutritive sweetening agents, fruit and fruit products, eggs and egg products, coffee, cocoa, chocolate, condiments, spices, ginger and nuts and it may also contain bakery products such as cake or cookies as a separate layer and/or coating. The said product may be frozen hard or frozen to a soft consistency; the said product shall have pleasant taste and smell free from off flavour and rancidity; the said product may contain food additives permitted in these regulation, the said product shall conform to the microbiological requirements specified in Appendix B of regulation. The said product shall conform to the following requirements, namely-

The Quality Standards of Ice cream, Kulfi, Chocolate and Softy Ice cream under Indian Standard Agencies *e.g.* FSSR-2011 and BIS (Ice cream-IS 2802:1964 and Kulfi- IS:10501-1983) are given below:

Table 16.59: BIS and FSSAI Quality Standards for Ice Cream, Kulfi *etc.*

Sl. No.	*Quality Parameters*	*Quality Norms*						
		FSSR-2011			*Under BIS*			
		Low Fat	*Medium Fat*	*Ice Cream*	*Plain*		*Fruits/Nut*	
					Ice Cream	*Kulfi*	*Ice Cream*	*Kulfi*
1	Milk Fat per cent Min	Not more than 2.5	In between 2.5 - 10	Min 10	8	8	10	6
2	Milk Protein per cent Min NX 6.38	3.0	3.5	3.5	-	3.5	-	3.5
3	Total Solids per cent by weight Min	26	30	36	36	35	36	30
4	Starch per cent Max			-	-	-ive	-	-ive
5	Sucrose per cent by weight Min			-	15	13	15	13
6	Acidity per cent (Lactic acid) Max.			-	0.25	0.3	-	0.3
7	Weight in gm per litre Min	475	475	525	525	-	540	-
8	Permitted Stabilisers/ Emulsifiers per cent Max.	-	-	As per appendix -A of regulation	-	0.5	-	0.5
9	Phosphatase Test	-	-	-	Negative	Negative		

Note: In case where Chocolate, Cake or similar food coating, base or layer forms a separate part of the product only the Ice Cream portion shall conform to the requirements given above. The type of ice-cream shall be clearly indicated on the label otherwise standard for ice-cream shall apply.

As per BIS

Bureau of Indian Standards has also prescribed quality standards for Ice Cream under IS 2802:1964 and Kulfi under IS:10501-1983 which are given in aforesaid table.

b. Dried Ice Cream Mix/Dried Frozen Dessert/Confection (hereafter referred to as the said product) means the product in a powder form which on addition of prescribed amount of water shall give a product conforming to the requirements of the respective products, namely - ice cream, medium fat ice-cream, low fat ice-cream as prescribed under regulation (as given in above table) and frozen confection, medium fat frozen confection and low fat frozen confection as prescribed under regulation (As given in following table) of these regulations except the requirement of weight/volume for both the products. The moisture content of the product shall not be more than 4.0 per cent. It may contain food additives permitted in these regulation. It shall conform to the microbio-logical requirements prescribed in Appendix B of the regulation.

c. **Frozen Dessert/Frozen Confection(hereafter referred to as the said product)** means the product obtained by freezing a pasteurised mix prepared with milk fat and/or edible vegetable oils and fat having a melting point of not more than 37.0 degree C in combination and milk protein alone or in combination/or vegetable protein products singly or in combination with the addition of nutritive sweetening agents *e.g.* sugar, dextrose, fructose, liquid glucose, dried liquid glucose, maltodextrin, high maltose corn syrup, honey, fruit and fruit products, eggs and egg products coffee, cocoa, chocolate, condiments, spices, ginger, and nuts. The said product may also contain bakery products such as cake or cookies as a separate layer/or coating, it may be frozen hard or frozen to a soft consistency. It shall have pleasant taste and flavour free from off flavour and rancidity and may contain food additives permitted in regulation. It shall conform to the microbiological requirements prescribed in Appendix B. It shall conform to the following requirements:

Table 16.60: FSSAI Standards for Frozen Desserts

Requirement	*Frozen Dessert/ Frozen Confection*	*Medium Fat Frozen Dessert/Frozen Confection*	*Low Fat Frozen Dessert/Frozen Confection*
Total Solid per cent	Not less than 36.0	Not less than 30.0	Not less than 26.0
Wt/Vol (gms/l)	Not less than 525	Not less than 475	Not less than 475
Total Fat per cent	Not less than 10.0	More than 2.5 per cent but less than 10.0	Not more than 2.5
Total Protein (N x 6.25) per cent	Not less than 3.5	Not less than 3.5	Not less than 3.0

Note: In case where Chocolate, Cake or similar food coating, base or layer forms a separate part of the product only the frozen dessert/confection portion shall conform to the requirements given above. The type of frozen confection shall be clearly indicated on the label otherwise, standards of frozen dessert/frozen confection shall apply and every package of Frozen Dessert/Frozen Confection shall bear proper label declaration under regulation 2.4.5 (41) of Food Safety and Standards (Packaging and Labelling) Regulations, 2011.

d. **Milk Ice or Milk Lolly (hereafter referred to as the said product)** means the product obtained by freezing a pasteurized mix prepared from milk and/ or other products derived from milk with or without the addition of nutritive sweetening agents, fruit and fruit products, eggs and egg products, coffee, cocoa, chocolate, condiments, spices, ginger and nuts; the said product may also contain bakery products such as cake or cookies as a separate layer and/ or coating; the said product shall have pleasant taste and smell free from off flavour and rancidity. It may contain food additives permitted in regulation. The said product shall conform to the microbiological requirements prescribed in Appendix B of regulation. The said product shall also conform to the following requirements, namely-

i. Total solids (m/m) Not less than 20.0 per cent.
ii. Milk Fat (m/m) Not more than 2.0 per cent and.
iii. Milk Protein (Nx6.38) Not less than 3.5 per cent.

As per Compendium Version-XIV of Above Regulation

Ice Cream, Kulfi, Chocolate Ice Cream, Softy Ice-Cream, Milk Ice, Milk Lolly and Dried Ice Cream Mix-

a. Ice-Cream, Kulfi, Chocolate Ice Cream or Softy Ice-Cream means the frozen milk product conforming to the composition specified in following table, obtained by freezing a pasteurized mix prepared from milk or other products derived from milk, or both, with or without addition of nutritive sweeteners and other permitted non-dairy ingredients. The said product may contain incorporated air and shall be frozen hard except in case of softy ice-cream where it can be frozen to a soft consistency.

Table 16.61: Quality Standards for Ice Cream, Kulfi *etc.*

Parameter	*Cce cream or Kulfi or Chocolate Ice Cream or Softy Ice Cream*	*Medium Fat Ice Cream or Kulfi or Chocolate Ice Cream or Softy Ice Cream*	*Low Fat Ice Cream or Kulfi or Chocolate Ice Cream or Softy Ice Cream*
Total Solids, minimum, per cent, (m/m)	36.0	30.0	26.0
Weight, minimum, g/l	525.0	475.0	475.0
Milk Fat, per cent, (m/m)	10.0 (minimum)	More than 2.5 and less than 10.0	2.5 (maximum)
Milk Protein*, minimum, per cent, (m/m)	3.5	3.5	3.0

* Protein content is 6.38 multiplied by the total nitrogen determined.

Notes- **i.** In case where coating, base or layer of non-dairy ingredients forms a separate part of the product, only the Ice Cream portion shall conform to the respective composition and **ii.** When any type of ice cream, kulfi, chocolate ice cream or softy ice cream is offered for sale in contravention of the requirements of sub-item (b) of item 6, (Pl.see below)the standards prescribed for the type ice cream, kulfi, chocolate ice cream or softy ice cream as per this sub-regulation shall apply.

Sub-item (b) of item 6-The type, (as given under specification of ice cream *etc.*), of ice cream, kulfi, chocolate ice cream or softy ice cream shall always be indicated on the label of the product. For softy ice cream offered for sale directly from the freezer without prepackaging, the type of product shall be displayed in a manner and at a place that is clearly visible to the consumer.

b. Milk Ice or Milk Lolly means the product conforming to the composition specified in following table, obtained by freezing a pasteurized mix prepared from milk or other products derived from milk with or without the addition of nutritive sweeteners and other permitted non-dairy ingredients. The said product shall be frozen hard.

Table 16.62: Quality Standards for Milk Ice or Milk Lolly

Parameter	*Milk Ice or Milk Lolly*
Total Solids, minimum, per cent, (m/m)	20.0
Milk Fat, maximum, per cent, (m/m)	2.0
Milk Protein*, minimum, per cent, (m/m)	3.5

* Protein content is 6.38 multiplied by the total nitrogen determined.

Note: In case where base or layer of non-dairy ingredients forms a separate part of the product, only the milk ice or milk lolly portion shall conform to the above composition

c. Dried Ice-Cream Mix means the product in a powder form which on addition of prescribed amount of water and freezing shall result in a product similar in characteristics to the respective product described in under standards of Ice cream (a).

The products (All mentioned above under a,b,c) shall conform to the microbiological requirements specified in Appendix 'B' of these regulations. For products covered under this standard, specific food additives specified in Appendix 'A' of these regulations may be used and only within the limits specified.

9. Milk Powder

A. FSSAI

As per FSS (Food Products Standards and Food Additives) Regulation-2011

Milk Powder means the product obtained by partial removal of water from milk of Cow and/or Buffalo. The fat and/or protein content of the milk may be adjusted by addition and/or withdrawal of milk constituents in such a way as not to alter the whey protein to casein ratio of the milk being adjusted. It shall be of uniform colour and shall have pleasant taste and flavour free from off flavour and rancidity. It shall also be free from vegetable oil/fat, mineral oil, thickening agents, added flavour and sweetening agent. It may contain food additives permitted in these regulations. It shall conform to the microbiological requirements prescribed in Appendix B. It shall conform to the following requirements:

Table 16.63: FSSAI Quality Standards for Milk Powder

Product	*Moisture*	*Milk Fat*	*Milk protein in Milk Solids on Not Fat*	*Titrable Acidity (ml 0.1N NAOH/10 gm Solids not fat)*	*Insolubility Index*	*Total Ash on Dry Weight Basis*
Whole milk powder	Not more than 4.0 per cent m/m	Not less than 26.0 per cent m/m	Not less than 34.0 per cent m/m	Not more than 18.0	Not more than 2 ml	Not more than 7.3 per cent
Partly Skimmed milk powder	Not more than 5.0 per cent m/m	Not less than 1.5 and not more than 26 per cent m/m	Not less than 34.0 per cent m/m	Not more than 18.0	Not more than 2 ml	Not more than 8.2 per cent
Skimmed milk powder	Not more than 5.0 per cent m/m	Not more than 1.5 per cent m/m	Not less than 34.0 per cent m/m	Not more than 18.0	Not more than 2 ml	Not more than 8.2 per cent

As per Compendium Version-XIV of Above Regulation

Milk powders milk products which can be obtained by partial removal of water from milk or cream. The fat or protein content, or both of the milk or cream may be adjusted, only to comply with the compositional requirements given in following table, by addition or withdrawal of milk constituents in such a way as not to alter the whey protein to casein ratio of the milk or cream being adjusted. Product shall be free from added whey and whey preparations.

Cream powder is also included. The standards for Milk protein, Insolubility index and Titrable acidity are same as prescribed under FSSR-2011. The standards for Moisture, Milk Fat, Total ash and Scorched particle are given in Table 16.64.

The products shall conform to the microbiological requirements specified in Appendix 'B' of these regulations. For products covered under this standard, specific food additives specified in Appendix 'A' of these regulations may be used and only within the limits specified.

B. *As per BIS*

Earlier BIS has prescribed quality norms of all above types of milk powders under IS: 1165-1975 but later on BIS has prescribed separate quality standards for **Skim milk powder- Standard grade(**IS: 13334 (Part-I)-1998), **Skim milk powder-Extra grade(**IS: 13334 (Part-II)-1998)**, Partly Skim milk powder Type I and II (**IS:14542-1998) **and Milk Powder** (IS:1165-1992/IS 1165:2002-fourth revision). The definition of above types of milk powders are given below:

Table 16.64: Quality Standards for Milk an Cream Powders

Parameter	Whole Milk Powder	Partly Skimmed Milk Powder	Skimmed Milk Powder	Cream Powder
Moisture*, maximum, per cent, (m/m)	5.0	5.0	5.0	5.0
Milk fat, per cent, (m/m)	Minimum 26.0 and less than 42.0	More than 1.5 and less than 26.0	1.5 (maximum)	42.0 (minimum)
Total ash, maximum, per cent (m/m), on moisture and fat free basis	9.3	9.3	9.3	-
Scorched particles, maximum	Disc B	Disc B	Disc B	Disc B

* The moisture content does not include water of crystallization of the lactose; the milk solids-not-fat content includes water of crystallization of the lactose. ** Protein content is 6.38 multiplied by the total nitrogen determined.

Note: Total sodium content in the milk powder shall not be more than 650 mg/100 gm SNF. The maximum level does not apply to sodium that could be present due to the use of sodium containing additives in milk powders.

a. *Skim Milk Powder-Standard Grade {IS: 13334 (Part-I)-1998}*

The product obtained from skimmed milk of cow or buffalo or a mixture there of by removal of water through spray or roller drying. It may contain Calcium chloride, Citric acid and sodium citrate, sodium salt of orthophosphoric acid and poly phosphoric acid not exceeding 0.3 per cent by mass of the finished product. **Extra Grade**{IS: 13334 (Part-II)-1998}-shall be material prepared by spray drying of fresh skimmed milk of cow or buffalo or a mixture there of. It shall be free from additives and shall be free from whey and butter milk.

b. *Partly Skim Milk Powder (IS:14542-1998)*

BIS has categorized partly skimmed milk powder in two types, the details of which are given below:

Partly Skim milk powder(Type-1)- It shall be the product obtained by spray drying or roller drying of Partly skimmed milk of cow or buffalo or a mixture there of.

Partly Skim milk powder(Sour, Type-2)- It shall be the product obtained by spray drying or roller drying of Partly skimmed milk of cow or buffalo or a mixture there of,primarily for use by industry like bakery.

c. *Milk Powder (IS:1165-1992)*

It shall be the material prepared by spray drying of Standardised milk obtained from fresh cow or buffalo or a mixture there of.

Table 16.65: BIS Quality Standards for different Types of Milk Powders

Sl.No.	Quality Standards	Skim Milk Powder		Partly Skim Milk Powder		Milk Powder (Spray)
		Standard Grade IS: 13334 (Par-t-I)-1998	Extra Grade * IS:13334 (Part-II)-1998	Type-I IS:14542-1998	Type-II IS: 14542-1998 Partly Skim Milk Powder (sour)**	IS:1165-1992/ IS-1165:2002 (Fourth revision)
1	Milk Fat per cent by mass	Not more than 1.5	Max 1.25	Min-1.5 Max-26	Min-1.5 Max-26	Not less than 26
2	Moisture per cent by mass Max,	4.0	3.5	4.0	4.0	4.0
3	Total acidity as Lactic acid per cent Max.	1.5	19.5 ml of 0.1N NaOH	1.5	1.5	1.2
4	Total Solids(Milk solids +Added solids) per cent Min	96	96.5	96	96	96
5	Total Ash on dry basis per cent Max	8.2	8.2	8.2	8.2	7.3
6	Insolubility Index Max.					
	Roller dried	15 ml	-	15 ml	-	-
	Spray dried	1.5 ml	0.5 ml	1.5 ml	2.5ml	2.0 ml
7	Lactate content mg/g, Max.	-	1.5	-	-	-
8	Scorched Particles Max.	-	15mg equivalent to disc - B			-

Sl.No.	Quality Standards	Skim Milk Powder		Partly Skim Milk Powder		Milk Powder (Spray)
		Standard Grade IS: 13334 (Par-t-I)-1998	Extra Grade * IS:13334 (Part-II)-1998	Type-I IS:14542-1998	Type-II IS: 14542-1998 Partly Skim Milk Powder (sour)**	IS:1165-1992/ IS-1165:2002 (Fourth revision)
9	Additives i. Calcium chloride, Citric acid and sodium citrate, sodium salt of Ortho phosphoric acid and phosphoric acid.	Not exceeding 0.3 per cent by weight of finished product.	-	Not exceeding 0.3 per cent by weight of finished product	Not exceeding 0.3 per cent by weight of finished product.	-
	ii. Butylated hydroxianisole (BHA)	-	-	Not exceeding 0.01 per cent by weight of finished product	Not exceeding 0.01 per cent by weight of finished product	Not exceeding 0.01 per cent by weight of finished product.
10	Microbiological quality					
	SPC/g Max	50000	40000	50000	50000	40000
	Coliform count/0.1gm.	Absent	Absent	Absent	Absent	Absent
	Coagulase positive Staph. aureus count in 0.1gm.	-	Absent	-	-	Absent
	Salmonella and Shigella in 25 gm	-	Absent	-	-	Absent

** It may also contain Sod. bicarbonate with "Unfit for direct consumption" label.

Partly Skim milk powder and milk powder may also contain BHA not more than 0.01 per cent by mass of the finished product. For improving the dispersibility of the product, lacithin to a maximum extent of 0.5 per cent by mass may be added and declared on the label as per the PFA/FSSAI rules.

All above types of powders shall be free from lumps except those that break up readly under slight pressure, dirt, other extraneous matter and added colour/flavour. It shall be white or white with greenish tinge to light cream in colour. It shall be reasonably free from scorched particles. The flavour of the aforesaid products except Partly Skim milk powder(Sour,Type-2 or of the reconstituted milk) shall be pleasant and clean. It may have slightly cooked but not be burnt flavour.

The details of quality norms under aforesaid BIS Standards are given in Table 16.65 given on Back Page.

C. *Under Codex Stan 207-1999 (Adopted in 1999. Amendment 2010)*

Milk powders are milk products which can be obtained by the partial removal of water from milk or cream. The fat and/or protein content of the milk or cream may have been adjusted, only to comply with the following compositional requirements by the addition and/or withdrawal of milk constituents in such a way as not to alter the whey protein to casein ratio of the milk being adjusted. Powders have made from **Raw materials** -Milk and cream. The following milk products are allowed for protein adjustment purposes:

i. Milk retentate Milk retentate is the product obtained by concentrating milk protein by ultrafiltration of milk, partly skimmed milk, or skimmed milk.

ii. Milk permeate Milk permeate is the product obtained by removing milk proteins and milkfat from milk, partly skimmed milk, or skimmed milk by ultrafiltration and.

iii. Lactose (as per quality allowed under codex).

Table 16.66: Codex Standards Milk Powders

Sl. No.	*Quality Standards*	*Milk Powders*		
		Skimmed Milk	*Partly Skimmed Milk*	*Whole Milk*
1	Milk Fat per cent m/m	1.5 Max.	1.5 to 26	Min. 26 and less than 42.
2	Water content* (Moisture) per cent m/m	5.0 Max	5.0 Max	5.0 Max
3	Milk Protein in milk solid not fat* per cent m/m	34 Min	34 Min	34 Min

Sl. No.	Quality Standards	Milk Powders		
		Skimmed Milk	Partly Skimmed Milk	Whole Milk
4	Food Additives i. Stabilizers 331 Sodium citrate 332 Potassium citrate	5 g/Kg singly or combination, expressed as an-hydrous substances	5 g/Kg singly or combination, expressed as an-hydrous substances	5 g/Kg singly or combination, expressed as an-hydrous substances
	ii. Firming agents- 508 Potassium chloride, 509 Calcium chloride	Limited by GMP	Limited by GMP	Limited by GMP
	iii. Emulsifiers-322 Lacithin	Limited by GMP	Limited by GMP	Limited by GMP
	471 Mono and diglycerides of fatty acids	2.5g/Kg	2.5g/Kg	2.5g/Kg
	iv. Antioxidants a. 300L-Ascorbic acid, 301Sodium ascorbate, 304Ascorbyl palmitate.	0.5g/Kg expressed as ascorbic acid.	0.5g/Kg expressed as ascorbic acid.	0.5g/Kg expressed as ascorbic acid.
	b. 320 Butylated hydroxyanisole (BHA)	0.01 m/m	0.01 m/m	0.01 m/m
	v. Acidity regulators 339Sodium phosphates, 340Potassium phosphates, 450Diphosphate, 451Triphosphate, 452Polyphosphate, 500Sodium carbonates, 501Potassium carbonates.	5g/Kg	5g/Kg	5g/Kg
	vi. Anticaking Agents Calcium carbonate, 170(i), Calcium carbonate, 341(iii), Tricalcium orthophosphate, 343(iii), Trimagnesium orthophosphate, 504(i) Magnesium carbonate, 530 Magnesium oxide, 551 Silicon dioxide, amorphous, 552 Calcium silicates, 553 Magnesium silicates, 554 Sodium aluminosilicate, 556 Calcium aluminosilicate, 559 Aluminium silicates.	10g/Kg singly or in combination	10g/Kg singly or in combination	10g/Kg singly or in combination
5	Contaminants			
	i. Heavy Metals	As per CAC	As per CAC	As per CAC
	ii. Pesticide Residue	As per CAC	As per CAC	As per CAC
6	Optional Standards			
	i. Titratable acidity** (ml of 0.1 N NaOH/10 gm SNF.)	Max.18	Max.18	Max.18
	ii. Scorched Particles***-	Max. Disc–B	Max. Disc–B	Max. Disc–B
	iii. Solubility Index****(ml)	Max.1.0	Max.1.0	Max.1.0

Note: *-The water content does not include water of crystallization of Lactose where as milk solid not fat content includes water of crystallization of Lactose. ** -IDF Standard 86:1981, IDF Standard 81:1981, *** -IDF Standard 107A: 1995, ****-IDF Standard 129A:1988 and Number given before the food additives is INS No.

The Contaminants, Hygiene, Labelling and Methods of Sampling and Testing are as per norms given in concerned codex booklet.

10. Cheese

A. *FSSAI*

As per FSS (Food Products Standards and Food Additives) Regulation, 2011

1. Cheese

It means the ripened or unripened soft or semihard, hard and extra hard product, which may be coated with food grade waxes or polyfilm, and in which the whey protein/casein ratio does not exceed that of milk. Cheese is obtained by coagulating wholly or partly milk and/or products obtained from milk through the action of nonanimal rennet or other suitable coagulating agents and by partially draining the whey resulting from such coagulation and/or processing techniques involving coagulation of milk and/or products obtained from milk which give a final product with similar physical, chemical and organoleptic characteristics. The product may contain starter cultures of harmless lactic acid and/or flavour producing bacteria and cultures of other harmless micro-organisms, safe and suitable enzymes and sodium chloride. It may be in the form of blocks, slices, cut, shredded or grated cheese.

i. **Ripened Cheese:** It is cheese which is not ready for consumption shortly after manufacture but which must be held for some time at such temperature and under such other conditions as will result in necessary biochemical and physical changes characterizing the cheese in question.

ii. **Mould Ripened cheese:** It is a ripened cheese in which the ripening has been accomplished primarily by the development of characteristic mould growth through the interior and/or on the surface of the cheese.

iii. **Unripened cheese:** Unripened including fresh cheese is cheese which is ready for consumption shortly after manufacture.

Cheese or varieties of cheeses shall have pleasant taste and flavour free from off flavour and rancidity.

2. Processed Cheese

It means the product obtained by grinding, mixing, melting and emulsifying one or more varieties of cheeses with the aid of heat and emulsifying agents. It may contain cream, butter, butter oil and other milk products subject to maximum 5.0 per cent lactose content in the final product and edible common salt, vinegar/acetic acid, spices and other vegetable seasoning and foods other than sugars properly cooked or prepared for flavouring and characterization of the product provided these additions do not exceed one sixth of the weight of the total solids of the final product on dry matter basis and cultures of harmless bacteria and enzymes. It shall have pleasant taste and smell free from off flavour and rancidity.

3. Processed Cheese Spread

It means the product obtained by grinding, mixing, melting and emulsifying one or more varieties of cheese with emulsifying agents with the aid of heat. It may contain Cream, Butter oil and other dairy products, subject to a maximum limit of 5.0 per cent lactose in the final product, salt, vinegar, spices, condiments and seasonings, natural carbohydrate sweetening agents namely sucrose, dextrose, corn syrup, corn syrup solids, honey, maltose, malt syrup and hydrolysed lactose and food properly cooked or otherwise prepared for flavouring and characterization of the product provided these additions do not exceed one sixth of the weight of total solids of the final product on dry weight basis and cultures of harmless bacteria and enzymes. It shall have pleasant taste and flavour free from off flavour and rancidity.

4. Cheddar Cheese

It means ripened hard cheese obtained by coagulating heated/pasteurised milk of Cow and/or Buffalo or mixtures thereof with cultures of harmless lactic acid producing bacteria, non-animal rennet or other suitable coagulating enzymes. It shall be in the form of hard pressed block with a coating of food grade waxes or wrapping of cloth or polyfilm. It shall have firm, smooth and waxy texture with a pale straw to orange colour without any gas holes.

5. Danbo Cheese

It means ripened semi hard cheese obtained by coagulating heated/ pasteurised milk of cow and/or Buffalo and mixtures thereof with cultures of harmless lactic acid producing bacteria, non-animal rennet or other suitable coagulating enzymes. It shall be smooth in appearance with firm texture and uniform yellow colour and may be coated with food grade waxes or wrapping of cloth or polyfilm.

6. Edam Cheese

It means the ripened semi hard cheese obtained by coagulating heated/ pasteurised milk of Cow and/or Buffalo or mixtures thereof with cultures of harmless lactic acid producing bacteria, non-animal rennet or other suitable coagulating enzymes. It shall have a firm texture suitable for cutting with a yellowish colour and a hard rind which may be coated with food grade waxes, wrapping of cloth, polyfilm or vegetable oil.

7. Gouda Cheese

It means ripened semi hard cheese obtained by coagulating milk of Cow and/or Buffalo or mixtures thereof with cultures of harmless lactic acid producing bacteria,non-animal rennet or other suitable coagulating enzymes. It shall have firm texture suitable for cutting, straw to yellowish colour and a hard rind which may be coated with food grade waxes, wrapping of cloth, or vegetable oil.

8. Havarti Cheese

It means ripened semi hard cheese obtained by coagulating milk of cow and/or Buffalo or mixtures thereof with cultures of harmless lactic acid producing bacteria, non-animal rennet or other suitable coagulating enzymes. It shall have firm texture suitable for cutting, a light yellow colour and may have a semi soft slightly greasy rind.

9. Tilsiter Cheese

It means ripened semi hard cheese obtained by coagulating milk of Cow and/or Buffalo or mixtures thereof with cultures of harmless lactic acid producing bacteria and cultures of Bacterium linens, non-animal rennet or other suitable coagulating enzymes. It shall have firm texture suitable for cutting with an ivory to yellow colour with a firm rind which may show red and yellow smear producing bacteria or coated with food grade waxes or wrapping of cloth or polyfilm after removal of the smear.

10. Cottage Cheese and Creamed Cottage Cheese

It means soft unripened cheese obtained by coagulation of pasteurised skimmed milk of Cow and/or Buffalo or mixtures thereof with cultures of harmless lactic acid bacteria with or without the addition of other suitable coagulating enzymes. Creamed Cottage Cheese is cottage cheese to which a pasteurised creaming mixture of cream, skimmed milk, condensed milk, non fat dry milk, dry milk protein, Sodium/Potassium/Calcium/Ammonium caseinate is added. It shall have a soft texture with a natural white colour. It may contain spices, condiments, seasonings and fruits pulp.

11. Cream Cheese (Rahmfrischkase)

It means soft unripened cheese obtained by coagulation of pasteurised milk of cow and/or buffalo or mixtures thereof and pasteurised cream with cultures of harmless lactic acid producing bacteria with or without the addition of suitable coagulating enzymes. It shall have a soft smooth texture with a white to light cream colour. It may contain spices, condiments, seasonings and fruits pulp.

12. Coulommiers Cheese

It means soft unripened cheese obtained by coagulation of milk of cow and/or buffalo or mixtures thereof with cultures of harmless lactic acid producing bacteria and non-animal rennet or other suitable coagulating enzymes and moulds characteristic of the variety. It shall have soft texture and white to cream yellow colour and may show presence of white mould including orange or red spots on the surface.

13. Camembert Cheese

It means ripened soft cheese obtained by coagulating milk of Cow and/or Buffalo or mixtures thereof with cultures of harmless lactic acid producing bacteria and cultures of Penicillium caseicolum and Bacterium linens,non-animal rennet or other suitable coagulating enzymes. It may be in the form of flat cylindrical shaped cheese covered with white mould (Penicillum caseicolum) with occasional orange coloured spots (Bacterium linens).

14. Brie Cheese

It means soft ripened cheese obtained by coagulating milk of Cow and/ or Buffalo or mixtures thereof with cultures of harmless lactic acid producing bacteria and cultures of Penicillium caseicolum and Bacterium linens, non-animal rennet and other suitable enzymes. It shall be white to creamy yellow in colour with a smooth texture showing presence of white mould (Penicillium caseicolum) with occasional orange coloured spots (Bacterium linens) on the rind.

15. Saint Paulin

It means ripened semi hard cheese obtained by coagulating milk of Cow and/or Buffalo or mixtures thereof with non-animal rennet, cultures of harmless lactic acid producing bacteria or other suitable enzymes. It shall have white to yellow colour with a firm and flexible texture and a hard rind which may be coated with food grade waxes or polyfilm.

16. Samsoe

It means hard ripened cheese obtained by coagulating milk of Cow and/or Buffalo or combination there of with non-animal rennet and cultures of harmless lactic acid producing bacteria or suitable coagulating enzymes. It shall be yellow in colour with a firm texture suitable for cutting and may have a rind with or without food grade waxes or polyfilm coating.

17. Emmentaler

It means hard ripened cheese with round holes obtained by coagulating milk of Cow and/or Buffalo or mixtures thereof with non-animal rennet, cultures of harmless lactic acid producing bacteria or other suitable coagulating enzymes. It may contain Cupric Sulphate not exceeding 15 mgm/Kg expressed as Copper. It shall have a light Yellow colour and a firm texture suitable for cutting and may have a hard rind.

18. Provolone

It means pasta filata cheese obtained by coagulating milk of Cow and/ or Buffalo or mixtures thereof with cultures of harmless lactic acid producing bacteria, non-animal rennet or other suitable coagulating enzymes. It may be smoked. It shall be white to yellow straw in colour with a fibrous or smooth body and rind which may be covered with vegetable fat/oil, food grade waxes or polyfilm.

All cheeses shall conform to the following requirements:

Table 16.67: A. Cheese and Processed Cheeses

Sl. No.	Parameters	Varieties of Cheese							
		Hard Pressed	Semi hard	Cheddar	Semi Soft	Soft	Extra hard	*Processed	Processed Spread
1	Moisture per cent Not More Than	39	45	39	52	80	36	47	60
2	Milk Fat per cent on dry basis Not less than	48	40	48	45	20	32	40	40

*: The processed cheese chiplets (packed sliced cheese) when sold in a package other than tin then it shall not contain more than 50 per cent moisture.

Table 16.68: B. Ripened Cheeses

Sl. No.	Parameters	Varieties of Cheese										
		Dambo	Edam	Gouda	Havarti			Tilsiter			Brie	Saint paulin
					Havarti	30 per cent	60 per cent	Tilsiter	30 per cent	60 per cent		
1	Moisture per cent Not More Than	39	46	43	48	53	60	47	53	39	56	56
2	Milk Fat per cent on dry basis Not less than	45	40	48	45	30	60	45	30	60	40	40

Table 16.69: C. Unripened and Ripened Cheeses

Sl. No.	*Parameters*	*Varieties of Cheese*											
		Unripened			*Ripened*						*Emmen-taler*	*Provo-lone-un soaked/ Soaked*	*Extra Hard Grating*
		Coulom-miers	*Cottage*	*Cream*	*Camembert*				*Samsoe*				
					30 per cent	*40 per cent*	*45 per cent*	*50 per cent*	*Samsoe*	*30 per cent*			
1	Moisture per cent Not More than	·56	80	55	62	56	56	56	44	50	40	47/45	36
2	Milk Fat per cent on dry basis Not less than	46	4 Creamed	70	30	40	45	50	45	30	45	45	32

19. Extra Hard Grating Cheese

It means ripened cheese obtained by coagulating milk of Cow and/or Buffalo, goat/sheep milk or mixtures thereof with cultures of harmless lactic acid producing bacteria, non-animal rennet, or other suitable coagulating enzymes. It may be white to light cream in colour with a slightly brittle texture and an extra hard rind which may be coated with vegetable oil, food grade waxes or polyfilm.

Common Requirements of different Cheeses and Varieties of Cheeses

All above mentioned cheeses shall also conform to the following requirements:

i. It may contain food additives permitted in these regulation including Appendix A. It shall conform to the microbiological requirements prescribed in Appendix B of FSSR-2011.

ii. Provided that cheese or varieties of cheeses coated with food grade waxes/or polyfilm/or wrapping of cloth shall bear proper label declaration as provided in regulation 2.4.5 (44) of Food Safety and Standards (Packaging and Labeling) Regulations, 2011.

As per Compendium Version-XIV of Above Regulation

Cheese- Same description as given in FSSR-2011 except the following:

i. Coagulating wholly or partly the protein of milk, skimmed milk, partly skimmed milk, cream, whey cream or buttermilk, or any combination of these materials, through the action of suitable enzymes of non-animal origin or other suitable coagulating agents, with or without use of harmless lactic acid bacteria and flavour producing bacteria, and by partially draining the whey resulting from the coagulation.

ii. All cheese shall be made from milk which is subject to heat treatment at least equivalent to that of pasteurization.

iii. **"Individual or Named Variety Cheese"** is a cheese, as defined in item 1 of this sub-regulation, that is designated with its well-established unique name as provided below:

Prescribed same definitions as given under FSSR-2011 for Cheddar Cheese, Danbo Cheese, Edam Cheese,Gouda Cheese, Havarti, Tilsiter, Cottage and Creamed Cottage,Cream cheese, Coulommiers, Camembert, Brie, Saint Paulin, Samsoe, Emmentaler (Emmental), Provolone and Extra Hard Grating Cheese with following slight changes:

a. **Type of milk: Cheddar Cheese, Danbo Cheese, Edam Cheese** -Heated/Pasteurised milk in place of heated/pasteurised milk of Cow and/or Buffalo or mixtures thereof.

Gouda, Havarti, Tilsiter, Coulommiers, Camembert, Brie Saint Paulin, Samsoe, Provolone Cheese- milk in place of milk of Cow and/or Buffalo or mixtures thereof.

Cottage and Creamed Cottage cheese- Pasteurised skimmed milk in place of pasteurised skimmed milk of Cow and/or Buffalo or mixtures thereof and with or without the addition of suitable enzymes of non-animal origin is added.

Cream Cheese- Pasteurized milk in place of pasteurised milk of Cow and/or Buffalo or mixtures thereof and with or without the addition of suitable enzymes of non-animal origin is added.

Emmentaler(Emmental)- milk in place of milk of Cow and/or Buffalo or mixtures thereof and "It may contain Cupric Sulphate not exceeding 15 mgm/Kg expressed as Copper" is deleted.

Extra Hard Grating Cheese- milk in place of milk of Cow and/or Buffalo, goat/sheep milk or mixtures thereof and "It may be white to light cream in colour" is deleted.

b. Suitable Enzymes of Non-Animal Origin in place of non animal rennet.

iv. **"Cheese Products"** are the products prepared from cheese(s) with other milk products and may contain permitted non-dairy ingredients. Processed cheese and Processed Cheese spread are categorised under this with the following description:

Processed Cheese means the product obtained by grinding, mixing, melting and emulsifying one or more varieties of cheeses with the aid of heat and emulsifying agents and may contain cream, butter, butter oil and other milk products. It may also contain non-dairy ingredients not exceeding one sixth of the weight of the total solids of the final product on dry matter basis.

Processed Cheese Spread means the product obtained by grinding, mixing, melting and emulsifying one or more varieties of cheese with emulsifying agents with the aid of heat and may contain cream, butter oil and other dairy products. It may also contain natural carbohydrate sweetening agents and other non-dairy ingredients not exceeding one sixth of the weight of total solids of the final product on dry weight basis.

v. Description/Definition of following cheeses are included:

1. **Mozzarella cheese:** It means unripened cheese obtained by coagulating milk with cultures of harmless lactic acid producing bacteria, suitable enzymes of nonanimal origin or by direct acidification. It is a smooth elastic cheese with a long stranded

parallel-orientated fibrous protein structure without evidence of curd granules. The cheese is rindless and may be formed into various shapes.

a. **Mozzarella** with a high moisture content is a soft cheese with overlying layers that may form pockets containing liquid of milky appearance. The cheese has a near white colour.

b. **Mozzarella with low moisture content** is a firm or semi-hard homogeneous cheese without holes and is suitable for shredding. Mozzarella is made by 'pasta filata' processing, which consists of heating curd of a suitable pH value kneading and stretching until the curd is smooth and free from lumps. Still warm, the curd is cut and moulded, then firmed by cooling.

2. **Whey Cheeses:** These are solid, semi-solid, or soft products which are principally obtained through either of the following processes-

    i. The concentration of whey and the moulding of the concentrated product.

    ii. The coagulation of whey by heat with or without the addition of acid.

    In each case, the whey may be pre-concentrated prior to the further concentration of whey or coagulation of the whey proteins. The process may also include the addition of milk, cream, or other raw materials of milk origin before or after concentration or coagulation. The ratio of whey protein to casein in the product obtained through the coagulation of whey shall be distinctly higher than that of milk. The product obtained through the coagulation of whey may either be ripened or unripened.

3. **Cheeses in Brine**: These are semi-hard to soft ripened cheeses. The body has a white to yellowish colour and a compact texture suitable for slicing, with none to few mechanical openings. The cheeses have no actual rind and have been ripened and preserved in brine until delivered to, or prepacked for, the consumer. Certain individual cheeses in brine contain specific herbs and spices as part of their identity.

vi. **Composition:** The product shall conform to the compositional specifications provided in the Table 16.70

Table 16.70: Composition of different Chesses

Sl.No.	Product	Moisture, Maximum, Per cent (m/m)	Milk Fat, Minimum, Per cent (dry basis)	Lactose, Maximum, Per cent (m/m)
1	Cheese			
	Hard- Pressed Cheese	39.0	48.0	-
	Semi Hard –Cheese	45.0	40.0	-
	Semi-Soft Cheese	52.0	45.0	-
	Soft Cheese	80.0	20.0	-
	Extra Hard Cheese	36.0	32.0	-
	Mozzarella Cheese	60.0	35.0	-
	Pizza Cheese	54.0	35.0	-
2	Extra Hard Grating Cheese	36.0	32.0	-
3	Named variety			
	Cheddar	39.0	48.0	-
	Danbo	39.0	45.0	-
	Edam	46.0	40.0	-
	Gouda	43.0	48.0	-
	Havarti	48.0	45.0	
	30 per cent Havarti	53.0	30.0	-
	60 per cent Havarti	60.0	60.0	-
	Tilsiter	47.0	45.0	-
	30 per cent Tilsiter	53.0	30.0	-
	60 per cent Tilsiter	39.0	60.0	-
	Cottage Cheese and Creamed Cottage Cheese	80.0	*	-
	Cream cheese	55.0	70.0	-
	Coulommiers	56.0	46.0	-
	30 per cent Camembert	62.0	30.0	-
	40 per cent Camembert	59.0	40.0	-
	45 per cent Camembert	57.0	45.0	-
	55 per cent Camembert	52.0	55.0	-
	Brie	56.0	40.0	-
	Saint Paulin	56.0	40.0	-
	Samsoe	44.0	45.0	-
	30 per cent Samsoe	50.0	30.0	-
	Emmental	40.0	45.0	-
	Provolone - Smoked	45.0	45.0	-
	Unsmoked	47.0	45.0	-

Sl.No.	Product	Moisture, Maximum, Per cent (m/m)	Milk Fat, Minimum, Per cent (dry basis)	Lactose, Maximum, Per cent (m/m)
4	Cheese products			
	a-Processed Cheese	47.0 (50 per cent for chiplets, packed sliced processed cheese), when sold in a package other than tin	40.0	5.0
	b-Processed Cheese Spread	60.0	40.0	5.0
5	Whey cheeses			
	a- Creamed whey cheese	-	33.0	-
	b- Whey cheese	-	**	-
	c- Skimmed whey cheese	-	***	-
6	Cheeses in brine			
	a- Soft cheese in brine	-	40.0	-
	b- Semi-hard cheese in brine	-	40.0	-

* Milk fat, Minimum 4 per cent (m/m) for creamed cottage cheese.; ** Milk fat in whey cheese shall be minimum 10 per cent and less than 33 per cent on dry basis. *** Milk fat in skimmed whey cheese shall be less than 10 per cent on dry basis.

vii. **Common requirements of different cheeses and Varieties of Cheeses:** All above mentioned cheeses shall also conform to the following requirements:

a. It may contain food additives permitted in these regulation under Appendix A. It shall conform to the microbiological requirements prescribed in Appendix B of compendium.

b. The details related to the Food Additives,Contaminants, Hygiene, Labelling and Methods of sampling and Testing are as per norms given in compendium.

B. *As per Codex*

Codex has given following Quality standards for different types of Cheeses -

a. *Individual Definition/Requirement*

i. **Mozzarella Cheese (CODEX STAN 262-2007)-** Mozzarella is an unripened cheese in conformity with the **General Standard for Cheese (CODEX STAN 283-1978)** and **the Standard for Unripened Cheese Including Fresh Cheese (CODEX STAN 221-2001).** It is a smooth elastic cheese with a long stranded parallel-orientated fibrous protein structure without evidence of curd granules. The cheese is rind less and may be formed into various shapes. Mozzarella with a high moisture content is a soft cheese with overlying layers that may form pockets containing liquid of milky appearance. It may be packed with or without the liquid. The cheese has a near white colour. Mozzarella with a low moisture content is a firm/semi-hard homogeneous cheese without holes and is suitable for shredding. Mozzarella is made by "pasta filata" processing, which consists of heating curd of a suitable pH value kneading and stretching until the curd is smooth and free from lumps. Still warm, the curd is cut and moulded, then firmed by cooling. Other processing techniques, which give end products with the same physical, chemical and organoleptic characteristics are allowed.

ii. **Cheddar Cheese (CODEX STAN 263-1966 Formerly CODEX STAN C-1-1966. Adopted in 1996. Revision 2007. Amendment 2008.**)- Cheddar is a ripened hard cheese in conformity with **the General Standard for Cheese (CODEX STAN 283-1978).** The body has a near white or ivory through to light yellow or orange colour and a firm-textured (when pressed by thumb), smooth and waxy texture. Gas holes are absent, but a few openings and splits are acceptable. The cheese is manufactured and sold with or without rind which may be coated. For Cheddar ready for consumption, the ripening procedure to develop flavour and body characteristics is normally from 5 weeks at 7-15 °C depending on the extent of maturity required. Alternative ripening conditions (including the addition of ripening enhancing enzymes) may be used, provided the cheese exhibits similar physical, biochemical and sensory properties as those achieved by the previously stated ripening procedure. Cheddar intended for further processing need not exhibit the same extent of ripening when justified through technical and/or trade needs

    Note: This is not to mean that the rind has been removed before sale, instead the cheese has been ripened and/or kept in such a way that no rind is developed (a "rind less" cheese). Ripening film is used in the manufacture of rind less cheese. Ripening film ay also constitute the coating that protects the cheese. For rind less cheese see also the Appendix to **the Codex General Standard for Cheese (CODEX STAN 283-1978).**

iii. **Danbo Cheese (CODEX STAN 264-1966,Formerly CODEX STAN C-3-1966. Adopted in 1966. Revision 2007. Amendment 2008)-** Danbo is a ripened firm/semi-hard cheese in conformity with **the General Standard for Cheese (CODEX STAN 283-1978).** The body has a near white or ivory through to light yellow or yellow colour and a firm-textured (when pressed by thumb) texture, suitable for cutting, with few to plentiful, evenly distributed, smooth and round pea sized (or mostly up to 10 mm in diameter) gas holes, but a few openings and splits are acceptable. The shape is flat squared or parallelepiped. The cheese is manufactured and sold with or without hard or slightly moist smear ripened rind, which may be coated. For Danbo ready for consumption, the ripening procedure to develop flavour and body characteristics is normally from 3 weeks at 12–20 °C depending on the extent of maturity required. Alternative ripening conditions (including the addition of ripening enhancing enzymes) may be used, provided the cheese exhibits similar physical, biochemical and sensory properties as those achieved by the previously stated ripening procedure. Danbo intended for further processing need not exhibit the same extent of ripening when justified through technical and/or trade needs. *m*)

Note: This is not to mean that the rind has been removed before sale, instead the cheese has been ripened and/or kept in such a way that no rind is developed (a "rind less" cheese). Ripening film is used in the manufacture of rind less cheese. Ripening film may also constitute the coating that protects the cheese. For rind less cheese, see also the Appendix to **the Codex General Standard for Cheese (CODEX STAN 283-1978).**

iv. **Edam Cheese(CODEX STAN 265-1966, Formerly CODEX STAN C-4-1966. Adopted 1966. Revision 2007. Amendment 2008)-** Edam is a ripened firm/semi-hard cheese in conformity with the **General Standard for Cheese (CODEX STAN 283-1978).** The body has a near white or ivory through to light yellow or yellow colour and a firm-textured (when pressed by thumb) texture, suitable for cutting, with few more or less round rice to pea sized (or mostly up to 10 mm in diameter) gas holes, distributed in a reasonable regular manner throughout the interior of the cheese, but few openings and splits are acceptable. The shape is spherical, of a flat block or of a loaf. The cheese is manufactured and sold with dry rind, which may be coated. Edam of flat block or loaf shape is also sold without rind. For Edam ready for consumption, the ripening procedure to develop flavour and body characteristics is normally from 3 weeks at 10-18 °C depending on the extent of maturity required. Alternative ripening conditions (including the addition of ripening enhancing enzymes) may be used, provided the cheese exhibits similar physical, biochemical and sensory properties as those achieved by the previously stated ripening procedure. Edam intended for further processing need not exhibit the same degree of ripening when justified through technical and/or trade needs.

v. **Gouda Cheese(CODEX STAN 266-1966 Formerly CODEX STAN C-5-1966. Adopted 2001. Revision 2007. Amendment 2008)-**Gouda is a ripened firm/semi-hard cheese in conformity with **the General Standard for Cheese (CODEX STAN 283-1978).** The body has a near white or ivory through to light yellow or yellow colour and a firm-textured (when pressed by thumb) texture, suitable for cutting, with few to plentiful, more or less round pin's head to pea sized (or mostly up to 10 mm in diameter) gas holes, distributed in a reasonable regular manner throughout the interior of the cheese, but few openings and splits are acceptable. The shape is of a flattened cylinder with convex sides, a flat block, or a loaf. The cheese is manufactured and sold with a dry rind, which may be coated. Gouda of flat block or loaf shape is also sold without rind. For Gouda ready for consumption, the ripening procedure to develop flavour and body characteristics is normally from 3 weeks at 10-17 °C depending on the extent of maturity required. Alternative ripening conditions (including the addition of ripening enhancing enzymes) may be used, provided the cheese exhibits similar physical, biochemical and sensory properties as those achieved by the previously stated ripening procedure. Gouda intended for further processing and Gouda of low weights (< 2.5 kg) need not exhibit the same degree of ripening when justified through technical and/or trade needs.

vi. **Havarti Cheese (CODEX STAN 267-1966 Formerly CODEX STAN C-6-1966. Adopted 1966. Revision 2007. Amendment 2008)-**Havarti is a ripened firm/semi-hard cheese in conformity with **the General Standard for Cheese (CODEX STAN 283-1978).** The body has a near white or ivory through to light yellow or yellow colour and a texture suitable for cutting, with plentiful, irregular and coarse large rice seed sized (or mostly 1–2 mm in width and up to 10 mm in length) gas holes. The shape is flat cylindrical, rectangular or of a loaf shape. The cheese is sold with or without a slightly greasy smear ripened rind, which may be coated. For Havarti ready for consumption, the ripening procedure to develop flavour and body characteristics is normally, depending on weight, 1–2 weeks at 14–18 °C (for smear development) followed by from 1–3 weeks at 8–12 °C depending on the extent of maturity required. Alternative ripening conditions (including the addition of ripening enhancing enzymes) may be used, provided the cheese exhibits similar physical, biochemical and sensory properties as those achieved by the previously stated ripening procedure. Havarti intended for further processing need not exhibit the same degree of ripening when justified through technical and/or trade needs. Havarti with a fat in dry matter content of minimum 60 per cent may alternatively be designated Cream Havarti.

vii. **Samso Cheese(CODEX STAN 268-1966 Formerly CODEX STAN C-7-1966. Adopted in 1966. Revision 2007. Amendment 2008)-** Samsø is a ripened hard cheese in conformity with **the General Standard for Cheese (CODEX STAN 283-1978).** The body has a near white or ivory through to light yellow or yellow colour and a firm-textured (when pressed by thumb) texture suitable for cutting, with few to plentiful, evenly distributed, smooth and round pea to cherry sized (or mostly up to 20 mm in diameter) gas holes, but few openings and splits are acceptable. The shape is a flat cylindrical, flat square or rectangular. The cheese is sold with or without a hard, dry rind, which may be coated. For Samsø ready for consumption, the ripening procedure to develop flavour and body characteristics is normally from 3 weeks at 8-17 °C depending on the extent of maturity required. Alternative ripening conditions (including the addition of ripening enhancing enzymes) may be used, provided the cheese exhibits similar physical, biochemical and sensory properties as those achieved by the previously stated ripening procedure. Samsø intended for further processing need not exhibit the same degree of ripening when justified through technical and/or trade needs.

viii. **Emmental Cheese(CODEX STAN 269-1967 Formerly CODEX STAN C-9-1967 – Standard for Emmentaler. Adopted 1967. Revision 2007. Amendment 2008)-**Emmental is a ripened hard cheese in conformity with **the General Standard for Cheese (CODEX STAN 283- 1978).** The body has a ivory through to light yellow or yellow colour and an elastic, sliceable but not sticky texture, with regular, scarce to plentiful distributed, mat to brilliant, cherry to walnut sized (or mostly from 1 to 5 cm in diameter) gas holes, but few openings and splits are acceptable. Emmental is typically manufactured as wheels and blocks of weights from 40 kg or more but individual countries may on their territory permit other weights provided that the cheese exhibit similar physical, biochemical and sensory properties. The cheese is manufactured and sold with or without a hard, dry rind. The typical flavour is mild, nut-like and sweet, more or less pronounced. For Emmental ready for consumption, the ripening procedure to develop flavour and body characteristics is normally from 2 months at 10–25°C depending on the extent of maturity required. Alternative ripening conditions (including the addition of ripening enhancing enzymes) may be used, provided a minimum period of 6 weeks is observed and provided the cheese exhibits similar physical, biochemical and sensory properties as those achieved by the previously stated ripening procedure. Emmental intended for further processing need not exhibit the same degree of ripening, when justified through technical and/or trade needs.

Note: The purpose of these criteria is to provide targets for the validation (initial assessment prior to the design of the manufacturing process), respectively, of (i) whether the intended fermentation and ripening conditions are capable of achieving the activity of propionic acid producing bacteria, and of (ii) whether the curd management and pH development are capable of obtaining the characteristic texture.

ix. **Tilsiter Cheese- (CODEX STAN 270-1968 Formerly CODEX STAN C-11-1968. Adopted in 1968. Revision 2007. Amendment 2008)-**Tilsiter is a ripened firm/semi-hard cheese in conformity with **the General Standard for Cheese (CODEX STAN 283-1978).** The body has a near white or ivory through to light yellow or yellow colour and a firm-textured (when pressed by thumb) texture suitable for cutting, with irregularly shaped, shiny and evenly distributed gas holes. The cheese is manufactured and sold with or without a well-dried smear-developed rind, which may be coated. For Tilsiter ready for consumption, the ripening procedure to develop flavour and body characteristics is normally from 3 weeks at 10–16°C depending on the extent of maturity required. Alternative ripening conditions (including the addition of ripening enhancing enzymes) may be used, provided the cheese exhibits similar physical, biochemical and sensory properties as those achieved by the previously stated ripening procedure. Tilsiter intended for further processing need not exhibit the same degree of ripening when justified through technical and/or trade needs.

x. **Saint-Paulin Cheese-(CODEX STAN 271-1968 Formerly CODEX STAN C-13-1968. Adopted in 1968. Revision 2007. Amendment 2008)-**Saint-Paulin is a ripened firm/semi-hard cheese in conformity with **the General Standard for Cheese (CODEX STAN 283-1978).** The body has a near white or ivory through to light yellow or yellow colour and a firm textured (when pressed by thumb) but flexible texture. Gas holes are generally absent, but few openings and splits are acceptable. The cheese is manufactured and sold with or without a dry or slightly moist rind, which is hard, but elastic under thumb pressure, and which may be coated. For Saint-Paulin ready for consumption, the ripening procedure to develop flavour and body characteristics is normally from 1 week at 10-17 °C depending on the extent of maturity required. Alternative ripening conditions (including the addition of ripening enhancing enzymes) may be used, provided the cheese exhibits similar physical, biochemical and sensory properties as those achieved by the previously stated ripening procedure. Saint-Paulin intended for further processing need not exhibit the same degree of ripening when justified through technical and/or trade needs.

xi. **Provolone Cheese-(CODEX STAN 272-1968 Formerly CODEX STAN C-15-1968. Adopted in 1968. Revision 2007. Amendment 2008)-**Provolone is a ripened firm/semi-hard cheese in conformity with **the General Standard for Cheese (CODEX STAN 283-1978).** The body has a near white or ivory through to light yellow or yellow colour and a fibrous texture with long stranded parallel-orientated protein fibres. It is suitable for cutting and, when aged, for grating as well. Gas holes are generally absent, but few openings and splits are acceptable. The shape is mainly cylindrical or pear-shaped, but other shapes are possible. The cheese is manufactured and sold with or without a rind, which may be coated. For Provolone ready for consumption, the ripening procedure to develop flavour and body characteristics is normally from 1 month at 10–20 °C depending on the extent of maturity required. Alternative ripening conditions (including the addition of ripening enhancing enzymes) may be used, provided the cheese exhibits similar physical, biochemical and sensory properties as those achieved by the previously stated ripening procedure. Provolone intended for further processing and Provolone of low weights (< 2 kg) need not exhibit the same degree of ripening when justified through technical and/or trade needs. Provolone is made by "pasta filata" processing which consists of heating curd of a suitable pH value, kneading and stretching until the curd is smooth and free from lumps. Still warm, the curd is cut and moulded, then firmed by cooling in chilled water or brine. Other processing techniques, which give end products with the same physical, chemical and organoleptic characteristics are allowed.

xii. **Cottage Cheese-(Codex Standard 273-1968 Formerly CODEX STAN C-16-1968. Adopted in 1968. Revision 2007)-**Cottage Cheese is a soft, rind less, unripened cheese in conformity with **the General Standard for Cheese (CODEX STAN 283-1978) and the Standard for Unripened Cheese Including Fresh Cheese (CODEX STAN 221-2001).** The body has a near white colour and a granular texture consisting of discrete individual soft curd granules of relatively uniform size, from approximately 3–12 mm depending on whether small or large type of curd is desired, and possibly covered with a creamy mixture.

xiii. **Coulommiers Cheese(CODEX STAN 274-1969 Formerly CODEX STAN C-18-1969. Adopted in 1969. Revision 2007. Amendment 2008)-**Coulommiers is a soft, surface ripened, primarily mould ripened cheese in conformity with **the General Standard for Cheese (CODEX STAN 283-1978)** which has a shape of a flat cylinder or sectors thereof. The body has a near white through to light yellow colour and a soft-textured (when pressed by thumb), but not crumbly texture, ripened from the surface to the center of the cheese. Gas holes are generally absent, but few openings and splits are acceptable. A rind is to be developed that is soft and entirely covered with white mould but may have red, brownish or orange coloured spots. Whole cheese may be cut or formed into sectors prior to or after the mould development. For Coulommiers ready for consumption, the ripening procedure to develop flavour and body characteristics is normally from 10 days at 10–16 °C depending on the extent of maturity required. Alternative ripening conditions (including the addition of ripening enhancing enzymes) may be used, provided the cheese exhibits similar physical, biochemical and sensory properties as those achieved by the previously stated ripening procedure. Coulommiers intended for further processing need not exhibit the same extent of ripening when justified through technical and/or trade needs.

xiv. **Cream CHEESE(Codex Standard 275-1973 Formerly CODEX STAN C-31-1973. Adopted in 1973. Revision 2007. Amendment 2008)-**Cream Cheese is a soft, spreadable, unripened and rindless cheese in conformity with **the Standard for Unripened Cheeses Including Fresh Cheeses (CODEX STAN 221-2001) and the General Standard for Cheese (CODEX STAN 283-1978).** The cheese has a near white through to light yellow colour. The texture is spreadable and smooth to slightly flaky and without holes, and the cheese spreads and mixes readily with other foods.

xv. **Camembert Cheese(CODEX STAN 276-1973, Formerly CODEX STAN C-33-1973. Adopted in 1973. Revision 2007. Amendment 2008)-**Camembert is a soft surface ripened, primarily mould ripened cheese in conformity with **the General Standard for Cheese (CODEX STAN 283-1978),** which has a shape of a flat cylinder or sectors thereof. The body has a near white through to light yellow colour and a soft-textured (when pressed by thumb), but not crumbly texture, ripened from the surface to the center of the cheese. Gas holes are generally absent, but few openings and splits are acceptable. A rind is to be developed that is soft and entirely covered with white mould but may have red, brownish or orange coloured spots. Whole cheese may be cut or formed into sectors prior to or after the mould development. For Camembert ready for consumption, the ripening procedure to develop flavour and body characteristics is normally from 10 days at 10-16 °C depending on the extent of maturity required. Alternative ripening conditions (including the addition of ripening enhancing enzymes) may be used, provided the cheese exhibits similar physical, biochemical and sensory properties as those achieved by the previously stated ripening procedure. Camembert intended for further processing need not exhibit the same extent of ripening when justified through technical and/or trade needs. Carré de Camembert is a soft surface ripened cheese with a square shape and which comply with all other criteria and requirements specified for Camembert.

xvi. **Brie Cheese(CODEX STAN 277-1973,Formerly CODEX STAN C-34-1973. Adopted in 1973. Revision 2007. Amendment 2008)-**Brie is a soft surface ripened, primarily white mould ripened cheese in conformity with **the General Standard for Cheese (CODEX STAN 283-1978),** which has a shape of a flat cylinder or sectors thereof. The body has a near white through to light yellow colour and a soft-textured (when thumbs-pressed), but not crumbly texture, ripened from the surface to the center of the cheese. Gas holes are generally absent, but few openings and splits are acceptable. A rind is to be developed that is soft and entirely covered with white mould but may have red, brownish or orange coloured spots. Whole cheese may be cut or formed into sectors prior to or after the mould development. For Brie ready for consumption, the ripening procedure to develop flavour and body characteristics is normally from 10 days at 10–16°C depending on the extent of maturity required. Alternative ripening conditions (including the addition of ripening enhancing enzymes) may be used, provided the cheese exhibits similar physical, biochemical and sensory properties as those achieved by the previously stated ripening procedure. Brie intended for further processing need not exhibit the same extent of ripening when justified through technical and/or trade needs.

General Standard for Cheese (CODEX STAN 283-1978, Formerly CODEX STAN A-6-1973. Adopted in 1973. Revision 1999, Amendments 2006, 2008)

This Standard applies to all products, intended for direct consumption or further processing, in conformity with the definition of cheese in Section 2 of this Standard. Subject to the provisions of this Standard, standards for individual varieties of cheese, or groups of varieties of cheese, may contain provisions which are more specific than those in this Standard and in these cases, those specific provisions shall apply.

Cheese is the ripened or unripened soft, semi-hard, hard, or extra-hard product, which may be coated, and in which the whey protein/casein ratio does not exceed that of milk, obtained by: (a) coagulating wholly or partly the protein of milk, skimmed milk, partly skimmed milk, cream, whey cream or buttermilk, or any combination of these materials, through the action of rennet or other suitable coagulating agents, and by partially draining the whey resulting from the coagulation, while respecting the principle that cheese-making results in a concentration of milk protein (in particular, the casein portion), and that consequently, the protein content of the cheese will be distinctly higher than the protein level of the blend of the above milk materials from which the cheese was made; and/or (b) processing techniques involving coagulation of the protein of milk and/or products obtained from milk which give an end-product with similar physical, chemical and organoleptic characteristics as the product defined under (a).

i. Ripened cheese is cheese which is not ready for consumption shortly after manufacture but which must be held for such time, at such temperature, and under such other conditions as will result in the necessary biochemical and physical changes characterizing the cheese in question.

ii. Mould ripened cheese is a ripened cheese in which the ripening has been accomplished primarily by the development of characteristic mould growth throughout the interior and/or on the surface of the cheese. Unripened cheese including fresh cheese is cheese which is ready for consumption shortly after manufacture.

Additionally, the following terms may be used for declaration of fat:

Table 16.71: Terms for Declaration of Fat

High fat	(if the content of FDM is above or equal to 60 per cent),
Full fat	(if the content of FDM is above or equal to 45 per cent and less than 60 per cent).
Medium fat	(if the content of FDM is above or equal to 25 per cent and less than 45 per cent)
Partially skimmed	(if the content of FDM is above or equal to 10 per cent and less than 25 *per cent*)
Skim	(if the content of FDM is less than 10 per cent)

xvii. **Cheeses in brine (CODEX STAN 208-1999,Codex Standard 208-1999 Adopted in 1999. Amendment 2001, 2010).** This Standard applies to Cheeses in Brine, intended for direct consumption or further processing, in conformity with the description in Section 2 of this Standard. Subject to the provisions of this Group Standard, Codex standards for individual varieties of Cheeses in Brine may contain provisions which are more specific than those in this Standard. Cheeses in Brine are semi-hard to soft ripened cheeses in conformity with **the General Standard for Cheese (CODEX STAN 283-1978)**. The body has a white to yellowish colour and a compact texture suitable for slicing, with none to few mechanical openings. The cheeses have no actual rind and have been ripened and preserved in brine until delivered to, or prepacked for, the consumer. Certain individual Cheeses in Brine contain specific herbs and spices as part of their identity. It has the following composition:

Parameters	*Soft*	*Semi-hard*
Minimum fat in dry matter, per cent	40	40
Minimum dry matter, per cent	40	52

Additionally, the following terms may be used for declaration of fat:

Table 16.72: Terms for Declaration of Fat (Cheese in Brine)

High fat	(if the content of FDM is above or equal to 60 per cent);
Full fat	(if the content of FDM is above or equal to 45 per cent and less than 60 per cent)
Medium fat	(if the content of FDM is above or equal to 25 per cent and less than 45 per cent)
Partially skimmed	(if the content of FDM is above or equal to 10 per cent and less than 25 per cent)
Skim	(if the content of FDM is less than 10 per cent)

xviii. **Group Standard For Unripened Cheese Including Fresh Cheese (CODEX STAN 221-2001, Adopted 2001. Amendment 2008, 2010)-** This Standard applies to unripened cheese including fresh cheese, intended for direct consumption or further processing, in conformity with the description in Section 2 of this Standard. Subject to the provisions of this Standard, Codex Standards for individual varieties of unripened cheese may contain provisions, which are more specific than those in this Standard and in these cases; those specific provisions shall apply. Unripened cheeses including fresh cheeses are products in conformity with the General Standard for Cheese, which are ready for consumption shortly after manufacture. Additionally, the following terms may be used for declaration of fat:

Table 16.73: Terms for Declaration of Fat (Unripened Cheese)

High fat	(if the content of FDM is above or equal to 60 per cent)
Full fat	(if the content of FDM is above or equal to 45 per cent and less than 60 per cent)
Medium fat	(if the content FDM is above or equal to 25 per cent and less than 45 per cent)
Partially skimmed	(if the content of FDM is above or equal to 10 per cent and less than 25 per cent)
Skim	(if the content of FDM is less than 10 per cent)

xix. **Whey Cheeses (CODEX STAN 284-1971, Formerly CODEX STAN A-7-1971. Adopted in 1971. Revision 1999, 2006. Amendment 2010)**

i. **Whey Cheeses** are solid, semi-solid, or soft products which are principally obtained through either of the following processes-

1. The concentration of whey and the moulding of the concentrated product;
2. The coagulation of whey by heat with or without the addition of acid.

In each case, the whey may be pre-concentrated prior to the further concentration of whey or coagulation of the whey proteins. The process may also include the addition of milk, cream, or other raw materials of milk origin before or after concentration or coagulation. The ratio of whey protein to casein in the product obtained through the coagulation of whey shall be distinctly higher than that of milk. The product obtained through the coagulation of whey may either be ripened or unripened.

ii. **Whey Cheese** obtained through the concentration of whey is produced by heat evaporation of whey, or a mixture of whey and milk, cream, or other raw materials of milk origin, to a concentration enabling the final cheese to obtain a stable shape. Due to their relatively high lactose content these cheeses are typically yellowish to brown in colour and possess a sweet, cooked, or caramelized flavour.

iii. **Whey Cheese** obtained through the coagulation of whey is produced by heat precipitation of whey, or a mixture of whey and milk or cream, with or without the addition of acid. These whey cheeses have a relatively low lactose content and a white to yellowish colour.

Permitted Nutrients for Whey Cheese

Where allowed in accordance with the **General Principles for the Addition of Essential Nutrients for Food (CAC/GL 9-1987),** maximum and minimum levels for minerals and other nutrients, where appropriate, should be laid down by national legislation in accordance with the needs of individual country including, where appropriate, the prohibition of the use of particular nutrients.

For cheeses obtained from the concentration of whey, the declaration of milk fat content may be combined with an indication of the fat content as follows:

a. *Fat on the Dry Basis*

Creamed whey cheese minimum -33 per cent

Whey cheese -minimum 10 per cent and less than 33 per cent

Skimmed whey cheese- less than 10 per cent

The dry matter content of whey cheese includes water of crystallization of the lactose.

b. *Composition*

The aforesaid types of Cheeses shall have the following composition in respect to Milk fat in dry matter, Fat Free dry matter/Dry Matter and Moisture on Fat free Basis and in respect to Dry Matter basis.

i. Milk fat in dry matter, Fat Free dry matter/Dry Matter and Moisture on Fat free Basis:

Table 16.74: Fat Free in Dry Matter, Milk Fat in Dry Matter (DM) and Moisture on Fat Free Basis

Type of Cheese	*a. Milk Fat in Dry Matter*		
	Minimum Content m/m	*Maximum Content m/m*	*Reference Level m/m*
Mozzarella			
1. With high moisture	20 per cent	Not restricted	40 per cent to 50 per cent
2. With low moisture	18 per cent	Not restricted	40 per cent to 50 per cent
Cheddar	22 per cent	Not restricted	48 per cent to 60 per cent
Danbo	20 per cent	Not restricted	45 per cent to 55 per cent
Edam	30 per cent	Not restricted	40 per cent to 50 per cent
Gauda	30 per cent	Not restricted	48 per cent to 55 per cent
Havarti	30 per cent	Not restricted	45 per cent to 55 per cent
Samso	30 per cent	Not restricted	45 per cent to 55 per cent
Emmental	45 per cent	Not restricted	45 per cent to 55 per cent
Tilsiter	30 per cent	Not restricted	45 per cent to 55 per cent
Saint-Paulin	40 per cent	Not restricted	40 per cent to 50 per cent
Provolone	45 per cent	Not restricted	45 per cent to 50 per cent
Cottage	0 per cent	Not restricted	4–5 per cent
Coulommiers	40 per cent	Not restricted	40 per cent to 50 per cent
Cream*	25 per cent	Not restricted	60–70 per cent
Camembert	30 per cent	Not restricted	45 per cent to 55 per cent
Brie	40 per cent	Not restricted	45 per cent to 55 per cent

Type of Cheese	*b. Moisture on Fat Free Basis*		
	Minimum Content m/m	*Maximum Content m/m*	*Reference Level m/m*
Cream	67 per cent	–	Not specified

Type of Cheese	*c. Fat Free Dry Matter/Dry Matter*		
	Minimum Content m/m	*Maximum Content m/m*	*Reference Level m/m*
Cottage- Fat Free dry matter	18 per cent	Restricted by the MFFB	-
Cream - Dry Matter	22 per cent	Restricted by the MMFBFF	Not specified

Note: *The cheese has been kept in such a way that no rind is developed (a “rind less” cheese).

ii. **Dry matter:** Depending on the fat in dry matter content, according to the table below:

Table 16.75: For Mozzarella, Cheddar, Danbo and Edam

Fat in Dry Matter Content (m/m)	*Corresponding Minimum Dry Matter Content (m/m)*				
	Mozzarella		*Cheddar*	*Danbo*	*Edam*
	With Low Moisture	*With High Moisture*			
Equal to or above 18 per cent but less than 30 per cent	34 per cent	-	-	-	-
Equal to or above 20 per cent but less than 30 per cent	-	24 per cent	-	41 per cent	-
Equal to or above 22 per cent but less than 30 per cent	-	-	49 per cent	-	-
Equal to or above 30 per cent but less than 40 per cent	39 per cent	26 per cent	53 per cent	44 per cent	47 per cent
Equal to or above 40 per cent but less than 48 per cent	-	-	57 per cent	-	-
Equal to or above 40 per cent but less than 45 per cent	42 per cent	29 per cent	-	50 per cent	51 per cent
Equal to or above 45 per cent but less than 50 per cent	45 per cent	31 per cent	-	-	55 per cent
Equal to or above 45 but less than 55 per cent	-	-	-	52 per cent	-
Equal to or above 50 per cent but less than 60 per cent	47 per cent	34 per cent	-	-	57 per cent
Equal to or above 48 per cent but less than 60 per cent	-	-	61 per cent	-	-
Equal to or above 55 per cent	-	-	-	57 per cent	-
Equal to or above 60 per cent	-	-	66 per cent	-	62 per cent
Equal to or above 60 per cent but less than 85 per cent	53 per cent	38 per cent	-	-	-

Table 16.76: For Gauda, Havarti, Samso, Emmental and Tilsiter

Fat in Dry Matter Content (m/m)	*Corresponding Minimum Dry Matter Content (m/m)*				
	*Gauda***	*Samso*	*Havarti*	*Emmental**	*Tilsiter*
Equal to or above 30 per cent but less than 40 per cent	48 per cent	46 per cent	46 per cent	-	49 per cent
Equal to or above 40 per cent but less than 48 per cent	52 per cent	-		-	
Equal to or above 40 per cent but less than 45 per cent	-	52 per cent	48 per cent	-	53 per cent

Fat in Dry Matter Content (m/m)	*Corresponding Minimum Dry Matter Content (m/m)*				
	*Gauda***	*Samso*	*Havarti*	*Emmental**	*Tilsiter*
Equal to or above 45 per cent but less than 50 per cent	-	-		60 per cent	55 per cent
Equal to or above 45 but less than 55 per cent	-	54 per cent	50 per cent	-	
Equal to or above 50 per cent but less than 60 per cent	-	-		62 per cent	57 per cent
Equal to or above 55 per cent but less than 60 per cent	-	-	54 per cent	-	-
Equal to or above 48 per cent but less than 60 per cent	55 per cent	-	-	-	
Equal to or above 55 per cent	-	59 per cent	-	-	
Equal to or above 60 per cent	62 per cent	-	58 per cent	67 per cent	
Equal to or above 60 per cent but less than 85 per cent	-	-	-	-	61 per cent
*Propionic acid in cheese ready for sale	Min. 150 mg/100g				
Calcium content	Min. 800 mg/100g				

** Gouda with between 40 and 48 per cent FDM and with a weight of less than 2.5 kg can be sold with a DM content of min. 50 per cent, provided that the name is qualified by the term "baby".

Table 16.77: For Saint-Paulin, Provolone, Cottage, Coulommiers and Cream

Fat in Dry Matter Content (m/m)	*Corresponding Minimum Dry Matter Content (m/m)*				
	Saint-Paulin	*Provolone*	*Coulom-miers*	*Camem-bert*	*Brie*
Equal to or above 30 per cent but less than 40 per cent	-	-	-	38 per cent	-
Equal to or above 40 per cent but less than 45 per cent	-	—	-	41 per cent	42 per cent
Equal to or above 40 per cent but less than 50 per cent	—	-	42 per cent	-	-
Equal to or above 40 per cent but less than 60 per cent	44 per cent	-	-	-	-
Equal to or above 45 per cent but less than 50 per cent	-	51 per cent	-	-	-
Equal to or above 45 but less than 55 per cent	-	-	-	43 per cent	43 per cent
Equal to or above 50 per cent but less than 60 per cent	-	53 per cent	46 per cent	-	-

Fat in Dry Matter Content (m/m)	Corresponding Minimum Dry Matter Content (m/m)				
	Saint-Paulin	Provolone	Coulom-miers	Camem-bert	Brie
Equal to or above 55 per cent but less than 60 per cent		-	-	-	48 per cent
Equal to or above 55 per cent	-	-	-	48 per cent	-
Equal to or above 60 per cent	54 per cent	60 per cent	52 per cent	-	51 per cent

c. ***Common Requirements of different Cheeses and Varieties of Cheeses-***

1. **Raw materials and Permitted ingredients** (Mozzerella,Cheddar,Danbo, Edam, Gouda, Havarti, Samso, Emmental, Tilsiter, Saint-Paulin, Provolone, Cottage, Coulommiers, Cream, Camembert, Brie,General standards of cheese, Unripened Chasse)- Cheeses are made from Raw materials-Cows' milk or buffaloes' milk, or their mixtures, and products obtained from these milks(for cream cheese, Cheese in Brine and Unripened Cheese - Milk and/or products obtained from milk), Whey Cheese-
   i. For products obtained through the concentration of whey-whey, cream, milk and other raw materials obtained from milk.
   ii. For products obtained through the coagulation of whey-whey, milk, cream and buttermilk have Permitted ingredients:
      a. Starter cultures of harmless lactic acid and/or flavour producing bacteria and cultures of other harmless micro-organisms.
      b. Rennet or other safe and suitable coagulating enzymes (Geotrichum candidum, Brevibacterium linens, and yeast for Coulommiers, Camembert, Brie),
      c. Sodium chloride and potassium chloride(Except Cheese in brine, General standards of cheese) as a salt substitute
      d. Safe and suitable processing aids(Except Cottage Cheese).
      e. Vinegar(for Mozzarella, Cream)
      f. Potable water.
      g. Safe and suitable enzymes to enhance the ripening process(Except for Mozzarella, Cottage, Cream),
      h. Herbs and spices where part of the identity of the Cheese in Brunei-(Except for Cheese in Brine and General Standards of Cheese)Rice, corn and potato flours and starches, Notwithstanding the provisions in the General Standard for Cheese (CODEX STAN 283-1978), these substances can be

used in the same function as anti-caking agents for treatment of the surface of cut, sliced, and shredded Mozzarella with a low moisture content only, provided they are added only in amounts functionally necessary as governed by Good Manufacturing Practice, taking into account any use of the anti-caking agents listed in section 4. 1 (Except Cottage Cheese), The cheese has been kept in such a way that no rind is developed (a "rind less" cheese).ix-(For Cottage,Unripened and Cream Cheese)- Gelatin and starches-These substances can be used in the same function as stabilizers, provided they are added only in amounts functionally necessary as governed by Good Manufacturing Practice taking into account any use of the stabilizers/thickeners listed in section 4.

Permitted ingredients for whey Cheese- Sodium chloride, Starter cultures of harmless lactic acid bacteria Only for use in products obtained through the concentration of whey by heat treatment, Sugars (limited by GMP).

For Edam, Gouda,Samso, Saint-Paulin, Tilsiter, Provolone Cheese- This is not to mean that the rind has been removed before sale, instead the cheese has been ripened and/or kept in such a way that no rind is developed (a "rind less" cheese). Ripening film is used in the manufacture of rind less cheese. Ripening film may also constitute the coating that protects the cheese. For rind less cheese, the Appendix to the Codex General Standard for Cheese (CODEX STAN 283- 1978) is also referred.

2. The details related to the Food Additives,Contaminants, Hygiene, Labelling and Methods of sampling and Testing are as per norms given in codex booklets.
3. Compositional modifications beyond the minima and maxima specified above for milk fat and dry matter are not considered to be in compliance with section 4.3.3 of the Codex General Standard for the Use of Dairy Terms (CODEX STAN 206-1999).

Modification/Addition in FSSAI Quality Standards [7]

Initially FSSAI has prescribed quality standards for different foods including milk and milk products under Food Safety and Standards (Food Product specification and Food Additives) regulation-2011 but later on as per the requirement, FSSAI has issued several amendments which are compiled as compendium (Different versions).

Note-Compendium encompasses all amendments to the individual regulations notified in the Gazette of India for the purpose of providing ready reference to food business operators and consumers; and can't be considered a legal document. Each amendment in compendium is marked with the amendment number and the same are listed at the end of document. In case of any errors/overlapping in compendium, the original notifications/amendments would prevail for the same.

- ✰ Under chapter -16, FSSAI Quality Standards for milk and milk products are mentioned as per FSSR-2011 and compendium (Version-XIV -25.03.2021).
- ✰ Latest amendments issued by FSSAI on the Quality Standards of milk and milk products, are compiled under Version-XXV (23.09.2022) of compendium and as per this version, the following major changes/additions (in comparison to version 25.03.2021) in quality standards of milk and milk products are made-

1. Milk

i. Quality standards for Low Lactose or Lactose free milk is introduced.

ii. "All milk and milk products, including composite milk products, as defined in subitem b, e, f, h and i of item 1 of this sub-regulation shall exclusively use the following logo on the product label."

2. Milk Fat, Anhydrous Milk Fat, Ghee

The following changes are made

Table 16.78: Changes in Milk Fat, Anhydrous Milk Fat and Ghee

Sl.No.	*Parameter*	*Milk Fat, Butter Oil*	*Anhydrous Milk Fat, Anhydrous Butter Oil*	*Ghee*
1	Butyro-refractometer Reading at 40°C	40.0 to 44.0	40.0 to 44.0	40-44
2	Reichert Meissl Value, minimum	24	24	24
3	Polenske Value	0.5-2.0	0.5-2.0	0.5-2.0

Sl.No.	Parameter	Milk Fat, Butter Oil	Anhydrous Milk Fat, Anhydrous Butter Oil	Ghee
4	FFA as Oleic Acid, maximum, %	0.4	0.3	2.0
5	Iodine Value	-	-	25-38
6	Saponification value	-	-	205-235
7	Presence of β-sitosterol	Absent	Absent	Absent
8	Fatty acid composition	-	-	The product shall meet the Requirement of Table 1

3. Milk Powder

The following additions are made:

Table 16.79: Additions in Milk Powder

Parameter	Whole Milk Powder	Partly Skimmed Milk Powder	Skimmed Milk Powder	Cream Powder
Titrable acidity, maximum (ml 0.1 NaOH for 10 g - solids-not-fat)	18.0	18.0	18.0	-
Insolubility Index, maximum, ml	2.0	2.0	2.0	2.0

Sources

1. *FSS (Food Products Standards and Food Additives) Regulation-2011.*
2. *Compendium version XIVo f FSS (Food Products Standards and Food Additives) Regulation -2011.*
3. *Agmark standards.*
4. *Different BIS Standards.*
5. *Different Codex Standards.*
6. *Different IDF standards.*
7. *FSSAI-Compendium version XXV (23-09-2022) related to FSS (Food Production Specification and Food Additives) regulation-2011.*

Chapter 17

Cream, Butter and Ghee

1. CREAM

Composition and Quality Standards of different Types of Cream

Given in chapter-16 under composition and quality standards.

Creaming [1,2]

Creaming is a natural process that occurs during the storage of raw whole milk. Due to the differences in densities between fat globules and the plasma phase of milk, fat globules tend to accumulate as a cream layer on top of the milk. However, natural creaming of fat highly depends on the storage temperature of milk [1].

- ☆ In the laboratory sense, creaming is the migration of the dispersed phase of an emulsion, under the influence of buoyancy. The particles float upwards or sink, depending on how large they are and how much less dense or more dense they may be than the continuous phase, and also how viscous or how thixotropic the continuous phase might be. For as long as the particles remain separated, the process is called creaming [2].
- ☆ Where it is important that either the form or the concentration of the emulsion should be stable, it is desirable that the continuous and the dispersed phases should have similar densities, and it also is desirable that the continuous phase should be viscous or thixotropic [2].
- ☆ Creaming is usually seen as undesirable because it causes difficulties in storage and handling, but it can be useful in special cases, especially

where it is desirable to concentrate an emulsion. A particular example is in the separation of dairy cream, either to achieve a desired concentration of butterfat, or to make butter [2].

Flocculation and Emulsion Breaking [2]

Flocculation is the process where particles clump and emulsion breaking is the process where particles coalesce. These processes differ from creaming in following ways-

- ☆ The process of migration is called creaming while the particles of the substance remain separated. In this it differs ideally from flocculation or emulsion breaking.
- ☆ Unlike flocculation and breaking, creaming of an emulsion is largely a simple process to reverse.

A creamed emulsion increases the likelihood of coalescence due to the close proximity of the globules in the cream. Factors that influence the rate of creaming are similar to those involved in the sedimentation rate of suspension particles, and are indicated by Stokes Law.

Emulsion Inversion

Creaming of an emulsion also increases the tendency of an emulsion to inversion. Emulsion inversion differs from emulsion breaking in that a breaking emulsion tends to separate the two phases into un-emulsified continuous phases. Inversion of an emulsion may or may not be difficult to invert, but generally more difficult than creaming.

- ☆ This class of process occurs mainly in special cases as given below:
  - **i.** When both the continuous and dispersed phases of an emulsion are liquid, as commonly is the state in dairy cream.
  - **ii.** It is common where the volume of the two fluid components is about the same or the volume of the dispersed phase is larger than that of the continuous phase.
  - **iii.** The process of emulsion inversion occurs when the dispersed droplets unite, but retain the formerly continuous material as droplets within the mass.
  - **iv.** The "invert emulsion" or "inverted emulsion", is the emulsion in which the formerly continuous phase has become the dispersed phase and vice versa. Inversion happens in dairy cream when the butterfat concentration is too high, and the resulting invert emulsion looks much like butter.
- ☆ Commonly invert emulsions look much like a paste or thick cream, and typical examples are mayonnaise, margarine (especially "low-fat" grades of margarine), similar creamy foods, *etc.* [2].

Factors Influencing the Fat Percentage of Cream [4,5,5a]

The important factors influencing the fat percentage of cream by centrifugal separation are given below:

1. Position of the Cream and Skim Milk Screw

Any of the screws can be moved IN or OUT. Thus,it comes nearer to or moves away from the centre of rotation. Adjusting the cream screw towards IN position or Skim milk screw towards OUT position yields high fat percentage in cream or vice-versa.

Cream Screw

The cream screw/outlet consists of a small, threaded, hollow screw pierced by a circular orifice through which the cream emerges. This screw can be driven IN or OUT, thus bringing it nearer to, or away from, the centre of rotation.

Skim Milk Screw

The skim milk screw/outlet is for the removal of skim milk.

Once the cream screw or skim milk screw has been adjusted, the cream separator delivers, under normal conditions, a definite ratio of skim milk and cream, which is usually 90:10 (or 85:15) by volume. Basically, any change in the separation procedures which alters the relative quantities of skim milk and cream will influence the fat test of the cream. By altering the position of the cream screw (or skim milk screw), the ratio of skim milk to cream changes.

- ☆ Thus, when the cream screw is moved IN towards the axis of rotation, a higher fat percentage in cream is obtained, and vice versa; this is because the force tending to discharge cream through the orifice is decreased ('R' in the formula $F = KWRN^2$ is decreased), A smaller proportion of cream is therefore discharged, which, contains the same quantity of fat, resulting in a higher fat percentage.
- ☆ Screwing OUT the cream screw produces thinner cream.
- ☆ Similarly, the skim milk screw OUT results in richer cream, and vice versa.

2. Fat Percentage in Milk

Milk containing high fat yields rich cream. The higher the fat percentage in milk, the higher the per cent fat in cream, and vice versa. Since practically all the fat in milk is contained in the cream, the cream from the separation of high-fat milk has a higher fat content than that from low-fat milk.

3. Speed of the Bowl

The higher and recommended speed of the bowl yields cream of higher fat content. The higher the speed of the bowl, the greater will be the centrifugal force and more rapidly the skim milk leaves the bowl with higher fat per cent in cream.

☆ The velocity of a fat globule is proportional to the square of the rotational speed so an increase in bowl speed will have a very major effect on separation efficiency.

i. An increase in bowl speed however requires an increase in energy input and a more robust design to withstand the large forces at the bowl periphery.
ii. The separator also generates more noise. For this reason bowl speeds have not increased significantly as skimming efficiency is quite adequate at moderate speeds of 4000-6000 rpm.
iii. The higher the speed of the bowl, the higher will be the fat per cent in cream.

4. Rate of the Milk Flow

Faster rate of inflow of milk to the separator produces cream with lower fat percentage, and vice versa. When the rate of inflow increases, the discharge from the cream outlet increases. As the skim milk discharge remains constant (with constant centrifugal force); more cream containing the same amount of fat results in a lower fat test, and vice versa.

5. Temperature of Milk [1,4,5]

Due to the differences in densities between fat globules and the plasma phase of milk, fat globules tend to accumulate as a cream layer on top of the milk. However, natural creaming of fat highly depends on the storage temperature of milk.

i. Creaming is more effective at room temperatures (15 °C) at extended creaming times rather than cool temperatures (4 °C) [1].
ii. An increase in temperature of milk leads to both an increase in density difference between milk fat and skim milk. So increase in temperature will lead to lower separation efficiency. Higher temperature will lead to disruption of fat globule which will result in heavy fat losses in skim milk.
iii. The fat losses are higher at 70°C than at 54.5°C. The optimum separation temperature is 40°C. Higher temperature leads to protein denaturation and phospholipids [4,5,5a].

☆ However, at very high temperatures (77 °C), creaming is less effective due to the denaturation of immunoglobulins in milk which encourage the flocculation of fat globules [1]. Lower temperature of milk during separation yields cream with higher fat percentage. Cold milk separators are also being used because of the followings [4,5,5a]-

- ☆ Cold milk separators (operate at temperatures less than 10°C) are available. These allow separation of milk as it is received at the factory, and although fat losses to skim milk are somewhat higher they do sometimes allow substantial savings in energy and capital costs.
- ☆ In some cheese making operations heat treatment of the milk is undesirable and cold milk separators offer some advantages.
- ☆ In addition cold milk separators produce cream with greater phospholipids content which gives better whipping properties.

☆ The major modification in a cold milk separator is wider disc spacing than in a conventional model to allow adequate flow of the more viscous cold cream [4,5].

6. Amount of Water or Skim Milk added to Flush the Bowl

If the amount of water or skim milk used to flush the bowl is more the fat percentage in cream will be low, and vice versa. The addition of more water or skim milk will cause an increase in the amount of cream produced, which, with the same fat content, will show a lower fat test.

Factors affecting Fat Losses in Skim Milk During Separation [4,5,5a]

1. Skimming Efficiency

The 'skimming efficiency (SE) of a cream separator can be defined in the following ways:

a. SE represents fat transferred from milk to cream and can be calculated by using the following formula [4a]-

$$Es = 1 - fs/fm$$

where, fs = skim fat as per cent w/w, fm = milk fat as per cent w/w

Separation efficiency depends on initial milk fat content and residual fat in the skim. Assuming optimum operation of the separator, the principal determining factor of fat loss to the skim is fat globule size. Modern separators should achieve a skim fat content of 0.04 - 0.07 per cent.

b. SE refers to the 'percentage total fat from milk recovered in the cream. The higher the fat per cent in milk, the greater the fat losses in the skim milk, the lower will be the skimming efficiency (SE) and vice versa. The best indication of higher SE is lower fat percentage in skim milk. SE of machine may be calculated using the following equation-

$$SE = \frac{C \times F/100}{M \times F_1/100} \times 100$$

where, C - Amount to cream (Kg), F - Fat percentage in cream, F_1 - Fat percentage in milk, M -Amount of milk (Kg)

Examples–The examples given in following table, help in understanding that high fat-test milk shows a higher SE than low-fat-test milk.

Table 17.1: Fat Percentage and SE

Sl.No.	*Particulars of Milk and Obtained Cream*	*Calculation of SE by Using above Formula*
1	100 kg milk testing 7.6 per cent fat; cream produced 14.2 kg cream testing 52.5 per cent fat.	$SE = \frac{14.2 \times (52.5/100)}{100 \times (7.6/100)} = 98.09$
2	100 kg milk testing 4.7 per cent fat produced 10.2 kg cream testing 45.0 per cent fat.	$SE = \frac{10.2 \times (45.0/100)}{100 \times (4.7/100)} = 97.66$

2. Factors affecting Fat Losses in Skim Milk [4,5,5a]

The important factors influencing the fat losses in skim milk obtained through centrifugal separation are given below:

i. **Temperature of milk-**The lower the temperature of milk, the higher the fat losses in skim milk and vice versa. For efficient separation, the temperature of milk should be above the melting point of fat, so that the milk fat in the fat globules is entirely in liquid form.

   - A satisfactory temperature for separation is around 40°C. The higher the temperature, the more efficient is the separation. The milk is heated before separation to 35-40 ° C in plate or tubular heaters for efficient separation. This is known as ' preheating/fore warming' of milk.
   - There is no marked increase in efficiency after 43-49°C.
   - On the other hand, separation at low temperatures (in warm-milk separators) may lead to partial clogging of the bowl due to high viscosity of cream at these temperatures, resulting in a greater fat loss in skim milk.

ii. **Speed of the separator bowl -** The lower the speed of the bowl, the higher the fat loss in skim milk, and vice-versa.

   - At below: rated speed there will be more fat loss in skim milk because insufficient centrifugal force is generated for efficient cream separation.
   - However, at above rated speeds, the skimming efficiency will not increase greatly.

iii. **Rate of milk in-flow-** The higher the rate of inflow of milk, the higher will be the fat losses in skim milk, and vice versa.

- ☆ If the rate of inflow is increased above the designed capacity of the separator, the milk passes through the bowl too rapidly and do not to allow for complete separation, thereby resulting in a higher fat loss in skim milk.
- ☆ On the other hand, underfeeding the separator does not greatly increase the efficiency of the separation.

iv. **Position of cream screw -** A good separator is designed to give efficient skimming within a fairly wide range of positions of the cream screw, so that the fat test of the cream can be varied without influencing the efficiency of skimming.

- ☆ With most separators, the position of the cream screw has little effect on the fat test of skim milk Until the cream test is up to 50 per cent.
- ☆ From this point(50 per cent) up to a 60 per cent fat test in cream, the fat content of the skim milk increases means there is greater fat loss in skim milk.
- ☆ If above 60 per cent of cream is obtained still higher fat loss in skim milk results.
- ☆ Separation of very thick cream at low temperatures may lead to higher losses due to clogging of the bowl with viscous cream.

v. **Mechanical condition of the machine-** Unsatisfactory mechanical condition of cream separator causes greater loss in skim milk. Vibration of the separator, conditions of discs, amount of separator slime in the bowl, are important unsatisfactory mechanical conditions of cream separator during cream separation.

a. **Vibration of separator-** This reduces the efficiency of separation by disturbing the currents of cream and skim milk. Vibration is caused by installation on an insufficiently firm foundation, the bowl being out of balance, bearings being worn out, the axis of rotation not exactly vertical, *etc.*

b. **Condition of the discs-** Discs in an unsatisfactory condition suffer a loss of skimming efficiency due to the uneven flow of the counter-current streams of cream and skim milk between them. An unsatisfactory disc is one which is out of shape, dirty, scratched or rough.

vi. **Amount of separator slime in the bowl-**If too much slime accumulates, the fat loss in skim milk increases. This is caused by the followings-

- ☆ A disturbance in the even flow of the counter-currents of cream and skim milk,
- ☆ By reduction in the centrifugal force (because of decrease in the 'effective' diameter of the bowl).

Separator slime (which is usually considered identical with clarifier slime) consists of the slimy mass which accumulates inside the bowl shell of the cream separator. It is made up of foreign matter, milk proteins, leucocytes, fragments of the secreting cells from the udder, fat calcium-phosphate and other minerals, bacteria and, occasionally, red blood corpuscles. The average composition of separator slime is given in Table 17.2

Table 17.2: Composition of Separator Slime [6]

Constituent	*Moist Slime Per cent*	*Dry Slime Per cent*
Water	68.2	-
Fat	1.4	4.4
Protein	25.3	79.6
Lactose	1.8	5.6
Minerals	3.3	10.4

vii. **Size of the fat globules -**The greater the number of fat globules of less than 2 μm size, the higher the fat loss in skim milk and vice versa. Fat globules of less than 2 μm size usually enter the skim milk, as they are not subject to sufficient centrifugal force(escape the centrifugal force) to be recovered in the cream.

viii. **Degree and temperature of separation-**The higher the degree and temperature of agitation, the greater will be the loss of fat in skim milk and vice versa. Agitation of hot milk causes the disintegration of normal fat globules in to smaller ones which escape the effect of centrifugal force there by leading to more fat loss in skim milk.

ix. **Presence of air in milk -**The greater the amount of air in milk the higher the fat losses in skim milk. If the milk delivered to the separator contains entrapped air bubbles, centrifugal force will disturb the counter-current streams of cream and skim milk between the discs, and lower the efficiency of separation. The effect of air in the milk is greater with hermetic than with non-hermetic cream separators.

x. **Acidity of Milk -**The higher the acidity of milk, the lower the efficiency of separation, the lower the stability of casein particles which in turn get precipitated and clog the bowl there by lowering the efficiency of separation.

Factors Reducing the Creaming [1,3]

Creaming is a reversible process, and although Stokes's law is inadequate in predicting creaming of concentrated emulsions, it predicts that creaming can be decreased by the following factors:

1. **Reduction of particle size [3]-** When the particle size distribution of milk fat globules, for example, is reduced to less than 1 μm by homogenization, creaming is practically eliminated due to Brownian motion which keeps the globules in suspension. Ideally, less than 3 per cent of all oil droplets should exceed 0.8 μm. Furthermore, the particle size should be monodisperse, which is not the case with food emulsions.
2. **Reduced density difference between the continuous and dispersed phase [3]-** This is not possible to obtain with food emulsions. Though the natural density difference between water and edible oils or fats may be decreased by the use of high-density additives, such additives are not permitted for use in foods. A thick adsorbed layer of proteins increases the density of small fat droplets (less than 0.3 μm), even to an extent where they may precipitate.
3. **Increased viscosity of the continuous phase**-The rate of creaming can be effectively reduced by adding hydrocolloids (gums) or carbohydrates, increasing the viscosity of the continuous phase. Chocolate milk is stable due to high zero shear viscosity created by a network of milk proteins, hydrocolloids and cocoa particles formed at high temperatures.
4. **Other factors [3]-** Other factors such as a high volume of dispersed phase, decrease the speed of creaming, *e.g.* mayonnaise is an example of an O/W emulsion, which does not cream off. On the contrary, polydispersity increases flocculation in low to medium concentration O/W emulsions, thus contributing to creaming.

Sources

1. *Mayumi Silva, Jayani Chandrapala, Fat Separation, Innovative Food Processing Technologies, 2021.*
2. *Wikipedia.*
3. *N. Krog, Emulsifiers and Emulsions in Dairy foods, Encyclopedia of Dairy Sciences, 2002.*
4. *Factors affecting Skimming Efficiency and Richness of cream, DT-2, Lesson 6, http: //ecoursesonline.iasri.res.in.*

4a. *Separation efficiency, Food Science, University of Guleh, https: //www. uoguelph. ca.*

5. *Y. Kotilinga Reddy, Yogesh Khetra and M.H. Sathish Kumar, Fat Rich Dairy Products, http: //www. Agrimoon.com.*

5a. *LPT 311: Milk and Milk Products Technology, http: //lms. tanuvas.ac.in.*

6. *Otto Frederick Hunziker, The Butter Industry - Prepared for Factory, School and Laboratory Hardcover, 1940.*

Neutralization and Ripening of Cream [1,2,3,4]

A. Neutralization of Cream

It is done for partial reduction in acidity of cream. It is understood that by neutralization of **cream acidity of cream is reduced**. Churning of High acid cream may cause high fat loss which can be prevented by neutralization. In pasteurization of sour cream, the casein curdles, by entrapping fat globules, as the bulk of curd goes in butter milk, causing high fat loss.

Objectives of Neutralization

The followings are the objectives of Neutralisation of cream:

a. The objectives of neutralization are to reduce the acidity in cream to a point (0.14 -0.16 per cent) which permits pasteurization without risk of curdling, to produce butter which keeps well in cold storage

b. To guard against undesirable flavours which may result when a cream of high acid which is subjected for pasteurization at higher temperatures.

c. To avoid excess loss of fat which result from the churning cream *i.e.* excessively sour.

d. To improve the keeping quality of butter from high acid cream. Salted-acid-butter develops a fish flavour during commercial storage at -23 to -29°C.

Theoretical Concept and Chemical Reaction of Cream Neutralization [3,4]

The concept of neutralisation mainly involves neutralisation of lactic acid developed in cream by Soda or Lime neutralisers. The details related to the different types of neutralisers, chemical reactions, molecular weights of neutralisers and lactic acid (quantitative relationship between reactants) and Neutralisation factor are given in Table 17.3 [3,4].

Table 17.3: Details of different Types of Neutralizers

Sl. No.	*Types of Neutralisers*	*Chemical Reaction*	*Molecular Weight*	*NF*
1	Soda neutralizers	**i. Sodium bicarbonate** $NaHCO_3 + CH_3\ CHOHCOOH \rightarrow CH_3CHOHCOONa + H_2O + CO_2$	$NaHCO_3 = 84$ $CH_3\ CHOHCOOH = 90$	84/90 = 0.933
		ii. Sodium carbonate $Na_2CO_3 + 2CH_3\ CHOHCOOH \rightarrow 2CH_3CHOHCOONa + H_2O + CO_2$	$Na_2CO_3 = 106$ $CH_3\ CHOHCOOH = 90$ 2x90=180	106/180 = 0.58
		iii. Sodium hydroxide $NaOH + CH_3CHOHCOOH \rightarrow CH_3CHOHCOONa + H_2O$	$NaOH = 40$ $CH_3\ CHOHCOOH = 90$	40/90 = 0.444

Sl. No.	Types of Neutralisers	Chemical Reaction	Molecular Weight	NF
2	Lime neutralizers	**i. Calcium hydroxide** $Ca(OH)_2$+2CH_3 CHOHCOOH → $(CH_3CHOHCOO)_2Ca + 2H_2O$	$Ca(OH)_2$ = 74 CH_3 CHOHCOOH=90 2x90=180	74/180 = 0.411
		ii. Magnesium hydroxide $Mg(OH)_2$+2CH_3CHOHCOOH → $(CH_3$ $CHOHCOO)_2$ Mg + $2H_2O$	$Mg(OH)_2$ = 58 CH_3 CHOHCOOH=90 2x90=180	58/180 = 0.322
Quantitative relationship between reactants		The quantitative relationship between amount of lactic acid present in solution and amount of pure neutralizer required to give exact neutralization is fixed and definite amount of CO_2 is driven off by air. The aforesaid table reveals that 90g of lactic acid requires 84g $NaHCO_3$ (or) 106g Na_2CO_3 (or) 40g of NaOH for neutralization.		

Notes: i. NF= Neutralization factor; ii. Sodium bicarbonate (0.83 kg up to 0.30 per cent cream acidity and 0.91 kg beyond 0.30 per cent cream acidity) [2] and iii. Calcium hydroxide (0.41 kg) [2].

Single Neutralization Procedure

Dissolve neutralizer in 10-15 times of its weight in clean water. Temp. of cream while addition of neutralizer should be 29-32°C.

Acidity in Cream [3,4]

Acidity in cream is present chiefly in the serum portion and not in fat. The acidity in cream as a whole known as cream acidity (C.A.).Cream acidity is calculated by using the following formula:

$$\text{Cream acidity \%} = \frac{\text{\% of serum in cream}}{\text{\% of serum in milk}} \times \text{\% of titratable acidity of milk}$$

Or

$$\text{Cream acidity \%} = \frac{\text{\% of Fat in cream}}{\text{\% of Fat in milk}} \times \text{\% of titratable acidity of milk}$$

The acidity per cent found in serum is known as cream serum acidity (C.S.A). Cream serum acidity is more reliable than cream acidity. There is a relationship between CA and C.SA as below:

$$\frac{\text{CA per cent}}{\text{CSA per cent}} = \frac{\text{Serum per cent in cream}}{100}$$

Factors affecting Neutralization of Cream [2,3,4]

Accurate neutralization of sour cream is important to get a desired quality product. The following factors effect the neutralization:

i. Accuracy in sampling.
ii. Accuracy in testing.
iii. Accuracy in estimation of amounts of cream and neutralizer,
iv. Careful weighing the quantity of neutralizer and.
v. Thorough mixing of neutralizer in cream prior to pasteurization.

Method of Neutralization of Cream [2,3,4]

There are following five important and essential steps to follow for cream neutralization-

i. **Adoption of definite standard of churning acidity [3,4]-**Acidity of cream at churning time controls the flavour and keeping quality of the butter. Hence, it is important to decide that at what acidity the cream shall be churned.

- ✰ Churning acidity should be kept upto that maximum acidity where freedom from chemical deterioration of butter (fishy flavour) with age can be ensured.
- ✰ For cream of average richness (about 30 per cent), fishy flavour can be prevented by keeping the churning acidity to 0.3 per cent maximum.
- ✰ The safe maximum limit of churning acidity varies with the richness of the cream. Since the acidity of cream is chiefly contained in the cream serum, cream serum acidity adjustment would give better results. Usually serum acidity is kept at 0.35 per cent to achieve best keeping quality of butter.

The following examples help in calculating the cream acidity and cream serum acidity:

a. **30 per cent fat cream, cream acidity 0.25 per cent -** Calculation of serum acidity is done by using the following formula:

$$\% \text{ serum acidity} = \frac{\%\ acid\ in\ cream}{\%\ cream\ serum}$$

$$\%\ serum\ acidity = \frac{0.25}{0.70} \qquad (100 - 30 = 70\%\ cream\ serum)$$

= 0.357 per cent

b. Cream test 20 per cent fat, serum acidity 0.35 per cent is desired what should be the acid test of the cream?

Per cent cream serum: 100-20 = 80

Per cent serum acidity = 0.35

Per cent acid in cream = per cent cream serum x per cent serum acidity

$= 0.80 \times 0.35$

$= 0.28$ per cent

ii. **Correct Estimation of Acidity** (Calculation of quantity of lactic acid to be neutralized per 100 kg cream) **[2,3,4]-**Representative sample should be taken. Weight of cream should be accurately measured as the final amount of neutralizer will be dependent on the weight of cream. Effect of CO_2 on acidity should be taken care especially in high acid cream (acidly > 0.65 per cent)

iii. **Calculating the amount of neutralizer to be added [3,4]-** The quantity of neutraliser can be calculated by using the following formula:

$$Quantity\ of\ Neutralizer\ (g)$$
$$= (\%\ initial\ acidity\ of\ sour\ cream - \%\ desired\ level\ of\ acidity)$$
$$\times \frac{qty.of\ cream}{100\ NF} = (a-b) \times \frac{w}{100} \times NF$$

N.F. for different neutralisers are given below:

a. Sodium bicarbonate ($NaHCO_3$) -1.1, b-Sodium bicarbonate (Na_2CO_3)-1.7, c- Calcium hydroxide $Ca(OH)_2$ - 2.43, d- Magnesium hydroxide $Mg(OH)_2$-3.1, e-Sodium hydroxide (NaOH)- 2.25.

iv. **Adding neutralizer in the correct manner [3,4]-**Neutralizer should be dissolved or emulsified in clean water, diluted to approx. 20 times its weight with water; the solution must be distributed quickly and uniformly throughout the entire batch of cream and mixed thoroughly with cream.

- ☆ For efficient mixing, neutralizer is usually sprayed onto the surface of well agitated cream.
- ☆ While the neutralizer is added, the cream should be agitated vigorously and continuously. Agitation of cream is preferable for 5-10 min after neutralization.
- ☆ Temperature at the time of neutralization should be 30°C. (High enough for smooth consistency and low enough to prevent abnormal heat curdling of the sour cream)

The aforesaid precautions are essential for assuring the followings-

a. efficiency of neutralization, b- protection of butter against neutralizer flavour, oily metallic flavour and mealy body, c-avoidance of pasteurizing difficulties and d-prevention of excessive fat losses in the buttermilk.

v. **Checking Results of Neutralization by Re-testing for Acidity [3,4]-**Acidity should not be checked immediately after neutralization because of the following reasons:

a. In case of lime and Magnesia neutralizers, the neutralizing action is slow. It completes after pasteurization and cooling.

b. In case of soda neutralizers, CO_2 is liberated and this reacts acid toward the phenolphthalein indicator. After pasteurization, expulsion of CO_2 is largely accomplished. Therefore, testing acidity after pasteurization would give correct results.

Role of Carbon Dioxide in Neutralization of Cream with Sodium Bicarbonate [3.4]

Fresh cream always contains some dissolved carbon dioxide (as carbonic acid) which reacts with sodium hydroxide during titration and shows a higher acidity test. But the carbon dioxide does not react with sodium bicarbonate neutralizer and consequently over neutralization results.

Double Neutralization with Lime and Soda [2,3,4]

The objectives of double neutralization are-

i. To avoid the intense effect on flavour of a large amount of any one neutralizer with high-acid cream.

ii. To avoid production of excessive carbon dioxide by the use of sodium bicarbonate with high-acid cream.

For this purpose the cream is first neutralized with lime neutralizer and brings the cream acidity down to 0.3-0.4 per cent. Next use soda neutralizer to bring the cream acidity down to the desired level.

Neutralizing Precautions [3,4]

In order to secure the desired results, *i.e.*, accurate acid reduction, absence of objectionable neutralizer flavour and of excessive fat losses, make sure of the correctness of weight of cream and acid test. Do not heat the sour cream above 85 to 90°F, before neutralization, use the correct amount of neutralizer in properly diluted form, distribute it evenly over the cream, and continue agitation of the neutralized cream for 5 to 10 minutes before starting to pasteurize.

Type of Neutralizers [3,4]

In order to accomplish the purpose, for which they are used in the creamery, neutralizers must have the following properties-

i. Neutralizers must have alkaline properties.

ii. They must be alkalis, alkaline earths or their substances. An alkali is a substance that has the property of neutralizing acids, forming salts with them.

The neutralizers used for reducing acidity in cream belong to either Lime Neutralizers or Soda Neutralizers. The details of these neutralisers are given below:

i. Lime Neutralizers

The principal constituent of the majority of lime neutralizers is calcium. Many of the lime neutralizers available for cream neutralization also contain some magnesium. The various commercial lime neutralizers differ from one another chiefly with respect to the proportion of calcium and magnesium they contain. On the basis of the magnesium content, these neutralizers can be classified in the following three groups:

Table 17.4: Classification of Magnesium Neutralizers

Sl.No.	Types of Mg. Limes	Per cent of Mg.	Other details
1	Low magnesium limes	5 per cent or less of Mg.	A well Known brand of creamery lime belonging to this group is peerless lime.
2	Medium magnesium limes	About 30-35 per cent Mg.	To this group belong such brands ad Kelly Island lime, Neutra-Lac and Neutra-Lime.
3	High magnesium limes	About 45 to 55 per cent Mg	All wood lime is an outstanding member of this group. All magnesium limes in the form of magnesium oxide and magnesium carbonate are also available. They are artificially prepared limes and demand a higher price than the natural limes. Their effect on the flavour of cream and butter however is outstandingly favourable.

In general the medium and high magnesium limes react somewhat more satisfactorily in the cream than the low magnesium limes. The higher the magnesium oxide content of lime the greater is its alkalinity and its neutralizing strength.

Calcium Carbonate-low solubility, low alkalinity unsuitable for cream neutralization and action is very slow.

ii. Soda Neutralizers

Soda neutralizers commonly used in the creamery are-

a. Bicarbonate of soda or baking soda
b. Sodium carbonate or soda ash
c. Mixtures off baking soda and soda ash, such as Sodium sesquicarbonate, Neutralene and Wyandotte.

Comparison between Lime and Soda Neutralizers [3,4]

The comparative study on the properties of Soda Neutralizers and Lime neutralizers reveals the followings:

Table 17.5: Comparison between Lime and Soda Neutralizers

Sl.No.	Properties	Soda Neut1ralizers	Lime Neutralizers
1	Purity	Soda Neutralizers will have less than 0.1 impurities.	Lime neutralizers will have sand and clay.
2	Solubility	Sodium Neutralizers are completely soluble.	Lime neutralizers are slightly soluble.
3	Action	Sodium Neutralizers act quickly.	Lime Neutralizers react slowly.
4	Action on casein	Soda neutralizers will have solvent and softening action and assist in minimizing clogging during processing.	Lime neutralizers tend to granulate (or) precipitate casein which results in bitter lime flavour on pasteurization.
5	Acid reaction	Soda Neutralizers act on serum acidity first.	In lime neutralizers the calcium has natural affinity towards casein, lime particles attach themselves mechanically to casein and it is not completely available for neutralizing the lactic acid. The neutralizing capacity of lime neutralizer is 80-85 per cent.
6	Foaming effect.	Soda neutralizers will produce violent foam when there is high acidity. $NaHCO_3$ has more effect than Na_2CO_3.	Lime neutralizers do not create foam.
7	Neutralizing strength	Soda neutralizers are weaker alkalis than lime neutralizers about twice as many kilograms of soda neutralizers are required to neutralize given amount of acid as compared to lime neutralizers.	Lime neutralizers are stronger alkalis than soda neutralizers about half quantity of lime neutralizers are required to neutralize given amount of acid as compared to soda neutralizers.
8	Material cost	Soda neutralizers are costlier than lime neutralizers.	Lime neutralizers are cheaper than soda neutralizers.
9	Effect on texture of butter	In case of high acid cream especially neutralizes to a low point there is a tendency to butter from lime neutralizer cream will give less smooth texture than soda neutralizers.	
10	Effect on flavour of butter	Soda neutralizers produce soapy type of flavour.	Lime neutralizers produces course lime flavour. In over all the butter will give neutralizer flavours.

Significance of Neutralisation of Cream [2]

Neutralization of cream has following significance:

a. To avoid excessive fat loss in buttermilk due to entrapping fat globules with casein.

b. To lower down the production of undesirable off flavour in cream.

c. To improve the keeping quality of butter.

B. Ripening of Cream [2,3,4]

Objectives/Purpose

The ripening of cream is done for following two purposes:

i. The fundamental objects of cream ripening are to produce butter with a pleasing, pronounced flavour and aroma, and to produce this flavour and aroma uniformly from day to day. Ripening also influences somewhat the exhaustiveness of churning and it affects the keeping quality of the butter variously, according to quality of original cream, churning acidity, and whether made into salted or unsalted butter.

ii. To obtain low fat loss in buttermilk.

Ripening refers to the process of fermentation of cream with the help of suitable starter culture. This step can be eliminated if sweet-cream butter is desired. The main object of cream ripening is to produce butter with higher diacetyl content. Ripening improves the keeping quality of salted butter but it reduces the keeping quality of a salted butter. The ripening of cream involves the following steps:

a. Addition of Starter Culture [2,3,4]

Starter culture consisting of a mixture of the following microorganism, is added-

i. Mixture of both acid producing organisms (Streptococcus lactis, S.cremoris, Lactococcus lactis,L. cremoris) and flavour producing organisms (S.lactis subsp. diacetylactis, Leuconostoc citrovorum and/ or Leuconostoc dextranicum).

or

ii. Butter starter culture containing lactic acid producer (Lactococcus lactis subsp. lactis and Lactococcus lactis subsp. cremoris with aroma (diacetyl) producers (Lactococcus lactis subsp. lactis, Leuconostoc mesenteroides subsp. dextranicum is added to standardized, pasteurized and cooled (20-22°C) cream at rate of 0.5-2.0 per cent [2]. Citric acid can be used to improve the flavour in cream.

b. *Incubation of Cream*

Amount of starter added depends on several factors. Usually starter culture is added at the rate of 0.5 to 2.0 per cent of the weight of cream and after proper mixing cream is incubated at about 21°C till desired acidity is reached. Usually it takes 15-16 hrs [2,3,4]. Cream is subsequently cooled to 5-10°C to arrest further acid development.

Biosynthesis of diacetyl is not sufficient above pH 5.2. Stopping fermentation of cream by cooling at pH 5.1-5.3, results in a milder flavour; whereas continuing fermentation upto pH 4.5-4.7 results in higher levels of both diacetyl and lactic acid, giving more pronounced flavour.

Effect of Cream Ripening on Butter [3,4]

The cream ripening has following effects on the quality of butter:

1. Effect on Flavour and Aroma

The typical butter flavour is due to the presence of diacetyl in combination with lactic acid, carbon dioxide, acetoin and intermediary products such as acetaldehyde, and probably other aromatic products as yet not definitely determined. These substances are the products of fermentation, brought about by the associative action of lactic acid-producing bacteria and citric acid-fermenting bacteria.

These bacteria are propagated and their flavour and aroma substances produced in the starter. During cream ripening the starter that is added to the cream, therefore, functions in following two ways:

i. It seeds the cream with species of bacteria that are capable of producing the desired aroma and flavour substances in the cream.

ii. It adds to the cream the aroma and flavour substances produced and already contained in the starter.

The mildly acid and pronounced nutty flavour that is characteristic of the pleasing flavour of good butter is usually accompanied by a high, attractive aroma.

a. *Factors which Influence the Diacetyl + Acetoin Content of Butter*

Cream ripened with a normal starter shows a varying ratio of diacetyl to acetoin.

i. In the case of 20 per cent cream, a ripening temperature of 17°C. (62.6°F) yields the largest amount of diacetyl. With increasing fat content the amount of diacetyl+ acetoin increases.

ii. Cream testing 40 per cent fat yields higher diacetyl content in the butter than 20 per cent cream, or whole milk.

The diacetyl content of butter to affect its flavour as follows:

Sl.No.	Amount of Diacetyl	Flavour
1	Absence of diacetyl	Flavourless
2	0.2 to 0.6 ppm diacetyl	Mild flavour
3	0.7 to 1.5 ppm diacetyl	Full flavour

b. Distribution of Diacetyl + Acetoin by the Churning Process

Butter,Fresh butter milk *etc.* have different quantity of diacetyl and acetoin content. The details are given below:

- ✰ Butter contains a relatively small proportion of the diacetyl and acetoin content of the cream from which it is made.
- ✰ Fresh buttermilk contains larger amounts of diacetyl and acetoin than the corresponding butter and than the original cream, at churning time.
- ✰ The serum of butter contains larger amounts of diacetyl + acetoin than the fat of the same butter.
- ✰ The wash water contains appreciable amounts of diacetyl + acetoin. Churning 2 liters of cream yields the following amounts of diacetyl in the buttermilk, wash water and butter-

Sl.No.	Butter milk	6.14 mg Diacetyl
1	First wash water	3.16 mg Diacetyl
2	Second wash water	0.00 mg Diacetyl
3	Third wash water	0.00 mg Diacetyl
4	Butter	0.30 mg Diacetyl

- ✰ The following amounts of diacetyl + acetoin are present in washed and unwashed butter:

Sl.No.	Type of Butter	Diacetyl (mg/kg)	Diacetyl + Acetoin (mg/kg)
1	Unwashed Butter	1.69	9.30
2	Washed Butter	0.86	3.78

- ✰ These findings indicate that in commercial manufacture of butter, particularly in the case of unsalted butter, excessive washing gives the finished butter, even when made from properly ripened cream, a disappointing washed-out flavour.

2. Effect of Cream Ripening on Keeping Quality of Butter [3,4]

The development of flavour and aroma in butter by cream ripening, or by any other process of manufacture, can be of value only, provided that it does not impair or destroy the keeping quality of the resulting butter. The ripening of cream affects the keeping quality of butter in following two fundamental ways-

i. By its control of age deterioration due to bacterial causes,
ii. By its influence on age deterioration due to chemical causes.

a. *Bacteriological Effect*

The ripening of cream improves the keeping quality of butter as far as keeping quality is dependent on freedom from age deterioration due to biological causes. Cream ripening assists in controlling bacterial deterioration in butter. In butter made from ripened cream there is a great prevalence of lactic acid bacteria and a relatively high acidity and probably an abundance of lactate salts. These agencies are antagonistic to the great majority of flavour-damaging organisms that may be present in the butter, thus retarding their action, preserving the fresh or desired flavour, and prolonging the keeping quality of the butter.

b. *Chemical Effect*

Cream ripening does not improve the chemical stability of butter. On the contrary, under average commercial conditions of manufacture, the ripening of cream to a full aroma and flavour shortens the life of salted butter.

- ✰ The usual flavour defects that develop with age in butter made from fully ripened cream are oily-metallic, fishy and sometimes tallowy flavour.
- ✰ There is a tendency also to intensify the well-known cold storage flavour. This is especially true of salted butter made from cream that arrives at the factory in sour fermented condition, is neutralized, pasteurized and re-ripened to a high acidity.
- ✰ It applies also, through to a somewhat lesser extent, to salted butter made from ripened sweet cream. It does not apply to unsalted butter.

Salted butter made from sweet, unripened cream, or from sour cream, neutralized and pasteurized, keeps better from the standpoint of absence of flavour deterioration due to chemical causes, than salted butter made from the same cream ripened to a full flavour and aroma.

Per cent Acid to which the Cream should be Ripened [3,4]

The ripening of cream to different acidities effects the quality of butter made from it. Therefore an approximate limits for acidities of cream, are recommended for maintaining the quality of butter.

i. For fresh consumption salted butter, cream of moderate richness (30 per cent fat) may safely be ripened to about **0.25 to 0.30 per cent acid**.

ii. For salted butter of commercial cold storage, it has been found preferable not to ripen the cream and to churn it at an acidity of about 0.21 per cent acid or lower.

iii. In case of unsalted butter, the cream may be ripened to any acidity without jeopardizing keeping quality.

Cooling and Ageing [3,4]

Cooling and ageing are processes which prepare the cream for subsequent operation of churning. When cream leaves the pasteurizer, the fat in the globule is in liquid form. When cream is cooled, fat crystallization starts, cream will not churn unless the butter fat is at least partially crystallized. If solidification of fat is not sufficient, the fat losses in butter are high.

- ✰ Rate of cooling has an important influence on the body and texture of butter. The temperature to which cream is cooled is chosen is such a way that the butter produced is of optimum consistency and cream churns to butter in a responsible time of about 35-45 minutes.
- ✰ Churning at too high temperature may give butter with greasy' body which may work up too quickly and become sticky.
- ✰ Generally cooling temperature in summer should be 7-9°C and that if in winter (10°-13°C)

Significance of Ripening of Cream [2]

Ripening of cream has following significance:

a. Typical flavour of butter from ripened cream is due to diacetyl and in some extent with acetic acid and propionic acid.

b. No diacetyl in sweet cream.

c. Normal diacetyl content in ripened cream butter 2.5 PPM and very rarely over 4 PPM.

Sources

1. *Wikipedia.*

2. *Butter- Classification, Neutralisation, and ripening of cream, Dairy Science and Technology, Agrojiva, 2021.*

3. *Process Outlines of Butter making, Lesson 18, http: //ecoursesonline.iasri.res.in.*

4. *Y. Kotilinga Reddy, Yogesh Khetra and M.H. Sathish Kumar, Fat Rich Dairy Products, http: //www. agrimoon.com.*

2. BUTTER [1,1a.3]

Butter is a dairy product made by churning fresh or fermented cream. Butter is a water-in-oil emulsion resulting from an inversion of the cream, an oil-in-water emulsion; the milk proteins are the emulsifiers.

- ☆ It is generally used as a spread and a condiment, as well as in cooking applications, such as baking, sauce making, and pan frying.
- ☆ Butter remains a solid when refrigerated, but softens to a spreadable consistency at room temperature, and melts to a thin liquid consistency at 32-35 °C.
- ☆ It generally has a pale yellow colour, but varies from deep yellow to nearly white. Its unmodified colour is dependent on the animals' feed and on food colourings in the commercial manufacturing process, most commonly annatto or carotene.

The composition, classification and quality standards under different quality/regulatory agencies are in chapter-16.

Churning of Cream [1,3]

Churning is the process of converting cream into butter through appropriate mechanical manipulations leading to the conversion of oil-in-water (O/W) emulsion of cream into water-in-oil (W/O) emulsion desired in butter. The emulsion change accompanied by removal of buttermilk and working of butter yields the desired structure and texture in the product.

- ☆ During the churning process, cream is converted into butter. Here the fat gloubles are disrupted under controlled conditions to destabilize o/w emulsion and bring about agglomeration of milk fat. The sequence of events that occur during churning is as follows:
  - **i.** Churning is initiated by agitation of cream causing incorporation of numerous air bubbles into the cream.
  - **ii.** With incorporation of air there is increase in the volume of cream and air plasma interface.
  - **iii.** Surface active (such as frictional, impact, concussion *etc.*) causes partial disruption of fat globule membrane.
  - **iv.** The fat film, thus formed, serve as a foam depressant causing the air bubble to burst.
  - **v.** The liquid fat also serves as cementing material causing fat globules to clump together and eventually butter grains are formed which floats in plasma *i.e.* butter milk.

Churning Efficiency (Ec) [2]

It represents fat transferred from cream to butter and can be calculated by using the following formula:

$$Ec = 1 - fbm/fc$$

where, fbm = buttermilk fat as per cent w/w,fc = cream fat as per cent w/w

- ☆ Maximum acceptable fat loss in buttermilk is about 0.7 per cent of churned fat corresponding to a churning efficiency of 99.3 per cent of cream fat recovered in the butter.
- ☆ Churning efficiency is highest in the winter months and lowest in the summer months.
- ☆ Fat losses are higher in ripened butter due to a restructuring of the FGM (possibly involving crystallization of high melting triglycerides on the surface of the globules).
- ☆ If churning temperature is too high, churning occurs more quickly but fat loss in buttermilk increases. For continuous churns assuming 45 per cent cream, churning efficiency should be 99.61 - 99.42 per cent.

Theories of Churning [1,3]

The conversion of oil-in-water (O/W) emulsion of cream into water-in-oil (W/O) emulsion to form butter has been explained by various theories of churning. These are discussed below:

1. The Phase Reversal Theory (Fischer and Hooker's Theory)

This theory was proposed by Fischer and Hooker in 1927. According to this theory churning is a process of phase reversal *i.e.* changing of oil in water emulsion (O/W) to water in oil emulsion (W/O). The stability of emulsion is related to the relative volumes of the two constituents present. When oil and water are mixed together, the resulting suspension may be a suspension of o/w or suspension of w/o. The type of emulsion obtained depends-

i. On the proportion of the two main constituents present,
ii. The order in which they are added and
iii. The type of emulsifier used.

- ☆ In churning cream, initially the ratio of surface area to volume (S/V) of the fat globules is large. When the churning proceeds, surface area decreases and with progressive churning, surface area keeps on decreasing. The reduced surface area can no longer hold all the butter milk so it breaks *i.e.* separates out.
- ☆ Agitation of cream during the churning process causes coalescence and clumping of fat globules until eventually the ratio of surface area to volume of the fat units becomes so small that the reduced surface area can no longer contain the butter milk in stable form. The O/W emulsion then suddenly breaks; giving butter grains consisting of an emulsion of W/O and free butter milk.

Supportive Evidences

a. The supportive evidence of this theory is established by the fact that in normal butter, water is not in continuous phase.

- It has been demonstrated that plastic cream, containing 80-82 per cent fat, conducts electricity and it responds to the pH determination showing water is in continuous phase.
- Butter is a very poor conductor of electricity and pH determination cannot be done on butter but only on serum separated from it.

b. Microscopic structural studies conducted by Rahn in1928 revealed that butter is not a true W/O emulsion. A proportion of globular fat are still intact in worked butter. He explained that since butter fat is cooled and largely crystallized before the start of churning, true W/O type emulsion is rarely possible.

2. The Foam Theory

This theory was put forward by Rahn in1928. According to this theory, cream (and also milk) contains a foam producing substance which gets solidified gradually when cream (or milk) is agitated.

- During churning first foam is produced. The fat globules then, due to surface tension, tend to concentrate on the foam bubble and thus are bought into such close contact that clumping of fat globules take place. Subsequently the foam producing substance assumes a solid character and the foam collapse. The fat globules then coalesce and butter is formed.
- According to this theory, fat in cream at churning time is completely crystallized and the pass in to butter with their membrane intact and thus butter is a compact mass of fat globules in which butter milk, water and air are distributed as small globules.
- Rahn's theory was based on his following findings that
  - **i.** Air was necessary for normal churning of butter. Application of normal amount of mechanical agitation, in the absence of air did not result in churning of cream.
  - **ii.** The effect of overloading of churn resulting in increased churning time supported this theory (in case of overloading the churn, there was no sufficient space in the churn for the formation of required amount of foam hence more time).

This theory was, however, subsequently criticized because of the fact that foam formation *i.e.* presence of air, is not required in some of the continuous butter making processes developed subsequently.

3. King's Theory (Modern Theory)

King's theory was proposed in 1930 and 1953 and it is regarded as the modern theory. According to this theory, during churning, the phenomenon happens is mid-way between the 'Phase Reversal theory' and Foam theory. The modern concept has been summarized by Mc Dowall as follows:

i. The fat in the cooled cream, at churning temperature, is present as clusters of fat globules. And within each globule it is present partly in solid and partly in liquid form.

ii. Agitation (churning) breaks up the clusters and causes foam formation. The globules become concentrated to some extent in the film around the air babble in the foam and thus are brought into close contact of each other.

iii. The movement of the globules over one another in the foam film and the direct concussion between them causes a gradual wearing away of the emulsion protecting surface layer (of phospholipid protein complex). The globules then adhere together to form larger and larger particles. Eventually these particles become visible as butter grains. The grains enclose some of the air from the foam. The fat still mainly remains in globular form.

iv. The working of the butter grains causes the globules to move over one another. Some of them, under the effect of friction and pressure cause some yields out a portion of the liquid fat, others are broken during working. Finally there is enough free liquid fat present to enclose the water droplets, air bubbles and intact fat globules.

Factors Influencing Churn Ability of Cream [1,1b,3]

The factors which influence the churnability of cream can be classified into two groups as:

i. **The factors related to the initial character of the cream-** The factors related to the initial character of the cream includes chemical composition of the butter fat, size of fat globules, richness of cream and viscosity of cream.

   a. **Chemical composition of butter fat:** Churnability of cream is greatly influenced by the proportion of soft fats (low melting point fat) and hard fats (high melting points). This proportion determines the degree of fat solidification in the cooled cream. If the proportion of soft fats is more than the churning period will be shortened, butter made will have less firmness and there will be more fat losses in butter milk. If the proportion of soft fats is low, it will prolong the churning period.

   b. **Richness of cream:** The amount of fat in cream affects its churnability considerably. The richer the cream the sooner will be the completion of the churning provided the cream is not rich enough to be so thick as to cause the cream to adhere to the inside of the churn and thus escape agitation.

      ☆ If rich cream is churned at a high temperature the butter will form in a remarkable short time, providing all other conditions are favourable. Thin cream churns much more slowly, and can be churned at high temperature than thick cream, without injuring the quality of butter when rich cream is churned at a high temperature and the butter forms in a short time (about 10 min), the butter will usually be greasy in body and will not contain a much of butter milk, which will be more or less difficult to remove on washing. When thick cream is churned, the butter does not break in the form of small round granules, as it does when thin cream is churned.

- ☆ When thick cream is (36 to 38 per cent fat) is churned at as high a temperature as is consistent with getting a good texture, the best result are obtained. This type of cream produces less butter milk and consequently less part loss in the butter milk and this will give increase over run and the breaking of the butter at the end of the churning will be such as to cause the granules to appear large and flaky, rather than small round granules. The more flaky granules of butter will retain more moisture than the small, harder granules under the same treatment.
- ☆ When thick cream is churned and the temperature is moderately high, it is almost impossible to churn the butter into granules. This condition causes butter from thick cream to contain more moisture than butter from thin cream.

c. **Viscosity of cream-**The more viscous the cream, more time is required to complete the churning process. More viscosity diminishes the freedom of movement of the fat globules, lessens their opportunity of being brought together and retard coalescence, thereby increases churning time.

d. **Size of fat globule-**Cream containing large fat-globules (avg. diameter 4.6µ) churn more quickly than cream containing small globules. Cream containing small fat globules (avg. diameter 3.4 µ) churn with difficulty and require twice as much time to break as the large globule cream. The butter made from such cream has short grains and crumbly character.

ii. **The conditions in the process of manufacturing-**The factors related to processing conditions are churning temperature, fullness of churn, speed of churn, design of churn *etc.*

a. **Churning temperature-**The temperature is one of the most influential factors in determining the churnability of cream. The higher the temperature of cream, the sooner the churning process will be completed.

- ✰ Too high a churning temperature is however not desirable. It causes the butter to contain soft lumps instead of in a flaky granular form. This is deleterious to the quality of the butter. It causes first a greasy texture of butter, and secondly, it causes the incorporation of too much butter milk in the butter.
  - i. This butter milk contains lactose, curd, and water, which when present together in butter, are likely to sour and in other ways deteriorate the butter.
  - ii. Curd and lactose should be excluded from butter as much as possible, in order to eliminate food for bacteria which may be present.
- ✰ Too low temperature is also undesirable although it is better to have the temperature a little low rather than too high.
  - i. Cream at low temperature becomes more viscous. On agitation in the churn such cream if it is very thick will adhere to the sides of the churn and rotate with it without agitating, consequently no churning will take place.
  - ii. Too low a temperature brings the butter in such a firm condition that it takes up salt with difficulty, and when this hard butter is being worked, a large portion of the water in the butter is expressed, and the overrun will be lessened to a great extent without increasing the commercial value of the butter.
- ✰ The degree of hardness of the fat in cream is the governing factor in deciding the temperature during churning. The hardness of the fat depends upon:
  - i. The season of the year.
  - ii. The individuality of the cow.
  - iii. The stage of lactation period.
  - iv. The kind of food fed for the cows.

All these factors influence the melting point of butter fat. The higher the melting point of butter fat, higher is the churning temperature and the lower the melting point of the fat; the lower is the churning temperature.

- ✰ During spring, the cows yield milk containing a longer proportion of soft fats; consequently the churning temperature is always lower in the spring than in the winter. During the winter, when the cows are fed on dry food chiefly the harder fat increases in quantity, therefore, a higher churning temperature is necessary during that time.

☆ The nature of food fed affects the melting point of butter to a considerable extent. Cotton seed will cause butter to become hard. When larger amount of cotton seed is fed, the butter assumes a crumby, hard condition.

It can be concluded that the churning temperature may vary between wide limits, but the average desirable churning temperature according to the season is-

Winter - 10-13°C, Summer- 7-9°C

b. **Amount of cream in churn (fullness of churn)-**For maximum agitation to take place during churning, the cream must dash from side to side or from top to bottom. Optimum load for maximum agitation should be one third to one half full.

   Overloading of the churn - It diminishes free space in the churn, diminishes concussions and leads to increase in churning time. The overloaded cream may be churned at higher temperature so that churning time is not prolonged but this is not recommended as higher temperature increases fat loss in butter milk and produces soft and leaky butter.

   Under loading the churn- It is not economical for the manufacturer and at the same time the cream will adhere to the inner side of the churn and delay churning.

c. **Nature of agitation-**Proper agitation is necessary for churning cream into butter. The speed of the churn provides agitation to cream. So, the maximum speed of the churn is the speed that yields the maximum amount of agitation. It is dependent on the ratio of centrifugal force and gravity force. Centrifugal force should be less than gravitational force. The speed of the churn is calculated by the following formula-

$$\text{Centrifugal force} < \text{Gravity force}$$

$$m\,\omega^2\,r < m.g$$

$$(2\pi n)^2.R < g$$

$$n < \sqrt{\frac{g}{R}\frac{1}{2\pi}}$$

$$\approx \frac{1}{2\sqrt{R}}$$

where, m=mass of the cream, N=speed of the churn, g= gravitational acceleration, R= distance from center of churn to the periphery.

Churning Difficulties [1,1b,3]

The causes of churning difficulties are usually associated with the peculiar character of the cream and particularly where the source of cream is confined to a single herd. Usual causes of prolonged churning time and difficulty in formation of butter are excessive hardness of fat, small fat globules, use of thin cream for butter making and high protein content in cream.

i. Excessive hardness of fat: Winter cream usually contains more hard fats. Use of such cream for butter making prolongs churning time because it diminishes the ability of fat globules to coalesce during churning.

ii. Small fat globules: cream that contains small fat globules takes more time for churning as the ratio of membrane material to fat increases and thus provides increased protection to fat globules.

iii. High protein content: such cream delays butter formation because of increased viscosity that minimizes the force of concussion between the globules.

iv. Use of thin cream: Such cream also have increased protection due to higher membrane protein to fat ratio and also due to intervening serum that keeps globules apart during churning.

Sources

1. *Churning of Cream, Lesson 19, http: //ecoursesonline.iasri.res.in.*

1a. *M.K. Srivastava, Handbook on Analysis of Milk Products. Write and Print Publications, New Delhi, 2016. ISBN: 9789384649258.*

1b. *Aneja, R.P., Mathur, B.N., Chandan, R.C. and Banerjee, A.K. 2002. Technology of Indian Milk Products. A Dairy India Publication, Delhi, India.*

2. *Churning Efficiency, Food Science, University of Guleh, https: //www.uoguelph.ca.*

3. *Y. Kotilinga Reddy, Yogesh Khetra and M.H. Sathish Kumar, Fat Rich Dairy Products, http: //www. agrimoon.com.*

Butter Colour [1,2,3,4]

Butter colours are required to manufacture uniform quality butter all the year round as the natural components (like carotene) responsible for colour in butter, varies according to many factors like season, breed, stage of lactation, feed among others. Therefore, a calculated amount of colour is added in butter so that it does not have influence of varying natural colouring components present in milk or cream.

Desirable Properties of Butter Colours [2,4]

The following desirable properties are required for butter colour:

i. It should be free from ingredients injurious to health.
ii. It should be free from undesirable oduors and flavours,
iii. Strength should be such that only a small quantity is required,
iv. It should have such permanency of emulsion as to prevent settling out upon standing and
v. It must be oil soluble.

Types of Butter Colours [2,4]

Butter colours are classified on the basis of their source as vegetable butter colours and mineral butter colours.

a. Vegetable Butter Colours

This class of butter colours is derived from plants. The most common is the colour obtained and extracted from the seed of the annatto plant (Bixa orellana). The extract of vegetable butter colour is made by boiling the annatto seed in oil for several hours. During the latter period of the process the heat is raised to a high temperature, about 115°C for extraction of annatto principle in permanent emulsion with oil. The mixture is then filtered through heavy canvas, either by gravity or under pressure.

b. Mineral Butter Colours

The colouring principle of this class of butter colours is derived from harmless oil soluble coal tar dyes. This coal tar dyes are mixed with the neutral oil, boiled and filtered. The coal tar dyes certified by USDA are:

i. Yellow A B (Benzeneazo- β- naphthlyamine).
ii. Yellow O B (Ortho- Tolueneazo- β- naphthylamine).

- ☆ Mineral colours have greater concentration of colouring principle and therefore less of the butter colour is needed to produce desirable colour in butter than is the case with vegetable colours. The emulsion of mineral butter colour in oil is more permanent as compared to vegetable butter colours.
- ☆ FSSR-2011 has permitted some food colours to be incorporated in butter. These are shown in Table 17.6.

Table 17.6: FSSR 2011 Permitted Colours in Butter [3]

Sl.No.	Colours (Natural : Singly or in Combination)	Quantity
1	Curcumin	100 ppm max.
2	β-carotene	100 ppm max.
3	Carotene (Natural extract)	100 ppm max.
4	Annatto extract on Bixin/Nor bixin basis (50:50 ratio)	20 ppm max.
5	β-apo-8 carotenal	35 ppm max.
6	Methyl ester of β-apo-8 Carotenoic acid	35 ppm max.

Quantity of Butter Colour to be Added [2,4]

The amount of colour that must be added varies greatly under diverse conditions and may vary from none to about 250 g for every 100 Kg of butter.

Manner of Adding Butter Colour [2,4]

The butter colour should be preferably added to the cream while loading the churn. Alternatively, it may be added to and mixed with salt just before final working of the butter. It is then worked into the butter and distributed uniformly.

Sources

1. *M.K. Srivastava, Handbook on Analysis of Milk Products. Write and Print Publications, New Delhi, 2016. ISBN: 9789384649258.*
2. *Y. Kotilinga Reddy, Yogesh Khetra and M.H. Sathish Kumar, Fat Rich Dairy Products, http: //www. agrimoon.com.*
3. *FSS (food product standards and Food additives) Regulation-2011.*
4. *Addition of Colour and Salting of Butter, Lesson 21, http: //ecoursesonline.iasri. res.in.*

3. GHEE [1,1a,2]

Ghee is a product of tradition with an established market prepared at household level, also at organized dairy plants. Ghee manufacture has great significance and relevance to Indian masses and the dairy industry there is sufficient recorded evidence to prove that the manufacture of ghee originated in India and it has been used extensively for dietary and religious purposes since Vedic times (3000-2000 B.C). Production of ghee is simple, convenient and affordable process to preserve milk fat would favour the milk producers and processors to a greater extent.

i. Ghee is a very popular product since from ancient time and have greater demand during festival and other ceremonial functions where use of ghee in food has been considered to delicacy due to its pleasing flavour and aroma.
ii. Best ways to salvage substandard and returned market milk at commercial dairy level is separation of fat from these and convert this fat to ghee.
iii. Ghee has a huge market demand in India, to meet this demand and also involvement of simple technology to produce ghee attracts the milk processor to have ghee making plant of variable capacity.

Compositional Changes during Manufacture [1,1a,2,3]

Ghee majorly consists of milk lipids and richest source of milk fat of all

Indian Dairy products. The constituents of ghee tend to vary with the method of its manufacture.

- ✰ Chemically ghee is a complete lipid of glycerides, 97-98 per cent triglycerides. Small amount of di-and mono-glycerides are also present in traces.
- ✰ cow milk ghee is different from buffalo milk ghee in terms of its composition. Chemical composition of cow milk ghee and buffalo milk ghee are given in composition of ghee under chapter-16.
- ✰ Fatty acid composition of buffalo milk ghee also varies from cow milk ghee.
    - i. The amount of butyric acid is significantly higher in buffalo than in cow ghee.
    - ii. The levels of short chain fatty acids caproic to myristic are significantly higher in cow than buffalo ghee where as levels of palmitic and stearic are higher in buffalo than in cow ghee.
- ✰ Ghee made from buffalo milk is white (lack of carotenoids) with greenish tinge and that made from cow milk is golden yellow. The characteristic colour of buffalo fat has been attributed to tetrapyrozole pigments- biliverdin and bilivubin, this pigment is conjugated to a protein in milk, but is released during the manufacturing process of ghee making. Thus, imparting yellowish-green colour to buffalo ghee.

Physico-chemical Changes during Manufacture [1,1a,2,6]

The quality of ghee depends on milk, cream, dahi or butter, methods of preparation, temperature of clarification, storage conditions, and type of animal feed. These factors in turn will determine the physicochemical characteristics of ghee. The principal measurements of ghee quality are: peroxide value, acidity, and flavour.

a. Flavour (Carbonyles, lactones)

Flavour formation in ghee happens during fermentation of cream and during clarification process.

- ✰ Much of the typical flavour of ghee is due to very complex mixture of compounds including carbonyls, lactones, FFA, and esters generated during processing.
- ✰ The typical flavour of burnt non fat milk solids combined with some component derived from oxidation of the milk fat is also involved. Lactones, which have a coconut-like flavour, seem important for the characteristic flavour of ghee. The levels of these are increased by increasing temperature during the production, and are often twice as high in ghee as in butter.

i. The most important factor controlling the intensity of flavour in ghee is the temperature of clarification.

  ☆ Ghee prepared at 120°C or above has an intense flavour which is usually referred to as cooked or burnt flavour,

  ☆ Ghee prepared at around 110°C has a somewhat mild flavour, often referred to as curdy.

ii. Ghee prepared from desi method generally produces the most desirable flavour.

iii. Flavour in ghee is retained longer when butter contains 1 per cent NaCI,

iv. The acidity of the cream or butter affects the flavour of ghee. Sweet cream-butter yields ghee with a fiat flavour whereas ripened cream or butter having an acidity of 0.15 to 0.25 per cent (lactic acid) produces ghee with a more acceptable flavour,

v. Cow ghee contains more volatile carbonyls. The total carbonyls of buffalo ghee is higher than that of cow ghee irrespective of the method of preparation and temperature of clarification.

b. During manufacturing, water gets evaporated and fat present in the cream or butter getting concentrated and curd particles (MSNF present in cream or butter) starts settling at the bottom during clarification process.

c. Colour

Development and granulation also happens during clarification for the subsequent packaging of ghee.

Physico-chemical Changes of Ghee during Storage [1,6]

The detailed related to Physical and Chemical changes during storage are given below:

a. Physical Changes during Storage

The following physical changes occur during storage:

i. Dependent primarily on the temperature of storage.

ii. Crystallization occurs with the formation of solid, semi-solid and liquid layers.

iii. Ghee prepared from cow and buffalo milk, kept either in a metal or glass container at 20°C or below, solidifies uniformly with fine crystal.

iv. Above 20°C and below 30°C, solidification is a loose structure.

v. To avoid layer formation it would be preferable to store ghee below 20°C.

b. Chemical Changes

The following Chemical changes occur during storage-

i. The liquid portion of ghee had a significantly higher RM value and iodine number than the granular solid or hard flaky portion of the same ghee.

ii. Ghee stored at high temperature is susceptible to oxidative deterioration, rancidity, and off flavour. Shelf life of ghee is also dependent on the method of preparation.

iii. The keeping quality of desi ghee is better than that of direct cream or creamery butter ghee.

☆ Peroxide value, TBA value and acidity increases with storage.

iv. Milk phospholipids in ghee improves its shelf life. Ghee having more residue, which is a source of phospholipids, has a longer storage life.

Hydrolytic and Oxidative Deterioration, their Causes and Prevention [1,2,5,6]

Normally this defect develops in ghee during storage, but in case the raw material used for ghee making is rancid, the freshly prepared ghee will also have this defect.

i. The nutritive value of ghee is also adversely affected due to rancidity in ghee.

ii. Milk fat hydrolysis is faster in liquid state than in solid state. Because of more solid fat in buffalo milk its rate of fat hydrolysis is slower than cow milk fat.

iii. Therefore, the cow ghee is more prone to developing rancid flavour during storage.

iv. Edible fats containing unsaturated molecules are susceptible to attack by molecular oxygen. Process is referred to as lipid oxidation and can give rise to undesirable volatile flavour compounds, potentially toxic oxidation products, and a general deterioration in the quality of the fat [1,6].

Hydrolytic Rancidity [1,5,6]

Hydrolytic rancidity develops when triglycerides in particular are hydrolyzed and free fatty acids are released; the process sometimes requires a catalyst. It is also the reason short-chain fatty acids, such as butter fats, are odorous. **It is also called hydrolysis or enzymatic oxidation**, occurs in the absence of air, but with the presence of moisture.

☆ This normally is accomplished through enzymatic peroxidation, where enzymes found naturally in fats (*i.e.*, lipoxygenase, cyclooxygenase) which can catalyze reactions between water and fat.

- ☆ Enzymatic rancidity is catalyzed by the presence of certain enzymes (examples are peroxidases and lipases). The heat treatment experienced during extrusion is usually sufficient to destroy these enzymes.

Oxidative Rancidity [1,5,6]

It is also known as autoxidation, occurs when oxygen is absorbed from the environment. In the presence of oxygen and/or ultraviolet (UV) radiation, most lipids breaks down and degrade, forming several other compounds.

- ☆ Unsaturated fatty acids, showing double bonds within the molecule, are susceptible to peroxidation. Such peroxidation could be called "oxidative deterioration" of lipids and can occur via a radical reaction in two ways. One way, and by far the most important one, is autoxidation. The other way, less important, is an enzymatic oxidation.
- ☆ Oxidation is a process where an oxygen ion replaces a hydrogen ion within a fatty acid molecule and higher numbers of double bonds within the fatty acid increase the possibility of autoxidation. Oxygen is eight times more soluble in fats than it is in water; it is the main cause of the autoxidation process, increasing the saturation of the oil.

Causes of Rancidity [5,6]

Rancidity in ghee is caused by the formulation of volatile compounds, which exhibit unpleasant odours even when present in small quantities. In ghee the chain reaction is catalyzed by heat, light, ionization reaction and trace metals (copper and iron), *etc.* The volatile flavour products of lipid oxidation include the followings:

i. Aldehydes.
ii. Ketones.
iii. Alcohols,
iv. Esters,
v. lactones,
vi. Hydrocarbons.

The unsaturated aldehydes and ketones are primarily responsible for undesirable oxidized (rancid) flavours.

Formation of flavour components are due to [2]:

i. Heat interaction between the native carbohydrates and protein system of cream;.
ii. Due to heat effect on the unfermented residue as well on fermented metabolic products formed by ripening process.

Cream constituents like lactose, citrate and glucose were responsible for the increase in ghee flavour components. Flavour in ghee is the resultant of following four different mechanisms [2]:

i. **Hydrolysis**- Free fatty acid formation.
ii. **Oxidation**-Saturated and unsaturated aldehyde, ketones, alcohols and hydrocarbons.
iii. **Decarboxylation** - Alkan-2-Ones.
iv. **Dehydration and Lactonization** - Lactones

Prevention of Hydrolytic and Oxidative Deterioration [1,6,7,7a,7b]

The following things/processes are to be carried out to prevent Hydrolytic and oxidative deterioration in ghee:

a. Use of Good Quality Raw Material [1,7,7a,7b]

i. Raw material used for the manufacture of ghee should be of good quality. Any off flavour, such as acidic, oxidized, and rancid present in raw material shall be carried over to the final product.
ii. The raw material should also be checked for the presence of copper and iron, which should not be more than permissible limits.

b. Method of Manufacture of Ghee [1,7,7a,7b]

i. Ghee prepared by desi method has higher moisture and higher acidity and thus lower keeping quality. If ghee is to be stored for longer time than this method should be avoided.
ii. The sulfhydryl and phospholipid contents have antioxidant properties in ghee. Those methods, which releases higher amounts of these natural antioxidant components should be adopted. Heating butterfat with higher amounts of solids-not-fat, as in case of direct cream method, at higher temperature of clarification will produce more sulfhydryl and thus better shelf life. Probably due to this reason the keeping quality of ghee is more than butter oil.
iii. The pre-stratification method produces ghee with higher amounts of phospholipids because its loss in ghee residue is minimum.
iv. Also the extraction of phospholipids from ghee residue and addition at 1 per cent to the ghee enhance its keeping quality.

c. Effect of Species of Mammals [7,7a,7b]

i. Cow ghee is apparently more shelf-stable than buffalo ghee due to the higher content of natural antioxidants the former product.
ii. Although buffalo ghee has been reported to be more resistant to lipolysis than cow ghee.

iii. Ghee prepared from cottonseed-fed animals showed that the fat had better keeping quality, presumably because of the antioxidant properties of gossypol, a phenolic substance in cotton seed observed.

iv. Ghee produced and packed in winter has longer shelf life than that packed in summer and rainy seasons.

d. Effect of Method of Preparation [7,7a,7b]

i. The keeping quality of ghee is affected by the method of manufacture. It is 9 months for DC method and 4 months for creamery butter method.

ii. Higher temperatures, or longer periods of heating at a particular temperature, have been shown to impart better oxidative stability because of greater liberation of phospholipids from phospholipid-protein complexes.

iii. It has been suggested that during heating, especially after most of the moisture has been evaporated, antioxidants are produced from phospholipids. The antioxidative properties of phospholipids in ghee have well established, and it has been shown that the presence of 0.1mg 100g phospholipids improves the keeping quality of ghee.

iv. Phospholipids may exhibit antioxidant activity by binding metals, regenerating other antioxidants and providing a synergism with phenolic antioxidant. The main fraction of phospholipids, which exerted antioxidant property, was found to be cephalin.

e. Addition of Antioxidants

The antioxidants are added universally to anhydrous butterfat and high fat food products. There are two sources of antioxidants, namely synthetic and natural antioxidant.

f. Packaging with Elimination of Oxygen [1,6,7,7a,7b]

i. Packaging under nitrogen - Nitrogen blanketing of any tanks where ghee is held is one method of preventing oxidation, but it is also important that pipe unions and pump seals that may allow air to be sucked into the process stream are suitable for the purpose. Small bubbles of nitrogen are streamed into the liquid ghee to scavenge any dissolved oxygen that may be present.

ii. packaging in vacuum.

iii. packaging with an oxygen scavenger.

iv. Tin cans are best to protect ghee against oxidative spoilage. The reason being that hot filling of ghee is possible in tin cans, which will exclude most of the oxygen from the product and also enable to replace oxygen with nitrogen gas. **The headspace in such containers can also be minimized.**

g. Decreasing the Rate of Oxidation [1,6,7]

i. Storage at low temperatures,

ii. storage in the dark.

Ghee should not be exposed to direct sunlight or irradiation. It should preferably be stored at about 22°C.

Role of Antioxidants [1,4,5,7,7a,7b]

Lipid oxidation is a chain reaction involving initiation, propagation and termination stages. The unsaturated fatty acids are oxidized to form odourless, tasteless hydro peroxides, which further degrade to yield carbonyls and other compounds, which give out the characteristic unagreeable oxidised flavour. The various factors which affect oxidation include, oxygen concentration, light intensity, contact with metals *etc.*,

- ☆ Anti-oxidants are substances used to inhibit the progress of lipid oxidation, and may be either
  - **i.** Natural (*e.g.*, Tocopherols, carotene).
  - **ii.** Synthetic substances (Butylated hydroxy anisole-BHA, Butylated hydroxy toluene-BHT, Hydroquinone, gallic acid esters *etc.*).
  - **iii.** Metal-chelating agents, such as citric acid and phosphoric acid, EDTA act in conjunction with anti-oxidants and inhibit fat oxidation.
- ☆ Oxidative rancidity is best controlled by the addition of antioxidants and 'oxygen scavengers'.
- ☆ There is a growing trend to utilize natural antioxidants such as mixed tocopherols, rosemary, and other unique ingredients.
- ☆ Often an 'antioxidant package' containing antioxidants and oxygen scavengers are used in synergy.
- ☆ Oxygen scavengers such as ascorbic and citric acid appear to increase the effectiveness of antioxidants.
- ☆ Keeping quality of Ghee can be extended by the addition of antioxidants. Under the PFA Rules (1976), BHA at not exceeding 0.02 per cent can be added. Under FSSR-2011,BHA and BHT are allowed. Naturally, betel leaves and curry leaves are added while ghee making at 1 per cent level, which gives good flavour to ghee as well as it acts as antioxidant.

Examples of Natural Sources of Antioxidants [1,7,7a,7b]

Some of the examples of natural antioxidants are as follows:

i. The seeds of soybean and safflower are rich source of phospholipids. Their addition to ghee and butter oil at 0.5 per cent level during boiling may delay the oxidative rancidity.

ii. Juices of Amla (Phyllanthus amblica) at level of 1.25 per cent in ghee can retard the fat oxidation possibly due to high content of ascorbic acid and gallate in amla.

iii. It has been found that addition of betel and curry leaves (at rate of 1 per cent of ghee) during heat clarification of butterfat improves not only the oxidative stability but also colour and flavour of ghee. The antioxidant properties of these plants are attributed to their phenolic compounds, predominately hydroxy charicol. The betel and curry leaves also contain carotene and ascorbic acid, which have tendency to undergo oxidation by consuming all free oxygen that may be present in the head space of the ghee container.

Defects in Ghee and their Prevention [2,3,4]

Possible defects in ghee with their causes and prevention, are listed in Table 17.7. Defects are categorized according to the sensory characteristics such as flavour related, texture related and appearance related.

Table 17.7: Different Defects in Ghee

1. Appearance Defects

Sl.No.	Defects	Causes	Prevention
1	Burnt colour	Excessively high temp. of clarification.	Optimum temp. of clarification of ghee.
2	Higher Sediment	Incorrect straining of ghee.	Correct straining of ghee.

2. Texture Defects

Sl.No.	Defects	Causes	Prevention
1	Greasy	Rapid cooling of hot ghee after clarification	Slow cooling of hot ghee after clarification
		Subjecting ghee to further heating and cooling treatment after preparation	Avoid further heating and cooling of ghee after preparation.

3. Flavour Defects

Sl.No.	Defects	Causes	Prevention
1	Smoky	Smoky fire used for boiling of milk/cream/butter.	Use non-smoky fire for boiling milk/cream/butter
2	Overcooked or burnt	Excessively high temperature of clarification.	Optimum temperature of clarification of ghee
3	Undercooked	Excessively low temperature of clarification.	Optimum temperature of clarification of ghee

Sl.No.	Defects	Causes	Prevention
4	Rancid	Fat hydrolysis due to lipase action in milk/cream/curd/ butter/ghee.	Inactivation of lipase by proper heat treatment.Clarify at optimum temperature which ensures minimum moisture content in ghee.
5	Oxidized/oily/ metallic	Fat oxidation due to direct contact of milk/curd/cream/ butter/ghee with copper or iron, exposure of these product to sunlight.	Storage of the milk/curd/ cream/butter/ghee in properly tinned or aluminium alloy/ ss vessels.Filling ghee upto the brim to minimize the headspace and use opaque containers. Avoid ghee storage at high temperature (>21°C) and avoid long storage.

Sources

1. *M.K. Srivastava, Handbook on Analysis of Milk Products. Write and Print Publications, New Delhi, 2016. ISBN: 9789384649258.*

1a. *Aneja, R.P., Mathur, B.N., Chandan, R.C. and Banerjee, A.K. 2002. Technology of Indian Milk Products. A Dairy India Publication, Delhi, India.*

2. *Y. Kotilinga Reddy, Yogesh Khetra and M.H. Sathish Kumar, Fat Rich Dairy Products, http: //www. agrimoon.com.*

3. *Quality and Grading of Ghee, Lesson 36, http: //ecoursesonline.iasri.res.in.*

4. *Ghee, LPT 311: Milk and Milk Products Technology (1+1).*

5. *Rancidity, Science digest, 2019, https: //www.science direct.com.*

6. *Soma Maji, Ghee, http: //courseware.cutm.ac.in.*

7. *Fat oxidation in ghee, Literature review about fat oxidation, 2016, https: //milkio. co.nz.*

7a. *Advanced dairy chemistry, P.F Fox, vol 2, lipids.*

7b. *Achaya, K. T. (1949). Rancidity in Indian butterfats (ghee), Biochemical Journal, 44(5), 561.*

Chapter 18

Traditional Dairy Products: Physico-chemical Changes in Milk Constituents during Manufacture and Storage

Traditional/Indigenous Products [1,1a]

India is emerging as the highest milk producing country in World. A large quantity of milk produced in the country amounting to over 54 per cent is being converted into various milk products; share of the traditional milk products comes to about 37 per cent.

- The market value of traditional products is considered to about to be 578 billion. The consumption of traditional milk products is likely to grow at an annual growth rate of more than 20 per cent.
- Traditional dairy products play a significant role in the economic, social, religious and nutritional wellbeing of the Indian masses.
- It is still mostly prepared, packed and or marketed by old traditional method without mechanization or well documented standardization of the process under unsanitary condition and have a poor shelf life.
- These products offer several technological benefits, *viz.* their mass appeal, lower cost of production, simple manufacturing technique, adopted at rural level/collection centre, high profit margin, generate lots of employment, value added products and nutritional status and have high export potential.

☆ The classification of these products based on the principle of their manufacture is given below:

Table 18.1: Classification of Traditional Dairy Products [1,1a]

Sl.No.	Type	Base Material	End Products
1	Heat and Acid coagulated	Chhana	Rosogolla, sandash, pantoa, chamcham, chhana mukri, chhanapodo, sitabhog
		Paneer	Culinary dishes/direct consumpton
2	Heat desiccated	Khoa	Peda, burfi, kalakand, milk cake, gulab jamun,
		—	Rabri
		—	Basundi
3	Fermented/cultured	—	Dahi, misti doi, lassi
		Chakka	Shrikhand, shrikhand wadi
4	Fat rich/Phase inversion	Ghee	Fat rich
5	Frozen	—	Kulfi/malai
6	Cereal based	—	Kheer, payasam, firni

Paneer [1,4]

Paneer is a heat-acid coagulated milk product obtained by coagulating standardized milk with the permitted acids at specified temperature. Paneer is an indigenous white (soft) cheese (unripened) which is used extensively as an ingredient in many vegetable delicacies.

☆ About 4 -5 per cent of the total milk produced in India is converted into paneer.Good quality paneer is obtained when standardised buffalo milk (Fat:SNF, 1:1.65) heated to 90°C and coagulating the milk with 1 per cent citric acid at 70°C and also characterized by a white colour, sweetish, mildly acidic, nutty flavour, spongy body and close knit texture.

Manufacturing Process

Milk (preferably buffalo milk) is heated to about 90°C and coagulated at 70°C using mostly citric acid as coagulant. Whey is then drained through cloth lined hoops and pressing the coagulated to form blocks. Blocks thus formed are floated in chilled water for firming their body. The cow and buffalo milk mixed in the ratio of 50:50 having a fat content of 5 per cent yields a superior product than cow milk alone.

Factors Influencing the Quality of Paneer [1,4]

i. Type and Quality of Milk

- ✰ Buffalo milk paneer possess desirable frying properties, body and texture as compared to cow milk. The cow milk paneer is soft, weak and fragile and during cooking it tends to disintegrate [4].
- ✰ Acidic milk having a titratable acidity of 0.2 to 0.23 per cent LA yields a product with inferior quality1,4]
- ✰ Homogenisation of cow milk is recommended to bring about improvement in the yield and to improve the organoleptic score of paneer [1].

ii. Heat Treatment of Milk [1,4]

- ✰ The objective of heating milk is to prepare it for rapid iso-electric precipitation, control the moisture content, develop typical body and texture, create conditions conducive to the destruction of pathogenic and other microflora present in milk and ensure safety as well as keeping quality of the final product.
- ✰ Heating milk to 80°C for 5 min and cooling to 70°C before coagulation affects the sensory and microbiological quality of paneer.

iii. Type of Coagulant [1,4]

Different coagulant like citric acid solution, one day old whey, 48 hr old sour whey, whey cultured with L acidophilus have been used to prepare paneer. Citric acid is generally used as a coagulant. Lemon or lime juice or vinegar imparts a typical flavour to the product. 1 per cent solution of citric acid yields good quality of paneer. In addition to this coagulant such as hydrochloric acid, phosphoric acid, tartaric acid and acetic acid have also been used in the manufacture of panner.

iv. Strength of Coagulant [1,4]

Best quality paneer can be obtained using 1 per cent citric acid. Two per cent citric acid also yielded good quality panner when cow milk is used. High acid concentration imparts acidic flavour, hardness and causes greater solids loss. Satisfactory quality products are obtained when nonconventional coagulants like hydrochloric acid and acidophilus whey (0.6 per cent) are used.

v. Amount of Coagulant [1,4]

For making good quality paneer from cow milk 2.34 g citric acid or lactic acid is required for 1kg of milk while for buffalo milk 2.1 g of citric acid or lactic acid per gram is required. Normally 1.8 to 2.0 kg citric acid is required for coagulating 1000lit of milk.

vi. Coagulation Temperature [1,4]

Coagulation temperature influences the moisture content of paneer. Coagulation temperature of 70°C has been suggested to be most effective. However good quality paneer can also be prepared at a coagulation of 80°C from both the milk.

vii. pH of Coagulation

The optimum pH of coagulation is 5.35 for buffalo milk and 5.0 for cow milk. The moisture retention in paneer decreases with the reduction in pH and consequently the yield also decreases. At pH more than 5.35 the paneer is very soft with fragile and crumbly body. Optimum pH for paneer preparation from cow milk is 5.2.

viii. Yield

The yield of paneer decreases with increase in the strength of coagulant, coagulation temperature and lowering the pH of coagulation in case of buffalo milk and it increases with increase in the coagulation temperature in cow milk paneer. The yield of paneer can also be increased by co-precipitation of casein and whey protein. Higher level of fat in paneer milk leads to lower yield of the product in case of buffalo and cow milk. Homogenisation of paneer milk improved the paneer yield.

Chemistry of Coagulation Process in Paneer [1,4]

The phenomena involves the formation of large structural aggregates of casein from the normal colloidal dispersion of discrete casein micelles, in which milk fat and coagulated serum proteins are entrapped together with whey. Acid and heat treatment(Coagulation) causes the following physical and chemical changes in casein-

a. Acidification

i. The progressive removal of tri calcium phosphate from the surface of casein and its conversion into mono calcium phosphate and soluble calcium salt.

ii. Progressive removal of calcium from calcium hydrogen caseinate to form soluble calcium salt and free casein.

iii. When the pH of milk system drops, the colloidal particles become isoelectric, *i.e.* the net electric charge becomes zero. Under such circumstances the dispersion is no longer stable; the casein gets precipitated and forms a coagulum.

b. Heating

Heating causes the interaction of β-lactoglobulin with 8-casein and complex formed between β-lactoglobulin and α-lactalbumin interact with 8-casein. The development of typical rheological characteristics of paneer could be due to preponderant and intensive heat induced protein- protein interaction. It is believed that β lactoglobulin and K- casein interact by sulphydryl disulphide inter change when heated together. It is initiated at 65°C, increases up to 83 per cent at 85°C and then decreases to 76 per cent at 99°C.

Khoa [1,4]

Khoa is a popular product throughout India and is called by different names in different regions like khoya, mawa, kava, palghova, *etc.* Khoa is a heat desiccated milk product.

- ☆ About 7-8 per cent of the total milk produced in India is converted into khoa.
- ☆ It is prepared traditionally by desiccating milk in an open shallow pan with continuous stirring till a coagulation mass is obtained which is worked up to form a pat.
- ☆ It is used as a base material for a large number of milk sweets such as burfi, peda, milk cake, kalakand, gulabjamun, *etc.* Three distinct types of khoa, namely, pindi, dhap and danadar are available in the country. These varieties differ in quality, texture, composition and specific usage.
- ☆ Khoa generally contains 17-20 per cent protein, 22-25 per cent lactose and 100-130 ppm of iron depending on whether it is prepared from cow, buffalo or mixed milk.

Different Type of Khoa [1,4]

The following three types of khoa are manufactured and available in the market:

i. **Pindi-** It is characterized as a circular ball of hemispherical pat with smooth and homogeneous body and texture. It shall not show any sign of fat leakage or presence of free water. This poses characteristic cooked flavour and is preferred for the preparation of burfi and peda. It contains 19 per cent moisture; 37 per cent fat; 18 per cent protein; 22 per cent lactose and 3.6 per cent ash.

ii. **Danedar-**This is characterized by the granular texture with hard grains of different sizes and shapes embedded in viscous serum. Usually 0.1 per cent of citric acid (on product basis) is added during the heat desiccation process in shallow pan to obtain such texture. It has 20-25 per cent fat and 35-40 per cent moisture. This forms base for the preparation of kalakand (granular texture), milk cake and granular milk solids.

iii. **Dhap-**It is characterized by loose and sticky body and smooth texture, attributable to higher moisture content and less intensive heat treatment. This khoa carries highest percentage of moisture over other varieties of khoa. This high moisture is necessary to provide adequate free water for soaking of maida (refined wheat flour) and semolina (suji) and for homogenous distribution of other ingredients in the preparation of smooth gulabjamun balls. It contains about 20- 23 per cent fat and 37-44 per cent moisture. It is preferred for the preparation of gulabjamun. This variety of khoa is used in the manufacture of gulabjamun, kalajamun, pantooa, carrot halwa, *etc.*

Chemistry of Khoa [1,4]

Khoa is essentially a matrix of heat denatured milk protein, in which fat, lactose, minor constituents and moisture are entrapped. Protein-protein reaction and reaction with other milk constituents occur during desiccation.

☆ Lactose and protein that undergo Maillard browning and caramelization are significant in imparting typical flavour and desirable sensory characteristics to the khoa.

Denaturation of Proteins

The main reaction in the preparation of khoa is the heat denaturation and coagulation of milk protein particularly the serum proteins and their interaction with casein and other constituents.

i. Most of the albumin and globulins are almost denatured; the protective properties of the other colloids are destroyed early in the boiling process.

ii. Casein is also irreversibly denatured from colloidal state to non-dispersible state.

iii. The denaturation is accelerated by the incorporation of air and frothing during stirring. Total heat coagulation of protein occurs when the boiling mixture thickens to a buttery consistency in the pan.

iv. Due to the combined action of heat and concentration coagulation of casein tends to increase logarithmically with milk solids concentration and forms a complex with denatured whey protein.

Subdivision of Fat

Fat globules in milk, undergo subdivision and disrupts the fat globule membrane as a result of heat, vigorous agitation and scraping during the manufacture of khoa.

i. Almost half of the globular fat is released as free fat, the extents of which depend upon the type and amount of fat content in the milk. Formation of free fat during transition from liquid phase to the semi-solid phase is very significant for the development of typical flavour, body and texture of khoa.

ii. The short chain fatty acids decreases, medium chain fatty acids increases and long chain fatty acids are unaffected during the preparation of khoa.

Lactose Crystals

i. From a dilute solution in milk, lactose is present in khoa as a super saturated solution. Most of the lactose is present as α-hydrate in khoa.

ii. The lactose is present in khoa mostly as supersaturated solution. Generally lactose does not crystallize out due to high viscosity and coagulations of proteins. However, large crystals of lactose are found in ordinary khoa.

Mineral Precipitation

A portion of milk salt is precipitated during preparation of khoa. Iron content, probably derived from heating pan, is raised considerably. From 2 to 4 ppm iron content in milk, the iron content of khoa increases more than 100 ppm due to the incorporation of additional quantities of iron from the karahi and the khunti by vigorous scraping.

Vitamins

Vitamin A, thiamine, riboflavin, ascorbic acid, nicotinic acid and folic acid are decreasing in the course of heating of milk for making khoa.

The Deterioration of Khoa

It is attributed to unhygienic methods of processing and subsequent improper handling, followed by lactose breakdown, proteolysis and lipolysis due to activities of the mixed flora during storage. Being aerobic in nature, moulds cover the entire surface of the product kept under ambient conditions producing an abnormal colour, flavour and appearance.

Physico-Chemical Changes in Milk during Khoa Making [1,4]

The changes in physico-chemical characteristics of milk during khoa making take place due to following three actions:

1. Concentration,
2. Heating and
3. Stirring and scraping.

Changes occurred during manufacture of Khoa are as follows:

a. Change in State- The removal of moisture from milk results into concentration of milk solids. This eventually changes the state of milk from liquid to solid/semisolid. All the constituents including lactic acidity increase in proportion to the degree of concentration. pH decreases.

b. **Change in intensity of colour-** The colour of khoa becomes intense with brownish tinge due to formation of melanoidins pigment. The browning reaction is Maillard type reaction due to interaction between aldose group of lactose and free amino group of casein.

c. **Change in milk fat-** As discussed above.

d. **Heat coagulation of milk protein**- As discussed above.

e. **Change in lactose-** As discussed above.

f. **Precipitation of milk salts-** As discussed above.

g. **Increase in iron content** - As discussed above.

h. **Changes in vitamins-** As discussed above.

Changes in Khoa during Storage [1,4]

The following chemical and microbiological changes occur during storage of khoa:

Chemical Changes

i. **Changes in organoleptic quality-**The changes in organoleptic quality of khoa during storage include development of off flavour (stale, sour, oxidized, *etc.*) hardening of body and texture and increase in intensity of colour. When stored at low temperature (5 - 10°C) for long time staleness developed.

ii. **Moisture-** Loss of moisture takes place resulting into dry and hard surface of khoa. The extent of evaporation however depends on the temperature of storage and packaging material. The higher the storage temperature the higher is the rate of evaporation. More the water vapour transmission rate (WVTR) of package faster the loss of water from khoa. Minimum loss of moisture takes place in case of tins and maximum loss in parchment paper.

iii. **Deterioration in fat-** Khoa has a low keeping quality at room temperature and develops rancid flavour. This may be due to fat hydrolysis which is due to lipase action. The peroxide value increases due to oxidative spoilage during storage of khoa, particularly at ambient temperature. The hydrolytic rancidity in khoa take place thereby increasing the free fatty acids measured as per cent oleic acid.

iv. **Breakdown in milk protein**- There is break down of milk protein in khoa as recorded by the increase in tyrosine value.

v. **Acidity-**Lactic acidity increases with the storage period. Higher percentage increase take place at higher temperature and vice-versa. The pH normally decreases during storage corresponding to the increase in acidity.

Microbiological Changes

- ☆ There is an increase of various types of microorganism during storage at room temperature. The shelf life of khoa is about 2-4 days under ambient condition and 3 weeks under refrigerated conditions.
- ☆ Mouldy surface developed during storage are due to high moisture content and excessively high humidity of storage room.
- ☆ Khoa manufactured by continuous methods in practically sterile when comes out of the machine. On the contrary khoa made by traditional methods contains comparatively high microbial load.

Factors affecting Composition of Khoa [1a,4]

The large variations in the market samples are attributed to the following factors-

a. Type and Quality of Milk Used [1a,4]

The quality of the products greatly depends on the type of milk used for their preparation.

i. Species of Milch Animals

The composition of cow and buffalo milks is different in respect of total solids, fat, proteins and lactose. Therefore, the use of these milks as such or their combination (without standardization) produces khoa of different composition.

- ☆ Buffalo milk is preferred over cow milk for khoa making because it yields a product with soft, loose body and smooth granular texture which is highly suitable for the preparation of sweets.
- ☆ The buffalo milk gives a greater out burn than that from cow milk.
- ☆ The average yield from buffalo milk is 21.6-23 per cent and from cow milk it is 18.3-18.5 per cent due to the higher T.S. Content.
- ☆ Khoa from cow milk is of inferior quality due to its dry surface, sticky and sandy texture and salty taste.

ii. Fat per cent in Milk

A minimum fat level of 4 per cent in cow milk and 5.5 per cent in buffalo milk is essential for production of khoa with desirable body and texture and to meet the FFA requirements also.

- ☆ A lower fat content results in undesirable hard body and coarse texture in the finished product which is not suitable for good quality sweets preparation.
- ☆ The fat level higher than the minimum improves the quality of the final product.

Buffalo milk khoa normally contains higher moisture and higher fat levels, but lower contents of protein, lactose and ash than cow milk khoa.

iii. Acidity of Milk [4]

Fresh sweet milk yields the best quality khoa, while developed acidity in milk produces an undesirable, coarse texture sour smell and bitter taste in to the khoa which is unsuitable for sweet preparation.

iv. Additives and Adulterants [1a,4]

- ✰ Adulteration of milk with water or removal of fat from milk also affects the composition of khoa. Adulteration of milk with water produces a brown khoa and a proportionate reduction in the yield.
- ✰ Neutralization of high acidic milk improves the texture, but does not improve the flavour of khoa. Further it gives saltish taste in the final product.
- ✰ Texture and flavour of khoa are not affected much when adulterated with starch the physical quality of khoa is considerably affected. It develops a harder body and a pasty sticky texture not suitable for the preparation of confections.

v. Presence of Colostrum

Colostrum in milk has a marked effect on the colour of khoa, which is deep yellow with cow colostrums. It also gives pasty texture in the final product making it not suitable for sweet making.

b. Method of Preparation [1a,4]

- ✰ In the conventional methods, adopted by halwais and unskilled workers, the desiccation of milk is done just by their past experience, which is bound to vary from lot to lot. Removal of more moisture during desiccation leads to decrease in moisture and increase in all other solids in khoa where as the removal of less moisture has opposite effect on the composition of khoa.
- ✰ The use of mechanized and continuous method produces khoa of more consistent quality.
- ✰ **Speed of stirring-**It should be optimum (Traditional method- 100RPM, Continuous systems - slightly more). The optimum speed of stirring prevents burning of milk solids and helps in developing desirable body and texture in khoa. Low speed results in to the burning of khoa. Higher speed makes the product pasty and sticky.

c. Degree of Heat Treatment [1a,4]

- ✰ Higher degree of heat treatment, results into very fast desiccation of milk solids particularly at the final stage. This may reduce the moisture content to a greater extent and increase all other milk solids in khoa.

- ✰ To obtain good quality khoa, milk should be maintained at the boiling temperature till it reaches a paste consistency and then temperature is lowered to 88°C till the pat formation stage.
- ✰ Continued heating at higher temperature at advanced stage of khoa making results into undesirable flavour (cooked) and texture (hands dry). The colour of such product is also brown. Slow heating is not only more time consuming but also produces sandy texture and brown colour.

d. Manner of Handling and Packaging [1a]

Leaving khoa on the hot surface of heating kettle for a longer time in the conventional method of preparation or in the open trays in continuous khoa making process, causes more evaporation of moisture thereby changing its final composition. Improper protection/package of khoa also leads to surface drying and decrease in the moisture content.

e. Storage Conditions [1a]

Storage of khoa in dry conditions or humid conditions, particularly when improperly protected, also results into compositional changes. The storage of khoa for longer time will have more drastic effect in these conditions.

f. Homogenization of Milk [4]

Homogenization of cow/buffalo milk produces a softer body in khoa, as against an homogenized milk. The khoa from homogenized milk also shows lower fat leakage, less browning, and a reduced patting tendency as compared to that obtained from unhomogenized milk.

Channa [1,4]

Chhana is a heat-acid coagulated milk product obtained by acid coagulation of boiled hot whole milk and subsequent drainage of whey. It is creamish-white in appearance with compact spongy body, sweetish-acidic flavour.

Chhana is mainly prepared from cow milk and used for preparation of varieties of Bengali sweets. Chhana is used as a base for the preparation of a variety of sweets like sandesh, rasogolla, chamcham, rasomalai, pantoa, chhana murki, *etc.*

Chemistry of Chhana Processing [1,2]

- ✰ The process of coagulation of milk in the preparation of chhana due to combined chemical and physical changes in casein brought about by the action of acid and high temperature of coagulation.
- ✰ This involves the formation of large structural aggregates of casein or curd from the normal colloidal dispersion of discrete casein micelles

in which fat and coagulated serum protein are entrained together with whey.

- ☆ Then the curd separates from the milk serum as a flocculent.
- ☆ A part of calcium and phosphate losses from the casein micelles during boiling and subsequent coagulation.

Factors Influencing Quality of Chhana [1,4]

Various factors influencing the quality of chhana-

a. Type of milk,
b. Fat content in milk,
c. Homogenization,
d. Calcium level,
e. Additives,
f. Type and amount of coagulant,
g. Temperature of coagulation,
h. pH of milk and
i. Speed of stirring.

a. Type of Milk [1]

Cow milk is preferred for chhana preparation since it yields moist surface, light yellow colour, a soft body, smooth texture, small grain and mildly acidic flavour which is more suitable for sweet preparation.

- ☆ The chhana from buffalo milk has a slight hard body, a greasy and coarse texture. Buffalo milk contain higher concentration of casein, mainly in the miceller form with bigger size of the micelles, higher proportion of high melting triglycerides and higher content of calcium may be responsible of hard body and coarse texture in buffalo milk chhana.
- ☆ A mixture of 75 per cent buffalo milk and 25 per cent cow milk produces suitable chhana, which offers good rosogolla.

b. Quality of Milk

i. **Fat content in milk [1,4]-** Fat level in milk affect the final quality of chhana. A minimum fat level of 3.2-4 per cent in cow milk and 5 per cent in buffalo milk is essential for satisfactory quality of chhana. Higher fat level would give rise to a softer chhana (with greasy surface) with higher fat loss in the whey. Lower than 3.5 per cent fat leads to hard body and coarse texture.

ii. **Acidity [1]-** Acidic milk produces chhana with sour smell and bitter taste hence unfit for sweet making. Addition of neutralizer to slightly acidic milk helps in getting chhana suitable for sandesh and not for rasogolla. Milk with 0.25-0.28 per cent LA can be used by adding 0.2 per cent Sodium Citrate followed by thorough washing of the coagulum.

c. Homogenization [3]

Homogenization of buffalo milk achieved only limited success in the preparation of good quality chhana and rasogolla.

d. Mineral Level of Milk [1]

Higher calcium content gave hardness in chhana. Addition of calcium or magnesium chloride produces a hard chhana whereas sodium chloride, sodium citrate and sodium acetate have negligible effect on the body and texture of chhana.

e. Presence of Colostrum [1]

The presence of colostrum in milk tends to produce a pasty texture in the coagulation mass. The adulteration of milk with starch results in a gelatinous mass on coagulation.

f. Type and Amount of Coagulants [1,2,4]

Coagulants generally added are organic acids like citric acid, lactic acid or their salts, lemon juice or sour whey.

- ✰ Lactic acid (0.5-0.75 per cent) tends to produce a granular texture chhana suitable for rosogolla making.
- ✰ Citric acid (1-2 per cent) produces doughy/pasty product suitable for sandesh making.
- ✰ Calcium lactate (4 per cent) produces soft body, smooth texture, and pleasant flavour chhana that is suitable for the preparation of sandesh.
- ✰ Generally low acid strength coagulant results a soft body and smooth texture required for rosogolla preparation and high acid strength results a comparatively hard body and less smooth texture required for sandesh preparation.
- ✰ **Quantity [4]**-The quantity of coagulant required is dependent on the type of milk. Generally, 2 to 2.5 g of citric acid per kg of fresh milk and 2.5 to 3.9 gm of lactic acid and 6 to 12 gm of calcium lactate per kg are required individually for complete coagulation.

g. Coagulation Temperature [1,2,4]

Temperature of coagulation exerts a significant influence on physico-chemical attributes. A temperature range of 80-85°C and pH 5.4 for cow milk and 70- 80°C and pH 5.7 for buffalo milk will give better yield and quality.

- ✰ Higher coagulation temperature results hard, granular and graininess texture chhana and lead to lower moisture content but higher percentage of milk solids comprising fat, protein, lactose and ash.
- ✰ The lower temperature imparts in a sticky chhana (with more moisture content in chhana)with slow drainage problem.

h. pH of Milk

pH 5.4 for cow milk and pH 5.7 for buffalo milk will give better yield and quality.

i. Stirring/Agitation

Generally slow stirring is preferred for avoiding foam formation and reduce the shattering of the coagulation mass, whereas fast stirring reduce the moisture content and increased hardness in chhana. Optimum speed of stirring is 40-50 rpm.

j. Method of Straining

- ✰ Again, delayed straining produced comparatively a soft body and smooth texture with higher yield chhana than immediate straining. Delayed straining is recommended for buffalo milk.
- ✰ Delayed drainage of whey following coagulation results in a softer body and smoother and sticky and less chewy texture and is more acceptable from overall quality.

k. Proteolysis

Proteolysis of buffalo milk with enzyme trypsin coagulation produced a chhana with very soft and smooth body having minutes sized micelles and suitable for rosogolla preparation. Addition of 0.3 per cent sodium citrate to buffalo milk produces a chhana similar to cow milk chhana.

l. Changes during Storage

- ✰ Post manufacture chemical reactions occurring during storage of chhana considerably alter its physico-chemical attributes and render it unfit for consumption.
- ✰ Shelf life of the product packaged in vegetable parchment paper was 2, 3 and 12 days at about 37°C, 24°C and 7°C, respectively.
- ✰ At low temperatures, there is a heavy growth of moulds on the surface together with the development of a stale flavour; on the other hand at high temperatures the stored product develops a sour smell and bitter taste together with moulds growth on the surface of chhana.

Fermented Milk [1,1b,4,5]

It is a milk product obtained by fermentation of milk, which may have been manufactured using other permitted raw material, by the action of suitable microorganisms and resulting in lowering of pH with or without coagulation (iso-electric precipitation). Fermented milk may be heat treated after fermentation.

Dahi [1]

The dahi may be classified into various types depending upon the type of milk used such as whole milk dahi, skim milk dahi and on basis of acidity and addition of sugar it can be classified as: sweet dahi (acidity < 0.7 per cent), sour dahi(acidity > 0.7 per cent) and sweetened dahi. The composition of dahi depends upon the type of milk and the manufacturing condition.

Type of Dahi Preparation1,1b]

The preparation of Dahi can be done in following two ways-

i. **Household preparation** -During preparation of dahi at home, cow or buffalo or mixed milk is boiled or simmered for a long time to concentrate it, allow cooling to room temperature and then transferred to earthenware pots or pan. A small amount of the previous day's curd is added as a starter culture and then allowed to set undisturbed overnight. The dahi is usually ready within 16 to 20 hours.

ii. **Commercial preparation**- Fresh, sweet, good quality milk (cow or buffalo or mixed) is received. It is then standardized to have 4-5 per cent fat and 10-12 per cent solids not fat, preheated to 60°C and homogenized at a pressure of around 176 kg/sq. cm. The milk is preheated to 80°C to 90°C for 15 to 30 minutes, cooled to 40 – 45°C and inoculated with 1 to 2 per cent of specific starter culture. It is then filled in a suitable containers using automatic filling and sealing machine and incubated at 40 -420C till curdling (3-4 hour). When a firm curd is formed, dahi cups are stored in a cold room temperature.

Biochemical Changes [1]

The following biochemical changes are done during dahi preparation:

i. **Break Down of Lactose:** It involves the following steps:

   a. **Hydrolysis of lactose** to glucose and galactose by starter organism is the first step in the process of fermentation during the preparation of dahi.

   b. The liberated glucose is then enters the classical Embden Meyer Hoff pathway (EMP). But the galactose produced cannot enter the EMP pathway directly but follows a longer route and ultimately enter the EMP pathway via glucose-1-phosphate.

c. Each of the simpler sugar is further acted upon by several enzymes successively before arriving at the lactic acid stage. There may be several form of acid end product.

d. The amount of lactic acid produced may range from 75 – 95 per cent of the total acidity, the rest being volatile acid.

ii. **Role of lactic acid-** The lactic acid cause at first an acid flavour combine with calcium of casein to form calcium lactate, then setting free the casein and coagulated it when isoelectric point (4.6) is reached. The fermentation is accompanied by gelling of the solids, principally the proteins and syneresis manifested by appearance of thin exudates of clear whey on the surface of dahi.

iii. **Nitrogen content**- Total nitrogen content of milk remains more or less unchanged during fermentation. Appreciable changes occur in non-protein nitrogen, albumin nitrogen, ammonia and dialyzable nitrogen during fermentation. The increase in the non- protein nitrogen during fermentation is attributed to the breakdown of protein.

iv. **Salt content**- Considerable changes occur in the mineral contents of milk. The total mineral content is not affected. The soluble calcium and phosphorus are increased during souring. Citric acid is completely disappears during fermentation.

v. **Vitamin content**- Vitamin content of dahi depends on the type of organisms used for fermentation.

- ✰ Fermentation of milk with streptococcus lactic and streptococcus cremoris results in increase in thiamine, riboflavin and folic acid and decrease in vitamin A.
- ✰ The incorporation of propionibacterium shermani in lactic culture would enrich in with riboflavin and folic acid and cynocobalamine (vitamin12).
- ✰ Fermentation of milk with streptococcus cremoris, streptococcus diacetylactis, streptococcus thermopilus and lactobacillus thermophilus results in increase in riboflavin and folic acid and decrease in nicotinic acid.

vi. During fermentation, the electrical charge on the fat particles is neutralized, causing the globule to coalesce and rise to top. The iso-electric point (pH 4.5 -4.6) of the fat globule in milk is slightly higher than that of washed globule pH 4.3) from cream. With respect to acidity, a pure culture is able to neutralize the charge on the fat globule at a lower level (1 per cent) of acidity than a mixed starter (1.3 -1.5 per cent).

Lassi [1b,4,4a,5]

Lassi, similar to butter milk, is a refreshing summer beverage popular in north India. Lassi is a white to creamy white viscous liquid with a sweetish, rich aroma and mild to high acidic taste. It is flavoured either with salt or sugar and other condiments, depending on regional preferences.

Product Description

Lassi can be described as a fermented milk beverage obtained after the growth of selected lactic acid bacteria in heat treated milk followed by sweetening with sugar. It is consumed as a cold refreshing beverage in summer. It is prepared by breaking the curd in to fine particles by agitation, addition of sugar, water and optionally flavour.

Table 18.2: Chemical Composition of Lassi [4,4a]

Sl.No.	*Components*	*Per cent in Lassi*
1	Milk Fat	1.5-3.8
2	Milk SNF	9.00
3	Sugar	13-15
4	Low Methoxy Pectin	0.5
5	Acidity	0.7 LA Min.

Production

Lassi is becoming popular and attracting demand throughout the year. To meet the consumer demand many dairies have started producing lassi on commercial scale.

- ☆ Similar to dahi, fresh, good quality milk is essential for production of good quality lassi. Curd is prepared from standardised milk(As discussed under dahi) and then the curd is broken with the help of a power driven agitator.
- ☆ Sugar syrup (25 per cent syrup) is added to the mix to give 12 per cent sugar concentration in the blend.
- ☆ Low methoxy pectin after making solution in water/syrup can also be added @ 0.5 per cent at this stage as a stabilizer to improve the appearance and mouth feel.
- ☆ The mixture can be flavoured with rose water and homogenized to improve body and texture. It is packed and stored at refrigeration temperature.

Shelf Life

Lassi packed in LLDPE pouch can be stored upto 7days at 5°C without any significant change in its sensory qualities. Lassi subjected to UHT treatment and packed aseptically has shelf life upto 120 days.

Chakka [1b,4,5]

Chakka is a fermented, intermediate dairy product obtained during the production of Shrikhand. Chakka can be described as strained dahi, in other words it is the curd mass obtained after removing whey from dahi, either through muslin cloth or basket centrifuge. Chakka is the base material for the production of shrikhand and shrikhand wadi.

Production of Chakka [1b,4,5]

It is prepared by traditional method and Industrial method.

Traditional Method

In this method, cow or buffalo or mixed milk is boiled thoroughly and cooled to room temperature (30°C). Previous day curd is added to this milk at the rate of 1 to 1.5 per cent. Milk is left undisturbed overnight at room temperature to set firmly. It is then stirred and hung in a muslin cloth for 10 to 12 hrs to drain off whey. The curd mass obtained after removal of whey is called as chakka.

Industrial Method

Skim milk is normally used in the commercial production of chakka. Low fat in the curd resulted with significant reduction in fat loss through whey, faster moisture expulsion and lower moisture retention in the final product.

Fresh, good quality skim milk is heated to 90°C for 10 min to kills the competitive microbes. Heated milk is cooled to 30°C and inoculated with LF-40 culture (*Lactococcus lactis* subsp. *lactis* and *Lactococcus lactis* var. *diacetylactis*) at the rate of 1.0- 1.5 per cent. This milk is incubated at 30°C for 10- 12hr. The curd (LA-0.8-0.9 per cent) is taken into basket centrifuge or quarg separator to remove whey from the curd. Use of quarg seperator for removal of whey has increased the chakka production to 8 tonnes/day.

- ☆ The yield of chakka produced traditionally is about 650g per 1000g of milk and yield of shrikhand is about 1.5 to 2.0kg per kg of chakka. Yield of Chakka is about 20 per cent more and shrikhand is 38.5 per cent more than that of traditional method, when produced commercially.

Factors effecting the fat losses and quality of Chakka [4]-

a. Quality of Chakka

i. **Body and Texture**- The chakka obtained from whole milk/standardized milk has smooth body, whereas the one obtained from skim milk is little rough and dry.

ii. **Moisture content-** Homogenization of milk leads to slow drainage of whey giving higher moisture content in chakka and a product with very soft consistency (not liked by the consumers).

iii. **Variation in moisture and acidity**-Conventionally made chakka varies from batch to batch with regard to moisture and acidity.

- Moisture content affects the yield, consistency and composition.
- Acidity affects the taste and quantity of sugar to be added.

b. Fat Losses

When whole milk is used for chakka making, high fat loss occurs in whey thereby affecting the recovery of fat in chakka. Therefore, it is preferred to use skim milk for chakka making and then mixing of cream or unsalted butter to adjust the fat in the finished product.

Shrikhand [1b,4,5]

Shrikhand is a popular fermented, sweetened, indigenous dairy product having semi solid consistency with typical sweetish-sour taste. It is very popular in the state of Gujarat, Maharashtra and part of Karnataka. It is prepared by mixing chakka with sugar, color, flavor, spices and other ingredients like fruit pulp, nuts *etc.* to form soft homogenous mass.

Production of Shrikhand [4,5]

It is prepared by traditional method and Industrial method. Use of skim milk has got many advantages as listed below:

- Fat losses are eliminated
- Faster moisture expulsion
- Less moisture retention.

Traditional Method

Prepare the chakka first as discussed above and Chakka is then added with calculated amount of sugar, colour, flavour and other optional ingredients like fruits, nuts, spices, herbs and served chilled.

Industrial Method

Shrikhand is the first traditional milk product for which large scale production technology was adopted. The first modern plant has been established at the Baroda District Cooperative Milk Producers Union ltd. Baroda Dairy has adopted a process.Prepare the curd mass or chakka as discussed under industrial preparation of chakka. The curd mass or chakka is taken into planetary mixer or scraped surface heat exchanger. Sugar at the rate of 80 per cent w/w, calculated amount of plastic cream (80 per cent fat) to give at least 8.5 per cent FDM in the finished product are added and mixed thoroughly. Optimal ingredients like colour, flavour, fruits, nuts *etc.* can also be added at this stage. Then it is packed at room temperature and stored at refrigeration temperature.

Physico chemical changes during Chaka and Shrikhand Manufacturing [6]

The preparation of shrikhand involves preparation of dahi,drainage of whey to form chakka and then conversion of chakka in to shrikhand. The following physico chemical changes occur during manufacturing of chakka and Shrikhand-

- ☆ **Dahi**: The progressive fermentation of milk into dahi brought about little or no change in the total solids, fat, proteins and milk salts, but significant decreases in the lactose content and pH value.
- ☆ **Chakka**: The conversion of dahi into chakka concentrated milk solids, separated out whey and advanced the fermentation of lactose.
- ☆ **Shrikhand**
    - i. Shrikhand prepared either from buffalo or cow's milk retained quantitatively all the constituents in proportion to the addition of sugar.
    - **ii. pH**-Little or no change in pH was seen during the conversion of chakka into shrikhand.
    - **iii. Minerals**-
        - **a.** The mineral make-up remained more or less unchanged, except citrate which disappeared completely at dahi stage of both buffalo and cow's milk.
        - **b.** Despite greater part of the solubilized minerals drained along with whey, considerable amounts of all the minerals remained aggregated into chakka from both milk systems.
        - **c.** All the minerals were quantitatively transferred from chakka into shrikhand for both buffalo and cow's milk. However, the minerals got diluted in proportion to the addition of sugar.
        - **d.** The fermentation of milk brought about highly significant changes in mineral contents during transition from colloidal to soluble phase, which was faster and greater for buffalo milk.

Sources

1. *Physico- chemical changes in milk constituents during manufacture and storage of Indigenous products, courseware.cutm.ac.in.*

1a. *BPVI-015, Dairy Products-II, IGNOU.*

1b. *Aneja, R.P., Mathur, B.N., Chandan, R.C. and Banerjee, A.K. 2002. Technology of Indian Milk Products. A Dairy India Publication, Delhi, India.*

2. *Jitendra Kumar, V. K. Gupta, Sanjeev Kumar and Suryamani Kumar, Effect of coagulants on the quality of chhana and rosogolla obtained from admixture of buffalo milk and butter milk, J Food Sci Technol 2015, 52(3): 1736-1741.*

3. *Soni K, Bandyopadhyay AK, Ganguli NC. Manufacture of rasogolla from buffalo milk. Indian J Dairy Sci. 1980, 33, 357-365.*

4. *M. Ranganadham, Sathish Kumar M. H., Devraja H. C. and F. C. Garg. Traditional Dairy Products, www.AgriMoon.com.*

4a. *Lassi and chhachh/mattha (country buttermilk) - methods of manufacture, packaging and shelf life, defects, Lesson 27, ecoursesonline.iasri.res.in.*

5. *M.K. Srivastava, Handbook on Analysis of Milk Products. Write and Print Publications, New Delhi, 2016. ISBN: 9789384649258.*

6. *Boghra Vitthal and O.N.Mathur, Physico-chemical Status of Major Milk Constituents and Minerals at Various Stages of Shrikhand Preparation, Journal of Food Science and Technology, 2000, 37(2): 111-115.*

Chapter 19

Chemistry of Cheese

Cheese is the generic name for a group of fermented milk-based food products, produced in a wide range off-Flavours and forms throughout the world. Although the primary objective of cheesemaking is to conserve the principal constituents of milk.

- ☆ The details related to the composition of different cheeses, nutritional values, quality standards *etc.* are given in chapter-16.
- ☆ Cheese manufacture and ripening involves the chemistry and biochemistry of milk constituents, fractionation and characterization of cheese constituents, microbiology, enzymology, molecular genetics, flavour chemistry, rheology and chemical engineering.

Chemistry of Cheese [1]

Milk is primarily water. Suspended in the water are solids; mainly proteins called casein and whey. To make cheese, separation of these two proteins is to be done. If the milk is not heated (unpasteurised), bacteria naturally present will start to turn it sour. Usually the milk is pasteurised, where it's heated to remove unwanted bacteria. Bacteria which gives cheese distinctive flavours such as Streptococci and Lactobacilli are then added, along with the enzyme rennet and the mixture is processed at 30-40 °C for a couple of hours. During this time, the natural sugar in the milk called lactose is fermented into lactic acid, reducing the pH and the rennet helps to coagulate the casein protein, forming curds. Curds have the consistency of a rubbery gel and they're allowed to set for an hour or two before the liquid whey is separated by cutting the curds into small pieces and draining the whey away.

- These drained cheese (Cottage cheese) curds have a soft consistency, with some of the whey still present. Cottage cheese tastes slightly sharp due to the pH change but is fairly tasteless as it hasn't had the time for the volatile flavours to develop.
- Some cheeses have additional bacteria injected into them or brushed onto their surface to mature into distinctive tastes.
- Others are warmed to release more whey, making them firmer and less rubbery. Softer cheeses ripen faster by the action of bacteria as they contain more water. Hard cheeses such as parmesan can last for weeks in fridge – it contains less water which bacteria find difficult to survive in.
- The longer a cheese is left before eating it, the more lactose is converted into lactic acid so older cheeses tend to have a sharper taste.
- In cheddar, the curds are mixed with salt before the whey is drained and this makes it a difficult breeding ground for bacteria. Extra-mature cheddar is much stronger in flavour than mild cheddar and needs to be matured for over a year. During this time, the proteins are chopped up into amino acids, which break down further to produce all the different flavour chemicals. Cheddar contains up to 500 compounds at detectable thresholds *e.g.* furaneol and homofuraneol (caramel), δ-dodecalactone (peach/coconut), butanedione (buttery), ethanoic acid (sharp, vinegar), (Z)-4-heptenal (creamy, biscuit) and methional (boiled potato).
- Gorgonzola and other blue cheeses get their distinctive taste from a class of chemicals called methyl ketones. In gorgonzola, 2-heptanone and 2-nonanone impart 'blue cheese' notes. This cheese can also have taste like 1-octen-3-ol (mushroom), 2-heptanol (fresh, lemongrass), ethyl hexanoate (apple), ethyl butanoate (pineapple), 2-nonanol (waxy) and 4-methylanisole (mothballs). All the flavour molecules found in cheese derive from just three starting materials -lipids in the milk fat, lactose (a sugar) and casein (a protein).

Milk Clotting Enzymes [1a,1c]

Milk clotting enzymes are important and essential for cheese making. A good milk-clotting enzyme is characterized by a high specific caseinolytic activity and a low general proteolytic activity, since the proteolysis strongly affects the sensory properties of cheese.

- Milk-clotting enzymes, obtained from animal, plant, and microbial sources have been used since antiquity for the manufacture of cheese. Until recently the calf gastric enzyme chymosin (rennin) in the form of a crude extract, paste or powder was used almost exclusively in commercial cheesemaking.

- ☆ Now as there has developed a substantial shortage of calf stomaches (veIls) and hence cheese makers have resorted to other enzyme preparations to meet the needs of an expanding cheese industry.
- ☆ Many proteolytic enzymes will clot milk, therefore, it is not surprising that milk-clotting enzymes have been obtained from virtually every class of living organism [1c].

Rennet

Rennet is a complex of enzymes produced in any mammalian stomach to digest the mother's milk. It contains many enzymes, including a proteolytic enzyme (protease) that coagulates the milk, causing it to separate into curd and whey. There are non animal sources for rennet also that are suitable for vegetarian consumption [1e].

Nomenclature of Rennet

A nomenclature system [1c] is adopted for rennet. The term "rennet" may be applied to all milk-clotting enzyme preparations (except porcine pepsin) used for cheesemaking.

i. Rennet is defined as "aqueous extracts made from the fourth stomaches of calves, kids or lambs."

ii. **Bovine rennet** is an "aqueous extract made from the fourth stomach of bovine animals, sheep or goats."

iii. **Microbial rennet**, "followed by the name of the organism" identifies milk-clotting preparations derived from microorganisms.

- ☆ Now-a-days,the important types of milk-clotting enzymes including animal rennet, microbial coagulants, recombinant coagulants, and plant-derived clotting enzymes are being used in cheese making.

A. The Traditional Enzyme [1a,1e]

The active enzyme in **rennet is called chymosin or rennin** but there are also other important enzymes in it, *e.g.* pepsin and lipase. Chymosin(rennin), is a proteolytic enzyme related to pepsin that synthesized by chief cells in the stomach of some animals.

- ☆ Renin is a naturally occurring, protein-digesting enzyme. Rennet, a commercial form of rennin, is used in the production of most cheeses. The practice of cheese making probably began when somebody discovered that milk stored in bags made from calf stomachs formed a sweet curd.
- ☆ Most cheese varieties, and about 75 per cent of total production, are produced by rennet coagulation but some acid coagulated varieties, *e.g.* Quark and Cottage cheese, are of major importance [2a].

- ☆ The enzyme most commonly used to coagulate milk is rennin. Milk proteins coagulate or precipitate to form a solid clot, or curdle under certain conditions. Rennin is milk-curdling enzyme, which is usually obtained from the fourth stomach (abomasum) of suckling calves. In other animals, the proteolytic enzyme, pepsin, substitutes rennin. Rennin is prepared commercially for use in cheese making as a salt extract of dried calf stomach. Such an extract containing the enzyme is called rennet or rennet extract [2a].
- ☆ Rennin is an extremely powerful clotting enzyme. One part of pure rennin can clot more than five million parts of milk. The optimum pH for rennin action on milk is 5.4. However, it can function very powerfully as a clotting agent at almost neutral pH (6.2-6.4) [1e,2a].
- ☆ The ratio of clotting to proteolytic power is very high in case of rennin. [1e,2a]
- ☆ For hard cheese such as Cheddar, usually about 2.5 g of commercial rennet powder is used for 100 L of milk. In case of Meito rennet it is 1.65 g/100 L milk.
- ☆ The rennet extract from the young milk-fed calf contains 88-94 per cent rennin and from 6-12 per cent pepsin, while in extracts from the older fodder eating bovine, it is almost reverse *i.e.* 90-94 per cent pepsin and only 6-10 per cent rennin [1e].

Preparation of Rennet

It is prepared by following traditional and modern methods [1e]:

In Traditional Method

In this method, dried and cleaned stomachs of young calves are sliced into small pieces and then put into saltwater or whey, together with some vinegar or wine to lower the pH of the solution. After some time (overnight or several days), the solution is filtered. The crude rennet that remains in the filtered solution can then be used to coagulate milk. About one gram of this solution can normally coagulate 2 to 4 litres of milk.

In Modern Method

In this method,the commercial enzyme is prepared from the fourth stomach (abomasums) of the calf, known as the vell. The lining of the stomach is washed, dried, cut into small pieces and macerated in water containing about 4 per cent boric acid at 30°C for about 5 days.

- ☆ Crude rennet extract contains active rennin as well as inactive precursor (prorennin). Addition of acid to the extract facilitates conversion of the prorennin to rennin and allows the extract to reach maximum activity.

- ☆ In one kg of rennet extract, there are about 0.7 g of active enzymes. The rest is water and salt and sodium benzoate. Typically, one kg of cheese contains about 0.0003 g of rennet enzymes.

Properties of Rennin [1e]

i. Rennin is a sulphur-containing protein.
ii. It is easily destroyed by heat, many chemical substances and many physical conditions.
iii. The isoelectric point of rennin is about 4.55 and it is easily destroyed by factors such as heat and high pressures.
iv. It is also digested by proteolytic enzymes and is readily adsorbed.
v. It is very sensitive to alkali, and heating at 70°C at pH 6.8-7.0 will destroy it in 14 min and.
vi. One part can clot about 50, 00,000 parts of milk.

B. Other Proteases [1a,1e]

Obtaining natural milk coagulants is a challenge since rennet availability is limited compared with the increasing demand of the dairy industry. Natural coagulants of animal origin are expensive, and their consumption has been restricted due to religious or dietary reasons. If calf rennet is used as a coagulant for cheese making, the product has to carry a tag of non-vegetarian which may lead to non-consumption of cheese by vegetarian population of the world. Plant proteases have been used as milk coagulants in cheesemaking for centuries either as crude extracts or in purified form.

a. Pepsins

The optimum pH for pepsin action on milk is 2.0. The ratio of clotting to proteolytic power is lower in case of pepsin and other proteolytic enzymes tried in cheesemaking. The latter type of enzymes result in a bitter product [1a,1e,2a].

b. Microbial Proteases/Microbial Rennet [1a,1e]

The growing demand for natural coagulants led to an increased necessity for rennet substitutes, promoting a search for new sources of proteases with coagulant properties.

- ☆ A large number of microorganisms are known to produce milk clotting enzymes. Milk clotting enzymes from bacteria (Streptococcus liquefaciens, Micrococcus caseolyticus, Bacillus cereus, B. polymyxa, B. mesentericus, B. coagulans and B. subtilis) have been used as coagulating enzymes.
- ☆ Milk clotting enzymes from fungi (*Aspergillus nidulans, A. galucus,*) are also used during cheese making. The details related to different microbial rennet/proteases are shown in Table 19.1 [1a,1e].

Table 19.1: Details of different Microbial Rennet/Proteases

Sl.No.	*Types of micro-organism*	*Activity and Uses*
1	Mucor miehei coagulants	i. The protease of these coagulants degrade casein fairly rapidly in the pH range 5.5-7.0; but inspite of this, the incidence of bitter cheese using this enzyme is low. ii. The enzyme is very sensitive to temperatures in the region 37-45°C and is destroyed at 70°C. iii. This enzyme is used successfully for many types of cheese.
2	Mucor pusillus extract	i. It is also used as microbial rennet. It is highly proteolytic than calf rennet or the *Mucor miehei* extract. ii. An increase in calcium ion concentration in milk decreases the clotting time but activity of this enzyme is not so pH-dependent as for other coagulants. iii. It tends to give hard curds because of its high proteolytic activity and the curd tends to lose fat into the whey, thereby giving lower yield as compared to others. iv. Thus, this enzyme is usually used in combination with other enzymes.
3	Endothia parasitica	i. The enzyme extracted from this, is more caseolytic than those from *Mucor* sp. or calf rennet. ii. Tends to produce more bitter flavors in high moisture cheese.
4	Protease obtained from *Bacillus* sp. P45	The milk-clotting ability of this was also established.
5	Commercial starter 'natto' (*Bacillus subtilis*)	This has been studied for its milk clotting activity and it was found to be a potential rennet substitute.

☆ The details related to the micro-organisms used in commercial production of microbial rennet are also shown in Table 19.2 [1e].

Table 19.2: Micro-organisms Used in Commercial Production of Microbial Rennet

Sl.No.	*Source of Enzymes*	*Name of Enzymes*
1	Mucor miehei	Hannilase, Rennilase, Fromase, Marzyme.
2	Mucor pusillus	Novadel, Noury, Emporase, Meito.
3	Endothia parasitica	Suparen, Sure curd.
4	Bacillus subtilis	Mikrozyme

C. Synthetic Chymosin [1a,1d,1e]

The imperfections in cheese production and scarcity of microbial and animal rennets compels cheese industry to seek replacements. In addition to the aforesaid reasons, many of these substitute rennets/proteolytic enzymes from microbial or plant origin have also following limitations-

- ☆ Cause flavor, texture and yield changes in certain types of cheese that are different than those produced by calf rennet.
- ☆ They are not suitable for long ripening cheeses as they have a different range of non-specific activities than chymosin and do not produce the correct flavours on prolonged ripening.

i. Recombinant Chymosin/Fermentation-Produced Chymosin (FPC) [1a,1d,ie]

Recombinant rennet/chymosin can be prepared by gene transfer technology in which the bovine chymosin is cloned in a suitable production strain and the enzyme is produced by fermentation. This enzyme can be used in cheese making as a coagulant having properties similar to that of calf rennet [1e].

- ☆ The genetic engineering, makes it possible to extract rennet-producing genes from animal stomach and insert them into certain bacteria, fungi or yeasts to make them produce chymosin during fermentation.
- ☆ The genetically modified microorganism is killed after fermentation and chymosin is isolated from the fermentation broth, so that the fermentation-produced chymosin (FPC) used by cheese producers does not contain any GM component or ingredient. FPC (considered the ideal milk-clotting enzyme), contains the identical chymosin as the animal source, but produced in a more efficient way and have been on the market since 1990. FPC was the first artificially produced and widely used enzyme [1d].

Synthetic chymosin by recombinant DNA techniques using strains of Eshericia coli or Klaveromyces lactis or Aspergillus niger as host organism is now available. The transferred genetic material exists in the host cell in the form of a plasmid and is used as a template for the production of an enzyme identical to chymosin [1a].

- ☆ Bovine chymosin is now produced recombinantly in E. coli, Aspergillus niger var awamori, and K. lactis as alternative resource.
- ☆ The most widely used fermentation-produced chymosin is produced either using the fungus Aspergillus niger or using Kluyveromyces lactis.

Benefits [1a,1d,1e]

Recombinant chymosin or FPC has following benefits:

i. Such chymosin can be produced in large-scale fermentors at low cost.

ii. Recombinant and highly pure chymosin also has some other advantages such as specific, low proteolytic activity, predictable coagulation behavior and vegetarian approval.

iii. FPC contains only chymosin B, achieving a higher degree of purity compared with animal rennet.

iv. FPC can deliver several benefits to the cheese producer compared with animal or microbial rennet, such as higher production yield, better curd texture and reduced bitterness.

D. Plant Coagulants for Cheese Making [1e]

Enzymes from many plant sources may be used as clotting enzymes in cheese making but most of the plant proteases are strongly proteolytic and cause extensive digestion of the curd, resulting in reduced yields, bitter flavours and pasty-bodied cheese. The optimization of using a plant rennet in a dairy application can be done by selecting the appropriate plant rennet with a consistent clotting efficiency.

Details related to the some available plant coagulants are shown in Table 19.3.

Table 19.3: Plant Coagulants

Sl.No.	*Plant Coagulants*	*Activities and Uses*
1	From *Cynara. cardunculus pistils*	The vegetable coagulant, extracted from *Cynara cardunculus* pistils, proved to be the most suitable milk-clotting enzyme for cheesemaking, since it possesses similar milk clotting properties to conventional calf rennet. Extracts from the flower petals of *Cynara cardunculus* were used for the manufacture of Serra cheese from sheep's milk by Portuguese farmers.
2	Ficin	Ficin is present in the latex of several species of the genus Ficus (fig) such as *Ficus glomarata, Ficus religiosa* and *Ficus carica* but the best source is *Ficus carica.* Cheese made with ficin develops bitter flavor which decreases in intensity during curing. *F. carica* latex, seemed to be a promising alternative coagulant at higher temperatures.
3	Papain	The latex of the plant *Carica papaya* yields papain and several other proteases. It has powerful milk clotting activity but it is also highly proteolytic. It requires a free sulfhydryl group for its catalytic activity.
4	Bromelain from pineapple	It has also been considered as a possible substitute for calf rennet.
5	Enzyme extracted from *Withania coagulans*	It was also used in the manufacture of Surati and Cheddar cheese.

Enzymatic Coagulation of Milk [1a,1b,1e]

The three stages of enzymic coagulation are given below:

a. Primary Stage

The hydrolysis of the κ-casein layer is called as the primary phase of rennet coagulation. The casein micelle is stabilized by κ-casein layer on the surface of the micelle.

- ☆ In the first stage, the proteolytic enzymes (chymosin, pepsin or microbial proteinases) contained in the rennet cuts off a specific fragment of one of the caseins, namely, κ -casein.
- ☆ The enzymes present in rennet (proteinases) hydrolyse κ-casein layer to form paracasein micelles. At the natural pH of milk, about 80 per cent of κ -casein must be cleaved to permit aggregation of the micelles to proceed.

b. Secondary Stage

The second stage is the subsequent clotting of the micelles (paracasein micelles which aggregate in presence of calcium) which have been destabilized by this enzymatic attack.

- ☆ This stage is the physical process of aggregation of casein particles (micelles) to form a gel. After losing its water soluble tail, κ -casein can no longer keep the casein particles separated, so they begin to form chains and clusters.
- ☆ The clusters continue to grow until they form a continuous, three dimensional network which traps water inside, and forms a gel, something like Jell-o.
- ☆ The aggregation of paracasein micelles in presence of calcium is called the secondary phase of rennet coagulation of milk.

c. The Third Stage

This refers to an ongoing development of the gel network. For some cheese the gel is cut as soon as it is firm enough to do so. For others, like soft ripened cheese, cutting is delayed while the gel continues to become firmer.

Chemistry Involved during Primary and Secondary Stages of Enzymatic Coagulation of Milk [1a,1e]

i. The amino acid chain forming the κ-casein molecule consists of 169 amino acids. Rennet enzymes act specifically at 105 (phenyl alanine)-106 (methionine) bond of this amino acid, thereby splitting it into following two parts-

- One part consists of amino acids from 1-105, called as para-κ-casein. This part is insoluble and remains in the curd together with αs and β-casein.
- The other part of amino acids from 106-169 is soluble part. These amino acids are dominated by polar amino acids and the carbohydrate, which gives this part its hydrophilic properties. This part of the κ-casein molecule is called the glycomacro-peptide and is released into the whey in cheese making.

ii. The formation of the curd is due to the sudden removal of the hydrophilic macropeptides and the imbalance in intermolecular forces caused thereby. Bonds between hydrophobic sites start to develop and are enforced by calcium bonds which develop as the water molecules in the micelles start to leave the structure. This process is usually referred to as the phase of coagulation and syneresis.

iii. Hydrolysis of κ-casein during the primary phase of rennet action releases the highly charged, hydrophilic C-terminal segment of κ-casein (macropeptide), as a result of which the zeta potential of the casein micelles is reduced from -10/-20 to -5/-7 mV and the protruding peptides are removed from their surfaces, thus destroying the principal micelle-stabilizing factors (electrostatic and steric) and their colloidal stability.

iv. When roughly 85 per cent of the total κ-casein has been hydrolyzed, the stability of the micelles is reduced to such an extent that when they collide, they remain in contact and eventually build into a three-dimensional network, referred to as a coagulum or gel.

Processing of Milk into Cheese [2,2a]

Cheese consists of a concentration of the constituents of milk, principally fat, casein and insoluble salts, together with water in which small amounts soluble salts, lactose and albumin are found.

- To retain these constituents in concentrated form, milk is coagulated either by means of lactic acid produced by bacteria or by the addition of rennet or by both.
- A portion of water is removed by cutting, cooking, stirring or draining the curd or by mechanical application of pressure. The cheese may or may not be ripened; the nature of the process depends upon the particular variety of cheese [2a].

Processing of Cheese [1a,2,2a,3]

Cheesemaking (or **caseiculture**) is the craft of making cheese. The processing of milk (an unstable, bulky but highly nutritious raw material) into cheese (a stable, flavorsome, and concentrated product that provides eating pleasure and has an extended shelf life) involves the following three major steps:

i. **Coagulation [1e,2,2a,3]**: The essential step in the manufacture of all cheese varieties involves coagulation of casein of milk to form a gel, which entraps the fat, if present. Coagulation may be achieved by the following three processes [2a]-

   - ✰ **Enzyme action(**Limited proteolysis by selected proteinases-rennets)- Rennet coagulation refers to the addition of enzymes to milk in order to make it clot. Many cheeses fall into this category-cheddar, gouda, queso fresco, and many others. Rennet enzymes act like a razor and shave off the κ-casein hairs. Without the hairs, the micelles can now stick, aggregate, and form the backbone of cheese structure.
   - ✰ Coagulation is a key step which involves the use of rennet and depends on several parameters (pH, calcium content, temperature, *etc.*) [2]. During the fermentation process, once sufficient lactic acid has been developed, rennet is added to cause the casein to precipitate. Rennet contains the enzyme chymosin which converts κ-casein to para-κ-caseinate (the main component of cheese curd, which is a salt of one fragment of the casein) and glycomacropeptide, which is lost in the cheese whey. As the curd is formed, milk fat is trapped in a casein matrix. After adding the rennet, the cheese milk is left to form curds over a period of time [3].
   - ✰ **Acid addition (**Acidification to pH 4.6) - Acid can either be added directly or can be produced by starter cultures. A few examples of acid coagulated cheeses include cottage cheese, quark, and chèvre. The addition of acid (**like adding positive charge**) neutralizes the micelle surface (Negatively Charged) and this allows them to bump into each other and stick. This effect is the most prominent at the isoelectric point of casein, pH = 4.6.
   - ✰ **Acid/heat addition (**Acidification to pH 5.2 and heating to 90°C)- In this process, acid and heat are used to clot milk. Examples of this include ricotta, mascarpone, and paneer. The effects of acid on casein is already discussed above. Whey proteins are denatured (unraveled) by heat exposing "sticky" portions of their structure. These sticky ends can bond to each other across whey proteins or bond to casein proteins. Acid can now also contribute to whey coagulation now that they have been denatured by heat. The result is a matrix of coagulated whey protein, and if casein is present, a matrix of coagulated whey/casein.

☆ Some milks never coagulate. To measure the coagulation ability of milk and identify different parameters in milk coagulation properties, the Formagraph, the computerized renneting meter and the Optigraph have been developed (reference methods). Equations have been developed using infrared spectrometry to predict the parameters obtained by the reference methods [2].

ii. **Dewatering [2,3]**: Once the cheese curd is judged to be ready, the cheese whey must be released. As with many foods the presence of water and the bacteria in it encourages decomposition. To prevent such decomposition it is necessary to remove most of the water (whey) from the cheese milk, and hence cheese curd, to make a partial dehydration of the curd. There are several ways to separate the curd from the whey and following four main ways of separating curds from whey, are discussed below:

   a. The first method is used for soft cheeses and involves curds being transferred into perforated molds.

   b. The second method involves removing supernatant whey after the curds have settled in the vat and is used for Gouda and Emmenthal cheeses.

   c. Next curds can be scooped from the vat and placed in molds, as is done with Parmigiano-Reggiano.

   d. Finally perforated screens can be used to drain the whey using gravity, as occurs with cheddar cheeses.

iii. **Refining** – Cutting, cooking,pressing of curd, **Scalding(**In making Cheddar -or many other hard cheeses, the curd is cut into small cubes and the temperature is raised to approximately 39 °C (102 °F) to 'scald' the curd particles),Maturation or aging,Ripening *etc.* are the steps which are done during this refining process.

Factors affecting Enzymatic Coagulation of Milk [1a,1e,2a]

The following factors are affecting the enzymatic coagulation of milk-

1. Effects of Processing Parameters on Enzymic Coagulation [1a]

Because rennet coagulation takes place in stages(three stages), it is necessary to understand the effect of processing on each stage. Only the first and second stages of coagulation, are discussed here. The primary and secondary phase of rennet coagulation are affected by the following factors – like milk composition, temperature, pH, calcium content, pre-heating of milk, rennet concentration *etc.*

a. **Effect of pH**: Lower pH increases enzyme activity and neutralizes charge repulsion between micelles. Hence, both primary and secondary stages of coagulation proceed more quickly at lower pH.

b. **Effect of Calcium**: Calcium is not required for the primary stage (*i.e.*, enzyme hydrolysis of 8-casein) but is essential to aggregation of the casein micelles. Increased calcium concentration is beneficial for coagulation of milk. For this reason, sometimes $CaCl_2$ is added to milk prior to cheese making. At low levels of calcium the primary phase goes to completion. Subsequently, instantaneous coagulation can be induced by adding sufficient calcium chloride. The addition of calcium promotes cheese making via following three beneficial changes [1e]-

   i. An increase in calcium ion concentration.

   ii. An increase in the concentration of colloidal calcium phosphate.

   iii. A concomitant decrease in pH (the addition of $CaCl_2$ to 0.02 per cent, *i.e.* 1.8 mM Ca, reduces the pH by ~ 0.05-0.1 units, depending on protein level).

c. **Effect of set temperature**: The principal effect of set temperature is on the secondary phase of enzymatic coagulation, which does not occur at temperatures below around 18°C. The optimum coagulation temperature for most cheese is 30-32°C, whereas Swiss set at 37°C.

   i. At temperatures less than 20°C coagulation does not occur, but the primary stage goes to completion and the milk will then coagulate quickly when warmed.

   ii. At temperature less than 30°C the gel is weak and difficult to cut without excessive yield loss due to fines.

2. Effects of Heat Treatments [1a,1e,2a]

Heat treatment affects the enzymatic coagulation in following ways-

i. **Mild heat treatment** (pasteurization) [1a,1e]–Heating milk to pasteurization temperature has beneficial effect on rennet coagulation due to heat induced precipitation of calcium phosphate and a concomitant decrease in pH.

   ☆ Pasteurization decreases the rate of the secondary stage. During heat treatment, the less calcium is available to assist with coagulation as calcium and phosphate move from soluble to colloidal (insoluble) form during heat treatment. This effect is reversed by cold storage or the addition of calcium chloride [1a].

   ☆ Pasteurization alters the indigenous microflora and facilitates the manufacture of cheese of more uniform quality, but unless due care is exercised, it may damage the rennet coagulability and curd-forming properties of milk.

   ☆ Even when properly pasteurized, Cheddar cheese (and probably other varieties) made from pasteurized milk develops a less intense flavour and ripens more slowly than raw milk cheese [2a].

ii. **Excess Heat treatment** -Heat treatment in excess of pasteurization results in increased clotting time and a weak gel. High heat treatments cause absorption of whey proteins onto the casein particles. The casein particles are then unable to form a strong gel. The deposition of heat-induced insoluble calcium phosphate leading to reduction in the concentration of native micellar calcium phosphate. This micellar calcium phosphate is important for cross linking para-κ-casein micelles and their aggregation during gel formation.

3. Effects of Homogenization [1a]

Homogenization primarily affects the secondary phase of aggregation. Some cheese quality effects are also noted. The following effects occur if the cheese milk is homogenized in its entirety:

i. Reduced aggregation of casein particles,
ii. Decreased syneresis,
iii. Finer gel network due to smaller fat globules,
iv. Improved texture of soft cheese,
v. Fat recovery is increased (Note: the same is true for acid and heat/ acid coagulated cheese),
vi. Hard cheese becomes rubbery and
vii. Makes cheese whiter because the yellow fat is masked by the artificial protein membranes on the homogenized fat globules.

As observed during treatment of milk for cheese making, some of these results may be different if only the cream is homogenized and then added back to the skim milk.

4. Composition of Milk [1e]

Variation in the composition of milk mainly affects the rate of coagulation and the curd firmness. Fast coagulation results in firmer curd.

- The rate of clotting is largely dependent on the nature of the casein micelles and the equilibrium with the calcium phosphate and calcium ions.
- The firmness of the curd are affected by pH value, calcium concentration, temperature, fat content and the ratio of rennin to casein.
- The rennet coagulation time (RCT) is affected by the protein content in milk. RCT decreases with protein content in the range of 2-3 per cent. Further increase in milk protein level (more than 3 per cent) result in a slight increase in gelation time. This is due to decrease in rennet:casein ratio, which necessitates an increase in the time required to generate sufficient hydrolysis of κ-casein to induce aggregation of paracasein

micelles. A minimum protein content of 2.5-3.0 is necessary to obtain gel in about 30-40 minutes during cheese making.

- ☆ Increase in fat content also results in decreased RCT but the effect is lower than that of the protein content.

5. Enzyme (Rennet) Concentration [1e]

The rate of enzymatic coagulation is directly related to the concentration of enzyme.

- ☆ Increase in concentration of rennet decrease RCT. During cheese making, rennet is added in such a concentration so as to coagulate the milk 30-40 minutes.
- ☆ More rennet concentrations can be used to shorten the coagulation time but it leads to retention of more rennet in the curd which has pronounced effect in ripening of the cheese, particularly proteolysis.

Requirements of Suitable Coagulating Enzymes/Rennet Substitutes [1a,1e]

The following are the requirements of suitable coagulating enzymes-

i. **Suitable ratio of clotting to proteolytic activity (C/P) [1a]-** This ratio is dependent on the specificity of the enzyme for the Phe105-Met106 bond of -casein. Most rennet substitutes are more proteolytic than rennet (*i.e.*, low C/P) and cause diminished yields of casein and fat, and bitterness during ripening.

   Rennet substitutes [1e]-The ratio of milk clotting activity to proteolytic activity should be high for Rennet Substitutes. This means that the milk should clot without much proteolysis *i.e.* breakdown of proteins to peptides.

ii. **Proteolytic specificity**-Structure and flavour of ripened cheese depends on the type of proteolysis caused by the coagulant during cheese curing. The exception is in cheese such as Swiss or Parmesan where most of the rennet activity is destroyed by the high cooking temperature. During ripening chymosin breaks down one of the caseins, namely s1-casein much more than other caseins.

iii. **High pH optimum-**

   Coagulating enzymes [1a]-Rennet activity is stable and able to coagulate milk at the normal pH of milk although its activity increases with decreasing pH. Most pepsins and microbial proteases are denatured at the pH of milk which has been a major difficulty in developing rennet substitutes.

iv. Denaturation temperature is important for two reasons-

- ✰ Ripening due to coagulating enzymes is not desirable in cooked cheese such as Swiss and Italian types. Rennet is eliminated during the high temperature cook in these cheeses but microbial coagulants are not.
- ✰ The coagulant must be eliminated by pasteurization to prevent proteolysis in products made from whey. Some microbial rennets survive pasteurization.

Rennet substitutes [1e]: It should be thermally stable (comparable to calf rennet) at pH and temperatures used during cheese making. It should have low thermal stability at temperatures of whey processing otherwise it may hinder the utilization of cheese whey in various products.

v. **Distribution between curd and whey [1a]**- Only 0-15 per cent of rennet remains in the curd, but small amounts of residual rennet are significant to ripening of aged cheese.

vi. The most important factors which determine rennet retention are following cooking treatments [1a]-

- ✰ As noted above, rennet does not survive in high temperature cooked cheese varieties.
- ✰ In low cooked cheese such as Cheddar, variations in cooking temperature and time influence rennet activity during aging.

vii. **The pH at draining [1a]**- Rennet is more soluble at low pH and, therefore, the amount retained in the curd increases with decreasing pH at draining. Retention of microbial rennets in the curd is independent of pH at draining.

viii. Changing rennet sources may also influence rennet retention and cheese ripening. Different rennets with the same coagulating properties may have different thermal tolerances and different proteolytic characteristics.

ix. **Standard and consistent activity [1a]**-Single strength rennet is standardized so that 200 ml coagulates 1,000 kg of milk in 30 - 40 min. Typical commercial rennet preparations are about 50 per cent chymosin (calf rennet) and 50 per cent bovine pepsin, so there is much opportunity for variation. Commercial calf rennet preparations are about 94-96 per cent chymosin. Using recombinant rennet it should be possible to produce commercial rennet preparations which are more consistent with respect to all of the properties listed above, including proteolytic specificity and heat tolerance.

x. **Flavour/Body and Texture**

Rennet Substitutes [1e]- It should impart desired flavour, body, and texture characteristics to the finished cheese

Sources

1. *Cheese chemistry, University of Sheffield, Cheesy Chemistry_EE.pdf.*

1a. *Enzymic Coagulation of Milk, Food Science, University of Guleh, https: //www. uoguelph.ca.*

1b. *D. G. Dalgleish, The Enzymatic Coagulation of Milk, Cheese: Chemistry, Physics and Microbiology pp 69-100.*

1c. *Rodney J. Brown and C.A.Emstrom, Milk-Clotting Enzymes and Cheese Chemistry, Part I-Milk-Clotting Enzymes, Fundamentals of Dairy Chemistry, pp 609-654.*

1d. *Wikipedia- Chymosin.*

1e. *S.K.Kanawjia and Yogesh Khetra, Cheese Technology, Agrimoon.com.*

2. *Thibault Troch, Émilie Lefébure, Vincent Baeten, Frédéric Colinet, Nicolas Gengler and Marianne Sindic, Cow milk coagulation: process description, variation factors and evaluation methodologies- A review, Biotechnology, Agronomy, Society and Environment, ISSN 1370-6233, E ISSN 1780-4507, Volume 21, 2017.*

2a. *Principles of Cheese manufacture, Lesson 5, http://ecoursesonline.iasri.res.in.*

3. *Wikipedia-milk coagulation.*

Biochemical Changes during Ripening [1-5]

Cheese ripening is basically about the breakdown of proteins, lipids and carbohydrates (acids and sugars) which releases flavour compounds and modifies cheese texture. The general principles of cheese ripening are given below [1,5]:

i. **Ripening Period-**Rennet-coagulated cheeses are ripened for periods ranging from nil for fresh cheese, about two weeks (mozzarella), two or more years (Parmigiano-Reggiano or extra-mature Cheddar) to 5 years for some hard ripened cheese depending on variety. The ripening process of cheese is very complex and involves microbiological and biochemical changes to the curd resulting in the flavour and texture characteristic of the particular variety.

ii. **Ripening processes Classification**- Ripening processes are broadly classified as interior and surface ripened.

   - ☆ **Interior Ripened**-Cheese which depend mainly on interior ripening (most hard ripened cheese - Cheddar and Italian types) may be ripened with rind formation or may be film wrapped before curing.
   - ☆ **Surface ripened** -Cheese which depend mainly on surface ripening include smear ripened and mould ripened.

iii. Three sources of cheese flavour-

   - ☆ Flavours present in the original cheese milk, such as natural butter fat flavour and feed flavour.

- ☆ Breakdown products of milk proteins, fats and sugars which are released by microbial enzymes, enzymes endogenous to milk, and enzyme additives.
- ☆ Metabolites of starter bacteria and other microorganisms. These include products from catabolism of proteins, fats and sugars.

iv. Flavour and texture development are strongly dependent on the followings-

a. pH profile,

b. Composition

c. Salting

d. Temperature

e. Humidity

v. As a general rule factors which increase the rate of ripening increase the risk of off flavour development, and reduce the period of time when the cheese is saleable.

1. Biochemical Changes

These changes in cheese during ripening may be grouped into the following-

A. Primary Reaction

(lipolysis, proteolysis and metabolism of residual lactose and of lactate and citrate) – These include the metabolism of residual lactose and of lactate and citrate (often, though erroneously, referred to collectively as 'glycolysis'), lipolysis and proteolysis.

a.

Most of the milk sugar has been removed in the whey or by fermentation, that is converted to lactic acid by the cultures. Residual lactose is metabolized rapidly to lactate during the early stages of ripening. Lactate is an important precursor for a series of reactions including racemization, oxidation or microbial metabolism. Citrate metabolism is of great importance in certain varieties.

Metabolism of Residual Lactose [1,5]

Milk contains no starch or fibre or any sugar other than lactose so all carbohydrate compounds in cheese are derived from lactose or produced by microorganisms. Relative to fat and protein lactose contributions to flavour are minimal. A key feature of cheese manufacture is the metabolism of lactose to lactate by selected cultures of lactic acid bacteria (LAB) known as starters.

- ☆ The rate and extent of acidification influence the initial texture of the curd by controlling the rate of demineralization.

- ☆ The pH of cheese affects the texture of curd directly by influencing the solubility of the caseins; all else being equal, high pH cheeses are softer than more acid cheeses.
- ☆ pH also affects texture and flavour indirectly by affecting the activity of enzymes important to ripening and, in the case of the coagulant, the retention of enzyme in the curd during manufacture.
- ☆ The complete fermentation of lactose is important in cheese to avoid the development of an undesirable secondary microflora.
- ☆ Residual lactose is metabolized quickly to L-lactate during the early stages of ripening at a rate largely determined by temperature and the salt-in-moisture (S/M) levels of the curd by the action of starter bacteria. Lactose that remains unfermented by the starter is probably metabolized by nonstarter lactic acid bacteria (NSLAB) flora. Some examples are:

  i. Lactose metabolism in **Swiss cheese** is quite complex. Residual lactose trapped in the curd after moulding is rapidly metabolized as the curd cools by Streptococcus thermophilus, which metabolizes glucose with the production of l-lactate. Galactose accumulates initially, but this sugar and any remaining lactose are metabolized by the lactobacilli present in the starter giving a mixture of D- and L-lactate.

  In well drained cheese such as **Swiss types**, lactose is completely used up in a few hours.

  ii. High salt in the moisture phase of **Cheddar** slows lactose metabolism so lactose content is 0.3 to 0.7 per cent at one day after manufacture and slowly declines to less than 0.1 per cent.

  iii. Residual lactose in **Camembert cheese** is used by Penicillium camemberti so it decreases quickly, especially on the surface, when the mould begins to grow.

  iv. In **washed cheese** varieties, lactose not leached by washing is quickly used up by the culture, especially for Dutch type cheese where salting is delayed. In Colby, early vat salting reduces the rate of utilization of residual lactose.

  v. Many organisms, including yeasts and moulds in mould and smear ripened cheeses utilize lactate and produce various flavourful compounds.

Metabolism of Lactate

Lactate produced from lactose by the growth of the starter is an important substrate for a range of reactions that occur in cheese during ripening. The rate at which l-lactate is racemized depends on the composition of the NSLAB flora. Racemization is probably faster in raw milk cheese than in cheese made from pasteurized milk. The pathway for lactate racemization probably involves oxidation of L-lactate by L-lactate dehydrogenase to form pyruvate, which is then reduced to d-lactate by the action of d-lactate dehydrogenase. Lactate can be oxidized by LAB in cheese to products including acetate, ethanol, formate and CO_2. However, the extent to which this pathway occurs in cheese depends on the NSLAB population and the availability of O_2. Anaerobic metabolism of lactate by Clostridium tyrobutyricum to butyrate and H_2 results in a defect known as **late gas blowing**, which causes cracks to appear in the cheese during ripening and the development of off-flavour.

- ✰ Metabolism of lactate is essential to the development of Emmental and related varieties characterized by the development of large eyes during ripening.
- ✰ Lactate metabolism is perhaps of most importance in the ripening of surface mould-ripened cheeses (*e.g.* Camembert and Brie).

Metabolism of Citrate

Milk contains approximately 1750 mg citrate per litre, most of which is in the soluble phase and thus is lost on whey drainage. Levels of citrate in cheese curd are approximately three times higher than in the whey. presumably due to the concentration of colloidal citrate. Cheddar cheese curd typically contains 0.2–0.5 per cent citrate.

Citrate is metabolized by citrate positive (Cit+) strains of lactococci, previously referred to as 'Streptococcus diacetylactis' or 'Lactococcus lactis ssp. lactis biovar diacetylactis', which contain a plasmid for citrate transport. Citrate is not metabolized by other LAB used as starters. The products of citrate metabolism include CO_2, which is responsible for the small eyes often found in Dutch-type cheeses, and important flavour compounds, particularly diacetyl, which contribute to the flavour of these cheeses.

b. Lipolysis (Fat Breakdown) [1,5]

The details related to the flavour given by milk fat with and without ripening of cheese, are given below:

- **i.** Dairy fat is a wonderfully rich source of flavours, because it contains an extremely diverse selection of fatty acids. In particular, butter fat is the only natural fat which is rich in short chain fatty acids. Butyric acid for example is a potent flavour compound. Dairy fat without any ripening during cheese making is an important contributor to cheese flavour and texture.
- **ii.** Lipolysis in cheese is catalysed by lipases from various source, particularly the milk and cheese microflora, and, in varieties where this coagulant is used, by enzymes from rennet paste. The fat derived flavours associated with cheese ripening result from the release of

fatty acids by lipolysis and further modification of fatty acids by microorganisms to other compounds.

- ✰ In cheese, oxidative changes are very limited due to the low oxidation/reduction potential (about –250 mV). However, triglycerides in all cheese varieties undergo hydrolysis by the action of indigenous, endogenous and/or exogenous lipases, which result in the liberation of fatty acids in cheese during ripening.
- ✰ The triglycerides of ruminant milk fat are rich in short-chain fatty acids that, when liberated, have low flavour thresholds that contribute significantly to the flavour of many cheese varieties.
- ✰ Lipolytic agents in cheese generally originate from the milk(lipoprotein lipase -LPL), the coagulant (in the case of rennet paste) and the cheese microflora.
- ✰ LPL is of most significance in raw milk cheeses as the enzyme is largely inactivated by pasteurization.
- ✰ Commercial rennet extracts used for the manufacture of most cheese varieties are free from lipase activity. However, rennet pastes used in the manufacture of certain hard Italian cheese varieties, such as Provolone and the various Pecorino cheeses, and often in traditional Greek feta, contain much lipase activity.
- ✰ LAB are present in cheese in high numbers, enzymes from these organisms are responsible for the liberation of significant levels of fatty acids during the long ripening period of many internal bacterially ripened cheeses.

c. *Proteolysis (Protein Breakdown) [1,5]*

It is the most complex and most important biochemical event that occurs during ripening and is catalysed by enzymes from residual coagulant, the milk (particularly plasmin) and proteinases and peptidases from lactic acid bacteria and, in certain varieties, other microorganisms that are encouraged to grow in or on the cheese. Because of its importance, proteolysis and the enzymes responsible for this process have been reviewed extensively. Protein degradation during cheese curing is a directed process resulting in protein fragments with desirable flavours. Protein breakdown involves three general types of processes:

- **i.** Proteases break proteins into smaller peptides, some of which are flavour compounds. For example, bitter and brothy flavoured peptides are well known to occur in cheese.
- **ii.** Peptidases further break down peptides to amino acids.
- **iii.** Further catabolism of amino acids by cheese microorganisms produces aldehydes, alchohols, carboxlic acids and sulfur compounds, many of which are flavourful.

- **Texture**-Proteolysis contributes to the softening of cheese texture during ripening due to hydrolysis of the casein matrix of the curd and through a decrease in the water activity (aw) of the curd due to changes in water binding by the new carboxylic acid and amino groups formed on hydrolysis. Protein breakdown causes shorter body which is less rubbery, less elastic, more meltable. For example, flavour and texture development in Cheddar are mainly dependent on protein breakdown and much less dependent on fat breakdown.
- **Flavour**-Proteolysis has a direct influence on flavour through the production of short peptides and amino acids, some of which are flavoured (often bitter), by facilitating the release of sapid compounds from the cheese matrix and, probably most importantly, by providing free amino acids that are substrates for a series of catabolic reactions that generate many important flavour compounds. Some off flavours associated with undesirable or excessive protein breakdown in cheese are bitter, stringent, putrid and brothy.
- The proteinases and peptidases that catalyse proteolysis in cheese during ripening originate from the coagulant, the milk, starter LAB, NSLAB, secondary starters and a complex Gram-positive bacterial microflora on the surface of smear cheeses.

B. Secondary Reaction

(Metabolism of fatty acids and of amino acids)- Secondary biochemical events are very important for the development of many volatile flavour compounds. These reactions lead to the production of volatile flavour compounds and pathways for the production of flavour compounds from fatty acids and amino acids are also reviewed.

Metabolism of Fatty Acids

While short-chain fatty acids contribute directly to cheese flavour, free fatty acids (FFAs) also contribute indirectly to cheese flavour by acting as precursors for the production of volatile flavour compounds through a series of reactions known collectively as metabolism of fatty acids.

- FFA metabolism is of most significance in blue-mould cheese in which FFAs are converted to 2-methyl ketones (alkan-2-ones) via a pathway corresponding to the early stages of β-oxidation caused by the action of spores and vegetative mycelia of P. roqueforti.

Metabolism of Free Amino Acids

Pathways for the catabolism of free amino acids during ripening produce many flavour compounds. The first pathway is initiated by the action of aminotransferases, pyridoxal-52 - phosphate (PLP)-dependent enzymes, which convert an amino acid to the corresponding α-ketoacid.

- ☆ The α-ketoacids produced are then degraded to a range of other compounds by enzyme-catalysed pathways or by chemical reactions. α-ketoacids, produced by the action of aminotransferases, are degraded further by cheese-related microorganisms to volatile flavour compounds

2. Microbiological Changes

These changes to cheese during ripening include the death and lysis of starter cells, the growth of an adventitious flora ('nonstarter lactic acid bacteria', principally facultatively heterofermentative lactobacilli) and, in many varieties, the development of a secondary microflora, *e.g.* Propionibacterium freudenreichii in Swiss cheese, moulds in mould-ripened varieties and a complex Gram-positive bacterial flora in smear cheeses, which is often of great importance to the flavour and, in some instances, the texture of these varieties.

Cheese texture softens during ripening as a consequence of hydrolysis of the casein micelle by proteolysis and changes to the water-binding ability of the curd and changes in pH (which in turn may cause other changes such as the migration and precipitation of calcium phosphate)

Sources

1. *Paul L H Mcsweeney, Biochemistry of cheese ripening, Department of Food and Nutritional Sciences, University College, Cork, Ireland.*
2. *McSweeney P L H (2004) Biochemistry of cheese ripening: introduction and overview. In Cheese: Chemistry, Physics and Microbiology, Vol 1: General Aspects, 3rd edn, pp 347–360. Fox P F, McSweeney P L H, Cogan T M and Guinee T P, eds. London: Elsevier.*
3. *McSweeney P L H and Fox P F (2004) Metabolism of residual lactose and of lactate and citrate. In Cheese: Chemistry, Physics and Microbiology, Vol 1: General Aspects, 3rd edn, pp 361–372. Fox P F, McSweeney P L H, Cogan T M and Guinee T P, eds. London: Elsevier.*
4. *McSweeney P L H and Sousa M J (2000) Biochemical pathways for the production of flavour compounds in cheese during ripening. Le Lait 80 293-324.*
5. *Ripening processes: chemical and physical changes, Food Science, University of Guelph, https: //www.uoguelph.ca.*

Chapter 20

Dried Milk Products: Concentrated and Dried Milk Products

Condensed Milk [1,2]

Condensed milk is a milk product obtained by evaporating part of water of whole milk, or fully or partly skimmed milk, with or without the addition of sugar.

- The term **condensed milk** is commonly used when referring to full cream sweetened condensed milk.
- The term **evaporated** milk is generally used while referring to full cream unsweetened condensed skim milk.

Sweetened Condensed Milks

Sweetened condensed milk (SCM) is one of the oldest industrially produced dairy products. The first record of an SCM-like product appears to be an English patent granted in 1810 to De Heine for preserving milk and sugar by heating the mixture in an open vessel. SCM is essentially a high-viscosity suspension of lactose crystals, fat globules, casein, and whey proteins in a saturated solution of lactose and sucrose. Its rheological behavior is complex.

- These milks are milk products which can be prepared by the partial removal of water from milk with the addition of sugar, or by any other process which leads to a product of the same composition and characteristics.
- The fat and/or protein content of the milk may have been adjusted, only to comply with the compositional requirements by the addition

and/or withdrawal of milk constituents in such a way as not to alter the whey protein to casein ratio of the milk being adjusted. Sweetened condensed milks are of following types:

i. Sweetened condensed milk.
ii. Sweetened condensed skimmed milk.
iii. Sweetened condensed partly skimmed milk and.
iv. Sweetened condensed high-fat milk.

The details related to the composition, Quality standards(FSSAI, BIS, Codex), nutritional values of condensed milks, are discussed under chapter-16.

Physico-chemical Changes during Preparation and Storage of Concentrated Milk [1,2,3a,4]

Physico-chemical changes during manufacturing and storage of concentrated/Evaporated milk are discussed below:

1. Changes during Manufacturing of Condensed Milk

There are numerous physico- chemical changes occurring in the condensed milks during their manufacture. These changes are because of the followings-

- ☆ Inherent properties of the milks.
- ☆ Any additives like sugars, stabilizers *etc.* added.
- ☆ The processing variables.
- ☆ Some of the product specific changes of commercial importance are discussed here.

Changes Caused by Concentration [2,2a,3a,4]

During concentration,there is removal of water from milk which increases the most of the solute concentrations, and hence causes numerous changes in properties, which often are approximately proportional to concentration factor.

- ☆ The changes also depend on other conditions, such as heat treatment and homogenization. Some important changes in properties are given below:

a. Changes in Physical Properties

i. **Density/Specific Gravity**- Removal of water in the manufacture of condensed milk increases the total solids content. Because of the total solids concentration the density/sp. gr. of condensed milk increases.

ii. Colligative properties(Osmotic pressure, freezing point depression, boiling point elevation), electrical conductivity, density and refractive index increase and heat conductivity decreases [2,2a]. Freezing point of milk is quite constant, however it is always lower than that of water and varies with composition of milk. Removal of water as well as addition of soluble substances, such as sugar, has direct effect on the freezing point of condensed milk [3].

- ☆ The details related to the change in density/Specific gravity and Freezing point of condensed and evaporated milks are given in Table 20.1.

Table 20.1: Changes in Density/Specific Gravity and Freezing Point of Condensed and Evaporated Milks

Sl. No.	*Type of Condensed Milk*	*Fa Per cent*	*Milk Solid not Fat Per cent*	*Sucrose Per cent*	*Specific Gravity*		*Freezing Point*	
					16°C	*49°C*	*°C*	*°F*
1	Condensed milk	9.0	22.0	43.0	1.303	1.2821	-14.9	5.1
2	Evaporated milk	9.0	22.0	-	1.083	1.072	-1.3	29.6
3	Sweetened Condensed skim milk	-	26.0	45.0	1.355	1.333	-	-

iii. **Water activity [2,2a]**-The water activity decreases. Increases of most of the solute concentrations.

iv. **Viscosity [2,2a,3]**-Viscosity of condensed milk varies greatly according to the composition, processing conditions and storage period. Increase in concentration increases the viscosity. The viscosity increases and the liquid become non-Newtonian and finally solid-like.

- ☆ Too high or low viscosity is not desirable in the product. Low viscosity tends to leave objectionable sugar sediment in condensed milk and fat separation in evaporated milk. Too high viscosity causes gel.

v. The diffusion coefficient of water decreases from approximately 10^{-9} $m^2.s^{-1}$ in milk to 10^{-16} $m^2.s^{-1}$ in skim milk powder with a small percentage of water [2,2a].

vi. **Changes in pH [3]**- A decrease in pH as compared to milk can be expected in condensed milks during manufacture because of the processing steps such as preheating, concentration and sterilization. Decrease in pH has definite effect on the stability of concentrated milk towards heat sterilization.

b. *Changes in Chemical Properties [2,2a]*

i. **Calcium ion activity**-The Calcium ions activity increases only slightly because calcium phosphate, which is saturated in milk, turns into an undissolved state. As the water content decreases, association of ionic species increases and also ionic groups of proteins are neutralized.

ii. **Protein concentration-**The conformation of proteins changes because ionic strength, pH, and other salt equilibria change.

- ✰ When milk is highly concentrated, the solvent quality decreases. Thus, the tendency of the protein molecules to associate and to attain a compact conformation is increased. Coalescence of casein micelles causes them to increase in size.
- ✰ This increase is smaller if the milk has been intensely preheated, presumably because β-lactoglobulin and other serum proteins have become associated with casein.

Changes in Evaporated Milk [2,2a,3a]

Some important changes in evaporated milks are observed during manufacturing and storage. The details of such changes are given below:

1. *Changes during Manufacturing [2,2a,3a]*

The changes occur in evaporated milk during its manufacturing are discussed below:

Viscosity

Viscosity is defined as resistance to the motion of the molecules of a fluid body among themselves caused by internal frictions as opposed to mobility and it is measured by viscosimeter.

i. **Effect of Sterilization on Viscosity of Product-** The condensing operation causes a slight but definite increase in viscosity (due to increase in concentration of milk solids). The increase in viscosity that gives the finished product its full body, however occur during process of sterilization.

- ✰ In general HTST yields a thin body and low viscosity, while LTLT process yield a full body and little more viscosity.

ii. **Increase in viscosity by Progressive Coagulation of Milk Proteins-**In sterilization process, the rate of thickening is greatest, shortly before the occurrence of a visible coagulation. It is observed that the thickening does not proceed rapidly until above 10 min before coagulation. This is because the thickening or increase in viscosity during sterilization appears to be a part of the function of coagulation. Thickening in sterilization is in fact the beginning of coagulation. It is very slow and gradual and in normal commercial process of sterilization, covering a period of approximately 20 minutes at the full sterilizing temperature. The increase in viscosity which milk undergoes toward the end of sterilizing process has to do with progressive coagulation of the milk proteins.

- ☆ It is observed further that for a heavy creamy body, the heat stability of milk should not exceed 30 - 40 minutes. Milks with heat stability in excess of 50 min will be exceedingly thin at the end of sterilization at 115°C for 30 min unless an increase in solids content is depended upon to build up body.
- ☆ The viscosity produced during sterilization is controlled mainly by the heat stability of milk. Therefore, factors which affect the heat stability of milk will certainly affect the viscosity of milk.
- ☆ The following factors provides low heat stability, high viscosity and very thick body and vice-versa-
- ☆ High acidity of fresh milk.
- ☆ Low fore-warming temperature.
- ☆ High concentration.
- ☆ High homogenizing pressure.
- ☆ Excessive holding of evaporated milk at ordinary temperature before sterilization

2. *Changes during Storage of Evaporated Milk [2,2a,3a]*

i. **Effect of Storage on Viscosity -** Evaporated milk becomes thinner with age. This loss of viscosity increases with the temperature of storage. The decrease in viscosity begins immediately. The rate at which viscosity is lost is accelerated at high storage temperature. Thus it is observed that:

- ☆ At or above 30°C, evaporated milk looses as much as 40 per cent of viscosity (original) during first 10 days of storage.
- ☆ At 15.6°C or below, the age thinning is very slight. It is reduced after which the rate of thinning is much more gradual. Age thinning of evaporated milk may therefore be definitely retarded and the attainment of final viscosity delayed by use of relatively low storage temperature.
- ☆ The finding observed during a study shows that viscosity loss upon storage for 110 days of 58.75 per cent at 26°C, 40 per cent at 15.6°C and 11.25 per cent at 7.2°C.
- ☆ The loss in viscosity depends upon quality of raw material (milk) used. Better grade milk lost less viscosity than the low grade initial milk with the same temperature and time of storage.

ii. **Gel Formation on Storage**-In some evaporated milks, the viscosity increases later in the storage period even to the gel formation. This tendency is greatest with milks that had received relatively light heat treatment and those of high solids concentration. It is observed that sterilisation processes effect the thickening and gel formation properties of evaporated milks.

- ☆ Evaporated milks that have been sterilized by the HTST process (135°C for 0.5 min) thicken quickly during storage. When the gel is broken before it becomes firm, the product wheyed off.
- ☆ When sterilized in the conventional manner (115.6°C for 20 min), gel formation is slow and holds the water better than from the HTST process.
- ☆ The more severe heat treatment of commercial process (115.6°C for 20 min) that retards thickening in storage is commercially preferable to the less drastic treatment of momentary exposure to high sterilizing heat.
- ☆ It may be due to insufficient time for the attainment of new solubility equilibrium of the Calcium and Magnesium salts when using HTST method of sterilization. When solutions of these salts are heated, it takes considerable time to attain that. Their peculiarity of being less soluble at higher temperature may be contributing to the reaction of these salts. In the HTST method of sterilization, conditions are thus less favorable than those of the LTLT procedure (115.6°C for 20 min). This suggests the possibility that, less of Ca and Mg content is rendered insoluble. This condition might encourage the protein particles to swell and to form a gel or it may be as low continuation of the coagulation process accompanied by an operation of caseinate molecule which finally produces an irreversible gel structure.

iii. **Protein Separation during Storage -**The proteins of some evaporated milks have a tendency to settle during storage. This is found more common in evaporated skim milk than evaporated whole milk.

- ☆ Homogenization retards protein separation in evaporated whole milk.
- ☆ It is observed that, fat in evaporated milk is a factor to maintain a normal dispersion of milk constituents during long period of storage.
- ☆ It is observed in this respect that fat acts to counter balance the protein settling to bottom.

iv. **Fat Separation in Storage-** During age thinning and before age thickening sets in, there is definite tendency to objectionable fat separation. This tendency emphasizes importance of efficient homogenization that ensure reduction of at least 90 per cent of fat globules to 2 μ or smaller. Such efficiency can be obtained by taking following care-

- ☆ Daily check up of homogenization Valve,
- ☆ Keeping the clearance surface of valve and valve seat smooth and intact by regrinding or resurfacing,

- ☆ Making sure of the continued accuracy of the pressure gauge and
- ☆ Examining fat globule size.

It is observed that HTST sterilization provides little opportunity to control body hence through a heavier body, control of the extent of fat separation can be achieved. Again here the LTLT procedure (115.6°C for 20 min) proves helpful. Fat separation in this more viscous body is slower and less intense. In case of LTLT, fat layer get re-emulsified easily than in case of HTST sterilization process.

Other Important Changes during Processing and Storage [1,2,4]

The other important changes such as change in colour, change in flavour (cooked flavour), deposit of sediments *etc.* are occurred in Condensed and evaporated milks.

1. Changes in Colour [2,3,4]

Objectionable brown colour is sometimes found in condensed milks. Exposure to high heat during manufacture and storage of condensed milk tends to darken their colour and develop a 'cooked' flavour.

- ☆ Various factors contributing to the brown discolouration and cooked flavour production, are heat treatment, total soils concentration, pH condition, addition of compounds like sugar stabilizers, storage temperature and time of storage [3].
  - i. **Relation of Caramelization to Discolouration [2,4]-** Browning of milk is attributed to reaction of sugar in milk. The colour changes in evaporated milk are not in direct relation of time, but it is the reaction of catalytic nature. Similarly, solution of sugar and proteins when heated, a marked brown colour develops and colour could not be removed by washing and also whey is clear showing no sign of any colour.
  - ii. **Relation of Amino Acid Sugar Complex to Browning** -It appears that in the lactose - protein reaction, responsible for the darkening of milk proteins, certain amino acids are involved. The sugar combines with NH_2 group which is alkaline, neutralizing the alkalinity and leaving the acid group free, thereby causing a reduction of pH accompanied by the formation of highly coloured product.
    - ☆ It is well known that sugar heated with an alkali turns brown; hence the amino acid - sugar complex which is alkaline would tend to assume a brown colour. This would be the case especially with a reducing sugar such as lactose, because of the availability of the aldehyde groups, which are naturally present in reducing sugar.
    - ☆ On addition formaldehyde, (being having a great affinity for amino acid) to evaporated milk at the time of sterilization, milk remains quite white.

iii. **Effect of steps of Processing on Colour-** It is found that time and temperature of heating is an important factor contributing towards colour of the milk. In addition this,effect of following steps, are also noted -

- ✰ Homogenization tends to diminish the colour of the product because of a finer sub division of the fat globules which prevent such penetration of the rays of light as it would reveal the butter fat more nearly in its natural colour which is yellow.
- ✰ Within the time of forewarming applied generally, the product suffers no decisive change in colour. If it is for long period (30′), effect is intense.
- ✰ Storage causes progressive darkening with increase in time and temperature. At (5°C) there is no change in colour. It is observed that for each 10 ° C decrease in forewarming temperature, the holding period is increased ~ 2.5 fold without causing an increase in colour.

2. Changes in Flavour-Cooked Flavour [2,4]

Heat treatment in processing tends to produce a cooked flavour in evaporated milk. It is generally observed that cooked flavour accompanies the darkening of colour.

- ✰ HTST process produces less cooked flavour than LTLT process.
- ✰ HTST sterilization treatment (135°C for 0.5 min) also reduces the extent of sterilization, decreases the viscosity and also fat separation *etc.* occur. So because of these limitation factors, LTLT process is important.

i. **Reaction Involved in Production of Cooked Flavour** -The reactions involved in production of cooked flavour are not sufficiently clear but the following facts help in understanding the possible reaction for producing cooked flavour-

- ✰ It is observed that when heating skim milk, cream, milk, *etc.*, -SH compounds are produced (provided heated to high temperature or at low temperature for longer time) from one or more proteins present and it is found that -SH compounds are wholly responsible for cooked flavour.
- ✰ Five hours heating at 70°C produces full cooked flavour.
- ✰ It is shown that in the reaction causing cooked flavour of milk by heat, O_2 is taken up and CO_2 is produced. Oxidation condition inhibits the formation of -SH group.

3. Sediments of Mineral Salts [2,4]

Some evaporated milk upon ageing show the tendency for granular deposit to form in the can. This deposit has a whitish colour, it is gritty and insoluble and seemingly of non-crystalline character. On analysis these granular structure are found to contain Tri Calcium Citrate - $Ca_3(C_6H_5O_7)_2$ and Tri Calcium phoshate - $Ca_3 (PO_4)_2$. These salts have peculiarity of being less soluble in hot rather than in cold solution.

i. **Effect of Processing on Mineral Salts-** With the increase in concentration of milk solids, there is an increase in the amount of sediment produced in evaporated milks. An increase in concentration necessarily increases Ca, Mg, citric acid and PO_4 content in the milk and thus provides possibility of precipitation. This can be controlled by addition of casein stabilizer.

ii. **Effect of Temperature of Storage on Sediment Formation** -The precipitation of mineral salts in the form of white sand like deposits is an age defect of Evaporated Milk. The temperature of storage appears to be a controlling factor in this respect. The Table 20.2 shows this effect.

Table 20.2: Effect of Temperature of Storage on Sediment Formation

Sl.No.	*Storage Temp./Storage time*	*Relative Amount of Ca-citrate Sediments*
1	29.4°c/30 days	Large amount of fine as well as fine particles.
2	20.0°C/78-110 days	Considerable amount of large particles.
3	7.2° C/4 months	No sediments

4. Effect of HTST Sterilization [2,4]

The impacts of HTST sterilization on concentrated milk are shown in Table 20.3.

Table 20.3: Effect of HTST Sterilization on Concentrated Milk

Sl.No.	*Properties/Physico-chemical Changes*	*HTST vs LTLT*
1	Flavour and colour change	The major advantages of HTST are to avoid cooked flavour and colour changes.
2	Retainment of natural property	By HTST process, natural property of milk is retained better than that with the process of low temperature long duration (LTLT) process.
3	Heat Stability	i. HTST sterilization gives heat stability better in concentrated product than with conventional method of sterilization. ii. The problems that are encountered in HTST process are insufficient viscosity and thin body which ultimately increase the tendency of fat separation on storage.
4	Self stability of product	Self stability of product is relatively small and gelation may occur during storage period.

Different types of HTST Processes

Some of the different types of processes developed are-

a. **Tin Sterilization**- Sterilization at 127°C-130°C for about 2 minutes to 40 seconds.

b. **Continuous Flow Sterilization**- Here temperature is 145°C or higher with only few seconds holding and with subsequent aseptic canning.

- ☆ For avoiding the problems due to this sterilization process, increase the viscosity and reduce the age thickening.
- ☆ Several methods are supplementing temperature treatment which may be given before or after sterilization. They may be given to the product during coming up period or cooking period and get the advantage of avoiding cooked flavour.

5. Effect on the Acidity of Concentrated Milk [2,4]

The acidity of the milk is affected by the heat treatment given to milk and storage time and temperature. The details on the effects on acidity during processing and storage, are given below:

a. Effect of Heat Treatment on the Acidity

When heat treatment is given to milk in manufacture of concentrated milk, the acidity measured in terms of H^+ concentration and titratable acidity is increased.

- ☆ The rate of formation of acid in concentrated depends upon the time and temperature treatment and lactose concentration. The rate of formation of acid in concentrated product is in proportion of time and temperature treatment.
    i. If the milk is heated in open pan, then the acidity will slightly drop in the beginning followed by an increase in acidity which will continue till the coagulation of casein. Initial decrease in acidity is due to loss of CO_2 dissolved in milk which usually contributes to 0.01-0.02 per cent of the total acidity expressed as lactic acid.
    ii. The increase in acidity on further treatment is due to breaking down of casein causing cleavage of phosphorus containing acid, probably nucleic acid and to a lesser extent to oxidation of lactose.
- ☆ The actual increase in acidity due to conventional temperature time ratio of forewarning in Sweetened condensed milk (SCM) and Evaporated milk is different.

SCM

The actual increase in acidity due to conventional temperature time ratio of forewarning in Sweetened condensed milk (SCM) is very slight. This slight increase in acidity is compensated by the initial loss of acidity due to expulsion of CO_2.

Evaporated Milk

In case of evaporated milk, the reaction of sterilization heat treatment in conventional type procedure is more and in addition, the milk is also heated at higher temperature of forewarming. Due to this treatment, usually there is increase in acidity approximately 0.05 - 0.1 per cent during the sterilization process than the normal acidity due to concentration.

b. Effect of Storage, Time and Temperature on Acidity

The acidity increase in concentrated milk is in direct relation with time and temperature of storage. The acid producing reaction due to heat treatment is continued even when the concentrated product is held in storage room.

- ☆ The rate of acid formation will be very slow depending upon the temperature of storage. SCM of 64.5 per cent sugar ratio having initial acidity of 0.43 per cent and the original bacterial count 4381/g when stored at different temperatures resulted in to increase in acidity as shown in Table 20.4.

Table 20.4: Storage Time/Temperature and Increase in Acidity

Sl.No.	Storage Temp./Storage time	Acidity Per cent
1	10.0°c/640 days	0.84
2	15.6°c/640 days	0.85
3	22.2°c/640 days	1.0

- ☆ This increase in acidity is mainly due to chemical changes as there was no increase in microorganisms during storage.

Changes in Sweetened Condensed Milk during Storage [2,3,4]

The following changes occur in SCM during storage-

a. Age Thickening

This is presumably the main change occurs in sweetened condensed milk during storage and, finally, gelation. Sweetened condensed milk markedly faster with age. It is usually assumed that added sugar (sucrose) inhibits age thickening. Sucrose increases the Ca^{2+} ? activity.

- ☆ There is difference in viscosity of SCM and Evaporated milks during before age thickening/during age thickening. In evaporated milk there is an initial decrease in viscosity before age thickening is not observed whereas the viscosity in sweetened condensed milk increases almost linearly with time. The following main factors affect.

- The age thickening in sweetened condensed milk is affected by the following main and important factors:
  - **i.** The variation in the type of milk and season occurs among batches of milk.
  - **ii.** Higher preheating treatment yields higher initial viscosity, and gel can form earlier. Hence, UHT heating is now generally applied.
  - **iii.** Higher concentration factor gives more age thickening.
  - **iv.** The addition of sugar at later stage in the evaporating process results in lesser age thickening.
  - **v.** The influence of added salts varies widely and depends on the stage at which it is added. Salts are added up to 0.2 per cent.
    - Adding a small amount of sodium tetra-polyphosphate (*e.g.* 0.03 per cent) mostly delays age thickening considerably, whereas adding more may have the opposite effect.
  - **vi.** Age thickening considerably increases with storage temperature.

b. Maillard Browning

Maillard reaction is chemical reaction between amino acids and reducing sugars that gives browned food and ongoing Maillard reactions are inevitable. Brown discoloration is stronger as:

- The storage temperature is higher,
- The milk is evaporated to a higher concentration and
- more intense heating is applied.
- Additional Maillard reactions occur if the added sucrose contains invert sugar.

c. Oxidative Changes

The oxidative changes effect the property/quality of SCM. These changes can be prevented by keeping head space oxygen to a minimum.

d. Lactose Crystallization [2,4,4a]

Good quality condensed milk should possess a smooth homogenous texture and be pleasant to the palate.

- Sometimes, however, the milk may be gritty and contain a large number of large lactose crystals. The solid particles are so large that the product lacks smoothness. This defect is readily detected by an average consumer. Sandy, rough, grainy, granular, and gritty are the terms used to describe this defect.
- Sweetened condensed milk (SCM) contains around 38 to 45 g lactose per 100 g water. Seeing the solubility of lactose in water at room

temperature, about 20 g of lactose is soluble in 100 g water, but in SCM this solubility is about half as much due to the presence of sucrose. It means that 75 per cent of the lactose tends to crystallize (meaning about 8 g per 100g SCM). Due to the high viscosity, nucleation will be slow and only a few nuclei would be formed per unit volume of milk, leading to large crystals.

i. The product (SCM) will obtain a relatively high quantity of large crystals and these crystals settle and are responsible to give sandy mouth feel. The crystals may not be so large as to be felt singly in the mouth, but they can be large enough to cause a non-smooth impression. The crystals smaller than about 8 μ m in length help in avoiding aforesaid problem.

ii. The prevention of crystallization in SCM is not possible and, accordingly, a large number of crystals should be obtained. Satisfactory results can be reached by applying followings:

   a. **By using seed lactose**- Adding 0.03 per cent seed lactose represents 0.004 times the amount of lactose to be crystallized. Correct cooling and induced rapid crystallization with the correct amount of seed lactose will help to avoid this defect.

   b. **By cooling-**The condensed milk must be cooled so that a large number of minute sugar crystals are formed, which results in a smooth texture. An optimum storage temperature is also essential to avoid this defect.

   c. **High viscosity-** High viscosity also delays crystal formation, therefore, it is necessary to rapidly cool the condensed milk in the initial stages.

Sources

1. *M.K. Srivastava, Handbook on Analysis of Milk Products. Write and Print Publications, New Delhi, 2016. ISBN: 9789384649258.*

2. *H.G. Patel, Condensed and Dried milks, www.Agrimoon.com.*

2a. *Soma Maji, Physico-chemical changes in concentrated milk, http: //courseware. cutm.ac.in/*

3. *Manish Kumar, Physico –chemical changes during preparation and storage of Concentrated milk, http: //www.basu.org.in.*

3a. *Physico-chemical changes taking place during manufacture of condensed milk-I, Lesson-13, http: //ecoursesonline.iasri.res.in.*

4. *Physico-chemical changes taking place during manufacture of condensed milk-II, Lesson-14, http: //ecoursesonline.iasri.res.in.*

4a. *S.D. Kalyankar and A.R. Sarode, Condensed Milk, in Encyclopedia of Food and Health, 2016.*

Dried Milk Products [1]

Milk powder is the cheapest source of milk solids for manufacturers. Both full cream and skimmed milk powder are used.

- ✫ Skimmed milk powder has two advantages in that not only does it allow the use of vegetable fat instead of milk fat but it also keeps better.
  i. A long shelf life is important in places where manufacturing milk is only available in the summer or where milk is not produced.
  ii. Almost all milk powder is now spray dried rather than roller dried. The spray drying process usually applies less heat than the old roller drying process. Low heat treatment makes the powder easier to reconstitute and improves the biological availability of the protein.
- ✫ The details related to the composition, Quality standards(FSSAI, BIS, Codex), nutritional values *etc.* of different types of milk powders, are discussed under chapter-16.

Physico-chemical Changes during Manufacture and Storage of Dried Milks [2,2a,2b]

Various physico-chemical changes occur in dried milk/milk powders during processing and storage. Hence effective and appropriate processing controls during processing and preventive measures during storage of milk powders need to be implemented. The various changes during manufacturing and storage of milk powders, are discussed below:

A. Changes during Manufacture/Processing [2,2a,2b]

During conversion from liquid milk to dry product, the milk powder manufactured undergo following changes-

- ✫ **Elimination of water**-The manufacturing process employed on milk (~87 per cent water, a_w=0.99) eliminates majority of water till it reaches the final stage of drying, the powder (~4 per cent moisture, $\mathbf{a}_w < 0.6$). In addition to the elimination of water, this process also changes the physical structure. In the process, the sol changes to a dust.
- ✫ **Dual Physical structure**-The dried milk/milk powder exhibits a well pronounced dual physical structure:
  1. **The Primary Structure**- It is the internal build-up of the powder particles from the milk solids and small amounts of moisture and air.
  2. **The secondary structure** -It represents a typical powder, a system of closely packed solid particles in a gas.
- ✫ **The structural elements** - The structural elements of dried milk particles are -

Lactose (either amorphous or partly crystalline), Proteins (casein micelles and whey protein particles), fat (in globular or non-globular -free fat forms) and air as spherical cells.

The fat, protein and air are presumably dispersed in continuous phase of amorphous lactose.

- **Protection against oxygen and moisture due to Fat content**- The protection against oxygen and moisture are different for whole milk powder and skimmed milk powder as these milk powders have different fat percentages.
  1. **Whole milk powder(WMP)and infant milk powders**- These powders are rich in fat (20-37 per cent) and hence need protection against oxygen and moisture, therefore, generally tinplate cans are used and the packaging is done under nitrogen cover to replace air with nitrogen in the tins.
  2. **Skim milk powders(SMP)**- SMP has very low fat (1.5 per cent) and as such a good moisture barrier material like multi wall paper sacks with one polythene liner is used.
- The milk powders are concentrated mass of milk components in the form of particles with small amount of moisture and air. If sufficient care is not taken to protect them from high humidity, high temperature of storage and the entry of air, the deterioration processes are accelerated.
- These changes while manufacture occurs in a sequence and they are discussed below:

1. *Changes of State of the Drying Droplets [2,2a,2b]*

- Atomizing pure water in a drying chamber in the usual way causes the water droplets to reach the wet-bulb temperature and to vaporize within 0.1s at this temperature.
- The presence of dry matter in the droplets, however, makes an enormous difference. The diffusion coefficient of water decrease substantially with increasing dry-matter content. Accordingly, the vaporization is significantly slowed down.
- In the drying droplet, temperature equalization occurs in less than about 10 ms. *i.e.* the temperature is virtually constant throughout a droplet during drying.

2. *Drying Stages [2,2a,2b]*

- Initially, the droplet has a very high velocity relative to the drying air.
- Therefore, there is a first stage during which circulation of liquid in the droplet occurs; this circulation greatly enhances transfer of heat and mass.

- ☆ For a droplet of 50 μ m diameter, this stage lasts for ~2 ms. In this time, the droplet covers a distance of about 10 cm and loses a small percentage of its water.
- ☆ Its velocity compared to the air decreases to the extent that the formed surface tension gradient of the drop surface arrests internal circulation of liquid.
- ☆ But in the second drying stage, the difference in velocity between drop and air is still great enough to accelerate water transport.
- ☆ The transport in the droplet occurs by diffusion but in the air by convection.
- ☆ After about 25 ms the relative velocity of the droplet has decreased so far that the water transport has become essentially equal to that from a stationary droplet.
- ☆ Relative to the air, the droplet then has covered a distance of a few decimeters and has lost about 30 per cent of its original water.
- ☆ In the third stage, lasting at least a few seconds, the droplet loses the rest of the water by diffusion.

3. *Drying Process [2,2a]*

- ☆ During drying the air temperature decreases and the humidity of the drying air increases. Moreover, the droplets vary in size and the smallest ones will dry fastest.
- ☆ In practice, the mixing is always between the two extremes.
  - **a.** For driers with a spinning disk, the situation tends to be fairly close to perfect mixing.
  - **b.** In most driers with nozzles, it tends to be closer to cocurrent flows.

  In all situations, the drying time may very by a factor of, say, 100 between the smallest and the largest drops. This is of great importance for fouling of the drying chamber; the largest drops have the greatest chance of hitting the wall and of being insufficiently dry to prevent sticking to the wall.
- ☆ Another factor that affects drying rate is the presence of vacuoles in the drying drops. It leads to significantly faster drying.
  - **i. Concentration Gradients [2,2a,2b]**-The difference in the concentration of a substance between two areas is called the concentration gradient. A rapid decrease in drying rate of the droplets occurs after the water content is reduced to 15 per cent. A strong concentration gradient forms rapidly during drying. The higher the drying temperature, the stronger the effect. Therefore, stronger gradients occur for cocurrent drying.

The concentration gradients are not of a lasting nature. The relatively dry outer layer of the droplet soon becomes firm and eventually glassy. This causes the droplet to resist further shrinkage. The droplet can react by forming vacuoles or by becoming dimpled. Especially at a low water content of the particles, so-called hair cracks may be formed.

ii. **Aroma Retention [2,2a,2b]-** Besides water, the drying droplets lose other volatile components, including flavour compounds (aroma). In many cases, however, the loss of flavour compounds is far less than expected, in spite of their volatility because of the following conditions:

   - ☆ The effective diffusion coefficient of most flavour components in the relatively dry outer layer of the droplet decreases far stronger with decreasing water content than the diffusion coefficient of water does due to the greater molar mass.
   - ☆ Thus, the aroma retention - retaining flavour compounds during drying - increases with:

     Droplet size because in larger droplets, the outer layer from which the flavour components get lost has a relatively smaller volume and

     Drying temperature because, at a higher temperature a solid rind forms more rapidly.
   - ☆ Formation of vacuoles diminishes aroma retention, especially if hair cracks develop in the particles and the vacuoles come into contact with the surrounding air.
   - ☆ Dissolved milk powder often has cooked flavour, which results during preheating and possibly during evaporation (flavour compounds formed). A cooked flavour mainly results from methyl ketones and lactones formed by heating of the fat and from Maillard products which are almost absent in skim milk powder.
   - ☆ During drying, conditions are mostly not such that off-flavours are induced. On the contrary, a considerable part of the volatile sulfhydryl compounds is removed.

iii. **Water Activity [2,2a,2b]**- The water activity (a_w) of a food is the ratio between the vapor pressure of the food itself, when in a completely undisturbed balance with the surrounding air media, and the vapor pressure of distilled water under identical conditions. Typically, as the water activity level is lowered, the rate of chemical degradative reactions decreases.

   - ☆ If the water content of a product decreases, its water activity (a_w) also decreases. Water activity is expressed as a fraction.

In pure water $a_w = 1$; in a system without water, $a_w = 0$. Water activity values of different dairy products are given in Table 20.5.

Table 2.5: Water Activity Values of different Dairy Products

Sl.No.	*Milk Products*	*Water Activity (a_w)*
1	Milk	0.995
2	Cheese	0.94-0.98
2	Ice cream mix	0.97
3	Concentrated milk	
	Evaporated milk	0.986
	Sweetened condensed Milk	0.83
4	Skimmed Milk Powder	
	1.50 per cent Water	0.02
	3.00 per cent Water	0.1
	4.50 per cent Water	0.2

Due to the very low water activity in milk powders, it exhibits following changes

a. **The hygroscopicity increases**-Usually, a (dry) product is called hygroscopic if a small increase of a w causes a considerable increase in water content. Obviously, this mainly concerns milk powder with very low water content.

b. **Diffusion coefficients decrease**- At low water content the effect is very strong. The diffusion coefficient of water decreases from approximately 10^{-9} $m^2. s^{-1}$ in milk to 10^{-16} $m^2. s^{-1}$ in skim milk powder with a small percentage of water.

B. Changes during Storage of Milk Powders [2,3]

It has been observed that the life of roller dried whole milk powder from cow's milk is about 10 months, the spray dried product however cannot be stored more than 6 months at 17°C and half of this time at 27°C. Although the milk powders are not sterile products, the low water activity does not permit microbial growth and microbial activity.

- ✰ The changes occurring in milk powders on storage are of physical, chemical and physico-chemical in nature.
- ✰ Both the lipids and the non-lipid constituents of milk undergo such changes and normally the rates of such reactions are accelerated with the elevation of storage temperature and levels of moisture and oxygen in the product.

- ✰ The quality of milk and the processing parameters adopted while manufacture of milk powders are also the major factors determining their storage stability.

  i. **Water Activity (aw) [2,3]-**The most important variable determining the rate of undesirable changes in milk powder is the water content. Moisture content defines the amount of water in food/milk powder and ingredients, but water activity explains how the water in food will react with microorganisms. The higher the water activity, the faster microorganisms like bacteria, yeast, and mold will be able to grow. Water activity is decreased by physically removing water, by adding solutes such as salt or sugar by freezing.Water activity of different milk powders, significance of a_w, effect on a_w *etc.* are given below:

  - ✰ The higher aw of whole milk powder as compared to skim milk powder of the same water content is caused by the fat not affecting a_w.
  - ✰ The a_w is considerably reduced if lactose crystallizes without absorption of water by the powder, at least if a w is less than about 0.5.
  - ✰ Crystalline lactose binds water very strongly, and that is also why the usual oven-drying methods to estimate the water content do not include the bulk of the water of crystallization. If the water content excluding the water of crystallization is taken as a basis, then a_w is even higher for the powder with crystallized lactose.
  - ✰ Whey powder has a_w slightly different from that of skim milk powder because in a dry product the soluble constituents (especially sugar and salts) decrease a_w somewhat less than casein. This is only true, however, as long as all lactose is amorphous, which often does not apply to whey powder.
  - ✰ It is thus advisable to make milk powder sufficiently dry and to keep it in that condition. If it is not hermetically sealed from the outside air, it will attract water in most climates.

    The higher the temperature, the higher the water activity. Because several reactions are faster at a higher a_w, this implies that a temperature increase may well cause an extra acceleration of deterioration.
  - ✰ The deterioration effect may be especially strong if the powder loses its glassy state. A lactose-water mixture will be at most ambient temperatures in the glassy state if its water activity is below 0.3. Because lactose is the dominant component of the amorphous material in a powder particle, about the same

relation is supposed to hold for the powder. This means that most dairy powders are in the glassy state (*i.e.*, the nonfat part of the material), except if the water content is high and the temperature is also high.

ii. **Colour, flavour and off flavour characteristics [2,3]**-The user of dried milk is concerned mainly with the flavour defects which arise during storage. The following important aspects affect the colour and flavour of dry milks:

- ✰ Some feed taints may be removed during the course of vacuum evaporation or drying, but not always.
- ✰ The use of poor quality milk is not desirable for manufacture of good quality dry milk.
- ✰ Acidity already developed in the milk will be retained in the powder, which acquires an unpleasant flavour.
- ✰ The flavour of the Milk powders produced from neutralised milk, is remains unsatisfactory and the colour of such milk powder may darken. First quality powder should not exceed an acidity of 0.15 (as lactic acid), and no powder should exceed 0.17 per cent. Very acid milk cannot be dried by the roller process.
- ✰ The most likely defect of freshly made powder from good quality milk, is some degree of burnt flavour.

  Roller Dried Powder-Burnt flavour may arise much more readily in roller powder owing to overheating on the rollers or faulty adjustment of the scraper knives. In any case, roller powders always possess a distinct cooked flavour.

  Spray dried powder-A distinct burnt flavour is only likely to occur in spray powder following considerable overheating in the drying chamber. The flavour of spray powder is influenced considerably by the milk preheating temperature used.

  i. "Low-heat" powder or powder from "ultra-high temperature" processed milk gives reconstituted milk having a flavour close to that of pasteurized milk.

  ii. "High-heat" powder when reconstituted possesses a more cooked flavour, which however, is not unpleasant and indeed is preferred by some consumers.

- ✰ In practice, storage changes produce various flavour defects which mask other factors and after 2-3 months the "low-heat" powder is commonly inferior if special precautions have not been taken.

- ☆ The colour and flavour changes will progress during prolonged storage if conditions like elevated temperature in combination with increased moisture content prevail. Two major types of deterioration are well recognized, affecting fat in one case and protein plus lactose in the other. Ultimately these affect the flavour, solubility, colour and nutritional value of the powder. Two major categories of reactions are:

  a-The oxidation of lipids and b- The non enzymatic browning commonly known as the Millard reaction.

  The former is associated with the tallowiness whereas presumably both these reactions contribute to stale and allied flavours.

  With the advancement of storage, the discolouration will also progress and fluorescence substances are also produced.

iii. **Fat Decomposition [2,3]** -Oxidation of the fat leading to the production of flat, and finally marked tallowy off-flavours is a major storage defect of full cream powder and some times occurs to a lesser extent in the residual fat content of separated milk powder.

The following observations were made by different researchers

- ☆ The chemical changes result from the addition of oxygen to the double bonds of unsaturated glycerides, giving peroxides which can be estimated by chemical means.
- ☆ At first, the peroxides accumulate but later they decompose to various aldehydes, ketones and acids which impart the unpleasant flavours.
- ☆ Peroxides are formed when the container contains an excess of free O_2, but later decomposition of peroxides exceeds the rate of their formation.
- ☆ In the oxidation of linoleic acid, the C-11 methylene group is preferential point of attack. Many other volatile aliphatic compounds such as alcohols, lactones, esters, acids and hydrocarbons arise from oxidation of unsaturated fatty acids.
- ☆ Aldehydes are detectable with colour reagents but the sense of taste is much more sensitive than chemical tests.
- ☆ The reaction commences with a slow linear induction phase, followed by a period of rapid exponential change and combination with oxygen, large quantity of O_2 can be absorbed. The break point at which the oxidation changes from first to second phase is generally found to be at 37 weeks in whole milk powder.

- ☆ The keeping quality of individual "low heat" powders varies considerably - from 4-12 months in temperate climates and perhaps half this time in hot climates.
- ☆ Though the development of poor flavour is parallel to O_2 absorption, the ratio is not constant and the most stable powders require the most O_2 in order to deteriorate to a given flavour value.
- ☆ The odour and flavour profiles of dried milk with addition of antioxidants (BHA/BHT and Maillard reaction mixtures from histidine and glucose) and stored under air and nitrogen are as follows:

  The Maillard reaction products are as effective as BHA/BHT in retarding the development of oxidative off flavours and off odours.

  Samples to which antioxidants are added or samples stored under nitrogen retain a higher acceptance for storage times as long as 84 weeks.

  The 2,3 butadione has a sweet and buttery odour and flavour at low concentrations and the branched chain aldehydes, the Strecker aldehydes are generally described as having fruity odours and flavours at low concentrations.

  The sulphur compounds are known to impart cooked odours and flavours.

  Some compounds, 1-octen-3-one and 1-octene-3-ol that are known to give rise to metallic and mushroom odour and flavour respectively in oxidatively deteriorated milk products.
- ☆ Factors affecting oxidative stability of lipids in dried milks include oxygen, moisture, temperature, light and metals like copper and iron. These influence the system depending upon their fat content, presence or absence of antioxidant and packaging under the nitrogen atmosphere.

The Use of Antioxidants

i. The application of antioxidants instead of nitrogen gas packaging, derived from various materials showing promise as practical antioxidants are- wheat germ oil, gum guaiac, rice bran concentrate mixed tocopherols, oat flour and special preparations made from ethyl gallate, ascorbic acid, nordihydroguaiaretic acid (NDGA), thiourea, hydroquinone monobenzyl ether, flavones, querectin, ascorbyl, palmitate, histide, tryptophan, phosphoric acid either alone or in combination of a few.

ii. For addition of any antioxidant legal acceptance is must and for the present in India, no antioxidant other than lecithin, ascorbic acid and tocopherol is permitted accepting BHA not more than 0.01 per cent by weight in milk powder. In infant milk no antioxidant is permitted.

iii. Inspite of some fairly promising indications that antioxidants retard tallowiness in dry milks, packaging in inert gas is the most efficient method at present available for preventing fat deterioration.

iv. **Fat Hydrolysis [2,3]**-True rancidity results from the hydrolysis of fat to free fatty acids such as butyric acid. The lipolytic enzymes involved can be derived from the milk itself or they may be of microbiological origin. This defect has become unusual since strict attention is now given to both milk production and plant hygiene and higher preheating temperatures which inactivate lipases.

v. **Other Forms of Fat Deterioration and their prevention [2,3]** -The other flavours are described as "stale" or most characteristically as like **"Coconut"**. A **toffee flavour** may also be a form of the same defect. They are produced only from milk fat and do not occur in dried separated milk.

- ✰ The defect is accelerated by high-temperature storage and by high moisture content, but is independent of O_2 and occurs in gas packed powder.
- ✰ This form of deterioration seems to occur at a fairly early stage during storage, it develops before the fat oxidation defect from which it also differs in flavour.
- ✰ There is strong evidence that the change is caused by a rearrangement of monoethenoid fatty acids in milk fat to form lactones and that δ-decalactone in minute quantities is the substance responsible for the coconut flavour. The origin of this lactone is thought to be 9-deconoic acid. These rearrangements involve no O_2 and the empirical formulae remain the same. No definite method of prevention has been suggested except low-temperature storage.
- ✰ In the absence of special precautions, oxidation occurs rather more rapidly in spray than in roller powders, but in recent years much progress has been made in extending the life of spray powders. To keep the autoxidation within reasonable limits for a long time, following measures should be taken-

a. **High Temperature Preheating [2,3]** The milk should be intensely heated to form antioxidants. The problem is, of course, that the heat treatment also causes a distinct cooked flavour.

- ✰ Compared with pre-heating the milk at 73.8 °C, preheating at 82.2 °C improves the keeping quality of the powder, whilst at 87.8 °C, the increase is 3-5 times. The mechanism is not fully understood, but in general it is associated with the formation of trace of protein decomposition products containing a sulfhydryl group which function as antioxidants and prolong the induction period.

- ☆ Natural milk proteins contain no free -SH groups, but in the denatured protein formed in heated milk, such groups become unmasked and reactive. The first action may be the formation of cysteine followed by further reaction at the exposed sulfur ends of the chains to split off H_2S or methyl mercaptan. Traces of H_2S are detectable in the heated milk. The action of -SH group is uncertain, but may be due in part to prevention of the catalytic action of heavy metals by removing them as sulfides.
- ☆ High temperature preheating may also prevent changes produced by enzymes derived from the raw milk or from bacteria, thus improves the bacteriological quality of the powder and preserves the vitamin content during storage.

b **The Rate of Autoxidation:** Strongly increases with decreasing a_w,however, to prevent other types of deterioration (especially Maillard reactions) a_w should be as low as possible. The effective Q 10 of the autoxidation reaction in milk powder is relatively low (about 1.5) because a higher temperature also causes higher a_w. The most suitable water content is generally 2.5 to 3 per cent [2,3].

c. **Packing in Inert Gas [2,3]:** Milk powders packed in inert gas have better quality and long shelf life. This is due to the prevention of the chemical reaction by removing the O_2 from the container and replacing it with N_2 or CO_2 usually the former. It is the only method which completely prevents oxidation for practical purposes and can extend storage life up to 7-10 years even at high temperature. Oxygen should be removed as effectively as possible by gas flushing.

- ☆ A problem is that the vacuoles in the powder particles contain some air, hence, O_2. Either the powder should contain hardly any vacuoles or the gas flushing should be repeated after a few days. Equilibrating the gas inside and outside the vacuoles by diffusion takes several days in whole milk powder (several weeks in skim milk powder). Equilibration is faster if the powder particles have a greater number of cracks.

d. **Addition of Antioxidants [2,3]:** Antioxidants play important role in enhancing the keeping quality of food products.

- ☆ Ascorbic acid added @0.03 per cent in the liquid milk prolongs powder keeping quality by several months but disappears during storage.

- ☆ The most effective antioxidants appear to be esters of gallic acid particularly Ethyl or Propyl gallate. The incorporation of as little as 0.07 per cent of ethyl gallate in the powder prolongs the keeping quality to about two years when only low temperature preheating has been used. Ethyl gallate is non-toxic and has no effect on flavour; it does not disappear during storage and appears to remain unchanged.
- ☆ Keeping quality may be extended still further by combining high-temperature preheating with the addition of ethyl gallate. In this way, it is possible to produce powders having a keeping quality of from 3-4 years in temperate climates.
- ☆ Nordihydroguaiaretic acid has also been found effective when added @ 0.04 per cent of the weight of the fat.

e. **The Rate of Deterioration:** It increases rapidly with rise of storage temperature. The keeping quality of an unstable powder may fall as low as 6 weeks in a tropical climate.

f. **Oxidation:** It is more rapid in Powder of high acidity.

g. The powder should be packaged in such a way that air and light are kept out. Generally, this implies packaging in cans.

h. Rigorous measures should be employed against contamination of the milk with heavy metals such as copper and iron which act as catalysts and shorten the induction phase. Stainless steel equipment should be used.

i. Other less certain factors are the possible presence of oxidizing enzymes of microbiological origin in the milk and seasonal variations in the glycerides-composition of the fat.

j. Intensive homogenization of the concentrate should be carried out.

vi. **Protein-Lactose/Maillard Reactions [2,3]** -Millard reactions between protein and lactose produce a series of storage defects which often occur together in a progressive sequence of stale flavours, increasing loss of solubility, unpleasant "glue" flavour and darkening or brown discoloration. These changes may occur in all types of powder but are associated predominantly with separated dried milk and tend to be most pronounced in roller dried powders.

- ☆ Milk powders contain relatively high concentrations of lactose and protein high in lysine (an amino acid) content. In the presence of moisture, these components readily participate in the Maillard reaction. This interaction may result in changes in protein quality that is accompanied or followed by undesirable colour changes. Typically both the temperature and humidity of storage conditions have an effect on the Maillard reaction.

- ☆ They are greatly accelerated by high-storage temperatures and in addition, the moisture content of the powder is a major factor. Critical moisture levels appear to be about 5 per cent for spray powder and 4 per cent for roller powder. This reaction in addition to accounting for the major part of browning, the protein carbohydrate complex or its decomposition products also result in production of reducing substances, fluorescent compounds and disagreeable flavour materials.
- ☆ About 80 different compounds have been isolated and identified which include furanics, lactones, pyrazines, pyridines, acetyl pyrrole, amides, pyrolidinone, succinate, glutarimides, carboxylic acids, acetone, 2-heptanone, maltol *etc.* Some of the findings in this regard are:
- ☆ Maillard reactions increase considerably with water content and with temperature. They lead to browning and to an off flavor. The gluey' flavor that always develops during storage of dry milk products with too-high water content is usually ascribed to Maillard reactions; the main component appears to be o -aminoacetophenone.
- ☆ If extensive Maillard reactions occur, they are always accompanied by insolubilization of the protein. Accordingly, the insolubility index increases when milk powder is stored for long at a high water content and temperature.
- ☆ One of the first reaction products detected in Maillard reaction is hydroxymethyl furfural (HMF), the concentration of which increases with storage time, temperature and moisture content.
- ☆ Many of the products of Maillard reaction are aroma substances (aldehydes, reductones, other furfurals *etc.*) which have an appetizing odour, but high levels of these compounds may cause considerable odours and off flavours. The ε-amino groups of lysine are mainly involved and because the compounds formed during the reaction such as fructoselysine, furosine *etc.* are resistant to enzymes, the content of available lysine is reduced, but usually only to a small extent.
- ☆ Browning reactions of the Maillard type occur if the storage of powdered dairy products at ambient temperature and having the moisture content ≥ 5 per cent. The HMF content is further increased in milk powder with added iron or added Vitamin A.
- ☆ With the progress of Maillard reaction, the decrease of amino-N is closely related to increase in combined sugar, increase in oxygen absorption and carbon dioxide production, increase in colour and reducing power and decrease in solubility.

- ☆ A number of compounds have been shown to inhibit the browning reaction.
    - **i.** In milk products, active -SH groups serve as natural inhibitors in retarding the heat induced browning, but the mechanism is not understood.
    - **ii.** Sodium bisulphite, sulphur dioxide and formaldehyde also inhibit browning in milk system as well as in simpler amino acid-sugar solutions.

vii. Deterioration of a high moisture powder [2,3]- It proceeds in a series of following stages:

- ☆ First the lactose absorbs moisture and changes to crystalline L-lactose monohydrate containing 5 per cent water of crystallization.
- ☆ In powders of low moisture content, the lactose is present in the amorphous α and β-forms (1: 1.5) with only 0.5 to 5.0 per cent of lactose hydrate.
    - i. If the moisture content and the temperature remain low, the lactose preserves its amorphous form and no deterioration occurs.
    - ii. If the moisture content is high, as much as 40-60 per cent of the lactose may be hydrated in a fresh powder and at levels of 6.5 to 7.0 per cent for full-cream powder and 7.5 to 8.0 per cent for separated milk powder crystallization becomes rapid and complete (dried milk can also absorb moisture very readily from the air). Such powders may show an apparent decrease in their moisture content.
- ☆ There is often a loss of free-flowing properties and possibly some definite "Caking" of the powder. The impervious lactose envelope around spray powder particles probably changes to a crystalline lattice.
- ☆ The free lactose content of the powder then falls and an insoluble protein-lactose compound is formed which contains lactose equivalent to the quantity of the free soluble lactose which has disappeared. The reaction is presumed to occur between the aldehyde group of the lactose and a protein-amino group. The amino group mainly involved is that of lysine, and about 40 per cent of the original lysine disappears, which considerably reduces the nutritional value of the powder. As the protein - lactose reaction proceeds, the remaining protein also becomes progressively insoluble and eventually the loss of solubility may be pronounced, particularly of roller -dried milk.
- ☆ The final stage involves a decomposition of the protein - lactose compound to form products which include substances having an

unpleasant glue-like flavour and a brown colour. The reaction involves absorption of O_2, and the evolution of CO_2, but the exact nature of the breakdown and the identification of the decomposition products are uncertain. In general, the O_2 reacts preferentially with the fat in full-cream powders, and tallowy flavours predominates, whilst the stale and gluey flavours are more characteristic of separated milk powder, but both forms of deterioration can proceed simultaneously.

- ✰ If the powder is gas packed, the protein lactose reaction and loss of solubility may still occur, but the final changes are much less serious. There may be some development of brown colour and a slight caramalize flavour, but unpleasant flavours are unusual. It would appear that development of the gluey flavour requires a supply of O_2. Gas packing is therefore partially effective but is not a real solution. These defects of browning can be avoided more simply and with greater certainty by producing powders
    1. By limiting heat treatments and time,
    2. By keeping low moisture content and
    3. Using moisture-proof packaging combined with low storage temperatures.

Sources

1. *M.K. Srivastava, Handbook on Analysis of Milk Products. Write and Print Publications, New Delhi, 2016. ISBN: 9789384649258.*

2. *H.G. Patel, Condensed and Dried milks, www. Agrimoon.com.*

2a. *Soma Maji, Physico-chemical changes in concentrated milk, http: //courseware. cutm.ac.in.*

2b. *The milk Powder System, Lesson 38, ecoursesonline. iasri.res.in.*

3. *Defects in dried milk during manufacture and storage, their causes and prevention-I, Lesson -42, ecoursesonline.iasri.res.in.*

Chapter 21

Ice Cream and Frozen Desserts

Ice cream and frozen desserts are popular throughout the world. Although ice cream is a popular frozen dessert in all parts of the world. Ice cream and frozen desserts are a food category that is characterized by normally being aerated and eaten while frozen [1,2,3].

- ☆ An ice-cream is made from milk fats while a frozen dessert is made from vegetable oil fats. Vanaspati is a hydrogenated vegetable oil used in cooking, a cheaper substitute for ghee or butter.
- ☆ The Food Safety and Standards Authority of India (FSSAI) has categorised ice-creams under "dairy based desserts/confections", further categorised as plain, medium fat and low fat depending on the percentage of total solids, milk fat and milk protein used.

The food regulator defines "frozen dessert/frozen confection" as a product obtained by freezing a pasteurized mix prepared with milk fat and/or edible vegetable oils and fat. The milk fat is used in such dessert for freezing.

- ☆ Ice cream is unique in that it is the only product that is consumed in the frozen state. Therefore, once it is manufactured it has to be stored, transported, distributed, and sold in the frozen state.
  - i. The imperfection in cold chain affects the quality of the product at the point of purchase and impacts consumer satisfaction.
  - ii. Factors affecting the shelf-life of ice cream are manufacturing procedures, warehouse equipment, warehouse handling practices, transportation, storage at retail premises, retail display equipment, and retail handling practice [1,2].

- Ice cream is a complex food colloid that consists of air bubbles, fat globules, ice crystals and an unfrozen serum phase. Ice crystals and air bubbles are generally in the range of 20–50 μm. The air bubbles are usually to some extent coated with fat globules and the fat globules are coated with a protein/emulsifier layer.
    i. Ice cream is a multiphase soft solid composed of ice crystals, casein micelles, air cells, and emulsion droplets (both individual and clustered), all embodied within an unfrozen serum phase that contains dissolved components.
    ii. These disparate parts interact as ice cream melts in the consumer's mouth, which defines the overall quality or acceptability of the product [1,2].
- Various steps in the manufacturing process(pasteurization, homogenization, ageing, freezing, and hardening) contribute to the development of this structure. Proteins and emulsifiers compete for inter facial space during the homogenization of the fat and the creation of the mix emulsion. Following homogenization, the emulsion is further affected by changes occurring during the ageing step, *viz.*, crystallization of the fat and rearrangement of the fat globule membrane to the lowest free energy state. This emulsion then undergoes both whipping and ice crystal formation during the dynamic freezing process, which contributes to the development of the four main structural components of the frozen product: discontinuous foam, a network of partially coalesced fat surrounding the air bubbles, ice crystals, and a continuous, freeze-concentrated, unfrozen aqueous solution [3].
- The details related to the composition, Quality standards(FSSAI, BIS, Codex), nutritional values of Ice creams and Frozen Desserts are discussed under chapter-16.

Sources

1. *M.K. Srivastava, Handbook on Analysis of Milk Products. Write and Print Publications, New Delhi, 2016. ISBN: 9789384649258.*

2. *H. Douglas Goff, Ice cream and in Frozen Desserts, Encyclopedia of Food Chemistry, 2019.*

3. *Stabilizers and Emulsifiers-classification, types, properties and role in ice cream, Lesson-7, http: //ecoursesonline.iasri.res.in.*

Physico-chemical Changes during Processing and Storage of Ice Cream and Frozen Desserts [1-4]

Various physico-chemical changes occurred during manufacturing and storage of Ice cream and Frozen desserts, are discussed on next page.

1. Physico-chemical Changes during Processing of Ice Cream and Frozen Desserts [1]

To produce ideal body and texture, the incorporation of air in ice cream during its processing, is essential. The amount of air that can be whipped into the mix will be influenced by the composition of the mix and by the regulations. In general, as a guide, maximum overrun should be 2.5 to 3 times the total solids content to avoid possible defects in the finished ice cream. Processing effects the physico-Chemical Properties(Whipping Ability of Mixes,etc.) of ice cream.

A. Whipping Ability of Ice Cream Mixes

The mix and air in-flow and ice cream out-flow rates determine the time that the mix spends inside the barrel (known as the residence time, typically 30-60 s), the overrun, the pressure inside the barrel(typically 5 atm.) and the throughput (which can be as much as 3000 lh-1in a large industrial freezer). All of these, together with the dasher rotation speed (typically 200 rpm), determine the outlet temperature.

- ☆ Modern factory freezers are computer-controlled, allowing easy monitoring and control of the process parameters.
- ☆ Air is injected into the barrel through a system of filters to ensure that it is clean, dry and free from microbiological contamination.
- ☆ Initially the air forms large bubbles. It is essential to create (and maintain) a dispersion of small air bubbles to obtain good quality ice cream. The beating of the dasher shears the large air bubbles and breaks them down into many smaller ones: the larger the applied shear stress, the smaller the air bubbles.
- ☆ Long residence times also lead to small air bubbles.
- ☆ It is easier to whip air into a foam that consists of a large volume fraction of liquid and a small volume fraction of air than vice versa. The high pressure inside the barrel reduces the volume of the air that has been introduced, and therefore makes it easier to aerate further.

  In a standard ice cream formulation, sufficient partial coalescence occurs to enable a stable air cell structure to be maintained at overruns up to about 120 per cent.
- ☆ It can be difficult to obtain overruns of more than about 60 per cent in products where fat and protein are not present, or only present in small quantities, such as sorbets.
- ☆ Similarly it is difficult to obtain high overruns if there is insufficient shear (for example because of a very short residence time) to produce partially coalesced fat. Extra shear, and hence increased de-emulsification, can be produced either by increasing the dasher speed or by using a closed dasher. Whilst the mix is aerated, it is simultaneously frozen.

Factors affecting Whipping Ability of Mix [1,2]

The factors affecting whipping ability of mix (Of great importance in batch freezing but not so important in continuous freezers) are discussed here. The factors can be grouped in Milk composition/Milk Solids, Additives and Processing factors.

a. Factors Related to the Milk Composition/Milk Solids

i. Total Solids [1,2]

Total solids includes the sum of all solid and dry ingredients of ice cream. Fat, sugar, MSNF, stabilizers and emulsifiers all contribute to total solids.

- ☆ Total solids in ice cream mix play important roles. They replace water in the mix, thereby increasing the nutritive value and mix viscosity and improving the body and texture of the ice cream.
- ☆ Increasing the percentage of total solids up to a certain extent decreases the percentage of frozen water and permits a higher overrun in ice cream.
- ☆ A heavy soggy product may result when the total solids content is too high *i.e.* above 40-42 per cent.

ii. Milk Fat/Butter Fat [1,1a,2]

Use of butter with emulsifier in mix has a positive influence on whipping ability of mix. Milk fat (MF) provides richness of flavor, carries fat-soluble flavors, and lubricates the mouth as well as the barrel of the freezer. Importantly, it partially coalesces during freezing to form a structure that helps to stabilize the foam structure and to give body (resistance) to the ice cream [1a].

- ☆ Limitations on excessive use of fat in a mix include followings:
  a. Cost,
  b. A hindered whipping ability,
  c. Decreased consumption due to excessive richness and
  d. High caloric value.
- ☆ A partially crystalline fat droplet is necessary for clumping to occur. This has been attributed to the protrusion of crystals into the aqueous phase, causing a surface distortion of the globule. The crystal protrusions can then pierce the film between two globules upon close approach. As the crystals are preferentially wetted by the lipid phase, clumping is thus inevitable. This phenomenon may account for partial clumping of globules under a shear force. The clusters thus formed actually hold the ice cream serum in their interstices, resulting in the observed dryness. These fat globule chains may also envelop the air cells, thus improving overrun, but fat crystals are also known to impair overrun development in whipped cream.

iii. Milk Solids-Not-Fat (MSNF) [1,1a,2]

The non fat portion of milk is composed of lactose, protein, minerals, and small amounts of vitamins, acids, and enzymes. Milk solids-not-fat (MSNF) or **serum solids** improve the texture of ice cream, aid in giving body and chew resistance to the finished product, are capable of allowing a higher overrun without the characteristic snowy or flaky textures associated with high overruns, and may be a cheap source of total solids.

- ☆ Less milk-solids not fat (MSNF) and higher fat contents caused rich mouth coating but textural attributes were reduced. As the MSNF contents are enhanced, the creamy, phenolic, buttery and caramel notes were also enhanced. On the other hand, ice crystals, coldness and meltdown were reduced with MSNF addition.

Lactose [1a]

Lactose can be undesirable in the diets of persons whose bodies produce insufficient β-D-galactosidase to hydrolyze the lactose. Therefore, it may be desirable to remove it by ultrafiltration and diafiltration or to hydrolyze it enzymatically.

Proteins [1,1a,2]

Along with their high nutritional value and favorable effect on flavor, milk proteins contribute desirable physical properties to ice cream. As good foaming agents, they are vital to the incorporation of air. They provide strength to the lamellae of the air cells [1a].

- ☆ Freeze-concentration of proteins in ice cream greatly increases the viscosity of the unfrozen phase, and this has a great effect on ice crystallization, ice crystal stability, and solute mobility [1a].
- ☆ Proteins contribute much to the development of structure in ice cream, including emulsification, whipping, and water holding capacity. The interfacial behaviour of milk protein in emulsions is well documented, as is the competitive displacement of proteins by small molecule surfactants [1,2].
    - ❖ Milk proteins are well known for their foaming properties and during the manufacture of ice cream, air is incorporated to about 50 per cent phase volume. Thus it should not be surprising that milk proteins contribute to stabilizing the air interface in ice cream. This air interface is very important for overall structure and structural stability. Loss of air can lead to a defect known as shrinkage, the occurrence of which is fairly common and very significant for quality loss and unacceptability of the product [1,1a,2].

- The proteins,which make up approximately 4 per cent of the mix, contribute much to the development of structure in ice cream including:
  a. Emulsification properties in the mix.
  b. Whipping properties in the ice cream.
  c. Water holding capacity leading to enhanced viscosity and reduced iciness.

iv. Minerals (Citrate, Phosphate, Calcium and Magnesium) [1,1a,2]

The minerals of ice cream, being mostly dissolved, reduce the freezing point, thus affecting the energy needed to freeze the product, rate of melting, and ease of dipping. They also contribute to the flavour and nutritional value [1a].

- These ions decrease tendency for fat coalescence (Sodium citrate, Disodium Phosphate). They prevent churning in soft ice cream for example, producing a wetter product.
- These salts decrease the degree of protein aggregation. Calcium and magnesium ions have the opposite effect, promote partial coalescence. Calcium sulfate, for example, results in a drier ice cream. Calcium and magnesium increase the degree of protein aggregation.
- Salts may also influence electrostatic interactions. Fat globules carry a small net negative charge, these ions could increase or decrease that charge as they get attracted to or repelled from the surface.

b. Factors Related to the Additives Added during Manufacturing [1,2]

i. **Stabilizers [1,1a,2]-** Stabilizers are added to ice cream for the primary purpose of affecting the texture and they give a velvety-smooth product with a moderate rate of melt, limited coldness, and a uniform distribution of particulates by absorbing water (therefore swelling) and limiting its migration, increasing viscosity, adsorbing to air cell lamellae, and, in some instances, forming a gel-like structure. Overuse of stabilizers can cause excessive mix viscosity, a heavy or soggy body (gumminess), interfere with flavour release and undesirable melting characteristics(poor melt down) [1a]. Initially stabilizers were given the name as holders, binders, fillers, improvers, hydrocolloids, gums and colloids due to their specific functionality of influencing the body and texture of the product.

   Basic Types of Stabilizers Added in Ice cream are of following two Sources

   - **Animal source**- Gelatin from bones and calf skin.
   - **Plant or vegetable source**-Gums (agar-agar, carboxymethyl cellulose, sodium alginate, acacia, oat, carrageenan and karaya).

Other details related to the Stabilisers are given under "Role and mechanism of Stabilisers and emulsifiers".

ii. **Emulsifiers [1,2]-** During manufacturing of ice cream, emulsifiers are added to smoothen the texture and thorough distribution of air cells. Mono and di-glycerides are most common emulsifiers used in ice cream plants. Emulsifiers which are used in ice cream are usually integrated with the stabilizers in proprietary blends, but their function and action is very different from the stabilizers.

- Emulsifiers are used for improvement of the whipping quality of the mix and Ice cream. The primary effect of emulsifiers in ice cream is related to their ability to de-emulsify the fat globule membrane formed during homogenization [1b].
- This de-emulsifications arising from the disruption of the fat globule membranes during freezing, facilitates the agglomeration and coalescence of fat globules, leading to partial churning out of the fat phase [1b]. Paradoxically an additional beneficial function of emulsifiers is to destabilize the fat globule. This permits a degree of clustering and prevents rapid melting [1c].
- The agglomerated fat globules stabilize air cells. Thus, emulsifiers are used to improve the whipping qualities of ice cream by producing smaller ice crystals and smaller air cells, resulting in a smoother ice cream texture and drier, stiffer ice cream [1b].

Other details related to the emulsifiers are given under "Role and mechanism of Stabilisers and emulsifiers".

iii. **Sugar [1,1a]-**Sucrose has been the major sweetener used in ice cream. It depresses freezing point and prolongs whipping. Sucrose performs following various functions in ice cream-

Optimum palatability, Maintain handling properties, Acceptability of ice cream, Sweetens the product, Increase pleasing sweet flavour, Enhance solid contents and viscosity and Texture and body improvement.

iv. **Fruits and flavours-**Depresses freezing point if they contain high sugar, hence prolong whipping.

c. *Factors Related to the Processing*

i. **Pasteurization** -Higher temperature enhances whipping.

ii. **Homogenization-**It Mixes up to 10 per cent fat content are subjected to homogenization to achieve optimal fat structuring and ice cream meltdown.

- The net effects of homogenization are in the production of a smoother, more uniform product with a greater apparent richness and palatability, and better whipping ability.

- ☆ Homogenization also decreases the danger of churning the fat in the freezer and makes it possible to use products that could not otherwise be used, such as butter and frozen cream.

iii. **Ageing** -An aging time of 4 h or greater is recommended following mix processing prior to freezing. This allows for hydration of milk proteins and stabilizers (some viscosity increase occurs during the aging period),crystallization of the fat globules, and a membrane rearrangement, to produce a smoother texture and better quality product.

- ☆ Non aged mix is very wet at extrusion and exhibits variable whipping abilities and faster meltdown. The appropriate ratio of solid: liquid fat must be attained at this stage, which depends on the temperature and the triglyceride composition of the fat used, as a partially crystalline emulsion is needed for partial coalescence in the whipping and freezing step.
- ☆ Emulsifiers generally displace milk proteins from the fat surface during the aging period.
- ☆ The whipping qualities of the mix are usually improved with aging. Aging is performed in insulated or refrigerated storage tanks, silos, *etc.* Mix temperature should be maintained as low as possible (≤ 5°C) without freezing.

iv. **Freezer** -Batch freezing processes differ slightly than the continuous systems.

Batch Freezing

The barrel of batch swept surface heat exchanger is filled to about one half volume with the liquid mix. The freezing unit and agitators are then activated and the product remains in the barrel for such time so as to achieve the proper degree of overrun and stiffness.

Whipping is not controlled but cannot exceed that which will fill the barrel with product *i.e.* 100 per cent overrun when starting half full. Whipping of air into ice cream mix without freezing will result in low overrun and large sized air bubbles.

Continuous Freezer

In a continuous scraped surface freezer, numerous processes take place that ultimately influence the overall quality of the ice cream.

- ☆ One of the most important steps, of course, is freezing water into ice. At the same time as ice is being formed, there is also air incorporation, leading to development of air cells and the desired overrun.
- ☆ At the same time that freezing is taking place within the barrel, changes are also occurring to the lipid phase and air component.

- ☆ In commercial scraped-surface freezers, filtered compressed air is injected under pressure through a diffuser at the end of the barrel where the mix is input. The fine air bubbles formed in the diffuser are incorporated within the mix as the dasher rotates within the barrel. The air cells are broken down into smaller and smaller bubbles based on the shear forces within the freezer as the ice cream is formed. Dispersion of air into fine bubbles (about 20-40 μm in size after drawing) requires that freezing occur at the same time to increase the shear forces within the freezer. Whipping air into ice cream mix without freezing results in lower amounts of overrun incorporated and larger air bubble sizes.
- ☆ In batch freezers, the mix is allowed to whip at atmospheric pressure. Hence whipping properties of the mix are very important, and overrun is more variable, being controlled simply by the head space remaining after the mix charge is put into the barrel.
- ☆ In the continuous freezer, air is injected through controlled valves, so whipping properties of the mix are perhaps less important, and overrun control is exact. Air distribution occurs under pressure in the continuous freezer, and it is the rapid expansion of the air bubbles at draw that establishes the air bubble interface.

B. Influence of Processing Parameters on the Microstructure Properties of Ice Cream [3]

The ice cream making process results in many complex interactions within the microstructure. Different microstructures can be generated depending on the processing parameters and the composition of the mix.

- ☆ During the freezing process, the speed of the dasher, amount of overrun, and draw temperature all play an important role in the development of the final microstructure.
  - i. **Dasher Speed-**The speed and design of the dasher during processing has great influence on the properties of ice cream products. In general, an increase in dasher speed tends to lead to an increase in partial coalescence. As dasher speed increases, air cell size has been known to decrease.
    - ☆ The size and amount of ice crystals can greatly be affected by the processing conditions. For example, as dasher speed increased, it was expected that mean ice crystal size would increase as the temperature inside the freezer increased due to greater frictional heat released, resulting in dissolution and recrystallisation.
  - ii. **Overrun-**Aeration is an important part of making ice cream. Without air incorporated, ice cream would be a densely-packed solid mass, resulting in little to no scoopability properties.

- ☆ As overrun is increased in the system, if affects other physical properties of ice cream and not just air cells. Partial coalescence increases as overrun is increased, as an increase in air content causes the repeated adsorption/desorption of fat to the air interface, which in turn increases the collision rate of the fat globules. As a result, the introduction of more air along with shear causes an increase in partial coalescence.
- ☆ In addition, researchers indicated that an increase in air lead to a decrease in overall size of ice crystals. Too much air can cause a fluffy and crumbly product, which are all defects in ice cream.

iii. **Draw Temperature-**The effect of draw temperature, the temperature at the outlet of the freezer, can influence the final viscosities as well as amount of total ice in the system. With warmer draw temperatures, less ice is formed, which results in lower viscosities, larger air cells, and lower amounts of partially-coalesced fat.

2. Physico-chemical Changes during Storage of Ice Cream and Frozen Desserts [4,5]

A. Changes Due to Storage Time [4]

With progress in the storage period, following changes observed in Ice cream and Frozen Desserts-

i. Total solid content increased due to loss of moisture during the storage of ice cream while no change in protein, fibre and ash content was observed.

ii. Acidity of all ice cream samples increased due to formation of lactic acid and pH, antioxidant activity and total phenolic content were found to reduce with progress in the storage.

iii. Increase in the storage time results in decreased overrun.

iv. Sensory scores of ice cream were found to decrease progressively with storage.

B. Changes due to Fluctuation in Storage Temperature [5,5a]

i. **Ice Recrystallization [5,5a]-**Temperature fluctuations during storage and distribution stages affect frozen food products such as ice cream, inducing physical and chemical changes.

- ☆ Ice cream is a very temperature sensitive product and temperature fluctuations during the storage and distribution steps of the food cold chain may result in changes in ice crystal size due to recrystallization phenomenon.

- ✰ Ice crystals are relatively unstable, and during frozen storage, they undergo changes in number, size, and shape, known collectively as recrystallization. This is probably the most important reaction leading to quality losses in all frozen foods.
- ✰ Some recrystallization occurs naturally at constant temperatures, but by far the majority of problems are created as a result of temperature fluctuations. Temperature fluctuations are common in frozen storage as a result of the cyclic nature of refrigeration systems and the need for automatic defrost. However, mishandling of product is probably the biggest culprit. The sight of ice cream sitting unrefrigerated on a loading dock, in the supermarket aisle, in a shopping cart, or in someone's grocery bag is too common. If one were to track the temperature history of ice cream during distribution, retailing, and finally consumption, one would find a great number of temperature fluctuations.
- ✰ Each time the temperature changes, the ice to serum content changes, and the smaller ice crystals disappear while the larger ones grow even larger. Recrystallization is minimized by maintaining low and constant storage temperatures.

ii. **Ice Cream Shelf-Life [5]-** The shelf-life depends entirely on its conditions of storage. It might be one year, or it might be two weeks or less. An excessive increase of the mean ice crystal size results in a reduction of ice cream quality and a shortening of its shelf life.

- ✰ Iciness is also the primary limitation to the shelf life of ice cream and probably also accounts for countless lost sales through customer dissatisfaction with quality.
- ✰ Ice crystals need to be numerous and of small, uniform size so they are not detected when eaten. It is heat shock, large temperature fluctuations, which is the greatest culprit to the loss of these small, uniform ice crystal size distributions and resulting coarse, icy texture.

Sources

1. *A. Jana, Suneeta Pinto and P.R.S. Moorthy, Ice Cream and Frozen Desserts, 2016, www.AgriMoon.com.*

1a. *Robert T. Marshall and Douglas Goff, Formulating and Manufacturing Ice Cream and Other Frozen Desserts, 2003, Volume 57, No. 5, Food Technology Magazine.*

1b. *Ice cream and Frozen desserts- Stabilizers and Emulsifiers - Selection, mechanism of action, influence on mix and ice cream, proprietary stabilizer blends, Lesson 8, http: //ecoursesonline.iasri.res.in.*

1c. *Soma Maji, Role and mechanism of stabilizers and emulsifiers in ice cream, http: //courseware.cutm.ac.in.*

2. *Syed QA, Anwar S, Shukat R, et al., Effects of different ingredients on texture of ice cream. J Nutr Health Food Eng. 2018; 8(6): 422-435.*

3. *Warren, M.M. and Hartel, R.W., 2014, Structural, compositional and sensorial properties of US commercial ice cream products. J Food Sci. 79(10).*

4. *Arashdeep Singh, Usha Bajwa and Rajpreet Kaur Goraya, Effect of Storage Period on the Physicochemical, Sensory and Microbiological Quality of Bakery Flavoured Ice Cream, 2014, International Journal of Bioinformatics Research and Applications 4(8): 80-90.*

5. *Ice Cream Shelf-Life, University of Guleph, https: //www.uoguelph.ca/foodscience.*

5a. *Physico-chemical properties of ice cream-II, Lesson-17, ecoursesonline.iasri. res. in.*

Role and Mechanism of Stabilizers and Emulsifiers in Ice Cream [1-4]

Stabilisers and Emulsifiers [1,2,3]

The stabilisers and emulsifiers play very important role in maintaining the valuable properties of Ice cream and Frozen desserts. Emulsifiers and stabilizers contribute to a great extent, to the desired body and texture characteristics of ice cream and frozen desserts. It is well recognized that individual compounds vary considerably in their emulsifying/stabilizing effects and often a single compound is not entirely satisfactory. Thus, a mixture of two or more emulsifiers/stabilizers is generally preferred to overcome the drawbacks of individual compounds.

Stabilisers

The stabilizers are a group of compounds, usually polysaccharides that are responsible for adding viscosity to the unfrozen portion of water and thus holding this water so that it cannot migrate within the product. This results ice cream that is firmer to the chew.

i. The excellent ice cream products can be made with only the natural stabilizing and emulsifying materials present in milk such as milk proteins and phosphates, additional stabilizers and emulsifiers have potential benefits.

ii. Without stabilizers, the ice cream would become coarse and icy very quickly due to the migration of the free water and growth of existing ice crystals [1].

   ✰ Stabilizers are a group of water-soluble or water-dispersible biopolymers used in small amounts (typically 0.2 per cent) in ice cream, sorbets, water ices and other foods. Nutritionally, stabilizers are a source of soluble fibre. Stabilizers are straight or branched polymers containing hydroxyl groups that can form hydrogen bonds to water molecules. Stabilizers are substances which make it possible to maintain the physico-chemical state of a foodstuff [1a].

Emulsifiers

Emulsifiers aid in the production of drier ice cream upon extrusion with smoother body and texture and good stand up or melt resistance. Milk also contains some naturally emulsifying compounds that aid in the manufacture of ice cream. These include milk proteins, phosphates and citrates.

- ☆ The emulsifying agents commonly used in ice cream are mono and diglycerides and polysorbate-80, polyoxyethylene sorbitan monooleate [1]. Added emulsifiers are more effective than milk proteins due to their smaller molecular size and greater mobility [4].

Selection Criteria for Stabilisers and Emulsifiers [1,2]

The suitability of emulsifiers and stabilisers for Ice cream preparation is essential and hence selection of these compounds to be added during processing, play significant role.

- ☆ The following criteria are used in choosing an emulsifier and a stabiliser for application during manufacturing of Ice cream and Frozen desserts:

Table 21.1: Criteria in Choosing an Emulsifier and a Stabiliser

Sl.No.	*Criteria for Stabilisers*	*Criteria for Emulsifier*
1	Ease of incorporation in mix	Fat percentage of mix
2	Effect on viscosity and whipping properties in mix	Type of frozen dessert
3	Ease of dispersibility in cold and hot mix	Effect on flavour of product
4	Type of body produced in the ice cream	Cost
5	Effect on meltdown characteristics	Composition of fat in mix
6	Ability to retard ice crystal growth	Compatibility with stabilizer used
7	Amount required to produce the stabilization	Type of freezer used
8	Perception as natural	Method of processing – homogenization
9	Effect on flavour perception	Formulation – effect of other ingredients
10	Cost	Legal standards

Roles of Stabilisers and Emulsifiers [1a]

The stabilisers and Emulsifiers are used in Ice cream for the purposes mentioned on next page.

Table 21.2: Purposes for which Stabilisers and Emulsifiers are Used

Sl.No.	*Stabilisers*	*Emulsifiers*
1	To increase the viscosity of the mix. To stabilize the mix *i.e.* to prevent wheying off.	Promote nucleation of fat during aging thus reducing aging time
2	To produce smoothness in texture during consumption. To reduce or slow down the growth of lactose crystals during storage mainly during temperature fluctuations. Mask the detection of ice crystals in the mouth during eating.	Improve the whipping ability of the mix due to their function at the air interface resulting in reduced air cell sizes and homogenous distribution of air in the ice cream.
3	To help in suspension of flavouring particles. To reduce moisture migration from the product to the package or the air.	Produce a dry and stiff ice cream as they enhance fat destabilization, facilitating molding, fancy extrusion and sandwich manufacture.
4	To produce a stable foam with desired stiffness at the time of packaging	Increase resistance to shrinkage and rapid melt down due to a combination of the above two factors.
5	To help prevent shrinkage of the product volume during storage. Prevent shrinkage and slow down moisture migration out of ice.	Increase resistance to the development of coarse/icy texture, due to the effect of fat agglomerates, more numerous air bubbles, and thinner lamellae between adjacent air bubbles on the size and growth of ice crystals.
6	To provide uniformity to the product and resistance to melting. Reduce the rate of meltdown (*i.e.* the rate at which ice cream melts).	Provide smooth texture in the finished product, due to fat structuring and interaction of fat agglomerates within the mouth during consumption.
7	Allow easier pumping and more accurate filling during processing. Facilitate the controlled in corporation of air in the freezer	-

Classification and Properties of Stabilizers [1a]

1. *Classification*

Stabilizers used in manufacturing of ice cream and frozen desserts are categorised in to the following groups:

Table 21.3: Classification of Stabilizers

Sl.No.	*Category*	*Examples*
1	Proteins	Gelatin
2	Plant exudates	Arabic, ghatti, Karaya and Tragacanth gums
3	Seed gums	Locust (carob) bean, guar, psyllium, starch and modified starches
4	Microbial gums	Xanthan

Sl.No.	Category	Examples
5	Seaweed extracts	Agar, Alginates, Carageenan
6	Pectins	Low and high methoxyl
7	Cellulose	Sodium carboxymethyl cellulose, Microcrystalline cellulose, Methyl and methylethyl cellulose, Hydroxypropyl and hydroxypropylmethyl cellulose.

2. Properties of Stabilisers

With the exception of milk proteins and gelatin, all the other stabilizers are carbohydrates. Casein being a hydrocolloid in nature also causes a appreciable stabilizing effect in ice cream. Therefore, the levels of milk solids, and so milk protein, will determine the amount of stabilizer to be added. The important properties of some commonly used stabilisers are shown in Table 21.4.

Table 21.4: Properties of Stabilizers

Sl.No.	Stabilisers	Properties
A	Carbohydrate based stabilisers (Plant source)	
1	Sodium alginate (E401)	☆ This is a polysaccharide of guluronic acid and mannuronic acid. It is extracted from brown seaweeds such as Macrocystispyrifera and Laminaria digitata. It consists of a negatively charged polymer chain with ionic bonds to positively charged sodium ions (Na+). ☆ In ice cream, alginates are blended with phosphate, citrate or tartrate ions to prevent premature gelation due to the calcium from the milk solids. ☆ The major advantage of alginate is its resistance to acid conditions, particularly when heated, whereas other stabilizers would lose their functionality. Usually, it is used at a level of 0.18-0.25 per cent.
2	Carrageenans (E407)	☆ Carrageenans are complex polysaccharides of esters of galactose and m-3,6-anhydrogalactose,found in red sea-weeds (Rhodophycae), such as Chondrus crispus (Irish Moss), Kappaphycus alverezii and Eucheuma denticu-laturn. ☆ It can have several structures. Classified as one of three types that have different properties- kappa (k),) iota (i) and lambda (l). ☆ Carrageenans can also form gels with both milk proteins and locust bean gum. ☆ In ice cream applications, k -carrageenan fractions are frequently used not as a primary stabilizer but as a secondary hydrocolloid to prevent wheying off of mix. ☆ k -carrageenan reduces phase separation of milk proteins and polysaccharides which could result in wheying off.

Sl.No.	Stabilisers	Properties
3	Guar gum (E412)	☆ It is extracted from the seeds of *Cyamposistragono-lobus,* an annual crop grown in the Indian subcontinent. ☆ Guar has a similar structure to LBG. It has a backbone of mannose units, about half of which have galactose side branches. ☆ It has a higher molecular weight than LBG and the side groups are more evenly spaced. The larger proportion of galactose units makes guar cold water soluble. ☆ Guar gum is significantly cheaper than LBG. Both these gums at 0.1-0.15 per cent impart a chewy body to the product. While locust bean gum may cause a comparatively short body, guar gum may impart stickiness to the product.
4	Locust bean gum (E410)	☆ Locust Bean Gum, also known as LBG, carob gum or St Johns Bread, is extracted from the seeds of the Mediterranean Ceratoniasiliqua tree. ☆ LBG is a polysaccharide consisting of a mannose backbone with galactose side branches on about a quarter of the mannose units. ☆ LBG is the best stabilizer for many ice cream applications and its ability to gel is crucial to some aspects of its use. It is also expensive stabilizer.
5	Pectin (E440)	☆ It is a polysaccharide consisting of linear chains of galacturonic acid and galacturonic acid methyl ester units. It is not a very satisfactory stabilizer for ice cream. ☆ It is extracted from citrus peel and apple pomace. ☆ Pectin is classified according to its degree of esterification. High methoxy (>50 per cent esterified)and low methoxy (< 50 per cent) pectins possess different properties. Low methoxy pectin requires calcium to gel.High methoxy pectin gels at low pH and in the presence of high concentrations of sugar. ☆ Pectin is the setting agent used in jam making. All fruits contain some pectin.
6	Sodium carboxymethyl cellulose (E466)	It is derived from purified cellulose from cotton and wood pulp. It is a sodium salt polymer of anhydroglucose residues. ☆ For use in ice cream, an average of 0.7 of the 3 hydroxyl groups in each glucose unit is substituted with a sodium carboxymethyl group. ☆ CMC hydrates at low temperatures. It has excellent water holding property but may cause wheying off. ☆ Usually used together with other gums, CMC at 0.1-0.2 per cent performs excellently as an ice cream stabilizer. ☆ Its perception as a 'chemical' has resulted in fairly low usage.

Sl.No.	Stabilisers	Properties
B	Protein based stabilisers (Animal Source)	
1	Gelatin	☆ It is a mixture of high-molecular-weight polypeptides derived from collagen from animal connective tissues, and was commonly used as a gelling and thickening agent. It is used at the rate of 0.2-0.3 per cent and requires ageing for at least 4 hours. ☆ Gelatin disperses in cold water and forms a gel upon heating the dispersion. ☆ It is a good ice cream stabilizer since forms a weak gel that melts readily in the mouth giving no impression of gumminess. ☆ It is not suitable for vegetarians, and has now generally been replaced by other stabilizers. ☆ The most important property of gelatin is its ability to form thermo-reversible gels. ☆ Gelatin's most important attribute is its gel strength and, when determined by the standard method, is called the 'Bloom strength' or 'Bloom value.' Commercial products normally have a Bloom value that falls between 50 to 280.

Functions of Stabilisers

Stabilisers have different functions in Mix and Ice cream. The details are given below [1,2,3]-

i. **Smoothens the texture and Provide body to the product-**Major functions performed by stabilizer in ice cream include contributing to body or 'substance' and creaminess, imparting a smooth texture [1-3]. Stabilizers interfere with ice crystallisation and limit the growth of large crystals. Use of stabilizer makes the smoother body. The effect of stabilizer is greater than expected and this may be due to the presence of sucrose as an additional solute [4].

ii. **Enhance viscosity and no effect on Freezing point -** Stabilisers enhance the viscosity and there is no effect on freezing point of Mix/Ice cream [1,2]. The viscosity of the mix is also increased as a result of interactions between gum type stabilizers and partially denatured proteins [4].

iii. **Prevent from texture coarsening -**The most important functions of stabilizers is to 'stabilize' the product texture *i.e.* to prevent or minimize detrimental effect of heat shock during storage and distribution.

iv. **Provide resistance during melting-** Addition of stabilisers providing melting resistance to ice cream.

Effects of Stabilisers on Mix and Ice Cream [1]

The following effects of Stabilisers on Mix and Ice cream are discussed here-

i. **Effect on texture and body characteristics-**The important involvement of stabilizers in promoting smoothness through control of ice crystal size begins after initial freezing and hardening. Once the ice crystals are formed, the influence of stabilizers is to do exactly what it implies *i.e.* stabilize the size of the ice crystals against growth as a result of temperature fluctuation (heats shock).

   ☆ Effect of Stabilisers on body characteristics is a major function at the time the product is drawn, during hardening and at the time of consumption. It involves the participation of the stabilizer ingredients in determining the cohesiveness of the product, particularly those aspects described by such adjectives as chewy, sticky, weak, gummy, *etc.*

ii. **Effect on melting rate and properties of melted product-**The structure resulted from the interaction of gums with water and other ingredients has a direct bearing on the rate of melting and appearance of the melted product. The meltdown appearance function is outcome to the role of stabilizer with respect to protecting against serum separation and fat destabilization in that by reducing the degree of destabilization of fat and protein, stabilizers can avoid the development of curdy appearance in melted product.

iii. **Effect on whipping and overrun retention-** Stabilizer are involved in the amount of air which is incorporated and the degree to which the air cells are stable.

iv. **Effect on sandiness-**Sandiness is caused by lactose. It is a problem of earlier days. This is because of the type of stabilizers used these days (CMC, natural gums, carrageenan, *etc.*), which functions in a manner similar to control of ice crystal size through decreasing the mobility of unfrozen water during heat shock.

   ☆ The supersaturated stages in which sugars responsible for sandiness are prone to crystallize in frozen desserts are reached at temperature associated with a high stabilizer concentration. Microcrystalline cellulose is particularly effective in this function, probably by providing an additional seeding function which encourages the development of small crystals.

Mechanism of Action of Stabilisers [1,2,3]

Water binding capacity of stabilizers is very high and added in small quantity hence effect on flavour and product value is inconsequential. Total solids, processing machine, stabilizers properties and some other factors affect the usage of stabilizers [2].

☆ Stabilizers are functionally also termed as hydrocolloids. These viscogenic compounds are primarily poly-saccharides although, gelatin a well known stabilizer is a protein in nature [1].

- ☆ The most apparent effect of stabilizers is the increased viscosity of the continuous liquid phase. The effect of stabilizers on viscosity exhibits considerable interaction with milk constituents. For instance, the basic viscosity of stabilizer solution is generally not affected by heat treatment in the absence of milk solids, but in their presence, within creasing total solids, the effect of stabilizer on heat induced increase in viscosity of the system becomes more pronounced [1].
- ☆ During the freezing/hardening process, as more and more water freezes (water content of ice cream falls considerably),the stabilizer and its complexes get concentrated, and these also provide followings-
  - **i.** Strength to the air cell wall. Therefore, the amount of air which is incorporated and the degree to which the air cells are stable is influenced by stabilizers. However, increased viscosity may prolong whipping time under certain circumstances [2,3].
  - **ii.** The increased stabilizer concentration together with decreased temperature greatly increases the viscosity, there by substantially reducing diffusion and mobility of the liquid in the frozen product. Also, gel formation induced by stabilizer in the mix is believed to effect, considerably, immobilization of the liquid. The movement of water is hindered partly by the ability of hydrocolloids to form hydrogen bonds, too. Thus, refreezing of water that originated from melting of ice due to temperature fluctuations would not permit the formation of large ice crystals [1].
- ☆ It should, however, be noted that merely by increasing the viscosity the desired stabilization of ice structure may not be achieved. Increasing the viscosity of a 36 per cent TS mix by addition of 5 per cent polyethylene glycol had no effect on ice crystal size in the frozen dessert [1].
- ☆ Certain stabilizers such as guar gum are effective thickening agents with no gelling ability. Further, the way stabilizers influence the body and texture perception in frozen ice cream could be related to their molecular structure and orientation besides their gel forming or viscosity building ability. However, extrapolation of results from model systems to ice cream is difficult as the later is a complex system with high concentrations of salt, sugar and stabilizers, which interact with other molecules, *e.g.* protein [1].
- ☆ The effect of stabilizers cannot be explained wholly in terms of the effect on ice growth. The three dimensional network formed binds, restricts movement of water in the frozen product and thus minimise the potential for formation of large crystal during temperature fluctuations [4].

- The following points also help in understanding the mechanism of action of stabilisers in Ice cream [4]-
  i. At a given rate of heat removal the coarseness of a population of crystals depends on the relative rates of nucleation and growth and it has been postulated that stabilizers act either by increasing nucleation or reducing the rate of growth.
  ii. It is observed that while stabilizers have no significant effect on rate of nucleation, the growth of ice, in a model system, is drastically retarded and the morphology altered by the presence of a stabilizer gel network.
  iii. The effect on ice growth rate and morphology has been explained qualitatively in terms of mechanical interference with the growth of ice.
  iv. It is probable that the gel fibres cause the ice interface to develop a radius of curvature of similar size of the mesh size of the network and thus to depress the freezing point. The gel fibres themselves are placed under stress and may rapture.

Emulsifiers [1,2,4]

Emulsifier are surface active agents known to improve the sensory quality of ice cream by aiding the whipping process, improving air cell distribution and enhancing the products heat shock resistance. Emulsifiers reduce the energy requirement to maintain the integrity of the fat globules [1,2].

- Generally, a mixture of high and low hydrophile-lipophile balance (HLB) emulsifiers(Mono- and diglycerides and polysorbate 80) are used. HLB concept was put forth by Griffin (1949) and HLB number ranges from 0 to 20. The HLB number is derived by calculating the proportion of the molecular weight of the emulsifier molecule represented by the hydrophilic portion and dividing that value by 5. The HLB system can be useful in describing an emulsifiers general characteristics; however, it is not precise enough [1]. In an emulsifier such as glycerol mono-stearate, the lipophilic portion of the molecule, the fatty acid chain enters the fat phase, while the hydrophilic portion the glycerol portion enters the water phase [4].

Functions of Emulsifiers [1,2,3]

The Functions of emulsifiers are discussed below:

i. **Stiffness to body and smoothness of texture-**They also impart a dry appearance and stiffness or 'stand up' property to the product being extruded from the freezer. These effects are brought about by the inter facial properties of emulsifiers.

ii. **Reduction in time of whipping-**Discussed earlier under "Factors affecting Whipping ability of Mix".

iii. **Even whip ability properties-** They are used for improvement of the whipping quality of the mix, for production of a drier ice cream to facilitate molding, fancy extrusion, and sandwich manufacture, for smoother body and texture in the finished product, for superior drawing qualities at the freezer to produce a product with good stand-up properties and melt resistance, and for more exact control of the product during freezing and packaging operations. Also discussed earlier under "Factors affecting Whipping ability of Mix".

iv. **Smaller uniformly distributed air cells-**Their ability to reduce surface tension seems to promote development of smaller but numerous air cells.

Classification and Properties of Emulsifiers [1,1a]

The classification of emulsifiers is done on the basis of the nature of the emulsifiers. The details related to the classification of emulsifiers and properties of individual emulsifier, are given below:

A. **Classification of Emulsifiers [1a]-** They are usually integrated with stabilizers in proprietary blends but their function and action differ remarkably from those of stabilizers. They can be classified in to the following two groups-

   a. **Hydrophobic emulsifier-**Examples include sorbitan esters, mono- and diglycerides of fatty acids, polyglycerol polyricinoleate, highly substituted sugars, polyglycerol esters, and propylene glycol esters.

   b. **Hydrophilic emulsifier-** Examples include ethoxylated sorbitan esters, monoglyceride derivatives such as lactates, tartarates, citrates, low-substituted polyglycerol esters, and monosubstituted sugar esters.

B. **Properties of Emulsifiers [1,1a]**-Different emulsifiers differ structurally, and so their action in ice cream differs, too. Obviously, therefore, the quantity of emulsifier required varies with the type of emulsifier used.

   ☆ The type of emulsifiers and their properties and roles in manufacturing of Ice cream and Frozen desserts, are shown in Table 21.5.

Mechanism of Action of Emulsifiers [1,2,3]

The mechanism of action of emulsifier can be summarized as follows:

a. Desired Product Structure [1]

The emulsion (two phase) is stabilized by casein micelles adsorbed at the fat globule serum interface in the homogenized ice cream mix.

Table 21.5: Properties and Roles of Emulsifiers

Sl.No.	Emulsifier	Properties/Roles of Emulsifier
1	Mono- and diglycerides (E471) [1,1a]	☆ The most commonly used emulsifiers in ice cream manufacture are mono-/diglycerides and the mono glycerides have a fatty acid molecule attached to a glycerol molecule. They are surface active because the glycerol end of the molecule is hydrophilic and the fatty acid end is hydrophobic. ☆ They are often used in combination, although the former is more effective. ☆ Fully saturated mono-/diglycerides that contain predominantly stearic and palmitic acids, such as glycerol monostearate, are often used for ice cream production. When unsaturated fatty acids such as oleic acid are present in the molecule, these emulsifiers promote better dryness at drawing from the freezer. ☆ The fat destabilization responsible for stiffness of the frozen mix is in the decreasing order for mono laurate, monooleate, and mono stearate in that sequence used at 0.1-0.2 per cent, mono- and diglycerides are less likely to cause churning. ☆ However, their drying and whipping ability is somewhat limited as compared to polysorbates.
2	Polysorbates i. Tween 80 or polysorbate 80 (polyoxyethylene sorbitan monooleate). ii. Tween 65 (polyoxyethylene sorbitan tristearate).	The sorbitan esters are similar to monoglycerides in that the sorbitan esters have a fatty acid molecule such as stearate or oleate attached to a sorbitol. They are water soluble components of sorbitol. To make the sorbitan esters water soluble polyoxyethylene groups are attached to the sorbitol molecule.
	iii. Tween 60 (polyoxyethylene monostearate). iv. Tween 40 (polyoxyethylene sorbitan monopalmitate). [1,1a]	a. Tween 80- It has a high drying power and aids in heat shock resistance, but the unsaturated fatty acid may undergo auto oxidation to develop an off flavour, particularly when a certain level is exceeded. Used usually at 0.02 to 0.06 per cent. Tween 80 in excess may cause churning especially in soft serve and high fat ice creams. b. Tween 65-Another polysorbate, Tween 65 has a little lower drying power as compared to Tween 80 but has excellent whipping properties. Because of its flavour stability, Tween 65 can be used at higher levels (e.g. 0.1 per cent) without any adverse effect on the products flavour. c. The fat destabilizing ability of different polysorbates is in the order Tween 80> Tween 40 > Tween 60> Tween.

Sl.No.	Emulsifier	Properties/Roles of Emulsifier
3-	Egg yolk [1a]	It contains several components with emulsifying properties, notably lecithin, is often used in 'all-natural', premium or homemade ice creams. ☆ Egg yolk has following approximate composition (by weight)-50 per cent water, 16 per cent protein, 9 per cent lecithin, 23 per cent other fat, 0.3 per cent carbohydrate and 1.7 per cent minerals.Lecithin consists of phosphatides and phospholipids. ☆ Egg yolk is usually used in ice cream manufacture either as pasteurized fresh egg yolk, frozen sugared pasteurized egg yolk or as dehydrated egg yolk. ☆ Egg yolk solids are normally used at about 0.5-3 per cent. High concentrations are only used for super- premium products, and cangive the ice cream an eggy flavour.

☆ When monoglyceride emulsifier is used, the fat globules are covered with an emulsifier layer and the milk proteins form an outer layer. Monoglycerides effectively compete with protein if they form crystals at the interface. Hence, the outer protein layer tends to be repelled from the fat globules. As a result, the emulsion is prone to destabilization by mechanical action during freezing.

☆ This destabilizing effect, in conjunction with other processing factors, is considered to be primarily responsible for development of the desired product structure.

b. *Dry Appearance [1,2,3]*

The dry appearance of ice cream coming out of the freezer is believed to be caused by several phenomena, one of which is an emulsifier induced clustering of fat globules at the liquid air interface. Also, more air cells of smaller size foaming on account of presence of emulsifiers appear to provide more surface with the available liquid, which implies that the liquid is spread over a larger area [1].

☆ They lower the fat/water interfacial tension in the mix, resulting in protein displacement from the fat globule surface, which in turn reduces the stability of the fat globule to partial coalescence that occurs during the whipping and freezing process, leading to the formation of a fat structure in the frozen product that contributes greatly to texture and meltdown properties [2,3]

☆ The extent of protein displacement from the membrane, and hence the extent of dryness achieved, is a function of the emulsifier concentration.

Egg yolk solids, like sweet cream buttermilk solids improve the whipping ability and shorten the freezing time [2,3].

c. *The De-emulsification Effect [1]*

This effect of emulsifiers is related to the quantity of emulsifiers used and also to the fat content of the product. With increasing emulsifier concentration, fat de-emulsification is enhanced and beyond a certain limit, greasy texture and short body result due to butter formation in the freezer.

- ✰ Higher fat levels magnify this de-emulsification phenomenon. Thus, an emulsifier level just enough to provide the correct amount of 'partial churning' is necessary.
- ✰ Less emulsifier is required in a high-fat ice cream than in a low fat one.
- ✰ Excessive emulsifier may promote slow melting and curdy meltdown.

Mix Processing in Relation to Action of Emulsifier and Stabiliser [1,2,4]

The mix processing (Aging, Homogenisation, Pasteurisation) effects the action of emulsifier and stabilisers.

a. *Effect of Homogenisation on Action of Emulsifier [1,2]*

Homogenisation of emulsifier is necessary for an emulsifier to be effective in ice cream mix. Milk protein(casein in particular) helps achieve size reduction of fat globules by homogenization and presence of emulsifiers promote greater size uniformity. Mix containing a monoglyceride displays upon homogenization, a very narrow band of size distribution (about 1μ) as compared to mix without monoglyceride (1-5 μ).

- ✰ Homogenization of ice cream mix helps to stabilize the oil-in-water emulsion. This stabilizing effect in conjunction with the destabilizing effect of the emulsifier results in the desirable texture quality.
- ✰ The greatly decreased fat globule size coupled with controlled de-emulsification caused by emulsifiers, in essence, brings about the development of the right kind of structure and corresponding texture in ice cream.

b. *Effect of Pasteurisation on the Action of Emulsifier and Stabiliser [4]*

The effect of pasteurisation on the role of stabilisers and emulsifiers are discussed below:

i. Microbiological safety.

ii. Whey proteins gets denatured and uncoiled which expose the lipophilic part of the molecule,

iii. Whey proteins starts to act as emulsifier and the water binding capacity increased,

iv. Increase the available binding site of protein: hydrocolloid interaction thus enhances the action of stabilizers like carrageenan,

v. During heating emulsifiers is melted and heat activated stabilizers are brought into colloidal solution and

vi. Excess heat treatment leads to unacceptable organoleptic deterioration.

c. *Effect of Aging on the Action of Emulsifier and Stabiliser [1,4]*

i. *On the Action of Emulsifier [1]*

Milk fat crystallizes during ageing of ice cram mix and much of crystallization taking place during the first hour.

- ✰ Presence of emulsifiers leads to more extensive fat crystallization.
- ✰ The emulsifier caused protein desorption from fat globules is time dependent and takes place during ageing. Further, the initial desorption from the surface of the fat globules, in the presence of emulsifiers, occurs as the removal of a coherent protein layer rather than individual casein particles.
- ✰ The process of de-emulsificaiton continues into the freezer to yield the desired body and texture.

ii. *On the Action of Stabiliser [1,4]*

Ageing of ice cream mix is an important processing step with regard to stabilizer action. Hydration of stabilizer is the most obvious effect of ageing, although not all stabilizers require ageing for complete hydration. The effects of ageing are more pronounced with gelatin stabilizers, and improvements in ice cream and freezing performance becomes more evident as the ageing time increases from 4 to 12 h or more. Even mixes containing stabilizers which require no ageing for complete activation benefit from a certain minimum ageing due to milk protein hydration and fat crystallization [1].

- ✰ The effect of aging on the role of stabilisers are discussed below [4]:

  i. Skim milk powder and stabilizers become fully hydrated.

  ii. Emulsion induced disruption of protein starts.

  iii. Substantial crystallization of fat. Triacylglycerols of the highest melting point are the first to crystallize and are situated closest to the surface of the fat globule.

  iv. The quantity of fat remaining liquid depends on the specific fat involved but balance between liquid and crystallised fat is essential for good quality.

  v. Use of unsaturated fat in ice cream results relatively low degree of fat crystallization.

Sources

1. *Ice cream and Frozen desserts- Stabilizers and Emulsifiers - Selection, mechanism of action, influence on mix and ice cream, proprietary stabilizer blends, Lesson 8, http: //ecoursesonline.iasri.res.in.*

1a. *Stabilizers and Emulsifiers- classification, types, properties and role in ice cream, Lesson -7, http: //ecoursesonline.iasri.res.in.*

2. *A. Jana, Suneeta Pinto and P.R.S. Moorthy, Ice Cream and Frozen Desserts, 2016, www.AgriMoon.com.*

3. *Syed QA, Anwar S, Shukat R, et al., Effects of different ingredients on texture of ice cream. J Nutr Health Food Eng. 2018; 8(6): 422-435.*

4. *Soma Maji, Role and mechanism of stabilizers and emulsifiers in ice cream, http: //courseware.cutm.ac.in.*

Index

D

E

I

K

L

M

N

R

S